Photo by William R. Anderson

STIGMAPHYLLON JATROPHIFOLIUM ADR. JUSS.

SYSTEMATIC BOTANY MONOGRAPHS

VOLUME 51

Monograph of Stigmaphyllon (Malpighiaceae)

Christiane Anderson

THE AMERICAN SOCIETY OF PLANT TAXONOMISTS

12 May 1997

SYSTEMATIC BOTANY MONOGRAPHS
ISSN 0737-8211

ISBN 0-912861-51-7

Printed in the United States of America

MONOGRAPH OF STIGMAPHYLLON (MALPIGHIACEAE)

Christiane Anderson
University of Michigan Herbarium
North University Building
Ann Arbor, Michigan 48109-1057

ABSTRACT. *Stigmaphyllon* is a genus of Malpighiaceae characterized by a vining habit, long-petioled leaves with broad or sometimes lobed laminas, interpetiolar stipules, inflorescences composed of umbels or pseudoracemes dichasially arranged, yellow petals, connate carpels, and a schizocarpic fruit splitting into three samaras. In most species the stamens are heteromorphic, the styles are apically foliolate, and each samara bears an elongate dorsal wing. The genus is distributed in the Neotropics from southern Mexico to northern Argentina, except Chile and the high Andes. One species, *S. bannisterioides*, is also found in coastal West Africa (Guinea Bissau, Guinea, and Sierra Leone). *Stigmaphyllon* is allied with *Banisteriopsis* and its relatives among the wing-fruited genera of Malpighiaceae. It is marked by a high incidence of parallelisms and/or convergences and, in some species, a great diversity in leaf shape. In this study, the species are circumscribed and provisionally grouped on the basis of morphological characters, particularly traits of the androecium and gynoecium. *Stigmaphyllon* comprises 90 species, including two varieties. Two species, *S. boliviense* and *S. carautae* are newly described. The often extensive nomenclature is reviewed for each taxon. All species are described and mapped, and many are illustrated. Two kinds of keys are provided, one for the whole genus and, in an appendix, nine for species from single countries or specific regions.

INTRODUCTION

Stigmaphyllon comprises 90 species that are widely distributed in the Neotropics in greatly diverse habitats and absent only from Chile and the high Andes. It is one of the wing-fruited genera of Malpighiaceae, in which the schizocarp splits into three samaras, each with an elongate dorsal wing. The species are typically vines with long-petioled, elliptical to cordate leaves and clusters of bilaterally symmetrical, yellow flowers borne in dichasially branched inflorescences. The androecium consists of ten usually unequal stamens. In most species, the three styles bear apical appendages, "folioles," for which the genus is named.

Originally included in Linnaeus's encompassing genus *Banisteria*, the distinctive *Stigmaphyllon* was readily segregated by Adrien de Jussieu (1833), whose generic circumscription remains generally unchanged. Jussieu published the name as *Stigmaphyllon*, but later authors at times rendered it as *Stigmaphyllum, Stigmatophyllon*, or *Stigmatophyllum*; the last version was adopted by Niedenzu (1896, 1899–1900, 1912, 1928) and thus became the most widely used orthographic variant. Traditionally, *Stigmaphyllon* has been allied with other similarly fruited genera, such as *Banisteriopsis* C. B. Rob. ex Small [="*Banisteria*" sensu Adr. Jussieu and Niedenzu], *Peixotoa* Adr. Juss., and *Heteropterys* H. B. K., which differ most notably in androecial and gynoecial details as well as vegetative characters. None have long-petioled, cordate or lobed leaves or foliolate styles; in *Heteropterys* the wing of the samara is thickened along the lower (abaxial) rather than the upper (adaxial) margin, as in *Stigmaphyllon, Banisteriopsis*, and *Peixotoa*. In some aspects of leaf morphology and inflorescence branching *Stigmaphyllon* is also similar to the Old World genus *Ryssopterys* Blume ex Adr. Juss. Currently, W. R. Anderson and M. W.

Chase are investigating the phylogeny of the Malpighiaceae by comparing evidence from chloroplast DNA and morphology, which will provide a better understanding of the phylogenetic position of *Stigmaphyllon*.

Although the genus is readily recognized, many species were little understood when Jussieu (1833, 1840, 1843) and Niedenzu (1899–1900, 1912, 1928) published their treatments, mainly owing to a lack of good specimens, particularly from western South America. The many collections that have accumulated since the 1920's permit a better delimitation of the species but also reveal the great variability and plasticity of many characters. Cladistic analyses were attempted but were unsuccessful in generating even a partially resolved phylogeny owing to extensive parallelisms. Most problematic are the widely distributed and greatly variable *S. sinuatum* and *S. tomentosum*. Also, the possible occurrence of hybridization and even apomixis in some species (*S. adenodon, S. bogotense, S. cardiophyllum, S. lindenianum, S. sinuatum, S. strigosum*), as suggested by pollen irregularity linked with intermediacy of morphological characters, remains to be investigated. The species are here grouped on the basis of morphological similarity without insistence that the arrangement necessarily reflects a pattern of ancestry/descent. Perhaps future work taking advantage of molecular techniques will succeed in revealing the phylogenetic relationships among species.

TAXONOMIC HISTORY

Linnaeus (1737) proposed the genus *Bannisteria* in his category Decandria Trigynia to accommodate species with a 3-parted schizocarp splitting into samaras, each with a large dorsal wing. In his *Hortus cliffortianus* (1738), he listed three species in *Bannisteria*, one now assigned to *Heteropterys* (W. R. Anderson, pers. comm.) and two now included in *Stigmaphyllon*. When Linnaeus published the *Species plantarum* (1753), he had expanded his genus *Banisteria* (now spelled with one "n") to comprise seven species, which are now assigned to eight taxa. To the two stigmaphyllons already described in the *Hortus cliffortianus* [*B. dichotoma* (=*S. dichotomum*) and *B. fulgens* (=*S. emarginatum*)] he added *B. angulosa* (=*S. angulosum*). The other four species include five elements: three species of *Heteropterys*, one of *Hiptage* Gaertn., and one of Rhamnaceae. Adrien de Jussieu, the first monographer of the Malpighiaceae, recognized that during the 80 years following the first edition of the *Species plantarum* the genus *Banisteria* continued to be a catch-all taxon for species with samaras bearing an enlarged dorsal wing. He proposed several new genera to accommodate the disparate elements he removed from a more narrowly circumscribed *Banisteria*. In his treatment of the Malpighiaceae for Saint-Hilaire's *Flora brasiliensis meridionalis* (1833) he published *Stigmaphyllon* and described 16 species, including 13 novelties. In 1838, he removed *S. paralias*, in which the units of the schizocarp consist of a nut bearing only an apical crest, to a new genus *Brachypterys* and renamed it *B. australis*. When Jussieu published his monograph of the family (1843), he recognized 48 species in *Stigmaphyllon* and two in *Brachypterys*. In the period of about 50 years that passed between Jussieu's work and the early publications on *Stigmaphyllon* by Niedenzu (1899–1900, 1912), the second monographer of the family, many new collections from South America became available to European botanists. The most notable contributions in *Stigmaphyllon* were those of Triana and Planchon (1862) on plants of Colombia and of Grisebach (1839) mostly on Sellow's Brazilian collections; Grisebach (1858) also treated the Malpighiaceae for Martius's *Flora brasiliensis*. Niedenzu

(1899–1900, 1912) included *Brachypterys* [comprising *S. paralias* and *S. ovatum* (=*S. bannisterioides*)] as a section in *Stigmaphyllon* and divided the genus into two subgenera (subg. *Eustigmaphyllon* and subg. *Baeopterys*), which he subdivided into sections, subsections, and series, mostly defined by the shape of the samara, particularly of the dorsal wing, and the stylar folioles; these groupings are artificial and are not recognized here. He recognized 60 species in his account of *Stigmaphyllon* for *Das Pflanzenreich* (1928). His admirable work was hampered by the paucity of collections, let alone good material, especially from western South America, and by the lack of opportunity to study holdings of the herbarium at Paris, particularly Jussieu's types, and the material in England, a reflection of the political realities of his time. Since Niedenzu's time and the beginning of my study, 22 new species were proposed, nine of which are recognized here. The most significant contribution was Cuatrecasas's excellent treatment (1958) of the Malpighiaceae of Colombia, which includes six new species of *Stigmaphyllon*, all recognized here. My own investigations have added 27 new species (C. Anderson 1986, 1987a, 1987b, 1989, 1990, 1992b, 1993b, 1993c), including *S. laciniatum*, previously recognized as a variety, as well as two additional species described here. *Stigmaphyllon*, including *Brachypterys*, is here proposed to comprise 90 species.

MORPHOLOGY

Vesture. The vegetative parts of nearly all species of *Stigmaphyllon* are at least moderately pubescent when very young, but the stems and sometimes also the petioles become glabrate to glabrous as they mature. The adaxial surface of mature laminas is commonly glabrous or glabrate, but the abaxial surfaces of the majority of species are variously pubescent; however, in a few species (e.g., *S. auriculatum, S. jatrophifolium*) the laminas are entirely glabrous, and even in species characterized by laminas abundantly pubescent at maturity very old leaves are sometimes only sparsely so. In a few species, e.g., *S. floribundum*, the abaxial laminar vesture is sloughed in patches as the leaves age.

The vesture is composed of medifixed unicellular "malpighiaceous" hairs, which mostly conform to a basic pattern: T-shaped, i.e., a cross-piece, the trabecula, borne on an erect stalk. Velutinous vesture, consisting of hairs in which the arms of the trabecula form a "V," common in many genera of Malpighiaceae, is found only in *S. gayanum* and *S. velutinum*. The T-shaped hairs vary in the length of the stalk, up to 0.5 (–0.6) mm long, and of the trabecula, 0.3–2.3 mm long, and in the shape of the trabecula, which may be straight (Fig. 7b) to wavy to crisped and curled (Fig. 29b). Pubescence of T-shaped hairs with the trabecula crisped and curled is here termed tomentose. The vesture is called sericeous if the stalk is so short or rudimentary that the trabecula is subsessile or even sessile. In hairs of sericeous vesture the trabecula may be terete and fusiform, or elliptical, flat, and closely appressed. Hairs of the latter type, if abundant, give a silvery metallic sheen to the surface. In appressed pubescence termed tomentulose, the trabecula is subsessile and crisped to curled. A type of T-shaped hair found on stems, inflorescence axes, and petioles but not on leaves has the trabecula flat and scalelike, 0.1–0.7 mm long and 0.1–0.4 mm wide, and sessile or borne on a stalk up to 0.3 (–0.4) mm long (Figs. 78a, 80b, 89b).

Habit. Most species are twining vines that become woody with age; plants of *S. bannisterioides* may also be twining shrubs. Only *S. harleyi* and *S. paralias* do not twine; *S. paralias* is a shrub up to 2 m tall, and *S. harleyi* is a twiggy, small-leaved subshrub ca. 50

cm tall. The genus may be characterized by woody rhizomes; however, the underground structures are only rarely collected and thus unknown for most species. Woody rhizomes are evident in collections of *S. calcaratum*, *S. diversifolium*, *S. emarginatum*, *S. paralias*, and *S. sagraeanum*, and woody tubers produced on the rhizomes are known for *S. arenicola*, *S. bonariense*, *S. ciliatum*, *S. finlayanum*, *S. jatrophifolium*, *S. panamense*, and *S. sinuatum*. The Sessé and Mociño drawing of "*Banisteria ternata*" [=*S. ellipticum*] includes a tuberous rhizome.

Leaves. In general, the leaves are opposite, but those associated with the inflorescence are sometimes alternate; only in *S. alternans* are the leaves apparently alternate throughout the plant. The laminas are borne on petioles usually several centimeters long (up to 25 cm long in *S. bogotense*) and associated with stipules; *S. crenatum* has sessile to subsessile leaves and apparently lacks stipules. Characteristically, each petiole bears a pair of glands at the apex; rarely the glands are borne at the base of the lamina (e.g., *S. bannisterioides*). These glands are commonly sessile though prominent but in a few species they are shallowly cupulate and borne on the petiole up to 6 mm below the base of the lamina (e.g., *S. puberulum*) or are peglike (e.g., *S. sagraeanum*). In some West Indian species, one or both glands are occasionally absent. *Stigmaphyllon harleyi* and *S. microphyllum* have unusually small leaves, the laminas less than 4 cm long borne on petioles (0.1–) 0.2–0.4 cm long. In most species, the petioles are free, but in some they are confluent across the node and form a band or ridge bearing the stipules. This fusion is most pronounced in *S. macedoanum* in which the bases of the petioles form a prominent corky ridge (Fig. 34b). The stipules are always borne on the axis between the petiole bases. They are inconspicuous and eventually deciduous, triangular or narrowly so, and generally free and eglandular. In a few species, the two stipules adjacent on the axis (i.e., not those of the same leaf) may be fused across the node into bifid structures, like the stipules in *Peixotoa*, e.g., in *S. echitoides*. In *S. adenophorum*, *S. bradei*, and *S. calcaratum*, the stipule consists of a basal circular gland and a triangular membranous tip; it is eglandular in all other species.

As in many vines, the size of the lamina may vary greatly along the length of one individual. Although some species have relatively small laminas, in others the laminas may be up to 31 cm long (*S. tomentosum*) and up to 24 cm wide (*S. bogotense*). The shape of the lamina is unusually diverse for Malpighiaceae and is commonly cordate but may be also linear, lanceolate, elliptical, ovate, oblong, oblate, reniform, or orbicular, or grades between these. Although the laminas are usually unlobed, in many species characterized by cordate or broader blades, some individuals may have palmately 2–5-lobed laminas as well. In *S. jatrophifolium* the large laminas are usually palmately to pedately lobed, and in *S. angustilobum*, *S. carautae*, *S. glabrum*, and *S. urenifolium* they are always pinnately (3–) 5–7-lobed. The most atypical laminas are those of *S. laciniatum*, which, as the name implies, are finely incised, and of *S. angulosum*, which are mostly sinuate-lobate with usually 5–7 (–9) lobes. Yet, although the laminas may be deeply lobed, they are never compound. *Stigmaphyllon* is the only genus in the Malpighiaceae in which such lobed laminas occur. In most species the shape is somewhat variable, e.g., from elliptical to ovate to cordate, but in a few species in the West Indies it is exceedingly diverse. In the well-named *S. diversifolium* and in *S. emarginatum* the laminas vary from linear, lanceolate, oblong, elliptical, or ovate to sometimes suborbicular; a single specimen may exhibit several different shapes.

The margin is generally entire though sometimes very shallowly and grossly crenate, as often in *S. sinuatum*, or irregularly and grossly dentate, as in *S. jatrophifolium* and *S. vitifolium*; the margin is deeply crenate with a sessile gland at each sinus abaxially in *S. crenatum*. It may be eglandular or beset with irregularly spaced filiform glands and/or with capitate glands. The filiform glands are best seen in young leaves or on the reduced ones associated with the inflorescence (e.g., Fig. 30c), since these glands are commonly broken off above the base in mature leaves, although in *S. ciliatum* even the mature leaves are evenly fringed. The capitate glands are mostly sessile but prominent and borne directly on the margin, though often hidden by the vesture. In a number of species of eastern Brazil (e.g., *S. tomentosum*) the glands are borne adjacent to the margin on the abaxial surface. Several species are characterized by stalked glands, borne on the margin, as in *S. adenodon*, or abaxially adjacent to the margin, as in *S. blanchetii*.

Inflorescence. The flowers are grouped into pseudoracemes (sensu Cuatrecasas, 1958, and W. R. Anderson, 1981) or umbels, which are often arranged into dichasia, compound dichasia, or small thyrses that may branch to the 5th order. The number of flowers per primary aggregate varies with each species: it may be as low as 3–8 (e.g., *S. bannisterioides*, *S. calcaratum*, *S. ciliatum*) to as many as 35–50 (e.g., *S. bogotense*, *S. lindenianum*, *S. sinuatum*). The number of flowers per umbel is fixed only in *S. microphyllum*, which has solitary 4-flowered umbels, and in *S. harleyi*, which has pairs of flowers borne in leaf axils. In species with pseudoracemes the peduncles are usually closely spaced, but in a few species they are more loosely arranged; in *S. bogotense* and *S. sarmentosum* the lowest pair of flowers of an aggregate is usually separated a short distance from the rest. Each flower is borne on a pedicel, subtended by two bracteoles, which itself is borne on a peduncle subtended by a bract (Fig. 1c); as described by W. R. Anderson (1981), this basic unit is a one-flowered cincinnus. In *Stigmaphyllon*, the bracteoles are always borne at the apex of the peduncle and the bract at its base. In some species (e.g., *S. diversifolium*), the peduncle is rudimentary or greatly reduced, and the pedicel is sessile to subsessile. In a few species (e.g., *S. ellipticum*), the pedicel is distally expanded and thus widest below the calyx. Whereas the bracts are eglandular, the bracteoles often bear 1–2 inconspicuous glands ca. 0.1–0.3 mm in diameter. In *S. adenophorum* and *S. singulare* each bracteole

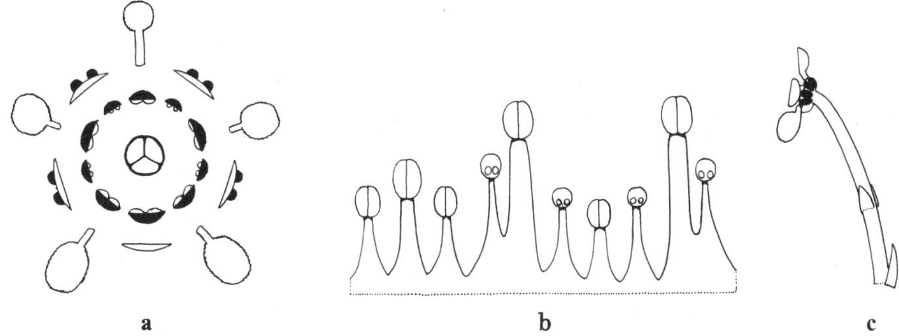

FIG. 1. Flower structure in *Stigmaphyllon*. a. Floral diagram. b. Androecium; stamen second from left opposes the anterior sepal, stamen fourth from right opposes the posterior petal (the "flag"), stamens with enlarged connectives and reduced locules oppose the lateral sepals. c. Flower borne on a pedicel subtended by two bracteoles; pedicel borne on a peduncle subtended by one bract.

bears a pair of prominent glands up to 0.8 mm in diameter, and in *S. aberrans* one bracteole of a pair has one prominent gland 0.5–0.8 mm in diameter.

Flowers. The terminology used here to describe the floral parts follows Niedenzu (1928), Cuatrecasas (1958), and W. R. Anderson (1981). The flowers are bilaterally symmetrical with the odd fifth petal (the "flag") and its opposing stamen in the posterior position and the fifth eglandular sepal and its opposing stamen in the anterior position; the plane of symmetry passes through the anterior sepal and posterior petal (frontispiece, Fig. 1a). The 3-carpellate, 3-locular, 3-stylar ovary is placed so that the odd style is aligned with the anterior sepal. The lateral sepals, petals, and stamens, and the two posterior styles are mirror images of each other.

Calyx. The calyx consists of five sepals, which are imbricate in bud. Each lateral sepal bears abaxially in the basal half a pair of oval glands; the anterior sepal is usually eglandular, although rarely it may bear one or even two, often rudimentary, glands. Only in *S. boliviense* and *S. coloratum* does the anterior sepal consistently bear a pair of glands equaling those on the lateral sepals. *Stigmaphyllon singulare* is noteworthy because its sepals are deciduous; in all other species, as is true for the family, the sepals are persistent.

Corolla. Each flower has five clawed petals with an orbicular or broadly ovate to obovate or sometimes elliptical limb. The limb of the anterior-lateral petals and posterior petal is basally truncate to cordate and abruptly set off from the claw, whereas the limb of the posterior-lateral petals is attenuate and gradually narrowed to the claw. The petals vary in size according to their position. In general, the anterior-lateral petals have the largest limbs and the posterior petal the smallest limb. The diameter of the limb varies with each species; the limb of an anterior-lateral petal may be as large as 18 mm (e.g., *S. cuzcanum*) or as small as 5 mm (e.g., *S. cardiophyllum*). The limb margin varies from erose to denticulate to fimbriate or digitate-fimbriate to sometimes lacerate. In a few species (e.g., *S. matogrossense*, Fig. 74c) the limb of the posterior petal bears at the base on the margin 1–3 stout gland-tipped fimbriae per side, 0.3–0.5 mm long and ca. 0.2 mm in diameter. The claw of the anterior-lateral petals is longer than that of the posterior-lateral petals; the claw of the posterior petal is longer and stouter than that of the lateral petals and is often constricted just below the limb. In old flowers, the claw of the posterior petal often bears the imprint of a pollinating bee's mandibles. In most species, the limbs are a pure yellow, but in some (e.g., *S. sinuatum*) the posterior petal only or all petals may be streaked or suffused with red. As a rule, the petals are glabrous, but in all of the few collections seen of *S. singulare* and in some specimens of *S. sinuatum*, primarily from Amapá (Brazil), the limbs are sparsely pubescent abaxially.

Androecium. The androecium consists of ten uniseriate stamens, which are heteromorphic in most species. They are basally connate generally in the pattern shown in Fig. 1b. Those associated with the three styles, i.e., those opposing the anterior sepal and the posterior-lateral petals, are usually the largest, and that opposing the posterior petal the smallest. Those opposing the lateral sepals not only vary in size but in most species are greatly modified. They consist of a very slender filament and an enlarged connective bearing 0–2 reduced locules. In some species (e.g., *S. sinuatum*) only the anterior-lateral stamens are modified, and the posterior-lateral stamens differ from the posterior stamen only in size, and in others (e.g., *S. bogotense*, Fig. 3f) the androecium is composed of subequal stamens, all with unmodified anthers. In most species the anthers are glabrous, but in some (e.g., *S. cardiophyllum*) they are pubescent along the margins of the locules and at the apex; in *S. tomentosum* the anthers may be glabrous or sparsely pubescent. Lowrie (1982), in his survey of pollen of the Malpighiaceae, described the pollen of *Stigmaphyllon* as "banisterioid." The pollen grains are spherical to cuboidal, hexaporate, and 12-rugate.

Gynoecium. The 3-carpellate, 3-locular ovary is densely pubescent. The carpels are completely connate in the ovary. Each locule is fertile and contains an anatropous, pendent ovule. The three styles are free to the base and persistent in fruit (though often broken off). They are terete but often distally laterally flattened, and glabrous or sparsely pubescent in the proximal 1/4–3/4; in a few species (e.g., *S. stylopogon*) the hairs are concentrated in an adaxial row. The two posterior styles are mirror-images of each other and differ from the anterior style. In most species the styles bear at the apex lateral folioles, for which the genus is named; each posterior style bears one foliole and the anterior style bears two. Instead of a foliole, the apex of the style may be only laterally expanded or perhaps have a narrow lip, or be linear and distally blunt or extended into a spur or short hook to long claw. The presence, as well as size and shape, of folioles varies with each species, although in a few species characterized by foliolate styles (e.g., *S. lindenianum*, *S. sagraeanum*) occasional individuals are encountered with styles bearing reduced folioles or lacking them. In *S. sinuatum*, a widespread and variable South American species, both foliolate and efoliolate anterior styles are common; in most of its extensive range the anterior styles are foliolate but the efoliolate form is more common in western Venezuela, Colombia, Ecuador, and northern Peru (C. Anderson 1993b). The stigma is circular, or horizontally or vertically elongate. It is always positioned at the apex of the style but on the adaxial angle, never terminally; this placement has traditionally been called "internal," i.e., facing the interior of the flower. Except for enlargement, the stance of anterior styles and of efoliolate posterior styles remains the same as the flower matures, but that of foliolate posterior styles changes. The latter are positioned in young bud so that the stigma faces the center of the flower and the foliole faces outward; as the flower expands, the distal portion of each style twists so that the stigmas face toward the posterior petal and the folioles face each other. Each style is closely associated with a stamen, particularly in species with foliolate styles. The anterior stamen is positioned so that the apex of the style fits between the locules and the folioles are spread over the apex of the anther. Each posterior style and the filament of the stamen opposing one of the posterior-lateral petals are closely apposed and often are twisted about each other; the anther fits into the axil formed by the distal portion of the style and the foliole, so that the foliole covers the anther. In most species the posterior styles are longer and stouter than the anterior style or the three styles are subequal in size; however, in a few species (*S. herbaceum*, *S. hypargyreum*, *S. maynense*, *S. orientale*, *S. puberum*) the anterior style and its opposing stamen are longer than the posterior styles and their opposing stamens.

Fruit. The fruit is a schizocarp splitting into three samaras, each suspended by a carpophore from a pyramidal torus. Generally, the samara consists of a spherical to ovoid nut with a large dorsal wing, thickened along the upper (adaxial) margin and often with a tooth near the nut. The areole, i.e., the adaxial face of the nut, is triangular to ovate and may be convex or concave at maturity. The nut may bear lateral winglets and spurs and/or crests or may be merely ribbed or smooth on its sides. The dorsal wing, which is generally elongate and distally flared, is of relatively constant size within each species and varies in length from 1.6 cm in *S. microphyllum* up to 6.2 cm in *S. saxicola*. The basic pattern is modified in various species. In some, especially those growing along watercourses, the nut contains air chambers surrounding the locule. The most extreme examples of this modification are found in *S. adenodon* var. *adenodon* (Fig. 27g, h), *S. lacunosum* (Fig. 27m), and *S. affine* (Fig. 80i, j) in which the nut is 12–19 mm in diameter and the dorsal wing is reduced. In other species from wet areas, the locule is surrounded by spongy tissue (e.g., *S. puberum*). The distinctive *S. ciliatum* has the nut laterally flattened (lenticular) and a triangular dorsal wing encircling it, and in *S. bonariense* the dorsal wing is broadly triangular to nearly square and encircles the nut. In *S. bannisterioides* (Fig. 11f, g), *S. paralias* (Fig. 58g), and *S. harleyi* (Fig. 58o) the dorsal wing is reduced to an apical crest. The most unusual "samara" is that of *S. calcaratum*, in which the dorsal wing is reduced to an erect crest and the nut is covered with numerous bulbous and warty excrescences composed of spongy tissue (Fig. 55h, i). In some species in which the samaras are not wind-dispersed the carpophore is absent (*S. bannisterioides*, *S. calcaratum*, *S. harleyi*, *S. lacunosum*, and *S. paralias*).

In most species, the embryo is ovoid and about twice as long as wide. The cotyledons are unequal, the inner often 2/3 as long as the outer. The cotyledons are straight or, especially in larger embryos, the distal portion of the outer cotyledon is folded over the inner one. Although the embryos, like the entire samara, may be quite small (e.g., ca. 3.7 mm long in *S. microphyllum*), they may also be more than 1 cm long (e.g., ca. 1.3 cm long in *S. affine*), especially in species likely to be water-dispersed. The embryos of *S. ciliatum* and *S. dichotomum* are laterally flattened and ca. 3 times as long as wide; the cotyledons are straight. In *S. bogotense* (Fig. 3k), *S. maynense* (Fig. 17h), *S. pseudopuberum*, *S. puberum* (Fig. 14m), and *S. sarmentosum* the embryo is nearly spherical and "brainlike" owing to the greatly convoluted cotyledons, which are folded within each other. The embryos of *S. bannisterioides* and *S. paralias* are circular to horseshoe-shaped. In *S. paralias* (Fig. 58h) the distal 2/5 of the outer cotyledon is folded over the inner cotyledon, but in *S. bannisterioides* (Fig. 11h) the inner cotyledon is rudimentary, only 1–2 mm long and wide.

HABITATS AND DISTRIBUTION

Stigmaphyllon occurs throughout most of the Neotropics in diverse habitats. Its range extends from southern Mexico to northern Argentina, and also includes the West Indies. It is absent only from Chile and the alpine Andes. One coastal species, *S. bannisterioides*, has become established in western Africa. Most species occur in wet or dry lowland forests or at the edges of thickets as well as in various types of secondary vegetation and disturbed areas. Although ca. 2/3 of the species occur exclusively below 1500 m, half of these below 500 m, a few are typically found mostly at altitudes above 1500 m in montane rainforests and cloud forests. Most of the high-elevation species are Andean, such as the widespread *S. bogotense*, which occurs from 1000 to 3200 m. The two exceptions are *S. pseudopuberum* of northern Guatemala and adjacent Mexico (1300–2700 m) and *S. cordatum*, also of Guatemala (1500–2500 m).

Two species are very widely distributed and commonly found in disturbed areas and secondary vegetation. *Stigmaphyllon ellipticum* is ecologically the most diverse; it occurs from Mexico to Peru along beaches and in various types of wet and dry forests from sea level to 2200 m. *Stigmaphyllon sinuatum* is a morphologically variable species of the Amazon region and the immediately adjacent areas up to 1000 m elevation. It is most common in wet primary and secondary forest but has also been recorded from white sand vegetation.

A number of species occur in very wet situations, such as lowland (often flooded) rainforest, or are associated with marshes and even swamps; they are commonly collected along water courses. Four species have been recorded from mangrove swamps (*S. affine*, *S. bannisterioides*, *S. ciliatum*) or nearby (*S. puberum*). In many of these, the nut of the samara exhibits modifications for dispersal by water, such as the presence of air chambers (e.g., *S. adenodon*, *S. affine*) or spongy tissue surrounding the locule (e.g., *S. bannisterioides*, *S. puberum*). Such modifications are sometimes associated with reduction of the dorsal wing.

Two species extend widely along the Atlantic Coast and also in the West Indies. *Stigmaphyllon bannisterioides* is found in seashore vegetation, including mangroves, from southern Veracruz, Mexico, to Maranhão, Brazil, from Cuba to Barbados, and, as noted above, along the coast of Guinea Bissau, Guinea, and Sierra Leone in western Africa. *Stigmaphyllon ciliatum* occurs in the same habitats, as well as on beaches, along rivers, and in roadside thickets more inland from Belize to Uruguay; it is naturalized in Barbados. There are unexpected gaps in the ranges of both species, which may reflect only the lack of collecting activity in suitable habitats; both have large showy flowers and are unlikely to be overlooked. They were only recently discovered in coastal Nicaragua by collectors associated with the Missouri Botanical Garden during field work in preparation for the Flora of Nicaragua. Neither species has been recorded from Costa Rica, and *S. ciliatum* is unknown from Panama and the Guianas, but both should be expected in those countries.

Four species apparently occur mainly on dry sandy substrates: *S. arenicola* on beaches, dunes, and in restingas; *S. auriculatum* in restingas and caatingas; *S. palmatum* in savannas and on beaches and sandbanks; and *S. paralias* in restingas and caatingas and on beaches. All of them are also found in dry woods or scrub forests, and *S. paralias* also grows in cerrado. *Stigmaphyllon lalandianum* and *S. tomentosum*, both of southeastern Brazil, are typically found in cerrado as well as in dry woods and secondary vegetation; *S. lalandianum* is also found in restingas.

The West Indian endemics *S. diversifolium*, *S. emarginatum*, and *S. sagraeanum* (and probably also *S. microphyllum*) occur in limestone and serpentine areas as well as in various disturbed situations. The Mexican *S. selerianum* also prefers limestone substrates.

The greatest diversity of the genus is in South America. Of the 90 species, only 15 occur in Mexico and Central America, and eight of these occur only there: *S. adenophorum, S. cordatum, S. lindenianum, S. panamense, S. pseudopuberum, S. retusum, S. selerianum,* and *S. tonduzii,* all with restricted ranges. Thirteen species occur in the West Indies, but of these *S. ciliatum* is naturalized (in Barbados only), *S. finlayanum* is recorded only once (from St. Vincent, probably introduced) and *S. convolvulifolium* only twice (from Martinique, probably introduced), and *S. adenodon* is found only on Grenada. Of the remainder, seven are endemic to the West Indies, and of these five are highly restricted (*S. angulosum* on Hispaniola; *S. floribundum* on Puerto Rico, Virgin Gorda, and St. John; *S. laciniatum* on Île de la Gonâve; *S. microphyllum* on Cuba; *S. sagraeanum* on Cuba and the Bahamas); *S. emarginatum* is found throughout the Antilles (except Cuba) and *S. diversifolium* on Cuba and the Lesser Antilles. As may be expected, none of the West Indian endemics are among the six species found in Trinidad and Tobago (*S. adenodon, S.*

bannisterioides, S. ciliatum, S. convolvulifolium, S. finlayanum, S. puberum); one of these, *S. finlayanum*, is only recorded from Trinidad and the adjacent Paria Peninsula of Venezuela.

In South America, *Stigmaphyllon* is most diverse in Colombia (25 species including 8 endemics), Peru (18 species including 5 endemics), and Brazil (45 species including 30 endemics). Three Amazonian species (*S. adenodon, S. lacunosum, S. sinuatum*) and the widespread *S. puberum* are found in all three countries. *Stigmaphyllon alternans, S. bogotense, S. ellipticum,* and *S. sarmentosum* occur in Colombia as well as in Peru, and four species of the Amazonian lowlands occur both in Brazil and Peru (*S. cardiophyllum, S. florosum, S. maynense, S. strigosum*).

In Colombia, *Stigmaphyllon* is found mostly in the Cordillera Central and in the area extending to the Pacific Coast; however, the eastern lowlands are botanically little explored. *Stigmaphyllon echitoides* and *S. velutinum* are the only widespread endemics, but whereas *S. echitoides* appears to be common, *S. velutinum* is known from only five collections. The other six species restricted to Colombia (*S. goudotii, S. orientale, S. romeroi, S. stenophyllum, S. suffruticosum, S. tergolanatum*) have been collected only once or very few times near their respective type localities. In Peru, the genus is represented both east and west of the Andes. Of the endemics, only *S. argenteum* has a broad range; the other four (*S. aberrans, S. cuzcanum, S. peruvianum, S. tarapotense*) are known only from the type or from a few, closely spaced localities. As may be expected, in addition to a few widespread species many of the 13 species found in Ecuador are also native to Colombia or Peru; three species (*S. ecuadorense, S. eggersii, S. nudiflorum*) are endemic to Ecuador. Bolivia shares three species with Peru, *S. cardiophyllum* (also in Brazil), *S. strigosum*, and *S. florosum,* which has a wide range in western South America. The other three Bolivian species, *S. boliviense*, *S. coloratum*, and *S. yungasense*, are known only from the departments of La Paz and Santa Cruz.

Within Brazil, the greatest diversity is concentrated in the eastern part of the country south of the Amazon Basin. Twenty-six species occur in the region extending from Ceará to Rio Grande do Sul, and of these only four occur elsewhere (*S. bonariense, S. ciliatum, S. jatrophifolium, S. paralias*). Of the remainder, only *S. auriculatum, S. blanchetii, S. lalandianum,* and *S. tomentosum* are moderately widespread. The planalto, a region characterized by high endemism in many groups (e.g., *Peixotoa*, C. Anderson 1982; *Paepalanthus* Kunth, Hensold 1988; *Mimosa* L., Barneby 1991; *Byrsonima* Rich. ex Kunth, W. R. Anderson, pers. comm.), harbors surprisingly few species of *Stigmaphyllon*: the common *S. lalandianum*, *S. paralias* on crystalline and white-sand substrates, *S. urenifolium* (also found in Bahia), and *S. macedoanum,* known only from the type locality near Capinópolis in Minas Gerais. Fourteen species occur in the Amazon region and the adjacent states of Maranhão and Piauí, and nearly all of them have broad ranges in South America. The exceptions are *S. jobertii*, known only from the type in Maranhão, and *S. calcaratum*. The latter is highly distinctive in both vegetative and reproductive characters, and has a disjunct distribution; it is found along the Rio Amazônas near Manaus and Santarém, and also in the Chaco region of northern Argentina, Paraguay, the Pantanal of southwestern Brazil, and adjacent Bolivia. Six species are known from Mato Grosso and Rondônia, including *S. matogrossense* and *S. stylopogon*, both only from these states.

Surprisingly, only ten species occur in Venezuela, and none are endemic there. Five species are recorded from the Guianas, four of them widespread (*S. bannisterioides, S. convolvulifolium, S. puberum, S. sinuatum*) and one, *S. palmatum*, extending from French Guiana to adjacent Brazil (Pará). In Argentina and Paraguay, in addition to *S. calcaratum*, mentioned above, only *S. bonariense* and *S. jatrophifolium* are found. These two species are most commonly collected along the Río Paraná, Río Uruguay, the southern Río Paraguay, and their tributaries, but only *S. jatrophifolium* extends into Uruguay. The coastal *S. ciliatum* is the only other species recorded from Uruguay.

GENERIC AND INFRAGENERIC AFFINITIES

The Malpighiaceae, noted for floral conservatism (W. R. Anderson 1979b), have traditionally been divided largely on the basis of fruit characters. Jussieu (1843), in the first monograph of the family, proposed four tribes based on stamen number and also presence/absence of fruit wings. Genera with an androecium composed of eight or fewer stamens he assigned to *Meiostemones*. Those with ten stamens he grouped into *Apterygieae* (genera with unwinged fruits), *Notopterygieae* (genera, including *Stigmaphyllon*, with mericarps bearing an enlarged dorsal wing and small lateral winglets or lacking lateral ornamentation), and *Pleuropterygieae* (genera with mericarps bearing well-developed lateral wings equaling or exceeding any dorsal ornamentation). Grisebach (1858) and Hooker (1862) accepted Jussieu's division of the family and his placement of *Stigmaphyllon*.

Nearly all wing-fruited genera have a pyramidal receptacle (the torus) to which the mericarps are attached; the genera lacking fruit wings generally have a flat receptacle. Niedenzu, who treated the Malpighiaceae for *Die natürlichen Pflanzenfamilien* (1896) and *Das Pflanzenreich* (1928), used the nature of the receptacle to define his two subfamilies, *Pyramidotorae* (*Pterygophorae*) and *Planitorae* (*Apterygiae*), and characters of the fruit to circumscribe tribes and subtribes as well as subgeneric divisions. Like Jussieu, he placed *Stigmaphyllon* nearest to *Banisteria* [=*Banisteriopsis*], *Peixotoa*, and *Heteropterys*, all New World genera that also have samaras with a large elongate dorsal wing. The samaras of *Heteropterys* differ from those of *Banisteriopsis*, *Peixotoa*, and *Stigmaphyllon* in that they lack a well-defined carpophore and are thickened along the lower (abaxial) margin; in the other three genera, the mature samaras are usually suspended by a carpophore from the torus, and the dorsal wing is thickened along the upper (adaxial) margin. *Stigmaphyllon* shares some characters of the inflorescence, flower, and fruit with *Banisteriopsis*, *Peixotoa*, and *Heteropterys*, and also a few attributes with the Old World *Ryssopterys*, notably cordate laminas borne on long petioles with a pair of glands at the apex, dichasially branched inflorescences, and a samara with an elongate dorsal wing.

Three additional New World genera have such dorsally winged samaras: *Cordobia* Nied. (a monotypic genus of Argentina), *Janusia* Adr. Juss., and *Ectopopterys* W. R. Anderson (a monotypic genus of Colombia, Ecuador, and Peru), none of them considered closely allied to *Stigmaphyllon*. In particular, *Ectopopterys* does not belong here, because the "dorsal" wing most likely is an enlarged lateral wing shifted to a functionally dorsal position (W. R. Anderson 1980). The New World genera with samaras bearing an enlarged dorsal wing can be separated with the following key.

1. Dorsal wing of samara thickened along the lower (abaxial) margin.
   2. Androecium homogeneous or nearly so; styles all free, unornamented or at most the apex with a dorsal hook. *Heteropterys.*
   2. Androecium strongly heterogeneous; posterior styles apically connate, anterior style foliolate. *Ectopopterys.*

1. Dorsal wing of samara thickened along the upper (adaxial) margin.
   3. Stigma internal, the styles commonly bearing apical folioles. *Stigmaphyllon.*

3. Stigma terminal (in *Cordobia* sometimes appearing slightly lateral), the styles never ornamented.
    4. Androecium of 10 stamens (and staminodes); gynoecium with 3 styles.
        5. Androecium of 10 fertile stamens; stipules distinct. *Banisteriopsis.*
        5. Androecium of 5 fertile stamens opposing the petals and alternating with 5 staminodes opposing the sepals; stipules of opposing leaves fused across the node into a heart-shaped structure.
            6. Staminodes consisting of a filament and a large glandular connective; flowers borne in umbels. *Peixotoa.*
            6. Staminodes consisting only of a filament; flowers solitary in the leaf axils. *Cordobia.*
    4. Androecium of 5 or 6 stamens and staminodes; gynoecium with 1 style. *Janusia.*

The subdivision of the family has since been reviewed by W. R. Anderson (1978), who proposed two differently defined subfamilies, the Byrsonimoideae and the Malpighioideae, based on habit, style and fruit morphology, and chromosome numbers. As summarized by W. R. Anderson (1993), chromosome numbers among the majority of wing-fruited genera of the Malpighioideae, including *Stigmaphyllon,* are not informative, because they are "monotonously uniform, with only rare departures from diploid (n = 10) to tetraploid (n = 20)." The few species of *Stigmaphyllon* that have been surveyed and are vouchered are all reported as n = 10 (Ormond et al. 1981; W. R. Anderson 1993). The answer to the question "which is the closest relative" must await the results from the current investigation of the phylogeny of the Malpighiaceae by W. R. Anderson and M. W. Chase. *Peixotoa* is a homogeneous group, highly specialized in its inflorescence and flowers, that probably arose from a banisterioid ancestor. Because it is not a likely sister-group to *Stigmaphyllon*, it is not considered further here. For additional information about *Peixotoa*, see C. Anderson (1982). The comments on *Banisteriopsis* and *Heteropterys* are based on the accounts by W. R. Anderson (1981) and Gates (1982).

Stigmaphyllon is characterized by a dichasial arrangement of the primary aggregates of flowers (pseudoracemes to umbels), an androecium with the stamens opposite the lateral sepals bearing an enlarged connective and reduced locules in most species, styles with an internal stigma and commonly folioles, and a samara in which the upper (adaxial) margin of the dorsal wing is thickened. The long-petioled heart-shaped leaves, so characteristic of many species of *Stigmaphyllon*, and the divided laminas of some are not found in any other New World genera of Malpighiaceae. In *Banisteriopsis*, the secondary inflorescence varies from racemose to cymose [but dichasial in *B. mathiasiae* (W. R. Anderson) W. R. Anderson], the stamens opposing the sepals of many species have enlarged connectives, and the dorsal wing of the samara is adaxially thickened, but the stigmas are strictly terminal and the styles lack apical ornamentation. *Heteropterys* differs from *Stigmaphyllon* in its racemose to paniculate secondary inflorescences and the samara, in which the lower (abaxial) margin is thickened. The androecium is similar to androecia in the species of *Stigmaphyllon* with subequal stamens, and the stigma is internal. The styles of *Heteropterys* are apically dorsally rounded, truncate, acute, or hooked, and resemble those of species of *Stigmaphyllon* that lack folioles.

As in many vines, the leaves of *Stigmaphyllon* are greatly variable, both in size and shape. The elliptical to lanceolate to ovate or obovate (sometimes orbicular) lamina borne on a short petiole (less than 3 cm long) that is so typical of the family occurs in *Stigmaphyllon* as well, but the large-cordate to reniform shape and the long petiole (e.g., in *S. bogotense* up to 25 cm long) is otherwise known only in *Ryssopterys.* Within the Malpighiaceae a lobed/divided lamina is unique to *Stigmaphyllon.* The lobed or divided leaf is clearly a derived condition, but in many species not only the outline of the lamina but also the expression of lobing is variable. In a few species (e.g., *S. laciniatum*, *S. angustilobum*) the leaves are always deeply lobed, but in many, seemingly characterized by entire laminas (e.g., *S. lindenianum*, *S. dichotomum*), occasional individuals with at least some lobed laminas are encountered. For species represented by few collections it is impossible to assess whether the tendency to produce lobed leaves is present. The vesture of the laminas (except in a very few species) is composed either of sessile or stalked hairs but no pattern is evident; both conditions are found in *Heteropterys* and *Banisteriopsis*, as well as in other genera, and thus are not informative. Only one leaf character is clearly derived within *Stigmaphyllon*. A group of species from eastern-southeastern Brazil and *S. alternans* (Colombia to Peru) differ in that the marginal glands are borne abaxially adjacent to the margin rather than on the margin, as in the other species. Some species lack marginal laminar glands, which may reflect an ancestral state or a loss.

The inflorescence unit in *Stigmaphyllon* varies from many-flowered pseudoracemes (15–50 flowers) to few-flowered umbels (3–8 flowers), which may be solitary but usually are grouped into dichasial inflorescences; in *S. harleyi* the flowers are grouped in pairs in the leaf axils. There is a loose correlation between size of petals and number of flowers per unit. Flowers with large petals, i.e., diameter of the limb 13–19 mm, are grouped into few-flowered umbels, e.g., *S. ciliatum* (3–8 flowers per umbel, limb diameter 12–18 mm), whereas in many-flowered pseudoracemes the limb of the petals is 8–11 mm in diameter, e.g., *S. dichotomum* (15–50 flower per pseudoraceme, limb diameter 7.5–9 mm); however, there are numerous exceptions to this generalization. *Stigmaphyllon microphyllum* has solitary and strictly 4-flowered umbels, and the petal limbs are 8.5–9.3 mm in diameter; *S. angustilobum* has compound inflorescences composed of 15–35-flowered pseudoracemes, and the petal limbs are 13–17 mm in diameter.

The peduncle and pedicel of *Stigmaphyllon* usually are similar in aspect, but in a few species that also share other character states (e.g., *S. ellipticum*, *S. echitoides*) the pedicel is distally expanded and differs in vesture or may be glabrous. In some species (e.g., *S. diversifolium*, *S. paralias*) the peduncle is much reduced or rudimentary; however, as also in *Banisteriopsis*, such a suppression of the peduncle is found in a number of diverse species and probably occurred several times. In *Stigmaphyllon*, a remnant of the peduncle is always present, whereas in *Peixotoa* and many species of *Banisteriopsis* the pedicels are sessile and subtended by the bract and bracteoles.

In *Stigmaphyllon*, three patterns are discernible in the androecium: stamens subequal or varying only slightly in size; stamens unequal in size, those opposite the anterior-lateral sepals with the connective enlarged and the locules reduced or absent; and stamens unequal in size, those opposite all lateral sepals with the connective enlarged and the locules reduced or absent. An unmodified androecium is clearly the ancestral condition in the family; however, it cannot be assumed that having only the stamens opposing the anterior-lateral sepals modified preceded having all stamens opposing the lateral sepals with the connective enlarged and the locules reduced or absent. The latter condition may well have arisen separately, i.e., directly from the unmodified androecium. In *Heteropterys* and *Banisteriopsis* subg. *Banisteriopsis* the androecium is homogeneous, but in many species of *Banisteriopsis* subg. *Hemiramma* (Griseb.) Nied. and subg. *Pleiopterys* Nied. the stamens have enlarged, glandular connectives and those opposing the sepals are larger than those opposing the petals. In a number of species of *Stigmaphyllon* in which all stamens opposing the lateral sepals are modified, these may be sterile (eloculate); this loss of locules is more common in those opposing posterior-lateral sepals. In a few species the presence of 1–2 locules or their absence is quite variable, which suggests that the reduction/elimination of locules in modified anthers is a change easily achieved and may have occurred several times.

Although unornamented styles may be assumed to represent the ancestral condition in *Stigmaphyllon*, absence of folioles may also indicate a loss. Several species characterized by foliolate styles include individuals whose styles are efoliolate or bear reduced folioles. Some of these variants are frequent enough that they have been named, e.g., "*S. lindenianum* var. *yucatanum*" for forms with an efoliolate anterior style. Representatives of *S. sinuatum*, a widespread and variable species in the Amazon and adjacent regions, have all three styles foliolate in the eastern and central part of the range, but in most western populations the anterior style has only a long claw and lacks folioles (previously recognized as "*S. brachiatum*" and "*S. monancistrum*"). Intermediates are infrequent but do occur throughout the range. In these individuals the anterior style may bear only one or two reduced folioles or two unequal folioles or only one large foliole (C. Anderson 1993b). Variation in the ornamentation of the posterior styles occurs in *S. harleyi, S. tarapotense*, and *S. sagraeanum*. In *S. harleyi*, known only from the type collection, the anterior style is always foliolate but the posterior styles are either foliolate or extended into a claw. In *S. tarapotense* the anterior style lacks folioles, but the apex of the posterior styles may vary even within the same umbel from bearing a lip to a foliole up to 0.8 mm long and wide. The anterior style of *S. sagraeanum* also bears only a short spur; in most collections each posterior style has a large foliole, but in some the folioles are reduced or represented only by a lateral lip or may even be absent. These variants, in all other respects agreeing with *S. sagraeanum*, have been recognized as "*S. sagraeanum* f. *primaevum*" and "*S. nipense*." Variation in both the anterior and posterior styles occurs in *S. lalandianum* (and to a lesser extent in the very similar *S. acuminatum*) in which the apical appendage of the styles varies from a very narrow lateral lip to a small foliole. In all these species, the pattern may vary among and within individuals of a population or sometimes even within the same umbel. Thus, it could be argued that the lack of folioles in species in which the style is extended into a claw reflects a secondary loss of folioles rather than an ancestral condition.

As described in "Morphology," a foliolate style is closely associated with an enlarged, fertile stamen, and the argument (untested) may be made that the folioles serve to prevent self-pollination and that changes in the androecium and gynoecium are somehow linked. For example, although in most species the anterior style and its associated stamen are shorter than or subequal to the posterior styles and their opposing stamens, in a few species, e.g., *S. puberum*, the anterior style and its stamen exceed the posterior styles and their stamens. In general, species with "*Heteropterys*"-like, efoliolate styles have an undifferentiated androecium, and species with foliolate styles have a modified androecium. Yet, exceptions exist, as the following examples demonstrate. In the West Indian endemics *S. emarginatum*, *S. diversifolium*, and *S. floribundum*, styles lack folioles or even a lateral lip but the stamens are heteromorphic. In the last two, the stamens opposing the sepals are sterile, i.e., they consist of a filament and a glandular connective lacking locules. In *S. aberrans* (Peru) and *S. bannisterioides* (Caribbean islands and Atlantic coast from Mexico south to northern Brazil), all styles have a long claw; in the former the androecium is modified, in the latter it is homogeneous. In *S. florosum* (Ecuador to western Brazil), the anterior style is dorsally blunt and the posterior styles have a very narrow lateral lip; all stamens opposing the lateral sepals are modified.

In some species of *Stigmaphyllon* the nut of the samara is enlarged and contains air pockets surrounding the locule, and the embryo is larger than that of species without such chambers in the samara; in a few cases, this is accompanied by a reduction of the dorsal wing (e.g., *S. adenodon*). Although this characteristic is clearly derived, it is found in diverse species associated with wet habitats in which a floating propagule is advantageous and is likely to have arisen several times. Such air pockets have been reported in few other genera of Malpighiaceae; similar devices for making the samara buoyant apparently occur in *Diplopterys* Adr. Juss. (Gates 1982), a genus of four riverine species closely allied with *Banisteriopsis*, and in *Jubelina* Adr. Juss. and *Lophanthera* Adr. Juss. (W. R. Anderson 1981). Two divergent patterns noted in embryos are clearly specialized within *Stigmaphyllon*. In most species the cotyledons are straight or folded back distally. In a few species, e.g., *S. bogotense*, the cotyledons are convoluted and folded within each other. [In Malpighiaceae an embryo with intricately folded cotyledons, though in a different pattern, has otherwise been reported only for *Mcvaughia* W. R. Anderson (W. R. Anderson 1979a).] The second type is found only in *S. bannisterioides* and *S. paralias*, two species also atypical for the genus in other characters. Their embryos are horseshoe-shaped instead of ellipsoid and consist of one very large cotyledon and a much smaller one; in *S. bannisterioides* the smaller cotyledon is rudimentary (1–2 mm long).

Although *Stigmaphyllon* is readily separated from other genera, few characters may be polarized with confidence; many appear to have arisen several times and several are variable within a species. Thus, attempts at cladistic analyses proved unsuccessful. In order to present a convenient overview of the genus, the species are here arranged in groups based primarily on similarities in the androecium and gynoecium, but without according these groups taxonomic recognition; they may well prove to be paraphyletic or perhaps even polyphyletic. In the section "Taxonomy," the sequence in which the species are listed matches these three groups and reflects morphological similarities. It is hoped that future attempts using molecular analyses may be useful in discerning the phylogenetic patterns in *Stigmaphyllon*. Because the nature of the androecium is not known for *S. jobertii* and *S. crenatum*, they are not assigned to any "group" below.

Group I: Stamens equal in shape and subequal or somewhat unequal in size, the anthers all with a small glandular connective and two full-sized locules. Half of the 16 species with such an androecium have styles without ornamentation or at most with a short spur (to ca. 0.2 mm long) (*S. bogotense, S. glabrum, S. pseudopuberum, S. sarmentosum, S. stenophyllum, S. suffruticosum, S. urenifolium,* and *S. yungasense*). Two of these, *S. glabrum* and *S. urenifolium*, stand out by their pinnately lobed leaves and large, lacerate-fimbriate petals (the limb of the lateral petals 14–16 mm in diameter). Of the others, only *S. yungasense* has the lateral petals with limbs 10–11.5 mm in diameter and fimbriae up to 0.6 mm long. The rest have small petals. Those of *S. bogotense*, *S. stenophyllum*, and *S. suffruticosum* are digitate-fimbriate and of *S. sarmentosum* short-fimbriate to erose-denticulate, all with the limb of the lateral petals 6.5–10 mm in diameter, and those of *S. pseudopuberum* are erose with the limb of the lateral petals 5.5–7 mm in diameter. *Stigmaphyllon bogotense*, *S. pseudopuberum*, and *S. sarmentosum* also have an unusual embryo in which the cotyledons are convoluted and folded within each other; the embryos of *S. stenophyllum* and *S. suffruticosum* are not known.

Of the remaining eight species, three have the anterior style extended into a claw; in *S. bannisterioides* the posterior styles are clawed as well, in *S. peruvianum* they are blunt, and in *S. cordatum* they vary from long-spurred to bearing a lip or a very small foliole. In *S. romeroi*, all styles have a narrow lateral lip. Only *S. orientale* and *S. harleyi* have all three styles bearing folioles, though in *S. harleyi* (known only from one collection) some flowers have the posterior styles only clawed. *Stigmaphyllon orientale* is unusual in that the anterior style exceeds the posterior two, a condition known otherwise only in four species of group II (*S. herbaceum*, *S. hypargyreum*, *S. maynense*, and *S. puberum*). In both *S. lalandianum* and *S. acuminatum*, which differ from each other most strikingly in leaf pubescence, the ornamentation of all styles varies from a narrow lip to a small foliole up to ca. 0.9 mm in diameter. The stamens opposing the lateral sepals often have the locules slightly reduced, but in neither species is the connective ever greatly enlarged. If laminar glands are present in *S. lalandianum* they are borne abaxially, adjacent to the margin; it shares this trait with the 14 species (nearly all, like *S. lalandianum* and *S. acuminatum*, from eastern Brazil) cited below in group III. The inflorescence unit is a 15–40-flowered pseudoraceme or umbel in four species (*S. acuminatum*, *S. cordatum*, *S. romeroi*, and *S. lalandianum*) but a few-flowered umbel in three others (*S. bannisterioides*, 3–6; *S. peruvianum*, 6–8; *S. orientale*, 8–12) and in *S. harleyi* an axillary pair of flowers.

Group II: Only the stamens opposite the anterior-lateral sepals with the connective enlarged and bearing two reduced locules. In 15 of the 20 species with such an androecium all styles bear large folioles. Two of the exceptions are noted above: in populations of *S. sinuatum*, particularly in the western part of its range, the anterior style lacks folioles but is extended into a long claw, and in *S. tarapotense*, in which the anterior style is efoliolate, the ornamentation of the posterior styles varies from a lip to a large foliole. *Stigmaphyllon carautae*, known only from the type, has all three styles efoliolate and is most similar in its flowers (except for the androecium) and its pinnately lobed leaves to *S. glabrum* and *S. urenifolium*, both included in group I. *Stigmaphyllon columbicum* is distinguished by its pairing of an anterior clawed style with foliolate posterior styles. In the enigmatic *S. velutinum*, the only species in the genus with mostly velutinous pubescence, all styles have a narrow lip that may be expanded into a tiny triangular foliole, up to 0.2 mm wide. Four species in this group (*S. herbaceum*, *S. hypargyreum*, *S. maynense*, *S. puberum*) stand out, because the anterior style and its opposing stamen are much larger than the posterior styles and their opposing stamens. *Stigmaphyllon orientale* (group I) is the only other species in which the anterior style exceeds the posterior two. In *S. goudotii*, *S. macedoanum*, and *S. singulare* the locules of the modified stamens are not as greatly reduced as in the other species but the connective is enlarged. The other species included in group II are *S. adenodon, S. adenophorum, S. angulosum, S. convolvulifolium, S. cuzcanum, S. ecuadorense, S. lacunosum*, and *S. nudiflorum.*

Group III: Stamens opposite the anterior-lateral and posterior-lateral sepals with the connective enlarged and bearing 0–2 reduced locules. (In *S. paralias*, a species atypical in many aspects, the stamen opposite the posterior petal and those opposing the posterior-lateral sepals lack locules.) This group comprises 60% of the species of *Stigmaphyllon*. Only six species in group III lack folioles: in *S. florosum* the anterior style is blunt and the posterior styles have only a narrow lip; in *S. diversifolium*, *S. emarginatum*, and *S. floribundum* the styles have a tiny spur; in *S. aberrans* all styles are clawed; and in *S. selerianum* the anterior style is extended into a claw and the posterior styles have either a narrow lip or a small triangular foliole. In *S. sagraeanum* the anterior style is always efoliolate; the posterior styles are usually foliolate, but in some individuals or populations they are efoliolate. *Stigmaphyllon diversifolium* and the very similar *S. floribundum* are also characterized by having the stamens opposing the lateral sepals sterile, i.e., the glandular connective lacks locules.

Five species have the anterior style with only a lip or a very small foliole on each side but the posterior styles with a large foliole. These folioles are broadly to narrowly lunate in *S. echitoides*, *S. ellipticum*, and *S. venulosum* but triangular to parabolic, the more common shape, in *S. eggersii* and *S. tergolanatum*. In a few collections of the very widespread *S. ellipticum* the posterior styles have the folioles reduced, sometimes to a narrow lip. These five species also share many other character states. The flowers are grouped in 8–15-flowered umbels, the pedicels are distally expanded, and the petals are large (the limb of the lateral petals 13–19 mm in diameter) and fringed with fimbriae 1–2 mm long. The leaves are never cordate or lobed and have relatively short petioles (up to 5.5 cm long, but usually shorter).

In all other species of group III all styles bear well-developed folioles. Among these several alliances may be discerned. In *S. boliviense*, *S. bradei*, *S. coloratum*, *S. puberulum*, and *S. vitifolium* the petiole glands are shallowly cupulate and borne on the petiole well below the base of the lamina instead of at the point of insertion. The stamens opposing the lateral sepals lack locules in *S. coloratum* and *S. puberulum* but are 2-loculate in *S. vitifolium;* in *S. boliviense* and *S. bradei* those opposite the anterior-lateral sepals lack locules but those opposite the posterior-lateral sepals have two tiny locules. *Stigmaphyllon vitifolium* also differs in its lobed or at least grossly dentate laminas, presence of the peduncle, and small triangular folioles of the anterior style; the other three species have broadly cordate to orbicular laminas, reduced to rudimentary peduncles, and the anterior style with large folioles.

Three species (*S. auriculatum*, *S. ciliatum*, *S. jatrophifolium*) have glabrous leaves and the other vegetative parts also glabrous or very sparsely sericeous; in *S. calcaratum* the herbage varies from sparsely sericeous to glabrate. Their flowers have large, fringed petals and are displayed in umbels. Only in *S. jatrophifolium* are the modified stamens always 2-loculate. In *S. calcaratum* and *S. ciliatum* they vary from sterile (eloculate) to 1–2-loculate, and in *S. auriculatum* the stamens opposing the posterior-lateral sepals show such variation, but those opposing the anterior-lateral sepals always bear two locules.

Fourteen species from eastern and southeastern Brazil and *S. alternans* (Colombia to Peru) differ from the others in bearing the laminar marginal glands abaxially adjacent to the margin instead of on the margin. In four (*S. blanchetii*, *S. gayanum*, *S. hatschbachii*, and *S. salzmannii*) the glands are stalked; in all of these the stamens opposite the posterior-lateral sepals lack locules. Among the other ten species, these stamens are also eloculate in *S. arenicola* (stamens opposite the anterior-lateral sepals sterile as well), *S. alternifolium* and *S. macropodum*, and usually only with one locule in *S. saxicola*. In *S. angustilobum* and *S. tomentosum*, the stamens opposing the lateral sepals are variable; they may be eloculate or bear one or two reduced locules. In *S. affine*, *S. alternans*, *S. bonariense,* *S. cavernulosum*, and *S. rotundifolium* they always have two reduced locules.

Among all but two of the remaining species that have laminas with eglandular margins or glands borne on the margin, the anterior-lateral stamens are 2-loculate. They lack locules in *S. laciniatum* and *S. matogrossense;* in the latter those opposite the posterior-lateral sepals are sterile (eloculate) as well. In six species (*S. dichotomum*, *S. lindenianum*, *S. microphyllum*, *S. panamense*, *S. retusum*, *S. strigosum*) the stamens opposing the posterior-lateral sepals have two reduced locules, in five species (*S. argenteum*, *S. cardiophyllum*, *S. finlayanum*, *S. stylopogon*, *S. tonduzii*) they have one reduced locule, and in *S. palmatum* they are sterile (eloculate). In *S. paraense* they vary from 0–2-loculate.

TAXONOMY

Stigmaphyllon Adr. Jussieu in St.-Hilaire, Fl. bras. mer. 3: 48. 1833 ["1832"]. *Stigmaphyllon* subg. *Stigmaphyllon* [subg. "*Eustigmatophyllum*"] sect. *Macropterys* Niedenzu, Ind. Lect. Lyc. Brunsberg. p. aest. 1900: 3. 1900. *Stigmaphyllon* subg. *Stigmaphyllon* sect. *Macropterys* subsect. *Machaeropterys* Niedenzu, Ind. Lect. Lyc. Brunsberg. p. aest. 1900: 19. 1900. *Stigmaphyllon* subg. *Stigmaphyllon* sect. *Macropterys* subsect. *Machaeropterys* ser. *Eumachaeropterys* Niedenzu, Verz. Vorles. Akad. Braunsberg W.-S. 1912–1913: 30. 1912.—LECTOTYPE, designated by Small, 1910: *Stigmaphyllon auriculatum* (Cavanilles) Adr. Jussieu.

Brachypterys Adr. Jussieu in Delessert, Icon. sel. 3: 20, t. 34. 1838 ["1837"]. *Stigmaphyllon* subg. *Stigmaphyllon* sect. *Brachypterys* (Adr. Jussieu) Niedenzu, Ind. Lect. Lyc. Brunsberg. p. aest. 1900: 30. 1900.—TYPE: *Brachypterys australis* Adr. Jussieu [=*Stigmaphyllon paralias* Adr. Jussieu].

Stigmaphyllon sect. *Baeopterys* Grisebach, Fl. Brit. W. I. 118. 1859. *Stigmaphyllon* subg. *Baeopterys* (Grisebach) Niedenzu, Ind. Lect. Lyc. Brunsberg. p. hiem. 1899–1900: 4. 1899. *Stigmaphyllon* subg. *Baeopterys* sect. *Eubaeopterys* Niedenzu, Ind. Lect. Lyc. Brunsberg. p. hiem. 1899–1900: 4. 1899. *Stigmaphyllon* subg. *Baeopterys* sect. *Eubaeopterys* subsect. *Odontopterys* Niedenzu, Ind. Lect. Lyc. Brunsberg. p. hiem. 1899–1900: 4. 1899.—LECTOTYPE, designated by Morton, 1968: *Stigmaphyllon emarginatum* (Cavanilles) Adr. Jussieu.

Stigmaphyllon subg. *Baeopterys* sect. *Eubaeopterys* subsect. *Homalopterys* Niedenzu, Ind. Lect. Lyc. Brunsberg. p. hiem. 1899–1900: 9. 1899.—LECTOTYPE, here designated: *Stigmaphyllon lanuginosum* Niedenzu [=*Stigmaphyllon bogotense* Triana & Planchon].

Stigmaphyllon subg. *Baeopterys* sect. *Monancistrum* Niedenzu, Ind. Lect. Lyc. Brunsberg. p. hiem. 1899–1900: 10. 1899. *Stigmaphyllon* subg. *Baeopterys* sect. *Monancistrum* subsect. *Eumonancistrum* Niedenzu, Ind. Lect. Lyc. Brunsberg. p. hiem. 1899–1900: 10. 1899.—LECTOTYPE, here designated: *Stigmaphyllon sagraeanum* Adr. Jussieu.

Stigmaphyllon subg. *Baeopterys* sect. *Monancistrum* subsect. *Hemiphyllum* Niedenzu, Ind. Lect. Lyc. Brunsberg. p. hiem. 1899–1900: 13. 1899.—TYPE: *Stigmaphyllon cordatum* Rose.

Stigmaphyllon subg. *Stigmaphyllon* sect. *Macropterys* subsect. *Probolopterys* Niedenzu, Ind. Lect. Lyc. Brunsberg. p. aest. 1900: 3. 1900. *Stigmaphyllon* subg. *Stigmaphyllon* sect. *Macropterys* subsect. *Probolopterys* ser. *Stenodoma* Niedenzu, Ind. Lect. Lyc. Brunsberg. p. aest. 1900: 3. 1900.—LECTOTYPE, here designated: *Stigmaphyllon incanum* Niedenzu [=*Stigmaphyllon velutinum* Triana & Planchon].

Stigmaphyllon subg. *Stigmaphyllon* sect. *Macropterys* subsect. *Probolopterys* ser. *Eurydoma* Niedenzu, Ind. Lect. Lyc. Brunsberg. p. aest. 1900: 7. 1900.—LECTOTYPE, here designated: *Stigmaphyllon convolvulifolium* Adr. Jussieu.

Stigmaphyllon subg. *Stigmaphyllon* sect. *Macropterys* subsect. *Prosodynamis* Niedenzu, Ind. Lect. Lyc. Brunsberg. p. aest. 1900: 22. 1900.—TYPE: *Stigmaphyllon puberum* (Richard) Adr. Jussieu.

Stigmaphyllon subg. *Stigmaphyllon* sect. *Eurypterys* Niedenzu, Ind. Lect. Lyc. Brunsberg. p. aest. 1900: 23. 1900. *Stigmaphyllon* subg. *Stigmaphyllon* sect. *Eurypterys* subsect. *Coelocarpium* Niedenzu, Ind. Lect. Lyc. Brunsberg. p. aest.

1900: 23. 1900.—LECTOTYPE, here designated: *Stigmaphyllon grenadense* Niedenzu [=*Stigmaphyllon adenodon* Adr. Jussieu].

Stigmaphyllon subg. *Stigmaphyllon* sect. *Eurypterys* subsect. *Pycnocarpium* Niedenzu, Ind. Lect. Lyc. Brunsberg. p. aest. 1900: 27. 1900.—LECTOTYPE, here designated: *Stigmaphyllon ciliatum* (Lamarck) Adr. Jussieu.

Stigmaphyllon subg. *Stigmaphyllon* sect. *Macropterys* subsect. *Machaeropterys* ser. *Xiphiopterys* Niedenzu, Verz. Vorles. Akad. Braunsberg W.-S. 1912–1913: 30. 1912.—LECTOTYPE, here designated: *Stigmaphyllon ellipticum* (H. B. K.) Adr. Jussieu.

Stigmaphyllon subg. *Stigmaphyllon* sect. *Eurypterys* subsect. *Pseudocoelum* Niedenzu, Verz. Vorles. Akad. Braunsberg W.-S. 1912–1913: 32. 1912.—TYPE: *Stigmaphyllon hasslerianum* Niedenzu [=*Stigmaphyllon calcaratum* N. E. Brown].

Perennial vines, becoming woody, or sometimes twining shrubs, or rarely shrubs or subshrubs (*S. paralias*, *S. harleyi*), with woody rhizomes, these sometimes bearing tubers. Leaves opposite (alternate in *S. alternans*), sometimes alternate near and in the compound inflorescence, petiolate (sessile to subsessile in *S. crenatum*), the petioles free or confluent across the node and forming a band or ridge bearing the stipules, usually with a pair of prominent but sessile glands at the apex of the petiole, or the glands sometimes shallowly cupulate and borne on the petiole, or the glands absent; laminas mostly entire, sometimes lobed (laciniate in *S. laciniatum*), commonly cordate but also linear to lanceolate to elliptical to ovate to oblong to oblate to reniform to orbicular, pubescent or glabrous, margin glandular or eglandular, usually entire or sometimes shallowly crenate or dentate (crenate in *S. crenatum*); stipules interpetiolar, inconspicuous, free or sometimes the opposing stipules fused across the node into a bifid structure, usually eglandular (glandular in *S. adenophorum*, *S. bradei*, *S. calcaratum*), eventually deciduous. Inflorescence an umbel (2-flowered in *S. harleyi*) or pseudoraceme, these sometimes solitary but more commonly borne in dichasia or compound dichasia or small thyrses; peduncles and pedicels present, peduncles sometimes reduced or rudimentary; bracts and bracteoles present, persistent. Sepals 5, imbricate in bud, persistent (deciduous in *S. singulare*), lateral sepals biglandular, anterior sepal eglandular (biglandular in *S. boliviense* and *S. coloratum*). Petals 5, clawed, the limb mostly orbicular or broadly ovate or obovate or sometimes elliptical, glabrous (pubescent abaxially in *S. singulare* and sometimes in *S. sinuatum*), yellow or sometimes yellow marked or suffused with red, erose to denticulate to fimbriate; the posterior petal with a longer and stouter claw, commonly constricted below the limb, and a smaller limb than the lateral petals, sometimes glandular at the base of the limb. Androecium uniseriate; stamens 10, connate proximally, sometimes subequal but usually unequal, those opposite the anterior sepal and the posterior-lateral petals (and the styles) usually the largest, those opposite the lateral sepals mostly with more slender filaments and the connective enlarged and the locules reduced or absent; anthers glabrous or pubescent. Styles 3, free to the base, glabrous or pubescent, stigmas internal; anterior style usually different from and shorter than the equal posterior styles (longer in *S. herbaceum*, *S. hypargyreum*, *S. maynense*, *S. orientale*, *S. puberum*), erect or slightly recurved, apex with two equal lateral folioles or only laterally expanded or linear and distally blunt or distally extended into a spur or hook; posterior styles mirror images of each other, lyrate or sometimes erect, apex with a lateral foliole or lip or linear and distally blunt or distally extended into a spur or hook. Ovary 3-carpellate, 3-loculate, the carpels connate. Fruit a schizocarp of 3 samaras borne on a pyramidal torus and usually suspended on carpophores (carpophore absent in *S.*

bannisterioides, S. calcaratum, S. harleyi, S. lacunosum?, *S. paralias*). Samara with a large dorsal wing thickened along the upper (adaxial) margin (the dorsal wing reduced to a crest in *S. bannisterioides, S. calcaratum, S. harleyi, S. paralias)*; the nut with small lateral winglets and/or spurs and crests or the nut only prominently ribbed or smooth (in *S. calcaratum* the nut covered with numerous bulbous and warty excrescences composed of spongy tissue); nut ovoid or spheroid (lenticular in *S. ciliatum*), the walls of the locule fibrous or sometimes woody or spongy, sometimes the locule surrounded by air pockets. Embryo usually ovoid (laterally flattened in *S. ciliatum* and *S. dichotomum*), cotyledons unequal, the distal portion of the larger outer cotyledon folded over the inner cotyledon or both cotyledons straight, or sometimes the embryo spherical and the cotyledons folded within each other and convoluted; in *S. bannisterioides* and *S. paralias*, the embryo circular to horseshoe-shaped, in *S. bannisterioides* the inner cotyledon rudimentary. Base chromosome number: x = 10.

KEY TO THE SPECIES OF STIGMAPHYLLON

Note: Measurements of floral characters are taken from material revived with Pohl's solution (1965). In some species with pubescent anthers the hairs may be present only on the unmodified anthers (i.e., the connective not very much enlarged nor the locules reduced or absent). The following key relies greatly on structural details of the flower, particularly the androecium. Readers may prefer to identify material by using the regional keys provided in the Appendix; the fewer species accounted for in each regional key permit the use of less technical characters.

1. Anterior style without folioles, the apex at most with a narrow lip on each side.
    2. Laminas subsessile (petiole up to 2 mm long), the margin deeply crenate, abaxially each sinus with a gland; Brazil (Espírito Santo). 45. *S. crenatum*.
    2. Laminas petiolate, the margin entire to subentire to dentate or very shallowly crenate, glandular or eglandular.
        3. Anthers all unmodified and subequal in shape, each anther with 2 locules about as long as the connective (in *S. acuminatum* and *S. lalandianum* anthers of stamens opposing the lateral sepals with the locules equally as long as the connective or slightly shorter, but the connective not greatly enlarged; in *S. peruvianum* anthers of stamens opposing the lateral sepals with the connective somewhat enlarged but the locules equally long).
            4. Laminas pinnately 5–7-lobed (those of the small leaves associated with the inflorescence 3-lobed or unlobed); limb of lateral petals 14–16 mm in diameter, the margin (lacerate-) fimbriate, the fimbriae up to 1 mm long.
                5. Laminas abaxially abundantly pubescent with T-shaped hairs; Brazil (Bahia, Minas Gerais). 7. *S. urenifolium*.
                5. Laminas abaxially glabrous; Brazil (Espírito Santo). 8. *S. glabrum*.
            4. Laminas unlobed; limb of lateral petals 5–13 mm in diameter, the margin erose to denticulate to fimbriate, the teeth/fimbriae up to 0.7 mm long.
                6. Mature laminas abaxially pubescent with T-shaped hairs to tomentose, the vesture evenly distributed.
                    7. Flowers (4–) 6–8 per umbel; even the mature laminas adaxially tomentulose; Peru (Cajamarca and adjacent Amazonas). 9. *S. peruvianum*.
                    7. Flowers 10–35 per umbel or pseudoraceme; mature laminas adaxially glabrous (at most with a few scattered hairs).
                        8. Laminas linear-lanceolate, up to 3 cm wide; petioles up to 2.8 cm long; Colombia (Antioquia). 5. *S. stenophyllum*.
                        8. Laminas elliptical to triangular to ovate to cordate to sometimes suborbicular, 4.3–21.5 cm wide; petioles 1.6–20 (–25) cm long.
                            9. Posterior styles with triangular to subsquare folioles 0.5–0.7 mm long, 0.5–0.9 mm wide; laminas with the margin eglandular or with glands borne abaxially adjacent to the margin; Brazil (Bahia, Minas Gerais, Rio de Janeiro). 15. *S. acuminatum*.

9. Posterior styles without folioles or a lateral lip; laminas with the margin eglandular or with glands borne on the margin.
   10. Lowest 2 (sometimes more) flowers of an umbel or pseudoraceme usually separated a short distance on the axis from the rest; petals suffused or margined with red, limb of the posterior petal 5–6.5 mm in diameter, limb of the lateral petals 6–10 mm in diameter; all with the margin digitate-fimbriate; margin of lamina with irregularly spaced sessile glands; Colombia, western Venezuela, Ecuador, Peru. 2. *S. bogotense.*
   10. Lowest 2 flowers of an umbel never separated from the rest; petals yellow, limb of the posterior petal ca. 8.5 mm in diameter, limb of the lateral petals 10–11.5 mm in diameter; all with the margin bearing fimbriae tapered to an acute apex; margin of lamina eglandular; Bolivia (La Paz). 6. *S. yungasense.*

6. Mature laminas abaxially sparsely to densely sericeous to glabrous or glabrate (with a few scattered sessile or T-shaped hairs, or the hairs more numerous but concentrated on or along the major veins).
   11. Flowers (3–) 4 (–6) per umbel; peduncles rudimentary to 2.5 mm long, pedicels 15–30 mm long; along the Atlantic Coast from southern Mexico (Veracruz) to northern Brazil (Maranhão), the West Indies, and coastal western Africa (Guinea Bissau, Guinea, and Sierra Leone). 10. *S. bannisterioides.*
   11. Flowers 12–50 per umbel or pseudoraceme; peduncles 2.7–14 mm long, pedicels 2.5–13.5 mm long.
      12. Mature laminas abaxially abundantly sericeous, the vesture evenly distributed.
         13. Limb of petals yellow but with a red center or suffused with red, margin of limb of posterior petal digitate-fimbriate; peduncles 0.2–0.3 times as long as the pedicels; laminas abaxially very densely white-sericeous or densely white sericeous-tomentulose, the epidermis hidden; Colombia (Cundinamarca), Ecuador (Loja, Pichincha), and Peru (Piura). 3. *S. sarmentosum.*
         13. Limb of petals yellow, margin of limb of posterior petal erose to erose-denticulate; peduncles 0.2–1.8 times as long as the pedicels; laminas abaxially sparsely to densely sericeous, the epidermis visible.
            14. Apex of the anterior style on each side with a lateral lip 0.1–0.4 mm wide, apex of posterior styles with a lateral lip or a small foliole; margin of lamina eglandular; petiole with a pair of prominent but sessile glands at the apex; flowers 15–40 (–50) per pseudoraceme; Brazil (Rio de Janeiro, Espírito Santo, Minas Gerais, São Paulo). 14. *S. lalandianum.*
            14. Apex of all styles without a lateral lip or small foliole; margin of lamina with irregularly spaced sessile glands; petiole with a pair of prominent to stipitate (peg-shaped) glands at the apex or up to 2 mm below the base of the lamina, each gland up to 1.3 mm long; flowers 12–20 per pseudoraceme; Mexico (Chiapas), and Guatemala (Alta Verapaz, Huehuetenango, Quetzaltenango, Suchitepéquez). 1. *S. pseudopuberum.*
      12. Mature laminas abaxially glabrous or glabrate (with a few scattered sessile or T-shaped hairs, or the hairs more numerous but concentrated on or along the major veins) or very sparsely sericeous but the vesture patchy.
         15. Petioles 0.9–2.2 cm long, with 2 (–4) peg-shaped glands at the apex, the glands 0.4–0.8 mm long; laminas 3.2–7 cm long, 1.7–4.2 cm wide; Colombia (Bolívar). 13. *S. romeroi.*
         15. Petioles 2.2–7.5 cm long, with 2 prominent but sessile glands at the apex; laminas 6.1–12 cm long, 4.5–8.8 cm wide.
            16. Laminas abaxially glabrous, the margin eglandular; limb of posterior petal 10–11 mm long, 8–9 mm wide, yellow; Guatemala (Guatemala, Huehuetenango). 12. *S. cordatum.*
            16. Laminas abaxially glabrate, with a few sessile or T-shaped hairs scattered on the surface, the hairs often more numerous and concentrated on or along the major veins, or abaxially very sparsely sericeous, the margin with sessile glands; limb of posterior petal ca. 7 mm in diameter, yellow suffused with red; Colombia (Caquetá, Valle). 4. *S. suffruticosum.*

3. Anthers of stamens opposing the anterior-lateral sepals and/or the posterior-lateral sepals with the connective enlarged and bearing 0–2 reduced locules.
  17. Anthers of stamens opposing the posterior-lateral sepals unmodified, i.e., like that of the stamen opposing the posterior petal.
    18. Laminas pinnately 5–7-lobed; the smallest pair of leaves below an inflorescence evenly fringed with filiform glands; Brazil (Rio de Janeiro). 21. *S. carautae.*
    18. Laminas unlobed or sometimes palmately 3–5-lobed; the smallest leaves below an inflorescence with marginal glands like the larger leaves or eglandular.
      19. Laminas abaxially sparsely to very densely sericeous, the hairs sessile or subsessile, the vesture appressed; lowlands of Colombia, Venezuela, the Guianas, northern Brazil, Ecuador, northern Peru, and Amazonian Bolivia. 33. *S. sinuatum.*
      19. Laminas abaxially pubescent with T-shaped hairs to tomentose.
        20. Laminas adaxially velutinous; lateral petals glandular-digitate-fimbriate; Colombia (Antioquia, Cundinamarca, Norte de Santander, Quindío, Risaralda). 22. *S. velutinum.*
        20. Laminas adaxially glabrous to glabrate; lateral petals erose or erose-denticulate to erose-fimbriate (fimbriae eglandular and tapered from the base).
          21. Laminas abaxially sparsely pubescent; peduncles (5.5–) 7.5–15 mm long, 1.2–3.3 times as long as the pedicels; Costa Rica (San José), northern and central Colombia, and Ecuador (Napo). 24. *S. columbicum.*
          21. Laminas abaxially very densely silvery-pubescent; peduncles 5.5–9.5 mm long, 0.75–1 times as long as the pedicels; Peru (San Martín). 23. *S. tarapotense.*
  17. Anthers of stamens opposing the posterior-lateral sepals with the connective enlarged and bearing 0–2 reduced locules.
    22. Stamens opposing the lateral sepals bearing only a glandular connective without locules, very rarely anthers of stamens opposing the anterior-lateral sepals with the connective enlarged and bearing 1–2 tiny locules.
      23. Flowers 8–18 (–27) per umbel (sometimes a pseudoraceme), the umbels usually solitary but sometimes borne in dichasia or compound dichasia or rarely in a small thyrse; apex of anterior style 0.9–1.7 mm long, 0.3–1.2 mm wide, linear with a spur 0.6–1.4 mm long or triangular or rhombic; laminas extremely variable, linear to suborbicular, 0.3–7 cm wide; Cuba and the Lesser Antilles. 40. *S. diversifolium.*
      23. Flowers (10–) 20–25 (–40) per congested or interrupted pseudoraceme (sometimes an umbel), the pseudoracemes usually in large compound inflorescences, rarely solitary; apex of anterior style 0.6–0.7 (–1.2) mm long, 0.1–0.2 mm wide, linear with a spur 0.2–0.3 (–0.6) mm long; laminas elliptical to broadly so to oblong or sometimes orbicular or lanceolate, 2.5–15.5 cm wide; Puerto Rico, Virgin Gorda, St. John. 41. *S. floribundum.*
    22. Anthers of stamens opposing the lateral sepals with the connective enlarged and bearing 2 reduced locules.
      24. Laminas abaxially pubescent with T-shaped hairs to tomentose.
        25. Anthers pubescent; apex of anterior style distally blunt or with a spur up to 0.3 mm long; bracteoles eglandular or each bracteole with a pair of inconspicuous glands (each 0.2–0.3 mm in diameter) or with a glandular area in the basal 1/3–1/2; eastern Ecuador (Napo) and eastern Peru to Bolivia and Brazil (Acre, Rondônia). 42. *S. florosum.*
        25. Anthers glabrous; apex of anterior style distally extended into a claw up to 1 mm long; only one of each pair of bracteoles with one lateral prominent gland, 0.5–0.8 mm in diameter; Peru (Junín, Pasco). 43. *S. aberrans.*
      24. Laminas abaxially glabrous to sericeous, the hairs sessile and appressed.
        26. Margin of lamina dentate or entire, with irregularly spaced sessile glands and/or filiform glands and/or stipitate glands.
          27. Laminas with the base truncate to cordate, the margin entire, abaxially moderately to densely sericeous, the vesture evenly distributed; anthers pubescent; flowers (9–) 12–35 per umbel; southern Veracruz, Mexico, to Panama and adjacent Colombia (Chocó). 63. *S. lindenianum.*

27. Laminas with the base deeply auriculate, the margin dentate or entire, abaxially glabrous or glabrate or very sparsely pubescent, especially on the major veins; anthers glabrous (rarely in *S. selerianum* the largest anthers with a few hairs); flowers 8–12 per umbel.

28. Laminas deeply 3–5-lobed or triangular to ovate, margin grossly dentate, each tooth ending in a filiform gland; anterior style with small triangular folioles, 0.2–0.3 mm long, 0.3–0.4 mm wide; peduncles 2.5–8 mm long, 0.5–1.3 times as long as the pedicels; limb of lateral petals 7–9 mm in diameter; Brazil (Rio de Janeiro). 51. *S. vitifolium.*

28. Laminas cordate or narrowly so, margin entire but with filiform glands up to 0.8 mm long (these often broken off in the larger leaves but the bases remaining); anterior style without folioles; peduncles 0.5–3 mm long, up to 0.3 times as long as the pedicels; limb of lateral petals 8.5–13 mm in diameter; Mexico (Chiapas, Oaxaca). 44. *S. selerianum.*

26. Margin of lamina entire, eglandular.

29. Petals with fimbriae up to 2 mm long (rarely fimbriate-dentate in *S. ellipticum*), limb of lateral petals 10–17 mm in diameter, limb of posterior petal (8–) 11–14.5 mm in diameter; peduncles present, pedicels distally expanded and widest below the calyx, unlike the peduncles.

30. Laminas abaxially glabrous or sometimes sparsely sericeous; petioles 0.6–2.8 cm long, free, not joined across the node; stipules free; flowers 3–9 (–12) per umbel; nut of samara without air chambers; southeastern Mexico to northern South America. 46. *S. ellipticum.*

30. Laminas abaxially abundantly sericeous; petioles 2.5–5.3 cm long, forming a band across the node bearing the stipules, the band becoming a ridge at older nodes; stipules commonly fused across the node to form a bifid structure; flowers 10–14 per umbel; nut of samara with air chambers surrounding the locule; Colombia (Antioquia, Caldas, Risaralda, Valle). 48. *S. venulosum.*

29. Petals erose, limb of lateral petals 7.5–11 mm in diameter, limb of posterior petal 6.5–9.5 mm in diameter; peduncles present or absent, pedicels of uniform diameter (like the peduncles, if present).

31. Flowers 4 per solitary umbel; laminas 0.8–3.7 cm long; anthers pubescent; Cuba. 38. *S. microphyllum.*

31. Flowers (5–) 15–50 per umbel or pseudoraceme, these solitary or in compound inflorescences; laminas 1–13 cm long; anthers glabrous.

32. Peduncles rudimentary to 5.5 (–9) mm long, up to 0.3 times as long as the pedicels; posterior styles terete, the apex with a foliole or sometimes efoliolate and extended into a spur; Cuba and the Bahamas. 39. *S. sagraeanum.*

32. Peduncles 1.5–25 mm long, 0.3–1 times as long as the pedicels; posterior styles canaliculate-complicate (Fig. 36m, n), the apex truncate, without a foliole (in dried material the posterior styles sometimes appearing flattened and seeming to have folioles); Jamaica, Hispaniola, Puerto Rico, Virgin Islands, and the Lesser Antilles south to Martinique except Dominica. 37. *S. emarginatum.*

1. Anterior style with well-developed folioles to 3.3 mm in diameter.

33. Flowering and fruiting plants leafless or rarely with a few very young leaves; Ecuador (Guayas). 32. *S. nudiflorum.*

33. Flowering and fruiting plants leaf-bearing.

34. Anterior and lateral sepals each with 2 glands.

35. Mature laminas abaxially very sparsely pubescent with T-shaped hairs to glabrous; Bolivia (La Paz, Santa Cruz). 60. *S. coloratum.*

35. Mature laminas abaxially densely pubescent with T-shaped hairs; Bolivia (Santa Cruz). 59. *S. boliviense.*

34. Anterior sepal eglandular, each lateral sepal with 2 glands.

36. Mature laminas abaxially glabrate to glabrous, sometimes with sessile or T-shaped hairs on the major veins, or sparsely to densely sericeous, the hairs sessile and appressed, or appressed-tomentulose (in *S. macropodum* and *S. stylopogon* some of the hairs with a stalk up to 0.1 mm long), or very sparsely pubescent and the hairs patchily distributed.
  37. Petiole glands peg-like, 0.8–1.5 mm long, borne up to 2 mm below the base of the lamina; Brazil (Piauí). 54. *S. jobertii.*
  37. Petiole glands sessile and borne at the apex of the petiole, or shallowly cupulate and borne up to 18 mm below the base of the lamina.
    38. Anthers all unmodified and subequal in shape, each anther with 2 locules about as long as the connective.
      39. Flowers grouped in 2's; peduncles rudimentary, to 0.8 mm long; laminas up to 2.8 cm long, 2.5 cm wide; petioles 0.2–0.3 cm long; twiggy subshrubs; Brazil (Bahia). 11. *S. harleyi.*
      39. Flowers 8–12 per umbel; peduncles 2–5 mm long; laminas 9–15 cm long, 7–10 cm wide; petioles 2.7–8.3 cm long; vines; Colombia (Meta). 16. *S. orientale.*
    38. Anthers of stamens opposing the anterior-lateral sepals and/or the posterior-lateral sepals with the connective enlarged and bearing 0–2 reduced locules.
      40. Anterior style and its opposing stamen longer than the posterior styles and their opposing stamens.
        41. Laminas abaxially very densely (white-) sericeous, the epidermis hidden; flowers (12–) 15–25 per umbel.
          42. Folioles of anterior style 1.3–1.6 mm long, 0.8–1.2 (–1.7) mm wide, narrowly rectangular or rarely subsquare but distally flared, the distal margin sagittate or coarsely erose; dorsal wing of the samara not abruptly narrowed at the nut, at least 0.8 cm wide, the nut with 3–4 lateral winglets per side; Ecuador (Napo, Pastaza, Zamora-Chinchipe), Peru (Amazonas, Huánuco, Loreto, San Martín, Pasco, Madre de Dios), and Brazil (southwestern Amazônas). 18. *S. maynense.*
          42. Folioles of anterior style 0.8–1.1 mm long, ca. 0.6 mm wide, oblong to triangular, the margin entire; dorsal wing of the samara abruptly narrowed at the nut to only 0.3–0.4 cm wide, the nut commonly with lateral spurs and/or crests or sometimes with a pair of lateral winglets; Panama and Colombia (Antioquia). 19. *S. hypargyreum.*
        41. Laminas abaxially glabrous to sericeous, the epidermis always visible; flowers 8–15 per umbel.
          43. Mature laminas abaxially sericeous, the hairs evenly distributed; margin of petals with fimbriae up to 0.6 (–0.8) mm long; the West Indies, Atlantic lowlands of Central America and also in the Pacific lowlands of Costa Rica (Golfo Dulce area) and Panama, lowlands of northern South America: Venezuela (Delta Amacuro, Monagas), the Guianas, Colombia (Antioquia, Chocó Putumayo), Peru (northern Loreto, Huánuco), Brazil (Amapá, Pará, Amazônas, Acre). 17. *S. puberum.*
          43. Mature laminas abaxially glabrate to glabrous, or very sparsely sericeous and the hairs unevenly and patchily distributed; margin of petals erose; lowlands of central and western Colombia and northern Ecuador (Esmeraldas). 20. *S. herbaceum.*
      40. Anterior style and its opposing stamen shorter than or subequal to the posterior styles and their opposing stamens.
        44. Pedicels glabrate to glabrous.
          45. Laminas ovate to cordate, the base deeply auriculate and the basal lobes overlapping in the larger laminas, the margin entire to shallowly dentate, evenly fringed with filiform glands; flowers 3–8 per solitary umbel; samara lenticular, the reduced dorsal wing encircling the laterally flattened nut; Atlantic lowlands from Belize to Uruguay, also in Trinidad and Barbados. 55. *S. ciliatum.*

45. Laminas palmately to pedately lobed or triangular to elliptical to ovate or rarely suborbicular, the base truncate to cordate to hastate or auriculate but the basal lobes not overlapping, the margin dentate or entire, eglandular or with irregularly spaced filiform glands or also with sessile glands; flowers 8–35 per umbel, usually borne in compound inflorescences; samara with an elongate dorsal wing, the nut ovoid.

46. Larger laminas palmately to pedately (2–) 5–7 (–9)-lobed, infrequently triangular to elliptical, or rarely suborbicular, 4.2–18.2 cm long, the margin shallowly to grossly dentate, the teeth ending in filiform glands, sometimes also with sessile glands in the sinuses; mostly along the Río Paraná and Río Uruguay and their tributaries in eastern Argentina, western Uruguay, southern Paraguay, and southeastern Brazil. 52. *S. jatrophifolium.*

46. Larger laminas triangular to ovate to elliptical to hastate or sometimes 2–3-lobed, 3.2–10.5 cm long, the margin entire, eglandular or with filiform and/or sessile glands.

47. Foliole of anterior style subsquare to rectangular, 1.8–2.3 mm long, 1.4–2 mm wide, folioles of posterior styles parabolic to subrectangular, 1.4–2.2 mm long, 1.3–2 mm wide; laminas ovate to elliptical to hastate or sometimes 2–3-lobed, the margin eglandular or with filiform and/or sessile glands; Brazil (Ceará to Rio de Janeiro). 53. *S. auriculatum.*

47. Foliole of anterior style triangular or narrowly so, 0.3–1 mm long, 0.5–0.8 mm wide, or the apex of the anterior style only distally laterally expanded into a broad lip on each side, folioles of posterior styles narrowly to broadly lunate to triangular, 1–1.1 mm long, 0.6–0.7 mm wide at insertion, or the apex only with a lateral lip; laminas unlobed, triangular to cordate to elliptical to sometimes lanceolate to ovate or rarely suborbicular, the margin eglandular.

48. Petioles 0.6–2.8 cm long, free, not joined across the node; stipules free; flowers 3–9 (–12) per umbel; peduncles 2.5–34 mm long, pedicels 2–13 mm long and glabrous, peduncles 0.5–5 times as long as the pedicels; nut of samara without air chambers; southeastern Mexico to northern South America. 46. *S. ellipticum.*

48. Petioles 2.5–5.3 cm long, forming a band across the node bearing the stipules, the band becoming a ridge at older nodes; stipules commonly fused across the node to form a bifid structure; flowers 10–14 per umbel; peduncles 3.2–5.5 mm long, pedicels 4–6 mm long and glabrous except for a row of hairs extending from the base of the pedicel to the base of the posterior petal, peduncles 0.6–1 times as long as the pedicels; nut of samara with air chambers surrounding the locule; Colombia (Antioquia, Caldas, Risaralda, Valle). 48. *S. venulosum.*

44. Pedicels pubescent, the hairs abundant and touching to overlapping.

49. Anthers of stamens opposing the posterior-lateral sepals unmodified, i.e., like that of the stamen opposing the posterior petal.

50. Petioles joined across the node and forming a prominent corky ridge bearing the stipules; larger laminas palmately 3–5 (–7)-

lobed to broadly ovate or broadly elliptical, the smaller 2–3-lobed to broadly ovate; Brazil (Minas Gerais). 36. *S. macedoanum.*

50. Petioles free, not forming a ridge across the node, the stipules borne on the stem; laminas unlobed (or rarely 3–5-lobed in *S. sinuatum*).
    51. Anthers pubescent; margin of lamina with raised or more commonly nail-like glands, i.e., stipitate and with a disk-like apex 0.3–0.5 mm in diameter; nut of samara 12–19 mm in diameter, the locule surrounded by air chambers.
        52. Laminas abaxially sericeous, the hairs sessile and evenly distributed; Amazon Basin in Brazil, Colombia, and Peru. 29. *S. lacunosum.*
        52. Laminas abaxially glabrate, with scattered T-shaped hairs; Amazon Basin and disjunct to the Paria Peninsula of Venezuela, Trinidad and Tobago, and Grenada. 28. *S. adenodon.*
    51. Anthers glabrous; margin of lamina with sessile and/or filiform glands; nut of samara 2.8–4.4 mm in diameter, without air chambers.
        53. Laminas abaxially appearing glabrous to the naked eye, but usually very sparsely sericeous, the hairs ca. 0.1 (–0.2) mm long and widely spaced, never touching; laminas ovate to cordate or narrowly so, the base cordate; Martinique, Trinidad, the Guianas, and northeastern Brazil (Amapá, eastern Pará). 34. *S. convolvulifolium.*
        53. Laminas abaxially sparsely to very densely sericeous, the hairs (0.2–) 0.3–0.5 (–0.7) mm long, usually touching to overlapping; laminas triangular to ovate to cordate to elliptical to broadly so to orbicular to oblate to reniform, rarely 3–5-lobed, the base acute to truncate to cordate to deeply auriculate; lowlands of Colombia, Venezuela, the Guianas, northern Brazil, Ecuador, northern Peru, and Amazonian Bolivia. 33. *S. sinuatum.*

49. Anthers of stamens opposing the posterior-lateral sepals with the connective enlarged and bearing 0–2 reduced locules.
    54. Anthers pubescent.
        55. Flowers 4 per solitary umbel; laminas 0.8–3.7 cm long; petioles (0.1–) 0.2–0.4 cm long; Cuba. 38. *S. microphyllum.*
        55. Flowers (9–) 12–35 per umbel, these borne in compound inflorescences or rarely solitary; laminas 5–19 cm long; petioles 0.8–8.5 cm long.
            56. Laminas abaxially evenly sericeous; anthers of stamens opposing the posterior-lateral sepals with the connective enlarged and bearing 2 reduced locules; Atlantic lowlands from southern Veracruz, Mexico, to Panama and adjacent Colombia (Chocó), in Costa Rica also reported from the Osa Peninsula, in Panama also in the Pacific lowlands. 63. *S. lindenianum.*
            56. Laminas abaxially glabrate to glabrous, sometimes with a few scattered sessile or T-shaped hairs on the major veins, or very sparsely pubescent and the hairs patchily distributed; anthers of stamens opposing the posterior-lateral sepals with the connective enlarged and bearing 0–1 reduced locule.

57. Limb of lateral petals 4–6.5 mm in diameter; marginal glands of lamina sessile and/or filiform and borne on the margin; Amazonian lowlands of Ecuador, Peru, Brazil, and Bolivia. 69. *S. cardiophyllum.*

57. Limb of lateral petals 10–16 mm in diameter; marginal glands of lamina stipitate, to 0.7 mm long, and borne abaxially adjacent to the margin; Brazil (Pernambuco, Bahia, Espírito Santo, southeastern Minas Gerais and adjacent Rio de Janeiro). 90. *S. salzmannii.*

54. Anthers glabrous (rarely with a few hairs in *S. macropodum*).

58. Laminas finely pinnately dissected; Île de la Gonâve (west of Haiti). 57. *S. laciniatum.*

58. Laminas linear, lanceolate, elliptical, ovate, cordate, orbicular, or palmately lobed.

59. Shrubs; anthers of stamens opposing the posterior petal and the posterior-lateral sepals consisting only of an enlarged connective without locules; stipules commonly fused across the node into a bifid structure; eastern Brazil from Maranhão to Rio de Janeiro (also recorded from Tucuruí, Pará, and once from Posse, Goiás). 62. *S. paralias.*

59. Vines; anther of stamen opposing the posterior petal unmodified, i.e., the connective not enlarged and bearing 2 unreduced locules, anthers of stamens opposing the posterior-lateral sepals with the connective enlarged and bearing 0–2 reduced locules; stipules free.

60. Flowers 4–6 (–8) per umbel, the umbels commonly solitary or sometimes in dichasia; petioles of even the largest leaves only up to 2.1 cm long; samara greatly modified, the nut with a dorsal crest and laterally covered with warty excrescences; northern Brazil along the Rio Amazônas near Manaus and Santarém, and in Argentina (Chaco, Corrientes, Santa Fe) and adjacent Paraguay, and the Pantanal (Mato Grosso) of Brazil. 56. *S. calcaratum.*

60. Flowers 10–35 per umbel or pseudoraceme, the umbels/pseudoracemes usually in compound inflorescences (mostly solitary in *S. bradei*); petioles of the larger leaves more than 2 cm long (up to 11 cm long); nut of samara with an elongate dorsal wing and laterally with winglets, crests and/or spurs, or unornamented (samara unknown in *S. bradei*).

61. Petiole glands of the larger leaves shallowly cupulate or discoid, borne 1–18 mm below the base of the lamina; peduncles rudimentary to 1.5 mm long; umbels solitary (rarely borne in dichasia); Brazil (São Paulo). 58. *S. bradei.*

61. Petiole glands prominent but sessile, borne at the apex of the petiole; peduncles (2.5–) 3–17 mm long, (0.4–) 0.5–2.3 times as long

as the pedicels; umbels/pseudoracemes usually borne in compound inflorescences.

62. Mature laminas abaxially glabrous to glabrate or very sparsely and patchily pubescent.
    63. Limb of lateral petals 8–9 mm in diameter, limb of posterior petal ca. 7 mm in diameter; anthers of stamens opposing the anterior-lateral sepals with the connective enlarged and always bearing 2 reduced locules, those opposing the posterior-lateral sepals always bearing 1 reduced locule; Trinidad and the adjacent Paria Peninsula (Venezuela). 68. *S. finlayanum.*
    63. Limb of lateral petals 10–15 mm in diameter, limb of posterior petal 9–11 mm in diameter; anthers of stamens opposing the lateral sepals with the connective enlarged and bearing 0–2 locules.
        64. Anthers of stamens opposing the lateral sepals consisting only of a glandular connective, without locules; laminas commonly 2–3-lobed, the marginal glands borne abaxially adjacent to the margin (or sometimes the margin eglandular); Brazil (Rio de Janeiro to Paraná). 86. *S. arenicola.*
        64. Anthers of stamens opposing the lateral sepals with the connective enlarged, those of the stamens opposing the anterior-lateral sepals with 2 reduced locules, those of the stamens opposing the posterior-lateral sepals with 1–2 reduced locules; laminas unlobed, the marginal glands borne on the margin (or sometimes the margin eglandular); central Panama and islands in the Gulf of Panama. 66. *S. panamense.*
62. Mature laminas abaxially moderately to densely pubescent, the hairs evenly distributed (in *S. rotundifolium* sometimes sparsely so and the lamina appearing glabrous to the naked eye).
    65. Mature laminas abaxially moderately or sparsely pubescent, not appearing white or silvery or golden but sometimes seeming glabrous to the naked eye; anthers of stamens opposing the posterior-lateral sepals with the connective enlarged and always bearing 2 reduced locules; Brazil (Bahia, Espírito Santo?). 84. *S. rotundifolium.*
    65. Mature laminas abaxially densely pubescent, the vesture appearing white to

silvery or golden; anthers of stamens opposing the posterior-lateral sepals with the connective enlarged and bearing 0–1 reduced locule (sometimes with 2 reduced locules in *S. paraense*).

66. Laminas abaxially very densely appressed-tomentulose to sericeous, the epidermis nearly or entirely hidden, at least some of the hairs with a tiny stalk up to 0.1 mm long, the trabecula 0.2–0.9 mm long.

67. Flowers (10–) 15–25 per umbel; anthers of stamens opposing the posterior-lateral sepals consisting only of an enlarged connective without locules; limb of posterior petal without stout gland-tipped fimbriae; anterior style glabrous, posterior styles with scattered hairs in the proximal 1/3; laminas up to 27 cm long, up to 22.5 cm wide; Brazil (Bahia).
85. *S. macropodum.*

67. Flowers 10–16 per umbel; anthers of stamens opposing the posterior-lateral sepals with the connective enlarged and bearing 1 reduced locule; limb of posterior petal near the base often with 2–3 stout gland-tipped fimbriae per side, these 0.6–1 mm long, ca. 0.2 mm in diameter; all styles with an adaxial row of hairs in the proximal 1/2–3/4 or nearly to the stigma; laminas up to 13.5 cm long, up to 10 cm wide; Brazil (Mato Grosso, Rondônia).
74. *S. stylopogon.*

66. Laminas abaxially sericeous but the epidermis always clearly visible, the hairs sessile.

68. Limb of lateral petals 6–7 mm in diameter, limb of posterior petal 5–5.6 mm long, 3.5–4.8 mm wide, the margin erose; nut of samara without lateral ornamentation or air chambers; lowlands of eastern Peru.
70. *S. argenteum.*

68. Limb of lateral petals 11–15 mm in diameter, limb of posterior petal 9–11 mm in diameter, the margin with fimbriae up to 1 mm long; nut of samara with lateral winglets and air cham-

bers surrounding the locule; Brazil (Goiás, Maranhão, Mato Grosso, western Piauí, Pará). 71. *S. paraense.*

36. Mature laminas abaxially pubescent with T-shaped hairs to tomentose, the vesture not appressed but spreading and evenly distributed.
    69. Anthers of stamens opposing the posterior-lateral sepals unmodified, i.e., like that of the stamen opposing the posterior petal.
        70. Laminas sinuate-lobate with 5–7 (–9) lobes (rarely ovate to suborbicular), base auriculate, deeply so in the larger laminas, margin sinuate; Hispaniola. 36. *S. angulosum.*
        70. Laminas unlobed, base cordate to truncate to subattenuate, margin entire or shallowly dentate.
            71. Limb of anterior-lateral petals 14–18 mm in diameter, limb of posterior-lateral petals 13–15 mm in diameter, limb of posterior petal 12–14 mm in diameter, the margin erose-denticulate to denticulate-fimbriate, the teeth/fimbriae up to 0.5 mm long.
                72. Margin of laminas shallowly dentate, each tooth terminated by a filiform gland or subsessile to stalked gland, the sinuses also with irregularly spaced subsessile to stalked glands; petals erose to erose-denticulate, the teeth up to 0.3 mm long; Ecuador (Manabí). 31. *S. ecuadorense.*
                72. Margin of laminas entire, eglandular or with sessile glands; petals fimbriate to denticulate-fimbriate, fimbriae up to 0.5 mm long; Peru (Cuzco) . 30. *S. cuzcanum.*
            71. Limb of anterior-lateral petals 9.5–12 mm in diameter, limb of posterior-lateral petals (6–) 8–10 mm in diameter, limb of posterior petal 7.5–10 mm in diameter, the margin erose to erose-denticulate, the teeth up to 0.2 mm long.
                73. Bracteoles each with a pair of prominent glands, each gland 0.5–0.8 mm in diameter; peduncles 4–8 mm long; flowers 10–21 per umbel.
                    74. Sepals deciduous; margin of laminas with stipitate to nail-like glands; anthers pubescent; stipules eglandular; Colombia (Norte de Santander) and Venezuela (Táchira, Zulia). 26. *S. singulare.*
                    74. Sepals persistent; margin of laminas eglandular; anthers glabrous; each stipule composed of a prominent, circular gland, ca. 0.8 mm in diameter, with a minute membranous acute tip; Costa Rica (Osa Peninsula). 27. *S. adenophorum.*
                73. Bracteoles eglandular or each bracteole with one or a pair of inconspicuous glands hidden by the vesture (each gland 0.2–0.4 mm in diameter) or with a glandular region in the basal 1/3–1/2; peduncles 3.5–21 mm long; flowers 15–50 per umbel or pseudoraceme.
                    75. Margin of laminas with nail-like glands, i.e., stipitate with a disk-like apex; anthers pubescent; peduncles 3.5–17.5 mm long, (0.8–) 1–2.75 times as long as the pedicels; Amazon Basin and disjunct to the Paria Peninsula of Venezuela, Trinidad and Tobago, and Grenada. 28. *S. adenodon.*
                    75. Margin of laminas eglandular or with irregularly spaced sessile to elongate glands (up to 0.2 mm long); anthers glabrous; peduncles 15–21 mm long, (1.5–) 3.4–4.2 times as long as the pedicels; Colombia (Antioquia, Tolima). 25. *S. goudotii.*
    69. Anthers of stamens opposing the posterior-lateral sepals with the connective enlarged and bearing 0–2 reduced locules.
        76. Laminas pinnately 5–7-lobed; Brazil (southeastern Minas Gerais, eastern Rio de Janeiro, and adjacent São Paulo). 76. *S. angustilobum.*
        76. Laminas unlobed or palmately 2–5 (–7)-lobed.
            77. Pedicels glabrate to glabrous or with a few scattered hairs, or glabrous except for a row of hairs extending from the base of the pedicel to the base of the posterior petal.

78. Petioles free, not joined across the node; stipules free; Ecuador (Guayas, Manabí). 47. *S. eggersii.*
78. Petioles forming a band across the node bearing the stipules, the band becoming a ridge at older nodes; stipules commonly fused across the node to form a bifid structure.
79. Laminas abaxially abundantly pubescent but the epidermis clearly visible; anterior-lateral petals ca. 15 mm in diameter, with fimbriae to 2.2 mm long; styles glabrous; western Colombia. 49. *S. echitoides.*
79. Laminas abaxially very densely woolly-tomentose, the epidermis nearly hidden; anterior-lateral petals ca. 19 mm in diameter, with fimbriae to 1 mm long; styles with scattered hairs in the proximal 1/4; Colombia (Huila). 50. *S. tergolanatum.*
77. Pedicels pubescent, the hairs touching and overlapping.
80. Marginal glands of lamina borne abaxially adjacent to the margin (sometimes hidden by the dense pubescence), the glands sessile to stipitate but never filiform.
81. Mature laminas adaxially pubescent, velutinous and/or with T-shaped hairs; petioles with a pair of glands borne at the apex or up to 7 mm below the base of the lamina; Brazil (Rio de Janeiro). 87. *S. gayanum.*
81. Mature laminas adaxially glabrate to glabrous; petioles with a pair of glands borne at the apex (or sometimes up to 0.2 mm below the base of the lamina in *S. affine*).
82. Anthers of stamens opposing the posterior-lateral sepals consisting of only an enlarged glandular connective without locules; unmodified anthers pubescent (usually glabrous in *S. tomentosum*).
83. Margin of petals digitate-fimbriate, limb of anterior-lateral petals 8.5–11 mm in diameter; larger laminas narrowly to broadly lanceolate to sometimes ovate, to 12 cm long and 5.5 cm wide; nut of samara 4–7 mm high, with a row of 5–7 lateral winglets on each side; Brazil (Rio de Janeiro, and adjacent Espírito Santo and Minas Gerais). 77. *S. alternifolium.*
83. Margin of petals erose to denticulate to fimbriate, the teeth/fimbriae tapered from the base to an acute apex, limb of anterior-lateral petals 11–15 mm in diameter; larger laminas triangular to elliptical to cordate to broadly ovate to orbicular or sometimes 2–5-lobed (rarely lanceolate), to 31 cm long and 19 cm wide; nut of samara 6.5–12 mm high, with 1–3 lateral winglets on each side and often also with spurs/crests or only with ridges/tubercles (samara unknown in *S. hatschbachii*).
84. Laminas abaxially densely pubescent with golden hairs (trabecula 1.1–2.3 mm long, straight, stalk 0.2–0.4 mm long), the epidermis nearly or completely hidden by the dense vesture; Brazil (Minas Gerais). 88. *S. hatschbachii.*
84. Laminas abaxially densely to sparsely pubescent with white or translucent hairs (trabecula 0.2–1.5 mm long, straight to curled, stalk 0.1–0.2 mm long or to 0.4 mm long in *S. tomentosum*), the epidermis never hidden.
85. Laminas with stipitate marginal glands to 0.4 mm long; unmodified anthers pubescent; limb of posterior petal without stout gland-tipped fimbriae; trabeculas of abaxial pubescence straight

to wavy, adjacent ones touching but not entwined to form a matted vesture, each hair readily discernible; Brazil (Paraíba, Pernambuco, Alagoas, Sergipe, Bahia, Espírito Santo, and adjacent Minas Gerais). 89. *S. blanchetii.*

85. Laminas with sessile to slightly raised marginal glands, 0.1 (–0.2) mm long; unmodified anthers usually glabrous (sometimes with a few hairs); limb of posterior petal at the base often with 1–3 stout gland-tipped fimbriae per side, these 0.3–0.5 mm long, ca. 0.2 mm in diameter; trabeculas of abaxial pubescence straight but more commonly wavy to curled, the adjacent ones overlapping to entwined and commonly forming a matted vesture, each hair not readily discernible; Brazil (Goiás, Minas Gerais, and adjacent Bahia, Rio de Janeiro, São Paulo, Paraná, Santa Catarina). 79. *S. tomentosum.*

82. Anthers of stamens opposing the posterior-lateral sepals with the connective enlarged and bearing 1–2 reduced locules; unmodified anthers glabrous or pubescent.

86. Mature laminas abaxially sparsely pubescent, the trabeculas straight or wavy, touching but not greatly overlapping and entwined, the individual hairs readily discernible; nut of samara ca. 20 mm high, 13–15 mm in diameter, the locule surrounded by air chambers; Brazil (Rio de Janeiro). 82. *S. affine.*

86. Mature laminas abaxially abundantly pubescent, the trabeculas straight to curled, greatly overlapping and entwined and thus the individual hairs not readily discernible (except rarely in *S. tomentosum*); nut of samara 4–12.5 mm high, 5–6.3 mm in diameter, without air chambers (with air chambers in *S. cavernulosum*).

87. Umbels in dichasia or solitary, borne on an axis, these and associated leaves arranged alternately on a primary axis.

88. Anthers glabrous; flowers 15–25 per umbel; nut of samara with air chambers surrounding the locule; Brazil (eastern Bahia and adjacent Minas Gerais). 83. *S. cavernulosum.*

88. Anthers pubescent; flowers 10–15 per umbel; nut of samara without air chambers; Colombia (Meta, Caquetá), Ecuador (Napo), and Peru (Loreto, San Martín). 78. *S. alternans.*

87. Umbels solitary or in dichasia or compound dichasia or thyrses, the units and associated leaves arranged oppositely on a primary inflorescence axis.

89. Limb of anterior-lateral petals 8–9 mm in diameter, limb of posterior petal 6–7.5 mm in diameter; dorsal wing of samara elongate, 5.5–6.2 cm long, ca. 1.8 cm wide, 3.5–4.7 times as long as wide; Brazil (Bahia, Espírito Santo, eastern Minas Gerais). 80. *S. saxicola.*

89. Limb of anterior-lateral petals 10.5–15 mm in diameter, limb of posterior petal 8–11 mm in diameter; dorsal wing of samara elongate or trian-

gular to subsquare, 1.6–5.2 cm long, 1.6–2.7 cm wide, 1–2.7 times as long as wide.

90. Limb of anterior-lateral petals 10.5–12 mm in diameter, limb of posterior petal without basal stout gland-tipped fimbriae; anthers of stamens opposing the lateral sepals with the connective enlarged and bearing 2 reduced locules; flowers 8–25 per umbel; samara with the dorsal wing triangular to subsquare, encircling the nut, 1.6–2.9 cm high, 1.7–2.7 cm wide; mostly along the Río Paraná and Río Uruguay and their tributaries in Argentina (Buenos Aires, Chaco, Corrientes, Entre Ríos, Misiones, Paraguay, Santa Fe), Brazil (Paraná, Rio Grande do Sul, São Paulo), Paraguay (Alto Paraná, Pilar), and Uruguay. 81. *S. bonariense.*

90. Limb of anterior-lateral petals 11–15 mm in diameter, limb of posterior petal near the base often with 1–3 stout gland-tipped fimbriae per side, these 0.3–0.5 mm long, ca. 0.2 mm in diameter; anthers of stamens opposing the lateral sepals with the connective enlarged and bearing 1 (–2) reduced locules; flowers 10–40 per umbel or pseudoraceme; samara with an elongate dorsal wing 3.6–5.2 cm long, 1.6–2.1 cm wide; Brazil (Goiás, Minas Gerais, and adjacent Bahia, Rio de Janeiro, São Paulo, Paraná, Santa Catarina). 79. *S. tomentosum.*

80. Marginal glands of lamina borne on the margin, the glands sessile and/or filiform and/or stipitate, or the margin eglandular.

91. Anthers of stamens opposing the posterior-lateral sepals with the connective enlarged and always bearing 2 reduced locules.

92. Margin of lamina eglandular; samara with a triangular to subsquare dorsal wing encircling the nut; along the Río Paraná, Río Uruguay, the southern Río Paraguay, and their tributaries in Argentina (Buenos Aires, Chaco, Corrientes, Entre Ríos, Misiones, Paraguay, Santa Fe), Brazil (Paraná, Rio Grande do Sul, São Paulo), Paraguay (Alto Paraná, Pilar), and Uruguay. 81. *S. bonariense.*

92. Margin of lamina with sessile and/or filiform glands; samara with an elongate dorsal wing.

93. Anthers pubescent; southeastern Mexico to northern Costa Rica. 64. *S. retusum.*

93. Anthers glabrous.

94. Limb of petals fringed with fimbriae up to 1 mm long, claw of posterior petal bearded; nut of samara with 3–4 lateral winglets per side; Peru (San Martín and adjacent Loreto, Huánuco, Junín, Cuzco), and adjacent Brazil (Acre) and Bolivia (Beni). 73. *S. strigosum.*

94. Limb of petals erose to denticulate or with fimbriae up to 0.5 mm long, claw of posterior petal glabrous; nut of samara with a pair of lateral winglets and/or crests and spurs or only ridges.

95. Limb of anterior-lateral petals 8–8.6 mm in diameter, limb of posterior-lateral petals ca. 7.5

mm in diameter, limb of posterior petal 6–6.5 mm long, 4–4.5 mm wide; peduncles 0.7–2.1 times as long as the pedicels; flowers 15–50 per umbel; Panama (Darién), northern Colombia, and western Venezuela. 65. *S. dichotomum.*

95. Limb of anterior-lateral petals 12–15 mm in diameter, limb of posterior-lateral petals 10–11.5 mm in diameter, limb of posterior petal 9–11 mm long, ca. 8 mm wide; peduncles 0.5–0.8 times as long as the pedicels; flowers 13–20 per umbel; central Panama and islands in the Gulf of Panama. 66. *S. panamense.*

91. Anthers of stamens opposing the posterior-lateral sepals with the connective enlarged and bearing 0–1 reduced locules.

96. Anthers of stamens opposing the anterior-lateral sepals consisting only of an enlarged connective without locules; peduncles rudimentary to 9 mm long, shorter than to equaling the pedicels.

97. Margin of lamina with stipitate or filiform glands; petiole of the larger leaves with a pair of shallowly cupulate glands usually borne up to 5 mm below the base of the lamina, each gland up to 1 mm high; posterior styles glabrous; Brazil (Bahia, Espírito Santo?, Pernambuco, not recorded from but to be expected also in Alagoas and Sergipe). 61. *S. puberulum.*

97. Margin of lamina eglandular; petiole with a pair of sessile glands borne at the apex; posterior styles bearded; Brazil (Mato Grosso). 75. *S. matogrossense.*

96. Anthers of stamens opposing the anterior-lateral sepals with the connective enlarged and always bearing 2 reduced locules; peduncles 2.5–17.5 mm long, 0.3–2.3 times as long as the pedicels.

98. Anthers pubescent; limb of petals digitate-fimbriate; nut of samara with a row of 5–7 lateral winglets on each side; Brazil (Rio de Janeiro, and adjacent Espírito Santo and Minas Gerais). 77. *S. alternifolium.*

98. Anthers glabrous; limb of petals erose or denticulate or with tapered fimbriae; nut of samara with 1 lateral winglet per side and/or crests and spurs or only ridges or unornamented.

99. Limb of all petals with fimbriae up to 1.8 mm long; laminas commonly palmately 3–5 (–7)-lobed, or lanceolate to ovate to cordate to suborbicular, the base of the larger laminas deeply cordate to auriculate; French Guiana, Brazil (Pará). 72. *S. palmatum.*

99. Limb of lateral petals erose (-denticulate), limb of posterior petal erose to denticulate or with fimbriae up to 0.5 mm long; laminas elliptical to ovate to cordate (in *S. tonduzii* sometimes palmately 3–5-lobed), the base truncate to cordate.

100. Limb of anterior-lateral petals 12–15 mm in diameter, limb of posterior-lateral petals 10–11.5 mm in diameter, limb of posterior petal 9–11 mm long, ca. 7.5 mm wide; central Panama and islands in the Gulf of Panama. 66. *S. panamense.*

100. Limb of anterior-lateral petals 9–11 mm in diameter, limb of posterior-lateral petals 7.5–9 mm in diameter, limb of posterior petal 6–8 mm in diameter.

101. Laminas abaxially pubescent with a mixture of sessile to subsessile hairs and some stalked hairs, the stalks up to 0.1 mm long, in mature leaves the vesture often sloughed off in patches; posterior styles and sometimes also the anterior style with scattered hairs; dorsal wing of samara 4–4.5 cm long; Trinidad and the adjacent Paria Peninsula (Venezuela). 68. *S. finlayanum.*

101. Laminas abaxially pubescent with T-shaped hairs to tomentose, the stalk of the hairs 0.1–0.3 mm long, the vesture abundant and evenly distributed; styles glabrous; dorsal wing of samara 2.3–3 cm long; Costa Rica (Nicoya peninsula of Guanacaste, northern Puntarenas). 67. *S. tonduzii.*

1. Stigmaphyllon pseudopuberum Niedenzu, Verz. Vorles. Akad. Braunsberg W.-S. 1912–1913: 28. 1912.—TYPE: GUATEMALA. Alta Verapaz: Cobán, Nov 1902, *von Türckheim 8385* (lectotype, designated by C. Anderson, 1987a: US!, photo: MICH!; isolectotypes: A! F! GH! M! NY! US!).

Vine. Stems and branches densely sericeous when young, eventually becoming glabrate to glabrous. Laminas 6–15 cm long, 3.5–12 cm wide, ovate to elliptical or broadly so or sometimes lanceolate, apex acuminate-mucronate, base attenuate to truncate or sometimes slightly cordate, glabrous adaxially, sericeous to densely so abaxially (trabecula 0.3–0.6 mm long, straight, sessile), margin with irregularly spaced sessile glands (0.3–0.6 mm in diameter); petioles 1–6.3 cm long, densely sericeous, not confluent across the node, with a pair of prominent to stout and stalked (peg-shaped) glands at the apex or up to 2 mm below the base of the lamina, each gland 0.6–2 mm in diameter, up to 1.3 mm long; stipules 0.6–1.6 mm long, 0.4–1 mm wide, free, triangular to narrowly so to sometimes linear, eglandular. Flowers ca. 12–20 per congested to interrupted pseudoraceme, these borne in dichasia or small thyrses (axes to the 3rd order, densely sericeous), rarely solitary. Peduncles 3–9 mm long; pedicels 2.5–9.5 mm long, terete; both densely sericeous, peduncles 0.2–1.8 times as long as pedicels. Bracts 0.8–1.8 mm long, 0.7–1.3 mm wide, triangular or narrowly so, apex acuminate; bracteoles 0.9–1.5 mm long, 0.6–1.4 mm wide, triangular or broadly so, apex obtuse, eglandular or each bracteole with a pair of glands (each gland 0.3–0.4 mm in diameter) or only a marginal glandular band in the basal 1/3–1/2; bracts and bracteoles sericeous abaxially. Sepals 1.5–2.5 mm long, 1.7–2.5 mm wide, glands 1.8–2.9 mm long, 1–1.6 mm wide. All petals with the limb orbicular, glabrous, yellow, margin erose or erose-dentate, teeth up to 0.3 (–0.4) mm long; anterior-lateral petals: claw 1.5–2.5 mm long, limb ca. 5.5–7 mm long and wide; posterior-lateral petals: claw 1.1–1.8 mm long, limb of posterior-lateral petals ca. 4–5.5 mm long and wide; posterior petal: claw 2.6–3.8 mm long, apex slightly if at all indented, limb ca. 3.5–5 mm long and wide. Stamens unequal in size but subequal in shape, that opposite the anterior style usually the longest or those opposite the anterior-lateral sepals equally as long; anthers all loculate, glabrous. Stamen opposite anterior sepal: filament 1.5–2.5 mm long, anther 0.8–1 mm long; stamens opposite anterior-lateral petals: filaments 1.3–2.4 mm long, anthers 0.7–1 mm long; stamens opposite anterior-lateral sepals: filaments 1.6–2.6 mm long, anthers 0.6–1 mm long; stamens opposite posterior-lateral petals: filaments 1.4–2.5 mm long, anthers 0.9–1 mm long; stamens opposite posterior-lateral sepals: filaments 1.5–2 mm long, anthers 0.7–0.9 mm long; stamen opposite posterior petal always shorter than the adjacent two: filament 1.3–1.8 mm long, anther 0.8–0.9 mm long. Anterior style 1.3–2.1 mm long, slightly shorter than or equal to the posterior two, terete, glabrous, erect or slightly recurved; apex 0.7–0.9 (–1.2) mm long including a spur up to 0.2 mm long, linear, 0.2–0.3 (–0.4) mm wide, folioles absent. Posterior styles 1.6–2.4 mm long, terete, glabrous, erect or slightly recurved; apex 0.8–1.3 mm long including a spur up to 0.2 (–0.3) mm long or the distal end blunt, linear, 0.3–0.4 (–0.5) mm wide, folioles absent. Dorsal wing of samara 4–5.4 cm long, 1.5–2.3 cm wide, upper margin with an obtuse tooth; nut bearing 1–3 triangular to rectangular lateral winglets on each side, these 1.5–5 mm long, 0.5–5.5 mm wide, or only with 1–2 spurs and/or crests or with a network of prominent ribs up to 0.5 mm high; nut 7–8 mm high, 5.5–6 mm in diameter, woody, without air chambers, areole 4.5–6 mm long, 4–4.5 mm wide, concave, carpophore up to 3 mm long. Embryo ca. 7 mm long, ca. 2 times as long as wide, ovoid, cotyledons convoluted and folded within each other, outer surface of outer cotyledon smooth but with a cleft in the basal half. Chromosome number unknown.

Phenology. Collected in flower from May through February, in fruit from September through May and in July.

Distribution (Fig. 2). Mexico (Chiapas) and Guatemala (Huehuetenango, Quetzaltenango, Suchitepéquez, and Alta Verapaz); upper elevation pine-oak forest and montane rain forest; (1000–) 1300–2700 m.

REPRESENTATIVE SPECIMENS. **Mexico.** CHIAPAS: Mpio. Zinacantán, along Mex. Hwy at paraje Sequentic, *Breedlove 10395* (DS, F, MICH, US, UTD); Mpio. Jitotol, 6.5 km N of Jitotol along rd to Pichucalco, *Breedlove 21417* (DS, ENCB, MICH, MO, NY); Mpio. Cintalapa, SE of Cerro Baul on the border with the state of Oaxaca, 16 km NW of Rizo de Oro, *Breedlove 27595* (DS, ENCB, MEXU, NY); Mpio. Angel Albino Corzo, above Finca Cuxtepec, *Breedlove 52070* (CAS); Mt. Ovando, *Matuda 2661* (A, DS, F, K, MEXU, MICH, NY, US, UTD); Mt. Male, near Porvenir, *Matuda 4588* (A, F, MEXU, MO, NY, US); Mpio. La Trinitaria, Lagos de Montebello, E side of Lago Tsikaw, *Shilom Ton 2643* (DS, ENCB, MICH, NY). **Guatemala.** ALTA VERAPAZ: Pansamalá, *von Türckheim 708* (GH, K, NY, P, US); vicinity of San Juan Chamelco, *Wilson 40900* (F), *Wilson 40983* (F).—HUEHUETENANGO: near Jacaltenango, *Steyermark 51840* (F, US).—QUETZALTENANGO: Chiquihuite, *Standley 68103* (F).—SUCHITEPÉQUEZ: flood plain of Río Mocá, *Skutch 1578* (A, F, US).

Stigmaphyllon pseudopuberum, of the pine-oak forests and montane rain forests of western Guatemala and adjacent Chiapas, Mexico, is one of the few species of *Stigmaphyllon* occurring at upper elevations (to 2700 m). It resembles *S. puberum* only in its ovate to elliptical leaves, which are sericeous abaxially, but the petiole glands are often peg-shaped. *Stigmaphyllon pseudopuberum* is distinguished by its small flowers, arranged in pseudoracemes, with a homogeneous androecium and efoliolate styles, aspects in which it is similar to *S. bogotense* (no. 2) and *S. sarmentosum* (no. 3), of upper elevations in western South America. As in these species, the cotyledons of *S. pseudopuberum* are also convoluted and folded within each other, but the outer surface of the outer cotyledon is smooth except for a cleft in the basal half.

2. Stigmaphyllon bogotense Triana & Planchon, Ann. Sci. Soc. Nat. Bot., sér. 4, 18: 320. 1862.—TYPE: COLOMBIA. Cundinamarca: "Entre Tenasuca et Tena, cordillère de Bogota," *Triana 5581-5* (lectotype, designated by Cuatrecasas, 1958: COL!, photo: MICH!; isolectotypes: BM! G! K!, photos of G isolectotype: F! GH! MICH!).

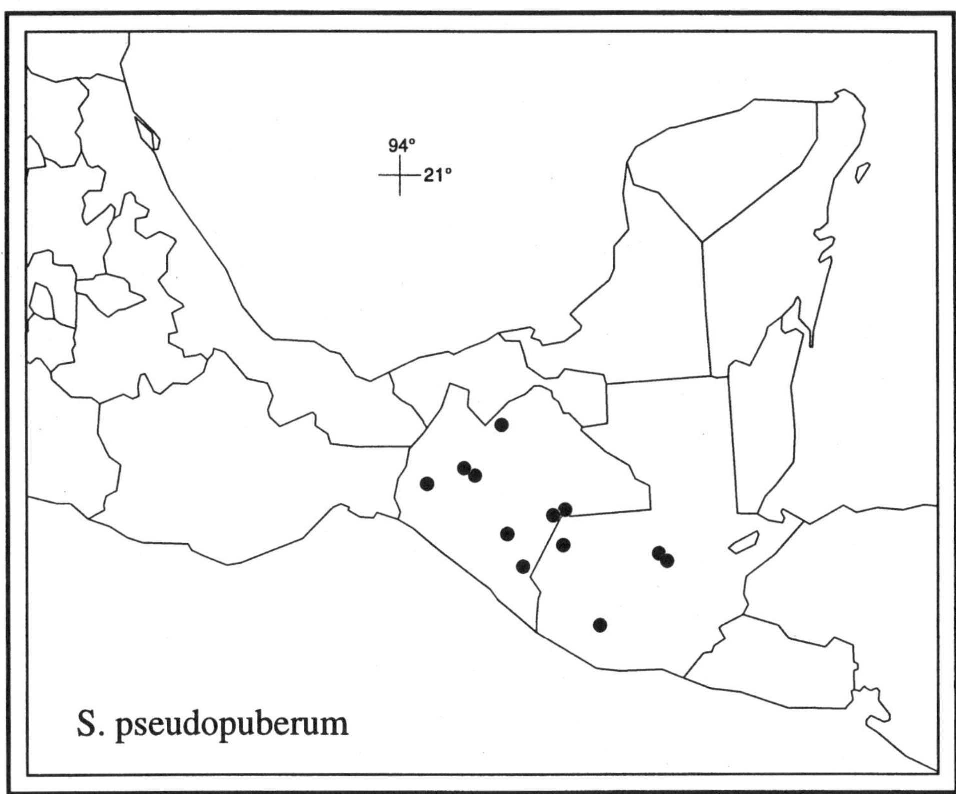

FIG. 2. Distribution of *Stigmaphyllon pseudopuberum.*

Stigmaphyllon lanuginosum Niedenzu, Ind. Lect. Lyc. Brunsberg. p. hiem. 1899–1900: 9. 1899.—TYPE: PERU. "ad Ponuzo," 1784, *Ruiz & Pavón s.n.*, distributed as no. 33/92 (holotype: B, destroyed; isotypes: F! MA!, photo of MA isotype: MICH!).

Stigmaphyllon ruizianum Niedenzu, Ind. Lect. Lyc. Brunsberg. p. hiem. 1899–1900: 9. 1899.—TYPE: VENEZUELA. Mérida, *Moritz 1196* (lectotype, here designated: K!, photo: MICH!; isolectotypes: BR! GH! GOET! W!).

Stigmaphyllon ruizianum var. *subglabratum* Niedenzu, Verz. Vorles. Akad. Braunsberg W.-S. 1912–1913: 26. 1912. *Stigmaphyllon bogotense* var. *subglabratum* (Niedenzu) Niedenzu, Pflanzenreich IV. 141(2): 481. 1928.—TYPE: COLOMBIA. Santa Marta, *H. H. Smith 2070* (holotype: B, destroyed; isotypes: CM! F! GH! K! MICH! MO! NY! US!).

Stigmaphyllon primaevum Niedenzu, Repert. Spec. Nov. Regni Veg. 26: 346. 1929.—TYPE: PERU. Cajamarca: Prov. Hualgayóc, montaña de Nanchó, 6000 ft, *Raimondi 5206* (lectotype, here designated: B, destroyed, photos: A! F! MICH! NY! US!).

Stigmaphyllon bogotense f. *renifolium* Niedenzu, Repert. Spec. Nov. Regni Veg. 26: 346. 1929.—TYPE: PERU. Amazonas: Chachapoyas, Valle de Huayabamba, *Raimondi 1865* (holotype: B, destroyed).

Vine to 25 m. Stems and branches sericeous or tomentulose when young, soon becoming glabrate to glabrous. Laminas 6.5–24 cm long, 4.3–21.5 cm wide, narrowly ovate to ovate to narrowly elliptical to elliptical to suborbicular, apex mucronate or emarginate-mucronate or sometimes acuminate, base cordate or sometimes truncate, glabrate to glabrous adaxially, sparsely to densely pubescent with T-shaped hairs to sparsely to densely tomentose abaxially (trabecula 0.9–1.3 mm long, wavy to crisped or curled, stalk 0.1–0.4 mm long), margin with irregularly spaced sessile glands (0.3–0.9 mm in diameter), these and the adjacent tissues sometimes drawn out into teeth; petioles 1.6–20 (–25) cm long, sericeous or tomentose, often glabrate in older leaves, not confluent across the node, with a pair of prominent but sessile glands borne at the apex, each gland (0.6–) 1–2 (–2.4) mm in diameter; stipules 0.3–1.2 mm long, 0.2–2.2 (–1.4) mm wide, free, triangular, eglandular. Flowers ca. 10–35 per condensed or interrupted pseudoraceme, usually the lowest two (sometimes more) flowers of an aggregate separated a short distance from the rest, the pseudoracemes sometimes solitary but usually borne in dichasia or compound dichasia or small thyrses (axes to the 3rd order, sericeous); subsidiary axes rare, 2–2.5 cm long. Peduncles 3.5–10.5 mm long; pedicels 6.2–15.5 (–19) mm long, terete; both sericeous or densely so (hairs subsessile), peduncles 0.3–0.8 (–1) times as long as the pedicels. Bracts 1.1–3 mm long, 0.4–1 mm wide, narrowly triangular, apex acute; bracteoles 1–2 mm long, 0.4–0.9 mm wide, narrowly triangular to almost linear, apex acute, eglandular or each bracteole with a pair of glands (each gland 0.1–0.3 mm in diameter) or with a glandular area in the basal 1/3–1/2; bracts and bracteoles sparsely to densely sericeous abaxially. Sepals 2–3 mm long and wide, glands 1.4–2.4 mm long, 0.6–1.4 mm wide. All petals with the limb orbicular (or that of the posterior petal sometimes obovate), glabrous, yellow but usually marked or suffused with red, margin digitate-fimbriate, fimbriae up to 0.8 mm long; anterior-lateral petals: claw (2.2–) 2.5–3.1 mm long, limb (6.5–) 7.5–10 mm long and wide; posterior-lateral petals: claw 1.8–2.2 (–2.5) mm long, limb (5–) 7–9 mm long and wide; posterior petal: claw (2.2–) 2.5–3.8 mm long, apex indented or slightly so, limb 5–6.5 mm long, 5–5.8 mm wide. Stamens unequal in size but subequal in shape, those opposite the anterior and anterior-lateral sepals usually with the longest filaments; anthers all loculate, glabrous. Stamen opposite anterior sepal: filament 2–3.6 mm long, anther 0.9–1.1 mm long; stamens opposite anterior-lateral petals: filaments 1.5–2.6 mm long, anthers 0.7–0.9 mm long; stamens opposite anterior-lateral sepals: filaments 2.3–3.5 mm long, anthers 0.8–1 mm long; stamens opposite posterior-lateral petals: filaments 2–3 mm long, anthers 0.9–1.1 mm long; stamens opposite posterior-lateral sepals: filaments 2–3 mm long, anthers 0.7–1 mm long; stamen opposite posterior petal always shorter than the adjacent two: filament (1.5–) 2–2.7 mm long, anther 0.7–1 mm long. Anterior style 1.8–3 mm long, subequal to the posterior two, terete but laterally flattened at the apex, glabrous, erect; apex 0.5–0.8 mm long including a spur up to 0.2 mm long or the distal end blunt, 0.2–0.3 mm wide, linear, folioles absent. Posterior styles 1.8–3.2 mm long, terete but the distal 1/4–1/3 laterally flattened, glabrous, slightly recurved toward the posterior-lateral sepals; apex 0.5–0.9 mm long including a spur up to 0.2 (–0.3) mm long or the distal end blunt, linear, folioles absent. Dorsal wing of samara 2.5–4.8 (–5.5) cm long, 0.9–2.1 cm wide, upper margin with or without a blunt tooth; nut commonly bearing lateral crests or ridges and/or small spurs and/or winglets up to 0.8 mm long and 0.4 mm wide, or with a pair of rectangular to broadly triangular coarsely dentate lateral winglets, these 3.5–7 mm long, 2–6.5 mm wide, or without lateral ornamentation; nut (2.8–) 3.5–7 (–8) mm high, 2.5–5.5 mm in diameter, without air chambers, areole 2.3–6 mm long, 2.5–4.5 mm wide, convex, carpophore up to 5 mm long. Embryo 4.2–7.3 mm long, ca. 1–2 times as long as wide, ovoid to spherical, "brainlike": cotyledons highly convoluted and folded, the inner nested in the outer, the distal 1/3 of the outer cotyledon folded over the inner cotyledon. Chromosome number unknown. Fig. 3a–k.

Phenology. Collected in flower and fruit throughout the year.

Distribution (Fig. 4). Colombia and western Venezuela to Ecuador and Peru; in montane evergreen rainforest and cloud forest, also at forest edge, in thickets, and at roadsides; (800–) 1000–3200 m.

REPRESENTATIVE SPECIMENS. **Venezuela.** BARINAS: Dtto. Bolívar, between Altamira and Santo Domingo, 08°51′N, 71°35′W, *van der Werff & Ortíz 5897* (MO).—DISTRITO FEDERAL: en la fila de San Isidro de Galipán, *Tamayo 340* (US, VEN).—FALCÓN: Sierra de San Luis, cerca de Parador Turistica, *van der Werff 3610* (MICH, U).—LARA: cerca de Sanare hacia el Parque Nacional Tacambú, *R. F. Smith 3134* (VEN).—MÉRIDA: 39 km W of Mérida along rd to La Carbonera, *Breteler 3141* (COL, G, MER, MO, NY, P, RB, S, U, US, VEN); Tabay, selva de La Isla y El Rincón, *Gehringer 432* (F, G, NY, US, VEN); Dtto. Tovar, carretera Tovar a Guaraque, parte alta de la Aldea San Francisco, *López-Palacios 9* (MER); Dtto. Campo Elías, bosque de San Eusebio, *Ruiz T. 1114* (CM, K, MER).—PORTUGUESA: Dtto. Unda, 20 km al N de Paraíso de Chabasquen, Fila de Helechal, 09°31′N, 69°59′W, *Aymard et al. 2043* (MICH).—TÁCHIRA: ad ripam dextram rivi Táchira ad limina reipublicae Colombiae, *Charpin & Jacquemoud AC13277* (G, MICH); Dtto. Cárdenas, entre Cordero y Páramo El Zumbador, *Romero 744* (MY); Quebrada La Lejia, S of Quebrada Agua Azul, 15–16 km SE of Delicias, 07°30′N, 72°24′W, *Steyermark 118529* (MICH).—TRUJILLO: carretera vieja entre Trujillo y Boconó, entre Urbina y San Rafael, *Steyermark 97226* (US, VEN); Vitú, Cerro El Zamuro, 09°28′N, 70°28′W, *van der Werff & Ortega 6112* (MICH). **Colombia.** ANTIOQUIA: 1 km al S de Hoyo Rico, *Barkley 18A176* (COL, NY, US); Mpio. Bolívar, carretera Medellín–Quindío, Km 93, *Forero et al. 9727* (MO); Mpio. Granada, rd between Granada and San Carlos, 18.6 km E of San Carlos, 06°09′N, 75°04′W, *MacDougal & Velásquez 4113* (MICH); Jericó, *Uribe U. 1419* (COL, US); Medellín, en Alto de Santa Elena, *Uribe U. 1856* (COL); Mpio. Sonsón, Km 16 of rd Sonsón–La Unión, 40 km from Unión, 05°46′N, 75°18′W, *Zarucchi et al. 5255* (K, MICH).—BOYACÁ: Valle de la Uvita, near Uvita, *Cuatrecasas & García Barriga 1874* (COL, F, US); region of Mt. Chapón, NW of Bogotá, *Lawrance 321* (BM, F, G, GH, K, MO, MT, NY, U).—CALDAS: Risaralda, Pan American Hwy between Anserma and Río Sucio, *Gentry 28800* (MICH, MO); Río Quindío, above Armenia, *Pennell et al. 8675* (NY, US); Manizales, *Yepes Agredo 662* (COL).—CAUCA: hoya del Río Palo, márgenes del río entre Tacueyó y La Tolda, *Cuatrecasas 19514* (A, F, MICH, US); Munchique, camino a la "Mina Tapada," *García Barriga et al. 12973* (COL, US); Popayán, *Lehmann 4930* (BM, F, G, GH, K, LE, MA, NY, S, U, US); ad pag. El Tambo, ad Quisquio, *von Sneidern 428* (A, F, MICH, NY, S).—CUNDINAMARCA: W of Bogotá on rd to El Colegio, ca. 1.5 km W of El Salto de Tequendama near Km 8, *Barclay et al. 3306* (COL, US); cuesta de Fusagasugá, *Cuatrecasas 8061* (COL, F, US); Estación Santana, arriba Sasaima, *Dugand & Jaramillo 3835* (COL, US); Santandercito, *Uribe U. 4033* (COL, NY); Viotá, vereda de Liberia, *Uribe U. 6017* (COL).—HUILA: Finca Merenberg, Cauca border E of Leticia, 02°16′N, 76°12′W, *Gentry 47713* (MICH); montes más arriba de Guadelupe, *Pérez Arbeláez & Cuatrecasas 8399* (COL, F, US); E of Neiva, *Rusby & Pennell 628, 649, 855* (NY).—MAGDALENA: Corrigimiento de Manaure, Finca Los Venados, *Romero Castañeda 7525* (COL, US).—NARIÑO: along Río Guisa on rd from Pasto to Ricuarte between Km 76–77, *Luteyn et al. 5081* (COL, MICH, MO, NY, US); Mpios. Túquerres y Guachavés, carretera de Túquerres a Samaniego, *Mora 342* (COL); region of Pasto, rd to Arandá, *Schultes & Villareal 7457* (COL, DS, F, IAN, K, NY, US).—NORTE DE SANTANDER: La Isla, *Killip & Smith 19810* (A, GH, NY, US); Culagá Valley, near Tapatá, N of Toledo, *Killip & Smith 20346* (A, GH, NY, US).—PUTUMAYO: entre Sachamates y San Francisco de Sibundoy, Planada de Minchoy, *Cuatrecasas 11437* (COL, F, US); carretera a Mocoa, entre El Pepino y Mirador, *Idrobo 2400* (COL); Sibundoy, *Schultes & Villareal 7625* (COL, F, GH, K, US).—SANTANDER: Valley of Río Minero, E of Florian, *Fassett 25827* (NY, US, WIS); vicinity of Charta, *Killip & Smith 18876* (GH, NY, US); Sucre, *Uribe U. 2428* (COL).—TOLIMA: Río Coello to San Miguel, new Quindío trail, *Hazen 9664* (NY).—VALLE: Mpio. Sevilla, vereda El Billar, *Devia 368* (MICH); hoya del Río Cali, Pichindé, *Duque Jaramillo 3802* (COL, NY); 2 mi from Bitaco on rd to Cali, Bitaco Valley, *Hutchison & Idrobo 3044* (COL, F, US); La Cumbre, *Pennell 5172* (GH, K, NY, US). **Ecuador.** AZUAY: Hwy Cuenca–Cola de San Pablo, *Boeke & Loyola 1053* (MICH, QCA).—BOLÍVAR: carretera Chillanes–Tiquibuso, sector de San José de Guayabal, 01°55′S, 79°05′W, *Zak & Jaramillo 2712* (K, MICH); carretera Chillanes–Tambillo–Trigoloma, quebrada de Atiacagua, 01°55′S, 79°05′W, *Zak & Jaramillo 2771* (K, MICH).—CARCHI: environs of Chical, 12 km below Maldonado on the Río San Juan, 01°04′N, 78°17′W, *Madison 4670* (F, MICH, SEL); Mira Cantón, El Carmen, camino hacia Chical, 00°51′N, 78°13′W, *Palacios et al. 9677* (MICH).—CHIMBORAZO: cañón of the

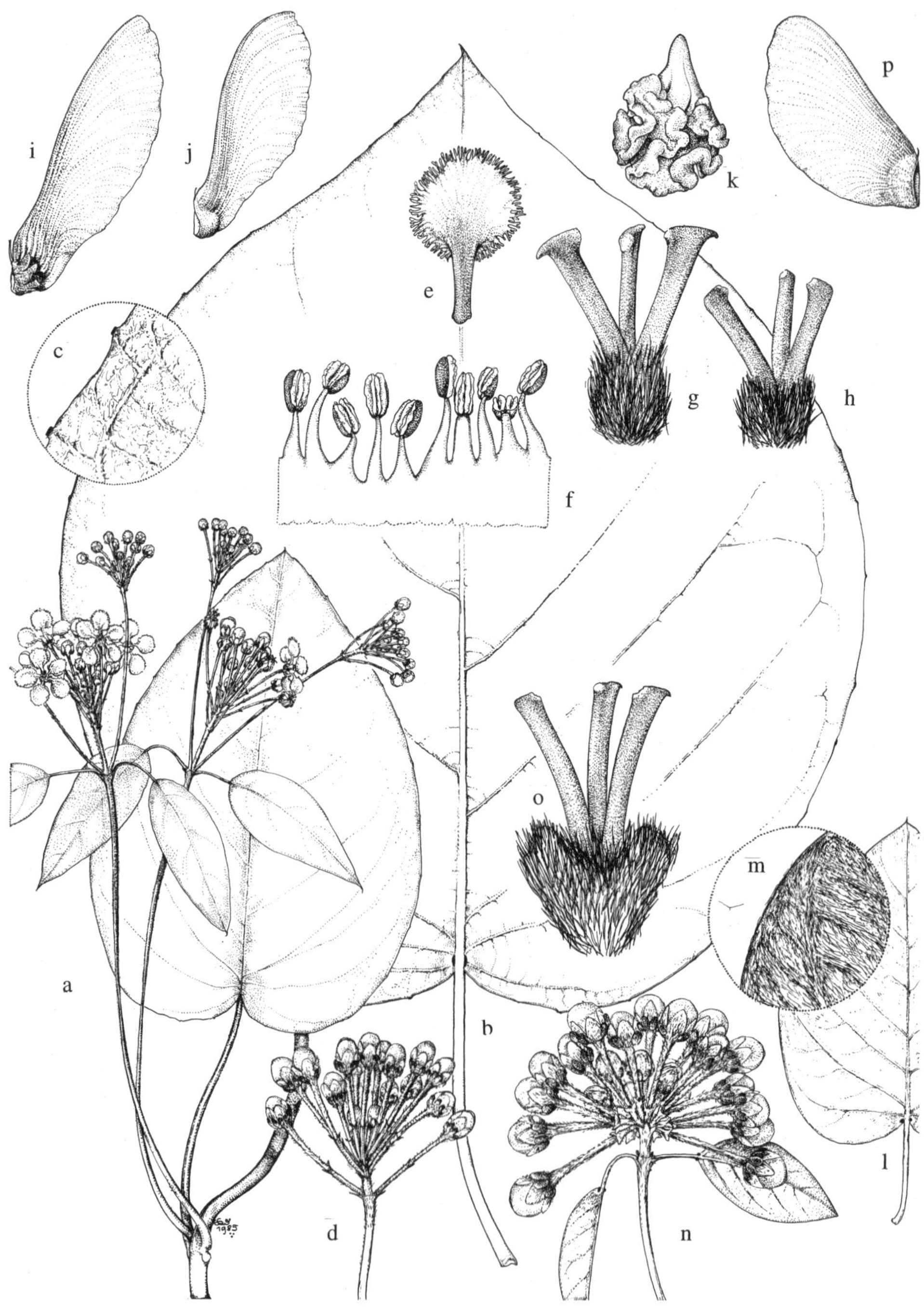

FIG. 3. *Stigmaphyllon bogotense* and *S. sarmentosum*. a–k, *S. bogotense*. a. Flowering branch. b. Large leaf. c. Detail of margin and abaxial surface of lamina. d. Umbel. e. Posterior petal (the "flag"). f. Androecium; stamen second from right opposes the posterior petal. g, h. Two gynoecia (anterior style to the left), illustrating variation in size and shape of style apex. i, j. Two samaras, illustrating variation in size and shape of dorsal wing and lateral ornamentation. k. Embryo. l–p, *S. sarmentosum*. l. Leaf. m. Detail of margin and abaxial surface of lamina. n. Umbel. o. Gynoecium, anterior style to the left. p. Samara. Scale: for a, b, l, bar = 4 cm; for c, bar = 1.3 cm; for d, n, bar = 2 cm; for e, bar = 8 mm; for f–h, o, bar = 2.7 mm; for i, j, p, bar = 2.3 cm; for k, bar = 6.7 mm; for m, bar = 4 mm. (Based on: a, d, *Uribe U. 1434*; b, c, *Triana s.n.*; e, f, h, *Steyermark & Liesner 118529*; g, *Harling & Andersson 13668*; i, *Hutchison & Idrobo 3044*; j, *Core 130*; k, *Woytkowski 7035*; l, m, *Steyermark 54395*; n, o, *Harling et al. 20419*; p, *Townsend A84*.)

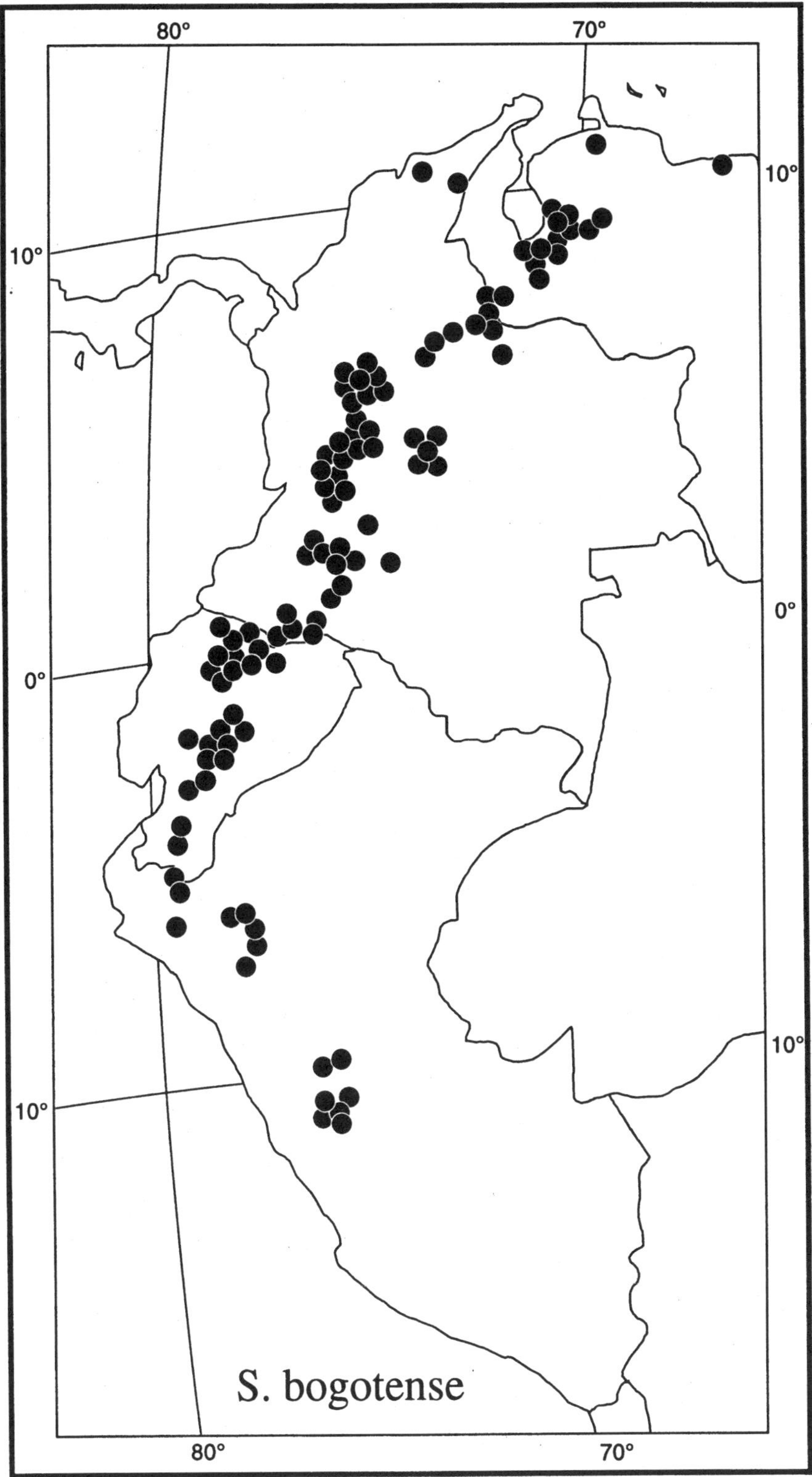

FIG. 4. Distribution of *Stigmaphyllon bogotense*.

Río Chanchan, ca. 5 km N of Huigra, *Camp E-3337* (F, GH, K, NY, US, W); 8.3 km E of Pallaganga, *MacBryde 950* (QCA).—ESMERALDAS: new rd from Lita to San Lorenzo, 00°58′N, 78°35′W, *Gentry 69969* (MO).—IMBABURA: carretera Apuela–Cemento Selvalegre–Otavalo, *Jaramillo 9293* (AAU); Cotachachi Cantón, Parroquia Apuela, Sector Cuellage, 00°15′N, 78°25′W, *Tipaz & Aulestia 1679* (MO).—MORONA-SANTIAGO: 9–10 km SE of San Juan Bosco, *Gentry 30844* (AAU, MICH, SEL); Cumandá, 6 km W of Mera, *Harling & Andersson 17288* (GB, MICH); along Río Palora, 1–4 km upstream from Arapicos, *Lugo S. 6069* (GB, MICH).—NAPO: Cantón Quijos, highest point on rd between Cosanga and Tena, 9 km SE of Cosanga, *Kirkbride & Chamba R. 4127* (MICH, US); 9 km NE of (above) Baeza on rd to Quito, 00°25′S, 77°57′W, *Molau & Eriksen 2161* (AAU, GB, MICH); Via Hollín–Loreto, entre Río Guamaní y Río Pucuno, Km 40, *Palacios 2213* (K, MICH).—PASTAZA: rd N from Mera, known as the "Via Anzu," 01°20′S, 78°10′W, *Boom et al. 7867* (MICH); Mera, *Harling et al. 9722* (GB, MICH, MO); Veracruz (Indillama), *Lugo S. 37* (GB, MICH, MO).—PICHINCHA: Reserva Geobotánica Pululahua, carretera Pululahua–Niebli, 00°05′N, 78°30′W, *Cerón & Iguago 5665* (MICH); Tandayapa–Nono, W of Nono, *Harling 14857* (GB, MICH); 13 km E of Tandapi on rd from Santo Domingo to Quito, *Wunderlin et al. 8686* (MO, SEL).—TUNGURAHUA: Baños, 4 km on the rd to Puyo, 0–3 km E of rd tunnel, 01°29′S, 78°29′W, *Madsen et al. 36490* (AAU); Hacienda La Merced, Pastaza Valley, *Mexia 7002* (F, US); between Machay and Río Mapoto, *Penland & Summers 289* (F, GH, US).—ZAMORA-CHINCHIPE: rd Loja–Zamora, above Tambo, *Harling & Andersson 13668* (GB, MICH); rd Loja–Zamora, Km 24–25, 03°59′S, 79°05′W, *Holm-Nielsen et al. 3465* (AAU, F, GB, MO, NY, S). **Peru.** AMAZONAS: Prov. Bongará, on rd to Pomacocha, *Hutchison & Wright 6824* (F, M, MICH, MO, NY, US); Mendoza, *Woytkowski 8119* (MO); Prov. Bongará, lower portion of Shipasbamba–Pomacocha trail, *Wurdack 1103* (F, NY, S, US).—CAJAMARCA: Prov. Santa Cruz, Dist. Catache, upper Río Zaña valley, ca. 5 km above Monte Seco on path to Chorro Blanco, *Dillon 4341* (MICH); Prov. Cutervo, entre Socota y La Pucarilla, a 9 km de Socota, *Sánchez V. & Miranda 6241* (MICH); Colasay, *Woytkowski 7035* (MICH, MO).—HUÁNUCO: Prov. Huánuco, rd Mirador to Chinchao, *Mexia 4143* (GH, MO); Muña, *Woytkowski 5191* (F, MO, P, S).—JUNÍN: 38.4 km NE of Tarma on carretera 20B to Oxapampa, *Jones & Davidson 9119* (MICH); Agua Dulce, ad Tarma, *Woytkowski 7466* (GH, MO, US).—PIURA: Huancabamba Prov., Canchaque, entre "Chorro Blanco" y "WarWar," *Díaz et al. 3194* (MICH); Huancabamba, arriba de Palambla, *Sagástegui 8806* (NY).

Stigmaphyllon bogotense is a widespread, common species of the uplands of northwestern South America. Although somewhat variable, it is easily recognized by its small flowers with digitate-fimbriate petals marked with red, homogeneous androecium, and efoliolate styles. The flowers are borne in a condensed or interrupted pseudoraceme; usually the lowest pair of flowers of an aggregate is separated a short distance from the rest. The distal end of the apex of the styles varies from blunt to extended into a spur up to 0.2 (–0.3) mm long. The leaves are usually abundantly pubescent abaxially but sometimes only sparsely so; Niedenzu recognized variants with sparsely pubescent leaves as var. *subglabratum*. The embryos are "brainlike," i.e., both cotyledons are convoluted and folded within each other.

Stigmaphyllon bogotense is most similar to the sympatric *S. sarmentosum* (no. 3) and to *S. pseudopuberum* (no. 1) of the highlands of Guatemala and adjacent Mexico. Both differ in that their leaves are sericeous to densely so abaxially rather than pubescent with T-shaped hairs. In *S. sarmentosum* the abaxial pubescence is so dense that the epidermis is obscured; the epidermis is always visible in *S. bogotense*. *Stigmaphyllon bogotense* has also been confused with *S. florosum*, a lowland species in which the styles also lack folioles. In *S. florosum*, the stamens are unequal and have pubescent anthers, the apex of posterior styles bears a narrow lateral lip, and the embryo is of the common type, i.e., ovoid and with the larger outer cotyledon folded over the smaller inner one, but not convoluted.

In a very few collections of *S. bogotense* the anthers of all or many flowers remain closed and contain abnormal pollen. In anthers of the following collections pollen is at least partly aborted: Colombia: *López F. 8001*, 85% normal; *Cuatrecasas 11531*, 95% normal; *Killip 979*, 55–85% normal; Ecuador: *Harling & Andersson 16284*, 0% normal;

Holm-Nielsen & Jeppesen 282, 0% normal; Peru, *Woytkowski 7791*, 66% normal. These collections may represent hybrids or perhaps apomictic populations.

3. Stigmaphyllon sarmentosum Cuatrecasas, Webbia 13: 529. 1958.—TYPE: COLOMBIA. Cundinamarca: Chinga valley, 3 km E of Gutiérrez, 50 km S of Bogotá, 2350 m, 22 Jul 1944, *Grant 9638* (holotype: US!, photo: MICH!; isotype: NY!).

Vine to 6 m. Stems and branches sericeous when young, soon becoming glabrate. Laminas 4.6–11.5 cm long, 2.5–7.4 cm wide, elliptical or narrowly ovate, apex mucronate, base cordate to truncate, sparsely sericeous to glabrate adaxially, densely white-sericeous-tomentulose abaxially (trabecula 0.3–0.6 mm long, wavy or crisped, subsessile or with a stalk up to 0.075 mm long); margin eglandular or with scattered sessile glands (0.3–0.4 mm in diameter); petioles 1.6–3.5 cm long, densely sericeous, not confluent across the node, with a pair of prominent glands at the apex, each gland 1–1.4 mm in diameter, up to ca. 1 mm high; stipules 0.6–1.1 mm long and wide, free, triangular, eglandular. Flowers ca. 13–17 per umbel, often the lowest pair of flowers of an aggregate separated a short distance from the rest, the umbels solitary or borne in dichasia or small thyrses (axes to the 2nd order, sericeous). Peduncles 1.7–10 mm long; pedicels 8.5–27 mm long, terete; both densely sericeous, peduncles 0.2–0.3 times as long as pedicels. Bracts 1–3 mm long, 0.7–1.3 mm wide, narrowly triangular; bracteoles 1–1.8 mm long, 0.7–1 mm wide, narrowly triangular to sublinear, eglandular or each bracteole with a pair of glands (each gland 0.3–0.4 mm in diameter) or with a glandular area in the basal 1/3–1/2 or only a marginal glandular band in the basal 1/3–1/2; bracts and bracteoles with the apex acute, sericeous abaxially. Sepals 2.5–3 mm long, ca. 2.5 mm wide, glands 1.8–2.1 mm long, 1–1.2 mm wide. All petals with the limb glabrous, yellow with a red center or suffused with red, marginal fimbriae up to 0.3 mm long, limb of lateral petals orbicular, margin proximally erose-denticulate and distally digitate-fimbriate; anterior-lateral petals: claw ca. 3 mm long, limb ca. 10 mm long and wide; posterior-lateral petals: claw ca. 2.5 mm long, limb ca. 8.5–9 mm long and wide; posterior petal: claw ca. 3.5 mm long, apex not indented, limb ca. 6 mm long, ca. 5 mm wide, elliptical, margin digitate-fimbriate. Stamens unequal in size but subequal in shape, those opposite the lateral sepals with the longest filaments; anthers all loculate, glabrous. Stamen opposite anterior sepal: filament 2.9–3.2 mm long, anther 1–1.1 mm long; stamens opposite anterior-lateral petals: filaments 2.2–2.7 mm long, anthers 0.8–0.9 mm long; stamens opposite anterior-lateral sepals: filaments 3.3–3.4 mm long, anthers 0.9–1.1 mm long; stamens opposite posterior-lateral petals: filaments 2.8–3 mm long, anthers ca. 1 mm long; stamens opposite posterior-lateral sepals: filaments 3–3.2 mm long, anthers 0.8–0.9 mm long; stamen opposite posterior petal always shorter than the adjacent two: filament 2.5–2.7 mm long, anther 0.7–0.8 mm long. Anterior style 2.8–3.1 mm long, slightly longer than the posterior two, terete proximally, laterally flattened in the distal 1/4–1/3, glabrous, erect proximally, the distal 1/2 recurved; apex ca. 0.6 mm long, 0.2–0.3 mm wide, linear, without a spur, folioles absent. Posterior styles 2.6–2.9 mm long, terete proximally, laterally flattened in the distal 1/3–1/2, glabrous, at least slightly recurved toward the posterior-lateral sepals; apex ca. 0.8 mm long including a spur up to 0.2 mm long, linear, folioles absent. Dorsal wing of samara 2.9–4.6 cm long, 1.6–2.1 cm wide, upper margin with or without a blunt tooth; nut without lateral winglets but rugose, 6.5–8.5 mm high, 4.5–5.5 mm in diameter, without air chambers, areole 4.5–5 mm long, ca. 3.5 mm wide, convex, carpophore up to 3 mm long. Embryo 5–7 mm long, ca. 1–2 times as long as wide, subspherical to ovoid, "brainlike": both cotyledons convoluted and folded, the inner nested in the outer, the distal 1/3 of the outer cotyledon folded over the inner cotyledon. Chromosome number unknown. Fig. 3l–p.

Phenology. Collected in flower in February, March, April, July, September, and December, in fruit in February, October, November, and December.

Distribution (Fig. 5). Colombia (Cundinamarca), Ecuador (Loja, Pichincha), and Peru (Lambayeque, Piura); in bosque húmedo montaño, dry forest, shrubby woods, matorral, and clearings; (800–) 1600–2900 m.

REPRESENTATIVE SPECIMENS. **Ecuador.** LOJA: Loja Cantón, Parroquia Vilcabamba, sector Yamburara, 04°17′S, 79°12′W, *Cerón & Ocampo 11865* (MICH); carretera El Cisne–Gualel, 03°52′S, 79°20′W, *Freire F. et al. 1022* (AAU); Las Chinches, on rd Catacocha–Loja, *Harling et al. 15423* (GB, MICH); Loja–Zamora rd, *Harling et al. 20419* (GB, MICH); Nudo de Sabanilla, W slope on rd to Yanaga, *Harling & Andersson 21822* (GB, MICH); NE slopes of Cerro Villonaco, 11 km from Loja along new rd to La Toma, 04°00′S, 79°15′W, *Molau & Eriksen 3084* (GB); Gonzanama Cantón, carretera Yangana–Valladolid, 04°25′S, 79°15′W, *Rubio et al. 2282* (MICH); desert country between Vilcabamba and Cachiyacu, *Steyermark 54395* (MICH, NY); Utuana and Colaisaga, *Townsend A84* (UC).—PICHINCHA: along N bank of river 3 km W of Alluriquín, *Werling & Leth-Nissen 513* (AAU, NY). **Peru.** LAMBAYEQUE: Prov. Lambayeque, El Protrero, Penachi, *Llatas Q. 1418* (F).—PIURA: Prov. Ayacaba, bosque de Huamba, *Cano 1547* (MICH).

Stigmaphyllon sarmentosum is readily recognized by its leaves, which are densely white-tomentulose-sericeous abaxially, and its flowers with their equal stamens and efoliolate styles. The embryos are "brainlike," i.e., both cotyledons are convoluted and folded within each other. This species is similar in its inflorescences, flowers, and embryos to the sympatric *S. bogotense* (no. 2), from which it is easily separated by its leaves. Those of *S. bogotense* are tomentose to sometimes sparsely so abaxially; the pubescence is never so dense as to obscure the epidermis, as in *S. sarmentosum*.

The paratype cited by Cuatrecasas [Colombia. Cundinamarca: Muchindote valley, Quebrada Seca and Quebrada Negra, 13 km NE of Gachetá, *Grant 9576* (NY, US, WIS)] differs from all other collections in that the abaxial leaf surface is only moderately sericeous. As Cuatrecasas noted, the specimens are incomplete; they consist of a few leaves, only some attached, a few loose samaras, and old inflorescences (the "flowers" represented only by the persistent sepals and torus). In other species, very old leaves or leaves on old plants do not retain the pubescence and, most likely, *Grant 9576* consists of old material of *S. sarmentosum* rather than an undescribed species.

4. Stigmaphyllon suffruticosum Cuatrecasas, Webbia 13: 528. 1958.—TYPE: COLOMBIA. Caquetá: Cordillera Oriental, vertiente oriental, bosque entre Sucre y La Portada, 1200–1350 m, 5 Apr 1940, *Cuatrecasas 9131* (holotype: US!, photo: MICH!; isotypes: COL! F! NY!).

Vine. Stems and branches sericeous when young, soon becoming glabrous. Laminas 7.2–12 cm long, 5.3–8.8 cm wide, ovate to cordate to sometimes suborbicular, apex abruptly acuminate, base cordate, glabrous adaxially, very sparsely pubescent with T-shaped hairs to glabrate abaxially (trabecula 0.8–2 mm long, straight, stalk 0.05–0.2 mm long), margin with irregularly spaced sessile glands (0.4–0.5 mm in diameter); petioles 3–7.5 cm long, sericeous, not confluent across the node, with a pair of prominent but sessile glands at the apex, each gland 1.3–1.7 mm in diameter; stipules 0.5–0.7 mm long and wide, free, triangular, eglandular. Flowers ca. 20–25 per pseudoraceme, these borne in dichasia or compound dichasia (axes to the 3rd order, sericeous). Peduncles 5–7 mm long;

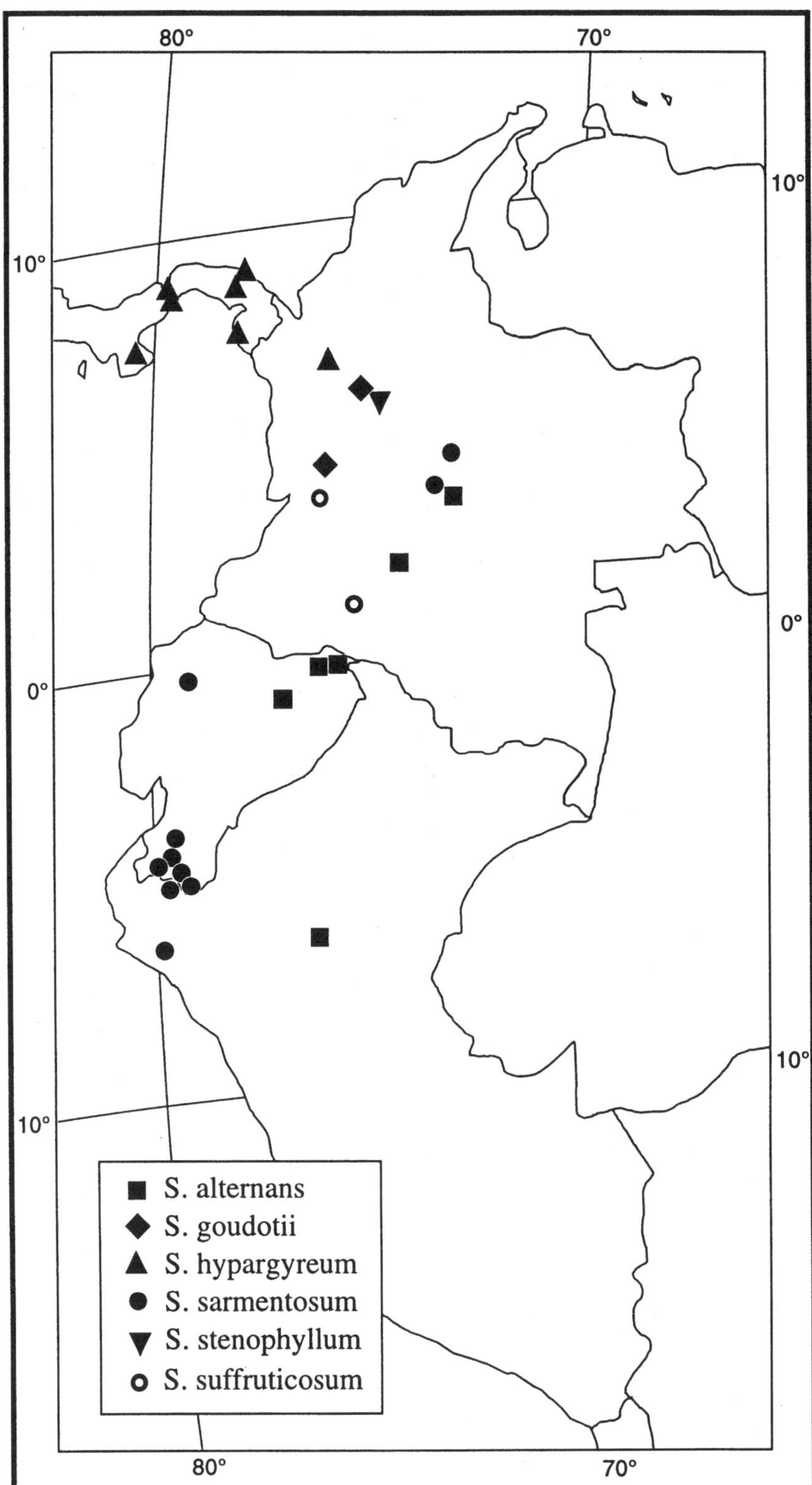

FIG. 5. Distribution of *Stigmaphyllon alternans*, *S. goudotii*, *S. hypargyreum*, *S. sarmentosum*, *S. stenophyllum*, and *S. suffruticosum*.

pedicels 7.5–8 mm long, terete; both sericeous, peduncles 0.6–0.9 times as long as the pedicels. Bracts 1.5–2 mm long, 1–1.5 mm wide, triangular, apex acute; bracteoles 1–1.5 mm long, 0.6–0.8 mm wide, narrowly triangular, apex obtuse, eglandular or with a glandular area in the basal 1/3–1/2; bracts and bracteoles sparsely sericeous to glabrate abaxially. Sepals ca. 2 mm long, ca. 2.2 mm wide, glands ca. 2 mm long, ca. 1 mm wide. All petals with the limb orbicular, glabrous, yellow but suffused with red, margin digitate-fimbriate, fimbriae up to 0.5 mm long, glandular; anterior-lateral petals: claw 2.5–2.7 mm long, limb ca. 9 mm long and wide; posterior-lateral petals: claw 1.8–2 mm long, limb ca. 8 mm long and wide; posterior petal: claw ca. 3.3 mm long, apex indented, limb ca. 7 mm long and wide. Stamens unequal in size but subequal in shape, those opposite the posterior-lateral sepals with the longest filaments; anthers all loculate, glabrous. Stamen opposite anterior sepal: filament ca. 2.9 mm long, anther ca. 0.9 mm long; stamens opposite anterior-lateral petals: filaments ca. 2.1 mm long, anthers ca. 0.7 mm long; stamens opposite anterior-lateral sepals: filaments ca. 2.6 mm long, anthers ca. 0.8 mm long; stamens opposite posterior-lateral petals: filaments ca. 2.6 mm long, anthers ca. 0.9 mm long; stamens opposite posterior-lateral sepals: filaments ca. 3.2 mm long, anthers ca. 0.5 mm long; stamen opposite posterior petal always shorter than the adjacent two: filament ca. 2.5 mm long, anther ca. 0.8 mm long. Anterior style ca. 2.7 mm long, subequal to the posterior two, terete, glabrous, erect; apex ca. 1 mm long including a spur ca. 0.2 mm long, linear, folioles absent. Posterior styles ca. 2.6 mm long, terete, glabrous, lyrate; apex ca. 1 mm long including a spur ca. 0.2 mm long, linear but somewhat incurved, folioles absent. Samara not seen. Chromosome number unknown.

Stigmaphyllon suffruticosum is known only from two flowering collections from Colombia (Fig. 5), the type and *Devia 2567* (US; Colombia. Valle: Mpio. Restrepo, Río Calima, Cusumbo, 680 m, 21 Feb 1981). It has small, digitate-fimbriate petals, a homogeneous androecium, and efoliolate styles. It is readily separated from other Colombian species with similar flowers by its ovate to cordate to suborbicular laminas, which are very sparsely pubescent to glabrate abaxially. In *S. bogotense* (no. 2) and *S. sarmentosum* (no. 3), the laminas are abundantly pubescent abaxially, and those of *S. stenophyllum* (no. 5) are linear-lanceolate.

5. Stigmaphyllon stenophyllum C. Anderson, Contr. Univ. Michigan Herb. 19: 428. 1993.—TYPE: COLOMBIA. Antioquia: Mpio. Medellín, Correg. Sta. Elena, Km 1 sobre la via Medellín–Ríonegro, 2000 m, 20 Jul 1986, *Callejas 2151* (holotype: MICH!; isotype: NY!).

Vine. Stems and branches sericeous when young, soon becoming glabrous. Laminas 9–11.5 cm long, 1.7–3 cm wide, linear-lanceolate, apex mucronate, base cordate, glabrous adaxially, sparsely pubescent with T-shaped hairs abaxially but densely so along the margins and on the midrib (trabecula 1.4–1.8 mm long, straight, stalk 0.2–0.3 mm long), margin eglandular; petioles 2–2.8 cm long, sericeous, not confluent across the node, with a pair of prominent but sessile glands at apex, each gland 0.6–0.7 mm in diameter; stipules ca. 0.5 mm long and wide, free, triangular, eglandular. Flowers ca. 15–20 per umbel (condensed pseudoracemes?), these solitary or borne in dichasia (axes sericeous). Peduncles 0.8–3 mm long; pedicels 10–15 mm long, terete; both densely sericeous, peduncles 0.1–0.2 times as long as the pedicels. Bracts 1.2–1.5 mm long, ca. 1 mm wide, triangular, apex acute, sparsely sericeous to glabrate; bracteoles 1.5–1.7 mm long, 0.7–1 mm wide, narrowly triangular, apex acute, sericeous abaxially but glabrous along the margins, eglandular. Sepals ca. 2.5

mm long and wide, glands ca. 2 mm long, ca. 1 mm wide. All petals with the limb glabrous, yellow but suffused with red, margin digitate-fimbriate, the fimbriae up to 0.5 mm long, glandular, limb of lateral petals orbicular; anterior-lateral petals: claw ca. 3 mm long, limb ca. 8 mm long and wide; posterior-lateral petals: claw ca. 2.5 mm long, limb ca. 6.5 mm long and wide; posterior petal: claw ca. 3.2 mm long, apex very slightly indented, limb ca. 5 mm long and wide, broadly elliptical. Stamens unequal in size but subequal in shape; anthers all loculate, glabrous. Stamen opposite anterior sepal: filament ca. 3.1 mm long, anther ca. 1.3 mm long; stamens opposite anterior-lateral petals: filaments ca. 2.5 mm long, anthers ca. 1.2 mm long; stamens opposite anterior-lateral sepals: filaments ca. 3.3 mm long, anthers ca. 1.1 mm long; stamens opposite posterior-lateral petals: filaments ca. 3 mm long, anthers ca. 1.4 mm long; stamens opposite posterior-lateral sepals: filaments ca. 3 mm long, anthers ca. 1.2 mm long; stamen opposite posterior petal always shorter than the adjacent two: filament ca. 2.5 mm long, anther ca. 1 mm long. Anterior style ca. 2.7 mm long, subequal to the posterior two, terete, with scattered hairs in the proximal 1/2, erect; apex ca. 0.8 mm long including a spur ca. 0.2 mm long, linear, folioles absent. Posterior styles ca. 2.7 mm long, terete, with scattered hairs in the proximal 1/2, erect; apex ca. 0.9 mm long including a spur ca. 0.2 mm long, linear, slightly incurved, folioles absent. Samara not seen. Chromosome number unknown. Fig. 6.

Stigmaphyllon stenophyllum is known only from the type, a flowering collection from Antioquia, Colombia (Fig. 5). It is readily recognized by its short-petioled, linear-lanceo-

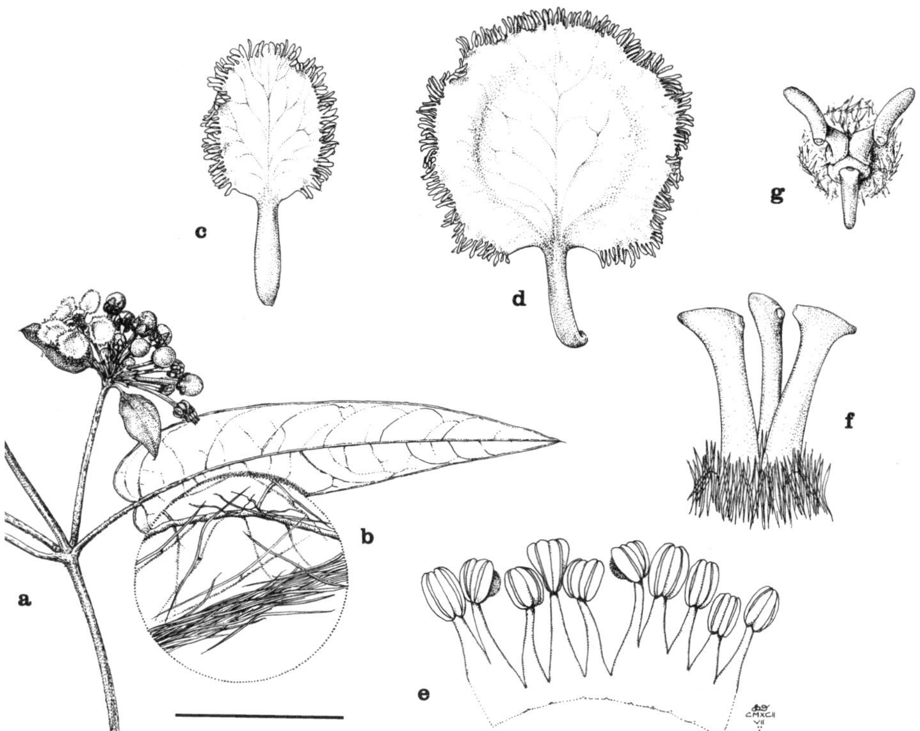

FIG. 6. *Stigmaphyllon stenophyllum*. a. Flowering branch. b. Detail of margin and abaxial surface of lamina. c. Posterior petal (the "flag"). d. Anterior-lateral petal. e. Androecium; stamen second from right opposes the posterior petal. f. Gynoecium, anterior style to the right. g. Gynoecium seen from above, anterior style at bottom. Scale: for a, bar = 4 cm; for b, bar = 1.3 mm; for c, d, bar = 5 mm; for e, bar = 4 mm; for f, g, bar = 2 mm. (Based on *Callejas 2151*.)

late leaves up to 3 cm wide, which are sparsely pubescent with T-shaped hairs abaxially (though densely so on the midrib and margin). The peduncles are very short (0.8–3 mm) and only 0.1–0.2 times as long as the pedicels. In other respects, this species most closely resembles the sympatric *S. bogotense* (no. 2), which differs in its long-petiolate leaves with elliptical to ovate to suborbicular laminas, abundantly pubescent abaxially, and longer peduncles and pedicels [peduncles 0.3–0.8 (–1) times as long as the pedicels]. In both species, the basal two flowers of an aggregate are separated a short distance from the cluster; the androecium is homogeneous, and the styles are efoliolate.

6. Stigmaphyllon yungasense C. Anderson, Contr. Univ. Michigan Herb. 16: 50. 1987.—TYPE: BOLIVIA. Depto. La Paz: Coripata, *Bang 2296* (holotype: NY!, photo: MICH!; isotypes: BM! F! G! GH! MICH! MO! NY! US! W!).

Vine to ca. 3 m. Stems and inflorescence axes with scalelike T-shaped hairs when young, soon becoming glabrous. Laminas 10–15 cm long, 7–11 cm wide, triangular to cordate, apex mucronate, base cordate, glabrous adaxially, pubescent with T-shaped hairs to tomentose abaxially (trabecula 0.6–2 mm long, wavy to crisped and curled, stalk 0.2–0.4 mm long), margin eglandular; petioles 2.8–5.2 cm long, with scalelike T-shaped hairs, not confluent across the node, with a pair of prominent but sessile glands at the apex, each gland 1.5–2.1 mm in diameter; stipules 0.6–1 mm long, 0.7–1 mm wide, free, triangular, eglandular. Flowers ca. (20–) 25–30 per umbel, these solitary or more commonly borne in dichasia or small thyrses (axes to the 3rd order, with T-shaped hairs). Peduncles 8.5–18 mm long; pedicels 5.5–12 mm long, terete; both densely pubescent with scalelike T-shaped hairs, peduncles 1–1.8 times as long as pedicels. Bracts 1.6–2.5 mm long, 1.3–2 mm wide, triangular, apex acute; bracteoles 1.4–1.9 mm long, 1–1.3 mm wide, oblong, apex obtuse, eglandular; bracts and bracteoles with scalelike T-shaped hairs abaxially. Sepals ca. 3 mm long, ca. 2.5 mm wide, glands ca. 1.8 mm long, ca. 1 mm wide. All petals with the limb orbicular or broadly obovate, glabrous, yellow, margin with fimbriae up to 0.5 (–0.6) mm long; anterior-lateral petals: claw ca. 2 mm long, limb ca. 11–11.5 mm long and wide; posterior-lateral petals: claw ca. 1 mm long, limb ca. 10 mm long and wide; posterior petal: claw ca. 3 mm long, apex slightly or not indented, limb ca. 8.5 mm long and wide. Stamens unequal in size but subequal in shape, those opposite the anterior-lateral sepals with the longest filaments; anthers all loculate, glabrous. Stamen opposite anterior sepal: filament ca. 3.2 mm long, anther ca. 0.9 mm long; stamens opposite anterior-lateral petals: filaments ca. 2.5 mm long, anthers ca. 0.8 mm long; stamens opposite anterior-lateral sepals: filaments ca. 4.1 mm long, anthers ca. 0.7 mm long; stamens opposite posterior-lateral petals: filaments ca. 3.5 mm long, anthers ca. 1.2 mm long; stamens opposite posterior-lateral sepals: filaments ca. 2.9 mm long, anthers ca. 0.7 mm long; stamen opposite posterior petal always shorter than the adjacent two: filament ca. 2.4 mm long, anther ca. 0.9 mm long. Anterior style ca. 3 mm long, equal or subequal to the posterior two, terete proximally, laterally flattened distally, glabrous, erect or slightly recurved; apex ca. 1.3 mm long, ca. 0.5 mm wide, triangular, obtuse, folioles absent. Posterior styles ca. 3 mm long, terete proximally, laterally flattened distally, glabrous, recurved; apex ca. 1.1 mm long including a spur ca. 0.2 mm long, ca. 0.3 mm wide, somewhat incurved, folioles absent. Mature samara not seen; dorsal wing of almost mature samara ca. 4.5 cm long, ca. 1.5 cm wide; nut with 1–3 lateral winglets per side, these 1–6.2 mm long, 1.2–2.5 mm wide, without air chambers. Chromosome number unknown. Fig. 7.

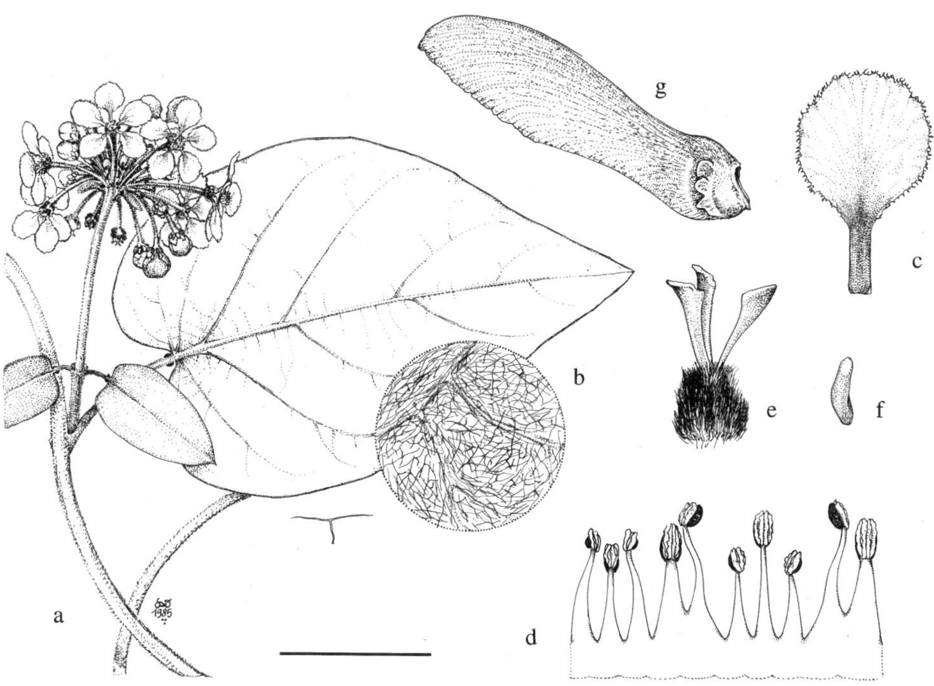

FIG. 7. *Stigmaphyllon yungasense*. a. Flowering branch. b. Detail of abaxial surface of lamina and one laminar hair. c. Posterior petal (the "flag"). d. Androecium; stamen second from left opposes the posterior petal. e. Gynoecium, anterior style to the right. f. Apex of a posterior style seen from above. g. Samara. Scale: for a, bar = 4 cm; for b, d, e, bar = 4 mm; for c, bar = 8 mm; for f and single hair, bar = 2 mm; for g, bar = 2 cm. (Based on: a–f, *Bang 2296*; g, *Beck 2251*.)

Phenology. Collected in flower and fruit in June and September, in young fruit in September.

Distribution (Fig. 8). Bolivia, known only from the Yungas region of La Paz; in forest; 1850 m.

ADDITIONAL SPECIMENS EXAMINED. **Bolivia.** LA PAZ: Nor Yungas, arriba de Puente Villa, Tarila Alto, [16°23′S, 67°37′W,] 1850 m, bosque natural fragmentario en depresión de ladera, 8 Mar 1979 (sterile), *Beck 390* (MICH, MO), 22 Sep 1979 (flowers, young fruits), *Beck 2251 p.p.* (MICH).

Stigmaphyllon yungasense, one of the few species in the genus with a homogeneous androecium and efoliolate styles, has been confused with *S. bogotense* (no. 2) and may also be mistaken for *S. florosum* (no. 42). *Stigmaphyllon bogotense,* a mostly upland species not yet reported from Bolivia, also has subequal stamens and efoliolate styles, but differs in its leaves with glandular margins, somewhat smaller digitate-fimbriate petals, and its inflorescences. The flowers are borne on pedicels that usually exceed the peduncles and are arranged in pseudoracemes; characteristically, the two lowermost flowers are separated on the axis a short distance from the rest of the cluster. In *S. yungasense*, the flowers are all grouped into umbels, and the pedicels are usually shorter than the peduncles. *Stigmaphyllon florosum,* which does occur in the Yungas region, is readily separated by its heterogeneous stamens bearing pubescent anthers and its glandular-margined leaves. It is a lowland species not recorded from above 1120 m.

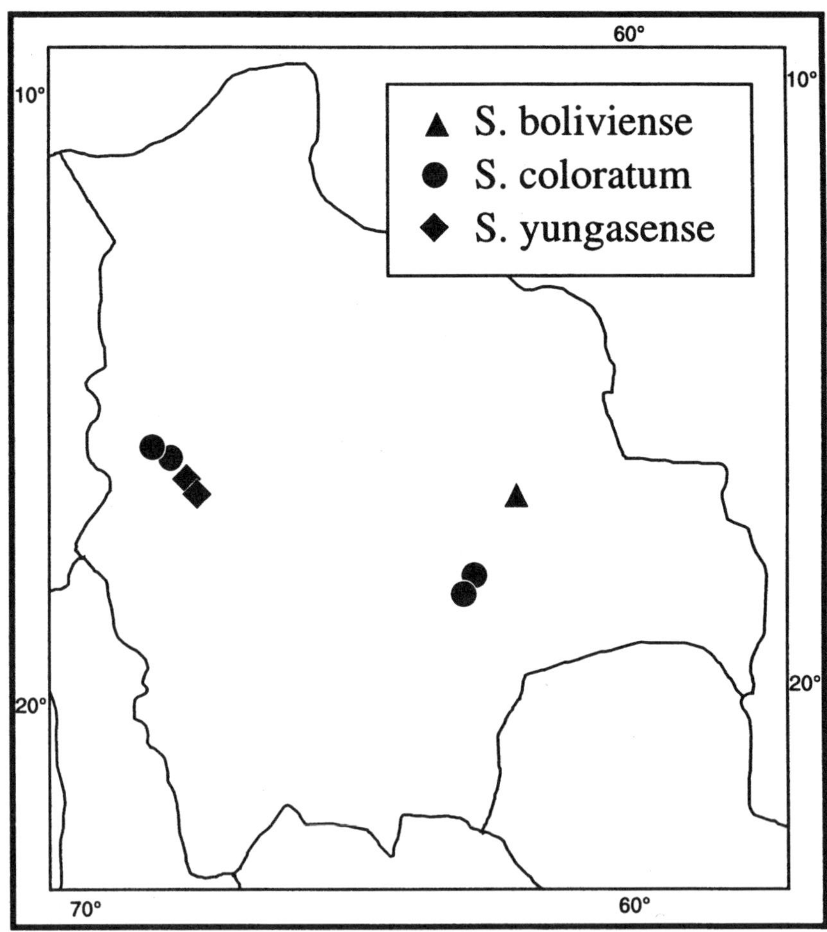

FIG. 8. Distribution of *Stigmaphyllon boliviense*, *S. coloratum*, and *S. yungasense*.

7. Stigmaphyllon urenifolium Adr. Jussieu in St.-Hilaire, Fl. bras. mer. 3: 52. 1833 ["1832"].—TYPE: BRAZIL. Minas Gerais: "prope pagum Sucurú, in parte Prov. Minas Geraes vulgo Minas Novas," *St.-Hilaire B1 1408* (holotype: P!, photo: MICH!, fragment: F!; isotypes: P!).

Stigmaphyllon anomalum Adr. Jussieu, Ann. Sci. Nat. Bot., sér. 2, 13: 288. 1840.—TYPE: BRAZIL. Minas Gerais: "in campis deserti Minarum novarum et ad Formigas," *Martius s.n.* (holotype: M!, photos: F! GH! MICH! NY! US!, fragment: P-JU!).

Vine to 6 m. Stems and branches with scalelike T-shaped hairs when young, soon becoming glabrous. Laminas 6.5–18 cm long, 6–16 cm wide, pinnately 3–7-lobed, the smaller leaves associated with the inflorescence sometimes unlobed and elliptical to ovate, apex of each lobe with a filiform gland (often broken off in older leaves), base auriculate, finely tomentulose to glabrous adaxially, pubescent with T-shaped hairs abaxially (trabecula 1.1–2 mm long, straight, stalk 0.1–0.2 mm long), margin with filiform glands (up to 3 mm long) and sometimes also with irregularly spaced sessile glands (each ca. 0.1 mm in diameter); petioles 3–5.6 cm long, pubescent with scalelike T-shaped hairs or sparsely so, not confluent across the node, with a pair of prominent but sessile glands at the apex or up to 6 mm below the base of the lamina, each gland 1.3–2 mm in diameter; stipules 0.6–1.2 mm long, 0.4–0.9 mm wide, free, triangular, eglandular. Flowers ca. 8–15 per umbel, these solitary or borne in dichasia (axes to the 2nd order, with scalelike T-shaped hairs). Peduncles 3.5–13.5 mm long; pedicels 5.5–12 mm long, terete; both densely pubescent with scalelike T-shaped hairs, peduncles 0.3–1.5 times as long as the pedicels. Bracts 1.6–2.5 mm long, 1–1.5 mm wide, triangular, apex acute, sericeous to glabrous abaxially; bracteoles 1.5–2.3 mm long, 0.9–1.5 mm wide, oblong, apex obtuse, sericeous abaxially, eglandular. Sepals 2.5–2.6 mm long, 2.5–2.7 mm wide, glands 1.8–2.5 mm long, 1–1.5 mm wide. All petals with the limb orbicular, glabrous, yellow and suffused with red, margin lacerate-fimbriate, fimbriae up to 0.9 mm long; anterior-lateral petals: claw 2–2.3 mm long, limb ca. 15 mm long and wide; posterior-lateral petals: claw 1.5–1.7 mm long, limb ca. 14 mm long and wide; posterior petal: claw 3.5–4 mm long, apex indented, limb 13.5 mm long and wide. Stamens unequal in size but subequal in shape, those opposite the anterior-lateral sepals the largest but those opposite the posterior-lateral sepals with the filaments equal or subequal; anthers all loculate, glabrous. Stamen opposite anterior sepal: filament 2.5–3 mm long, anther 1.5–1.8 mm long; stamens opposite anterior-lateral petals: filaments 2–2.5 mm long, anthers 1–1.5 mm long; stamens opposite anterior-lateral sepals: filaments 3.5–3.6 mm long, anthers 1.5–1.7 mm long; stamens opposite posterior-lateral petals: filaments 2.5–3 mm long, anthers 1.2–1.3 mm long; stamens opposite posterior-lateral sepals: filaments 3.2–3.5 mm long, anthers 0.8–1.2 long; stamen opposite posterior petal always shorter than the adjacent two: filament 2.5–3 mm long, anther 0.7–1 mm long. Anterior style 2.6–3.2 mm long, shorter than the posterior two, laterally compressed, glabrous or the proximal 1/3 pubescent, erect; apex 1.1–1.2 mm long including a spur ca. 0.2–0.3 mm long, linear, folioles absent. Posterior styles 3.2–4 mm long, laterally compressed, the proximal half pubescent, recurved at a ca. 75° angle; apex 0.8–1 mm long including a spur ca. 0.2 mm long or blunt distally, linear, folioles absent. Dorsal wing of samara ca. 3.3 cm long, ca. 1.5 cm wide, upper margin without a tooth; nut smooth or with a few prominent veins, 9–11 mm high, 6.5–7.2 mm in diameter, wall containing spongy tissue, without air chambers, areole ca. 6 mm long, ca. 4.5 mm wide, convex, carpophore up to 2.3 mm long. Embryo ca. 7.4 mm long, ca. 2 times as long as wide, ovoid, outer cotyledon ca. 10.5 mm long, ca. 4.4 mm wide, the distal 1/3 folded over the inner cotyledon, inner cotyledon ca. 7 mm long, ca. 3 mm wide, folded at the distal 2/3. Chromosome number unknown.

Phenology. Collected in flower in January and April; date of fruiting collection unknown.

Distribution (Fig. 9). Brazil (Bahia, Minas Gerais); at forest margin and roadsides; ca. 600 m.

REPRESENTATIVE SPECIMENS. **Brazil.** BAHIA: Chapadão Ocidental da Bahia, ca. 45 km N of Santa Maria da Vitória, on the rd to Serra Dourada, 13°03′S, 44°12′W, *Harley 21607* (MICH).—MINAS GERAIS: 22 km by rd W of Januária on rd to Serra das Araras, *Anderson 9208* (COL, F, MICH, MO, NY, RB, UB); Formigas, *Gardner 4474* (BM, K); entre Caraça et Mendanha, *Glaziou 12482* (BR, C, F, G, K, LE, P, RB); Presidente Soares, Serra do Caparó, *Heringer 10295* (UB); prope Sabará, *Martius s.n.* (M).

Stigmaphyllon urenifolium is readily distinguished by its pinnately lobed leaves, homogeneous androecium, and efoliolate styles. The androecium is unusual in that the sta-

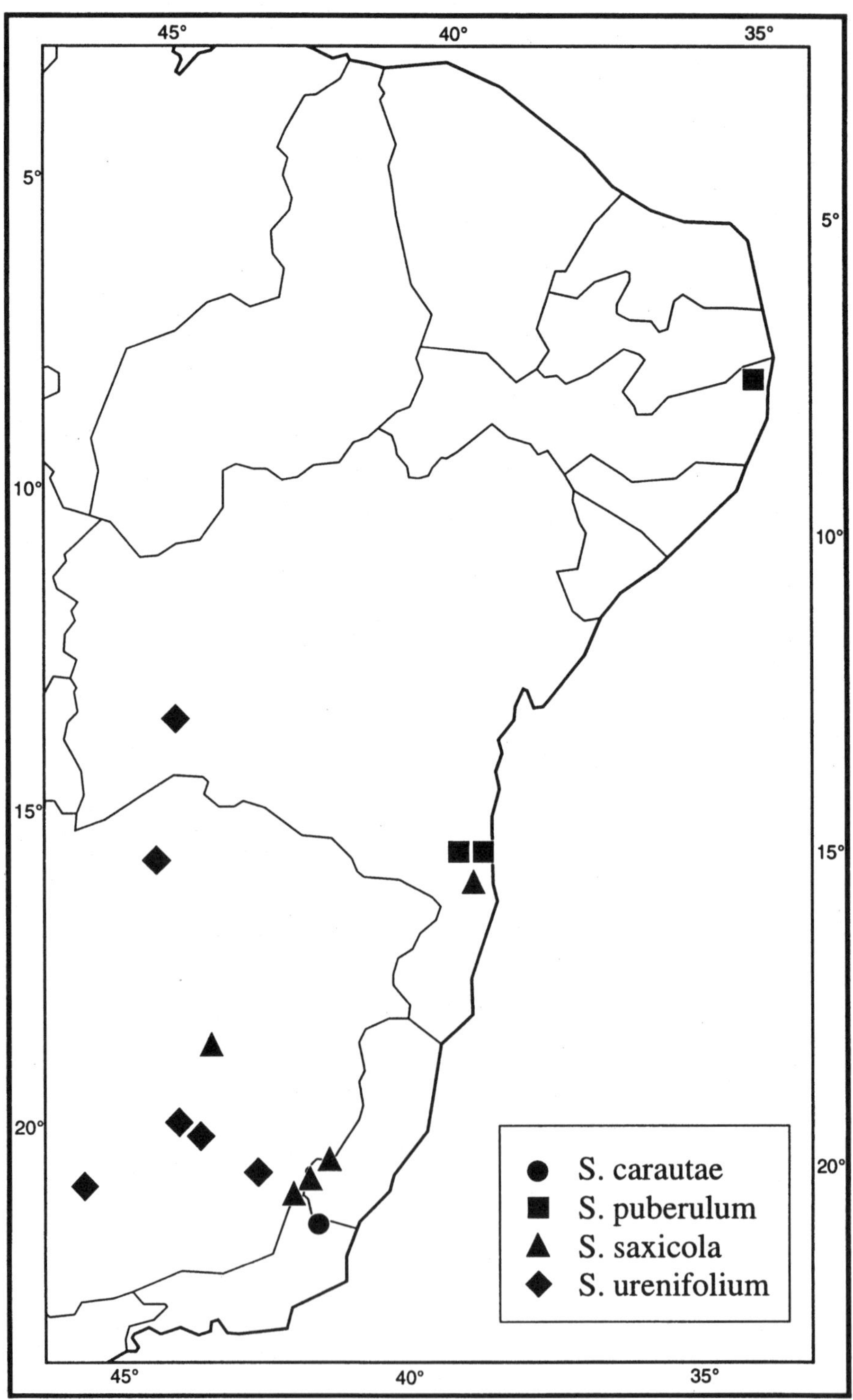

FIG. 9. Distribution of *Stigmaphyllon carautae, S. puberulum, S. saxicola,* and *S. urenifolium.*

mens opposite the anterior-lateral sepals have longer filaments than the ones opposite the posterior-lateral petals; in most species, the reverse is true. *Stigmaphyllon urenifolium* most closely resembles *S. glabrum* (no. 8), which differs in its glabrous leaves and sericeous stem pubescence, and *S. carautae* (no. 21), which has a heterogeneous androecium. *Stigmaphyllon urenifolium* might be confused with *S. angustilobum,* which also has pinnately lobed laminas abaxially pubescent with T-shaped hairs and occurs in southeastern Minas Gerais, eastern Rio de Janeiro, and adjacent São Paulo; however, *S. angustilobum* is readily separated by its heterogeneous androecium and foliolate styles.

Jussieu based *S. anomalum* on a Martius collection with entire rather than lobed leaves, but he suggested that it might prove to belong to *S. urenifolium.* The type includes only the smaller leaves associated with the inflorescence. In *S. urenifolium*, such distal laminas are sometimes unlobed even though the larger laminas borne on the same axis are typically lobed.

8. Stigmaphyllon glabrum C. Anderson, Contr. Univ. Michigan Herb. 19: 424. 1993.—TYPE: BRAZIL. Espírito Santo: Mpio. Castelo, estrada do Bocó, 20 Aug 1987, *Hatschbach 51343* (holotype: MBM!; isotype: MICH!).

Vine. Stems and branches sparsely sericeous when young, soon becoming glabrous. Laminas 14.2–15.3 cm long, ca. 12 cm at widest part, pinnately 5–7-lobed, the smaller leaves associated with the inflorescence unlobed and elliptical to ovate, apex of each lobe terminating in a filiform gland, base auriculate, glabrous adaxially and abaxially, margin with filiform glands and rarely also with a few irregularly spaced sessile glands (ca. 0.4 mm in diameter); petioles ca. 6 cm long, glabrous, not confluent across the node, with a pair of prominent stalked glands at the apex, each gland 0.5–0.7 mm long, 1.5–2.2 mm in diameter; stipules not seen. Flowers ca. 15–20 per umbel, these borne in compound dichasia (axes to the 3rd order, sericeous). Peduncles 5–6 mm long; pedicels 7–9.5 mm long, terete; both sericeous, peduncles 0.6–0.8 times as long as the pedicels. Bracts 1.3–1.7 mm long, 0.9–1 mm wide, triangular, apex acute; bracteoles 1.5–1.8 mm long, 1–1.3 mm wide, triangular, apex obtuse, eglandular; bracts and bracteoles sericeous abaxially but with a glabrous marginal band 0.2–0.3 mm wide. Sepals ca. 3 mm long, ca. 2.8 mm wide, glands ca. 1.5 mm long, ca. 1 mm wide. All petals with the limb orbicular, glabrous, yellow and suffused with red (?), margin with fimbriae up to 1 mm long; anterior-lateral petals: claw ca. 2.5 mm long, limb 15–16 mm long and wide; posterior-lateral petals: claw ca. 1.5 mm long, limb 14–15 mm long and wide; posterior petal: claw ca. 4 mm long, apex indented, limb ca. 10 mm long and wide. Stamens unequal in size but subequal in shape, those opposite the anterior-lateral sepals the largest; anthers all loculate, glabrous. Stamen opposite anterior sepal: filament ca. 3 mm long, anther ca. 1.9 mm long; stamens opposite anterior-lateral petals: filaments ca. 2 mm long, anthers ca. 1.5 mm long; stamens opposite anterior-lateral sepals: filaments ca. 3.5 mm long, anthers ca. 1.6 mm long; stamens opposite posterior-lateral petals: filaments ca. 3 mm long, anthers ca. 1.5 mm long; stamens opposite posterior-lateral sepals: filaments ca. 2.7 mm long, anthers ca. 0.6 mm long; stamen opposite posterior petal subequal to the adjacent two: filament ca. 2.6 mm long, anther ca. 0.8 mm long. Anterior style ca. 3.2 mm long, slightly shorter than the posterior two, terete, glabrous, slightly recurved; apex ca. 1 mm long including a spur ca. 0.1 mm long, linear, folioles absent. Posterior styles ca. 3.5 mm long, terete, glabrous or with a few scattered hairs in the proximal 1/2, lyrate; apex ca. 0.7 mm long, blunt distally, folioles absent. Immature samara with a large flared dorsal wing, nut without lateral winglets or ornamentation. Chromosome number unknown.

Stigmaphyllon glabrum is only known from the type from Espírito Santo, Brazil (Fig. 82). It is distinguished by its pinnately lobed and glabrous leaves, long-fimbriate petals, and efoliolate styles. The androecium is unusual in that the stamens opposite the anterior-lateral sepals have longer filaments than the ones opposite the posterior-lateral petals; in most species the reverse is true. *Stigmaphyllon urenifolium* (no. 7) of Minas Gerais and central Bahia and *S. carautae* (no. 21) are very similar, but in both the laminas are abaxially pubescent instead of glabrous.

9. Stigmaphyllon peruvianum Niedenzu, Arbeiten Bot. Inst. Königl. Lyceums Hosianum Braunsberg 8: 61. 1926.—TYPE: PERU. Cajamarca: Prov. Jaén, near the mouth of the Chinchipe, 600–800 m, May 1912, *Weberbauer 6216* (holotype: B, destroyed, two branchlets from holotype: NY!, photo: MICH!; isotypes: F! G! GH! S! US!).

Vine. Stems and branches densely beset with T-shaped hairs when young, becoming glabrate. Laminas 2.2–6.8 cm long, 1.8–5.3 cm wide, lanceolate to elliptical to suborbicular, apex mucronate or emarginate-mucronate, base truncate to cordate, tomentose to sparsely so adaxially, densely tomentose abaxially (trabecula 0.5–1.3 mm long, crisped and curled, stalk 0.1–0.2 mm long), margin eglandular; petioles 0.4–1.8 cm long, densely beset with subsessile T-shaped hairs, not confluent across the node, with a pair of prominent but sessile glands at apex, each gland 1–1.5 mm in diameter; stipules 0.4–0.6 mm long, ca. 0.4 mm wide, free, triangular, eglandular, obscured by the dense stem vesture. Flowers (4–) 6–8 per umbel, these solitary or borne in dichasia (axes to the 2nd order, densely pubescent with subsessile T-shaped hairs), the umbels sessile or borne on secondary peduncles up to 11 mm long. Peduncles 2–7.5 mm long, densely sericeous; pedicels 3.5–7.5 mm long, distally expanded, sericeous or sparsely so, peduncles 0.5–1.3 times as long as the pedicels. Bracts 1.2–2 mm long, 1–1.5 mm wide, triangular, apex acute; bracteoles 1.3–2 mm long, 0.8–1.2 mm wide, broadly triangular, apex obtuse, eglandular; bracts and bracteoles densely to sparsely sericeous abaxially. Sepals ca. 2.5 mm long and wide, glands 1.5–2 mm long, ca. 1 mm wide. All petals with the limb orbicular, glabrous, yellow, margin with fimbriae up to 0.7 mm long; anterior-lateral petals: claw 2.6–4.5 mm long, limb 11–12 mm long, 12–13 mm wide; posterior-lateral petals: claw 2.2–3 mm long, limb 10–11 mm long, 11–12 mm wide; posterior petal: claw 3–4 mm long, apex indented or only slightly so, limb ca. 8 mm long and wide. Stamens unequal in size but subequal in shape, although those opposite the sepals with the connectives somewhat enlarged (but the locules equally long) and with the longest filaments; anthers all loculate, glabrous or pubescent. Stamen opposite anterior sepal: filament 3–3.2 mm long, anther 0.8–0.9 mm long; stamens opposite anterior-lateral petals: filaments ca. 2.4 mm long, anthers ca. 1 mm long; stamens opposite anterior-lateral sepals: filaments 2.8–3.5 mm long, anthers 0.5–0.7 mm long; stamens opposite posterior-lateral petals: filaments 2.5–2.8 mm long, anthers 1–1.2 mm long; stamens opposite posterior-lateral sepals: filaments 2.9–3.7 mm long, anthers 0.5–0.6 mm long; stamen opposite posterior petal always shorter than the adjacent two: filament 2.3–2.5 mm long, anther ca. 0.8 mm long. Anterior style 2.1–2.8 mm long, shorter than or equaling the posterior two, terete, glabrous, erect or slightly recurved; apex 0.9–1 mm long including a spur 0.6–0.7 mm long, 0.1–0.2 mm wide, linear, folioles absent. Posterior styles 2.4–3 mm long, terete, glabrous, lyrate;

apex ca. 1 mm long including a spur 0.5–0.6 mm long, 0.2–0.3 mm wide, incurved, folioles absent. Dorsal wing of samara ca. 2.3 cm long, ca. 1.2 cm wide, upper margin with a blunt tooth; nut with 1–2 prominent veins, ca. 4.5 mm high, 3.5 mm in diameter, without air chambers, areole ca. 3.5 mm long, ca. 2.3 mm wide, concave (?), carpophore up to 1 mm long. Mature seed not seen. Chromosome number unknown.

Phenology. Collected in flower in May, June, and October, in young fruit in June.

Distribution (Fig. 10). Peru (northern Cajamarca and adjacent Amazonas, Marañón River Valley); dry thorn scrub (*Gentry 22784*); 470–800 m.

ADDITIONAL SPECIMENS EXAMINED. **Peru.** AMAZONAS: 5 km E of Bagua on rd to La Peca, Marañón Valley, *Gentry 22784* (AMAZ, MICH, MO); Prov. Bagua, "St. Julian" hill, on the Río Utcubamba, Hacienda Marelilla near Bagua Grande, 600 m, *Hutchison 1492* (US).

Stigmaphyllon peruvianum, known from only three collections, is an attractive species characterized by abundant pubescence on all its vegetative parts; even the upper leaf surfaces are at least sparsely tomentose. The large petals are fimbriate, the androecium is homogeneous, and the styles lack folioles. The pedicels are distally expanded and usually more sparsely sericeous than the peduncles. In some details of the inflorescence, *S. peruvianum* resembles *S. ellipticum* (no. 46) and *S. eggersii* (no. 47), but these species differ in their heterogeneous androecium and glabrous anthers. Also, in *S. ellipticum* the laminas are glabrate to glabrous or at most sparsely sericeous abaxially, and in *S. eggersii* all styles are foliolate.

10. Stigmaphyllon bannisterioides (L.) C. Anderson, Taxon 41: 328. 1992. *Malpighia bannisterioides* L., Pl. surinam. 9. 1775.—TYPE: SURINAME. *Dahlberg s.n.* (holotype: LINN 588.13, photo: MICH!, microfiche!; isotype: S-L, microfiche!).

Banisteria ovata Cavanilles, Diss. 9: 429. 1790. *Brachypterys borealis* Adr. Jussieu, Ann. Sci. Nat. Bot., sér. 2, 13: 291. 1840, nom. superfl. *Stigmaphyllon ovatum* (Cavanilles) Niedenzu, Ind. Lect. Lyc. Brunsberg. p. aest. 1900: 31. 1900. *Brachypterys ovata* (Cavanilles) Small, N. Amer. fl. 25(2): 138. 1910.—TYPE: DOMINICAN REPUBLIC. *Surian 828* (lectotype, designated by C. Anderson, 1987a: P-JU!, photos: A! MICH!).

Banisteria maritima Richard, Actes Soc. Hist. Nat. Paris 1: 109. 1792.—TYPE: FRENCH GUIANA. *Leblond 45* (holotype: G!, photo: MICH!).

Banisteria picta H. B. K., Nov. gen. sp. 5: 160. 1822 ["1821"].—TYPE: COLOMBIA. "Crescit locis humidis fluminis Sinu, inter Carthagenam et Isthmum Panamensis," *Humboldt & Bonpland s.n.* (holotype: P-HBK!, photos: F! MICH! US!; isotype: P!).

Banisteria brachyptera DC., Prodr. 1: 591. 1824.—TYPE: FRENCH GUIANA. Cayenne, *Perrottet s.n.* (holotype: G-DC!, microfiche: MICH!).

Banisteria calcitrapa Hamilton, Prodr. pl. Ind. occ. 40. 1825.—TYPE: *Desvaux s.n.* (holotype: P!, photo: MICH!).

Stigmaphyllon heringerianum de Paula & Alves, Rodriguésia 46: 165. 1978.—TYPE: BRAZIL. Maranhão: Rosário, cachoeira de Miranda, estuário do Rio Itapecuru, 12 Jan 1976, *de Paula 741* (holotype: UB!, photo: MICH!).

Vine or twining shrub to 3 m. Stems and branches sericeous when young, soon becoming glabrate to glabrous. Laminas 4–13 cm long, 1.2–5.5 cm wide, narrowly elliptical

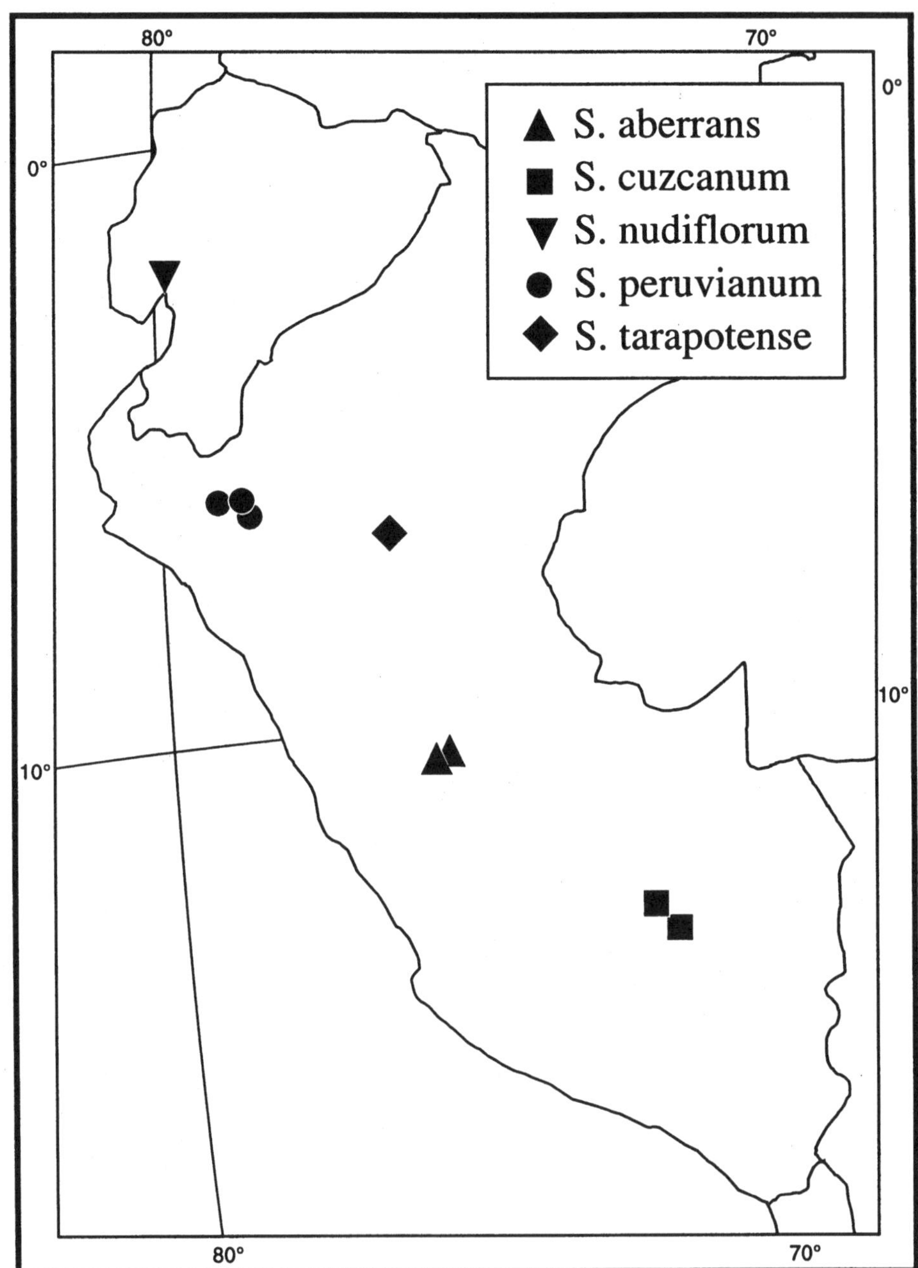

FIG. 10. Distribution of *Stigmaphyllon aberrans, S. cuzcanum, S. nudiflorum, S. peruvianum*, and *S. tarapotense*.

to lanceolate to ovate or rarely linear-lanceolate, apex acute, obtuse, or sometimes apiculate, base attenuate or truncate, glabrate to glabrous adaxially, sparsely sericeous abaxially (trabecula 0.2–0.3 mm long, straight, sessile), margin eglandular; petioles 0.4–1.8 cm long, sericeous to glabrate, not confluent across the node, with a pair of glands at the apex or the base of the lamina, these flush with the epidermis but with a raised margin, each gland 0.5–1 mm in diameter; stipules 0.2–0.8 (–1) mm long, 0.3–0.9 mm wide, free, triangular, eglandular; the pair of leaves subtending an umbel usually abruptly smaller and with the laminas broadly elliptical to broadly ovate to orbicular. Flowers (3–) 4 (–6) per umbel, these mostly solitary or borne in dichasia or sometimes in compound dichasia or rarely in small thyrses (axes to the 3rd order, sericeous or sparsely so). Peduncles 0.2–2.5 mm long; pedicels 15–30 mm long, terete; both sparsely sericeous to glabrous, peduncles up to 0.2 times as long as the pedicels. Bracts 1–2.3 (–5.3) mm long, 0.9–1.5 (–3.4) mm wide, ovate or elliptical; bracteoles 0.8–1.6 mm long, 0.6–1.1 (–1.9) mm wide, ovate to elliptical or triangular, eglandular; bracts and bracteoles with the apex acute to obtuse, sericeous abaxially. Sepals 2–3.2 mm long, 2.3–3.3 mm wide, glands 2–3 mm long, 1–1.4 mm wide. All petals with the limb orbicular or broadly obovate, glabrous, yellow but suffused with red in age, margin erose; anterior-lateral petals: claw 2.5–3.5 mm long, limb 11–12 mm long and wide; posterior-lateral petals: claw 2.5–4 mm long, limb 9–12 mm long and wide; posterior petal: claw 3.5–4.5 mm long, apex rarely slightly indented, limb ca. 8.5–10.5 (–11) mm long and wide. Stamens unequal in size but subequal in shape, those opposite the anterior-lateral sepals usually the longest, sometimes those opposite the posterior-lateral sepals equally long; anthers all loculate, glabrous, 0.8–1.2 mm long, anther of stamen opposite posterior petal usually a little smaller, (0.7–) 0.9–1 mm long. Stamen opposite anterior sepal: filament 2.5–3.4 mm long; stamens opposite anterior-lateral petals: filaments 2–2.5 mm long; stamens opposite anterior-lateral sepals: filaments 2.7–3.2 mm long; stamens opposite posterior-lateral petals: filaments 2.2–2.8 mm long; stamens opposite posterior-lateral sepals: filaments 2.4–3 mm long; stamen opposite posterior petal always shorter than the adjacent two: filament 1.8–2.5 mm long. Anterior style 2.5–3.7 mm long, equal or subequal to the posterior two, terete, glabrous, erect; apex 1–1.7 mm long, linear, 0.2–0.3 mm wide, folioles absent. Posterior styles 2.6–3.7 mm long, terete, glabrous, erect or distally slightly recurved; apex 1–1.2 mm long, usually 0.1–0.3 mm shorter than apex of anterior style, linear, 0.2–0.3 mm wide, folioles absent. Dorsal wing of samara reduced to an apical crest 4–9 mm high, 5.5–7.5 mm wide; nut bearing 4–6 ridges or winglets (up to 7 mm long and 2 mm wide) per side; nut 9–11 mm high, 8–11 mm in diameter, wall containing spongy tissue, without air chambers, areole 4.5–6 mm long and wide, slightly convex, carpophore absent. Embryo 7.2–8 mm long, circular to horseshoe-shaped, outer cotyledon 12.5–15.5 mm long, 4.6–5.7 mm wide, inner cotyledon rudimentary, 1–2 mm long and wide, straight. Chromosome number unknown. Fig. 11.

Phenology. Collected in flower and in fruit throughout the year.

Distribution (Figs. 12, 48). Along the Atlantic Coast from southern Mexico (Veracruz) to northern Brazil (Maranhão), not recorded from Honduras and Costa Rica but to be expected there; in the West Indies on Cuba, Jamaica, Hispaniola, Puerto Rico, Guadeloupe, Martinique, St. Lucia, and Barbados; also along the coast of West Africa in Guinea Bissau, Guinea, and Sierra Leone; along seashores and beaches, in mangrove swamps and salt marshes; sea level to 50 m.

REPRESENTATIVE SPECIMENS. **Cuba**. ORIENTE: Baracoa, *Ekman 4106* (G, S), *Shafer 3912* (A, F, NY, US), *Wright 2157* (G, GH, GOET, MO). **Jamaica**. PORTLAND: *Hunnewell 15299* (GH).—ST. MARY: Annotto Bay, *Proctor 23755* (LL, NY).—ST. THOMAS: Morant Point, *Webster & Wilson 5229* (A, BM, G, MICH, S). **Haiti**. Dept. du Nord, Bayenne, *Ekman H2668* (A, F, G, LL, NY, S, US); vic. of St. Louis du Nord, *Leonard & Leonard 14099* (A, NY, US); Bayeux, near Port Margot, *Nash & Taylor 1016* (NY). **Dominican Republic**. Samaná Peninsula, vic. of Sánchez, *Abbott 516* (GH, US); Prov. Puerto Plata, Puerto Plata, *Ekman 14357* (S, US); Prov.

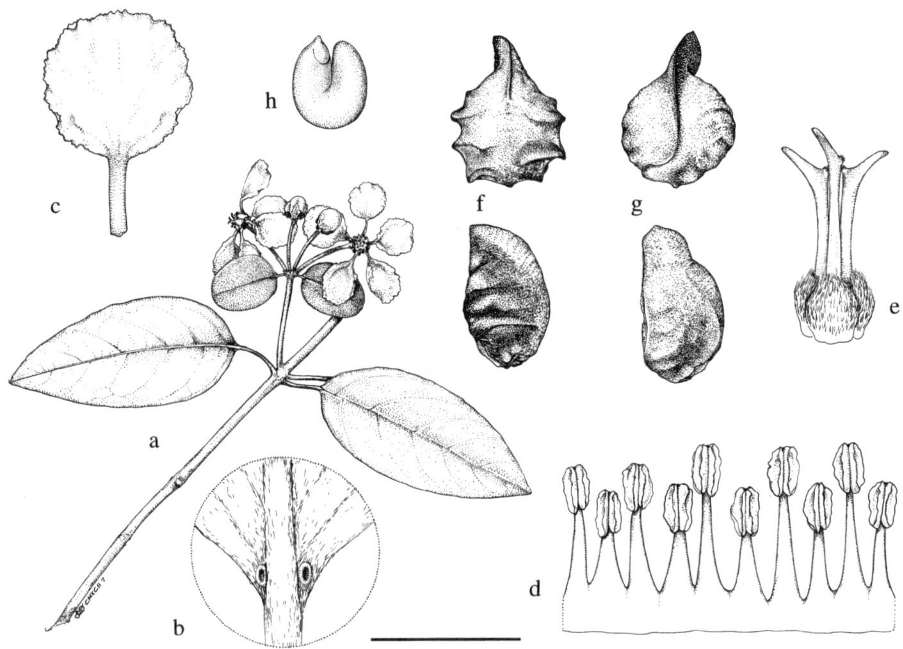

FIG. 11. *Stigmaphyllon bannisterioides*. a. Flowering branch. b. Base of leaf, abaxial view. c. Posterior petal (the "flag"). d. Androecium; stamen second from left opposes the posterior petal. e. Gynoecium, anterior style in the center. f, g. Adaxial views (above) and lateral views (below) of two samaras, illustrating variation in size and shape of dorsal wing and lateral ornamentation. h. Embryo. Scale: for a, bar = 4 cm; for b, d, e, bar = 4 mm; for c, bar = 1 cm; for f–h, bar = 1.3 cm. (Based on: a, b, *Crosby 42*; c–e, *Cremers 7812*; f, h, *Feuillet 898*; g, *Zanoni et al. 20185*.)

Barahona, *Fuertes 313* (A, BM, F, G, GH, K, MO, NY, P, S, US, W). **Puerto Rico**. Naguabo prope Río Blanco, *Eggers 407* (BR, G, GH, GOET, K, M, P, W); mouth of Río Santiago, *Liogier & Liogier 28999* (NY, UPR); Vieques Island, *Woodbury V-14* (UPR). **Guadeloupe**. *Questel 453, 576, 4993* (P, US); *Stehlé et al. 5515* (US). **Martinique**. *Sieber 125* (BR, G, GOET, K, M, MO. NY, P, W); *Duss 1414* (F, GH, MO, NY, US). **St. Lucia**. *Proctor 18024* (A, BM); *Slane & Boatman 249* (GH); *Sturrock 270* (A). **Barbados.** *McIntosh 65* (P); *Bovell & Freeman 210* (NY). **Tobago**. Bacolet, *Broadway 4396* (BM, F, G, GH, M, NY, P, S); W of Scarborough, *Webster & Miller 9785* (A, MICH, S). **Trinidad**. Point Cumana, *Britton & Hazen 804* (NY, US); *Broadway 5831* (BM, F, K, MO, S, US); Caroni swamp near Caroni Road, *Crosby 42* (GH, LL, MICH, MO); San Juan to Port of Spain, *J. Johnston 64* (GH, NY).

Mexico. TABASCO: Mpio. Frontera, Paso San Román, *Ventura A. 20404* (MICH); Frontera, *Rzedowski 30035* (MEXU, MICH, MO, P, SD).—VERACRUZ: antes de puente de Alvarado, *Calzada 435* (CAS, GH, MEXU). **Guatemala**. IZABAL: Puerto Barrios, *Deam 384* (GH, MICH, NY, US). **Belize.** Belize, *Kellerman 5737* (LL, US), *Lundell 4087, 4089* (MICH). **Nicaragua**. RÍO SAN JUAN: San Juan del Norte (Greytown), *Stevens 20824* (MO).—ZELAYA: Bahía de Bluefields, *Molina R. 2060* (F, GH); La Barra de Punta Gorda, *Moreno 13196, 13201* (MO); Río Kuawantla, 3 km W of Puerto Isabel, *Neill 4576* (MICH); El Bluff, N de El Muelle, *Sandino 2227* (MICH). **Panama**. COLÓN: Chagres, *Fendler 49* (GH, K, MO, US); Miguel de la Borda, *Croat 10073* (F, MO); trail above Río Indios, *Sullivan 122* (MO).

Colombia. ANTIOQUIA: Mpio. Turbo, bocas del Atrato, Vereda Turbito, en Bahía Colombia (Golfo de Urabá), *Callejas et al. 5019* (MICH); ca. 1 km W of Turbo, ca. 08°05′N, 76°43′W, *Feddema 2033* (MICH, NY, S, US).—ATLÁNTICO: Baranquilla, *Paul C-5* (US).—BOLÍVAR: Baranquilla, *Bro. Paul C-5* (US); Canal de Dique, a poco kilómetros de su desembrocadura en la Bahía de Cartagena, *Uribe U. 3185* (US).—CHOCÓ: between Punta Las Barcas and Turbo, *Duke 9709* (MO). **Venezuela**. DELTA AMACURO: Depto. Pedernales, along Caño Angosturita, SE of Pedernales, 09°N, 62°08′W, *Steyermark 114285* (MO).—MIRANDA: Higuerote, *Badillo*

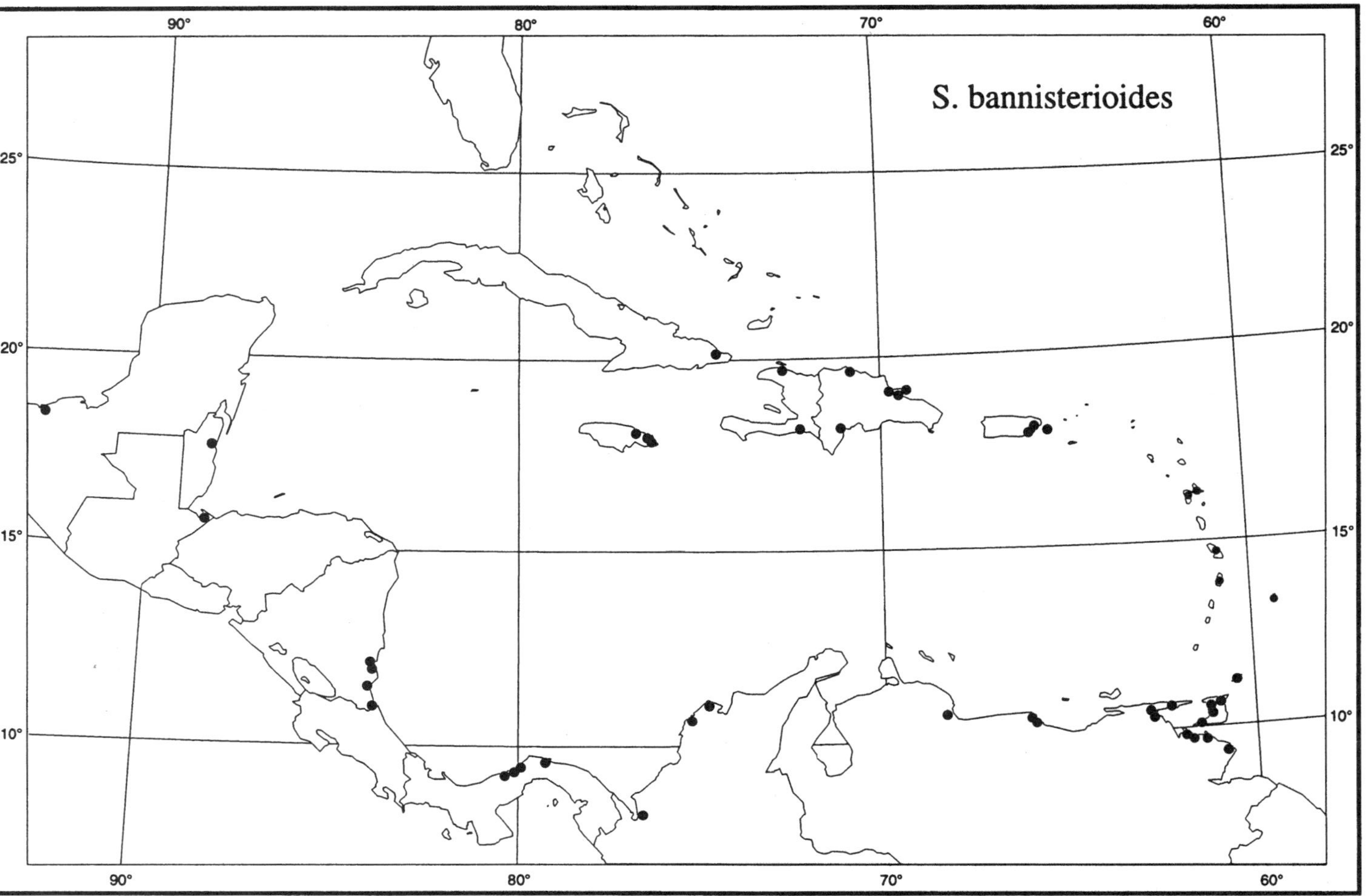

FIG. 12. Distribution of *Stigmaphyllon bannisterioides* in the West Indies, Central America, and northern South America; see Fig. 48 for distribution in the Guianas and Brazil.

715 (VEN).—SUCRE: vic. of Cristóbal Colón, Aricana beach, *Broadway 596* (GH, NY, US); Quebrada de El Purgatorio, 30 km W de Trapa, *Lasser & Vareschi 3862* (VEN); Peninsula de Paria, Ensenada de Patao, E de Puerto de Hierro, E de la boca del Río Patao, *Steyermark 91317* (US, VEN).—YARACUY: Caripito, Río San Juan, *Trujillo 1489* (MY). **Guyana**. Abary River mouth, 06°30′N, 57°45′W, *Boom 7165* (MICH); Pomeroon River, Pomeroon Dist., *de la Cruz 3078* (F, GH, K, MICH, MO, NY, US); Barima River, NW Dist., 08°20′N, 59°50′W, *de la Cruz 3370* (F, GH, MO, NY, US); Amakura River, NW Dist., 08°10′N, 60°W, *de la Cruz 3465* (F, GH, MO, NY, US); Demarara-Mahaica Region, Georgetown, 06°50′N, 58°10′W, *Hahn 4814* (MICH); Georgetown, vic. of Peter's Hall, *Hitchcock 16656* (GH, NY, S, US). **French Guiana**. Kourou, *Benoist 1337* (P); plage de Susini entre Bourda et la plage de Montjoly, *Cremers 9522* (CAY, MICH, US); Sinnamary, route de l'Anse, *Feuillet 1578* (MICH); Grand Crique–Bassin de l'Approvague, 04°07′N, 52°24′W, *de Granville 8353* (CAY, MICH, US); Ile de Cayenne, Anse de Montabo, 04°57N, 52°18′W, *Hoff 5166* (CAY); Iles du Salut, *Sagot 98* (BM, G, K, P, S, W). **Suriname**. Without locality, *Hostmann 278* (BM, G, GH, K, LE, MO, P, W); via secta ab Wia Wia bank ad Grote Zwiebelzwamp, *Lanjouw & Lindeman 1090* (K, IAN, NY); Coronie, *Lanjouw & Lindeman 1530* (K, NY); Nickerie, along rd behind Blufpunt, *Lanjouw & Lindeman 3069* (K, IAN, NY); rd to Carl Francois along Saramacca River, 79 km from Paramaribo, *Maguire & Stahel 23598* (NY, US); near Boskamp at mouth of Coppename River, *Wessels Boer 508* (LL, US); Paramaribo, *Wullschlägel 51* (BR, GOET, NY, W). **Brazil**. AMAPÁ: beira do Rio Oiapoque, *Black 49-8413* (IAN); Mpio. Amapá, São Joaquim, Rego Fundo, *Black 50-9422* (IAN); Macapá, *Ducke 1977* (MG); Rio Araguarí, Uruguaína, *Fróes & Black 27632* (IAN).—CEARÁ: margem do Rio Cocó (mangue), Parque Ecológico Adail Barreto, Fortaleza, *Fernandes EAC 20118* (MICH).—MARANHÃO: Maracassumé River region, Candido Mendes, *Fróes 1779* (A, G, K, MICH, MO, NY, P, S); Rio Itapecuru, *Schwacke 17* (RB).—PARÁ: Ilha do Marajó, Soure, *Black 48-3609* (P); Vigia, *Black 52-14233* (NY, P); Pará [Belém], *Burchell 9560* (GH, GOET, K, LE); Viseu, boca de Rio Gurupí, *Cavalcante 1923, 1924* (MG); Marajó, Rio Pacovalinho, *Huber 254* (INPA, MG, UB); Ilha de Paquetá, *Jobert 42* (P, R); Belém, *Murça Pires & Black 758* (GH).

Guinea Bissau. Rio Camocote, 11°14′N, 15°02′W, *Malaisse 14507* (MO); Bombadina, Ponta Inglês, *Pereira 2561* (K). **Guinea.** *Roberty 17885* (K). **Sierra Leone.** *Deighton 2768* (K); 5 mi above Mahela, *Elliot 4089* (BM); Mahela, *T. S. Jones 49* (K); Rokupr, *Jordan 54* (K); Tombo, Kent, *Tindall 21* (K).

Stigmaphyllon bannisterioides is an atypical, easily recognized species found along the Atlantic Coast from southern Mexico to northern Brazil and in the West Indies; it is the only species of the genus to occur in the Old World. The somewhat succulent, usually lanceolate or narrowly elliptical laminas are borne on short petioles (up to 1.8 cm long). Each of the pair of basal laminar glands is flush with the epidermis, rather than prominent, but has a raised margin. The inflorescence unit is an umbel of (3–) 4 (–6) large flowers; the umbels are usually solitary but sometimes borne in dichasia or rarely in compound dichasia. Typically, each umbel is subtended by a pair of leaves abruptly smaller than the foliage leaves and with the laminas broadly elliptical to broadly ovate to orbicular, although sometimes these leaves equal the foliage leaves or rarely are even larger (*Wullschlägel 51*, P). The peduncles are at most 2.5 mm long. The flowers differ from those of most species in their subequal stamens and subequal, efoliolate but hooked styles. Most unusual is the highly modified samara. The enlarged nut (8–11 mm in diameter) bears a much reduced dorsal wing, an apical crest 4–9 mm high, and is usually ribbed but lacks lateral winglets. The embryo is circular to horseshoe-shaped and consists mostly of the unusually large outer cotyledon; the inner cotyledon is rudimentary.

Stigmaphyllon bannisterioides and *S. paralias* (no. 62) are both atypical species, somewhat similar in habit and foliage, in which the "samara" consists of a nut with an apical crest and lacks a carpophore. Jussieu placed them in their own genus, *Brachypterys*, as *B. borealis* and *B. australis*, respectively. (A third species, unknown to Jussieu, with a "samara" like that of *S. paralias* is the distinctive *S. harleyi* of Bahia, Brazil.) Although the two are easily separated, they have been confused where their ranges overlap in northern Brazil. *Stigmaphyllon bannisterioides* is a vine or twining shrub found in wet and brackish places along the Atlantic coast and in estuaries, but *S. paralias* is always a shrub on sand, on beaches, and in restingas as well as in caatingas and other sandy inland habitats. Both species have large flowers in few-flowered umbels in which the peduncles are rudimentary to very short; however, in *S. bannisterioides* the androecium is composed of subequal stamens and the apex of the styles is extended into a claw but lacks folioles, whereas in *S. paralias* the stamens are highly modified and the styles all bear large folioles.

Stigmaphyllon bannisterioides is well-known species that has been most commonly cited as *Stigmaphyllon ovatum* or *Brachypterys ovata*, because an earlier name for it, *Malpighia bannisterioides* L., had been overlooked (C. Anderson 1992a).

11. Stigmaphyllon harleyi W. R. Anderson, Bol. Mus. Bot. Mun. Curitiba 43: 1. 1981.— TYPE: BRAZIL. Bahia: Serra di Açuruá, São Inácio, on rocky hillside called Pedra da Mulher just S of town, 11°07′S, 44°44′W, ca. 500–600 m, 25 Feb 1977, *Harley 19026* (holotype: CEPEC!; isotypes: K! NY! MICH!).

Twiggy subshrub to 50 cm. Stems and branches densely tomentose when young, becoming glabrate to glabrous. Laminas 0.8–2.8 cm long, 0.8–2.5 cm wide, orbicular or the largest sometimes reniform, apex obtuse, base shallowly cordate to truncate, glabrous adaxially and abaxially, margin eglandular; petioles 0.2–0.3 cm long, densely tomentulose, not confluent across the node, with a pair of prominent but sessile glands at the apex, each gland 0.3–0.8 mm in diameter; stipules 0.2–0.3 mm long and wide, free, triangular and scalelike, eglandular, hidden by the stem pubescence. Flowers grouped in 2's, axillary. Peduncles 0.3–0.8 mm long; pedicels 9.5–11.5 mm long, terete; both densely tomentulose, peduncles less than 0.1 times as long as the pedicels. Bracts 0.9–1.2 mm long, 0.6–0.9 mm wide, triangular, apex acute; bracteoles 0.8–1 mm long, 0.5–0.7 mm wide, triangular, apex obtuse, eglandular; bracts and bracteoles tomentulose or sparsely so abaxially. Sepals ca. 2.5 mm long, ca. 1.8 mm wide, glands 1.4–1.5 mm long, ca. 0.8 mm wide. All petals with the limb orbicular, glabrous, yellow, margin erose and often with 1–2 prominent glands near the base, each gland 0.1–0.3 mm in diameter; anterior-lateral petals: claw ca. 1.5 mm long, limb ca. 9 mm long and wide; posterior-lateral petals: claw ca. 1 mm long, limb ca. 8 mm long and wide; posterior petal: claw ca. 1.5 mm long, apex not indented, limb ca. 5 mm long and wide. Stamens unequal in size but subequal in shape, those opposite the anterior-lateral sepals and posterior-lateral petals the largest; anthers all loculate, glabrous. Stamen opposite anterior sepal: filament ca. 2.3 mm long, anther ca. 1.2 mm long; stamens opposite anterior-lateral petals: filaments ca. 2 mm long, anthers ca. 0.9 mm long; stamens opposite anterior-lateral sepals: filaments 2.4–2.5 mm long, anthers 1.1–1.2 mm long; stamens opposite posterior-lateral petals: filaments 2.4–2.5 mm long, anthers ca. 1.2 mm long; stamens opposite posterior-lateral sepals: filaments ca. 2.1 mm long, anthers ca. 0.9 mm long; stamen opposite posterior petal always shorter than the adjacent two: filament ca. 1.7 mm long, anther ca. 0.8 mm long. Anterior style 2.6–2.8 mm long, equaling the posterior two, terete, glabrous, erect, apex 1–1.3 mm long, each foliole ca. 0.8 mm long, 1–1.2 mm wide, subtrapezoidal. Posterior styles 2.8–3 mm long, terete, glabrous, erect; foliole 0.6–0.7 mm long, ca. 0.8 mm wide, subtrapezoidal, or the apex only laterally expanded but without a foliole, apex 1.4 mm long, ca. 0.5 mm wide, oblong. Dorsal wing of samara reduced to an apical crest ca. 0.5 mm high, nut bearing lateral crests up to 0.1 mm high or only prominently ribbed, nut ca. 4.2 mm high, ca. 2.2 mm in diameter, without air chambers, areole ca. 2 mm long, ca. 1 mm wide, concave, carpophore absent. Embryo ca. 3.2 mm long, ca. 2 times as long as wide, ovoid, outer cotyledon ca. 3.1 mm long, ca. 2.4 mm wide, straight, inner cotyledon ca. 2.6 mm long, ca. 1.8 mm wide, straight. Chromosome number unknown. Fig. 58i–p.

This distinctive species is known only from the type collection from Bahia, Brazil. (Fig. 82). It differs from all other species in its twiggy habit; small, subsessile, orbicular, glabrous leaves; and most strikingly by its 2-flowered inflorescences. Although glands may be found at the base of the limb of the posterior petal in a number of species, including the sympatric *S. paralias* (no. 62), *S. harleyi* is the only species in which the lateral petals may bear such glands as well. The anterior style has folioles, but the posterior styles are variable; they may be foliolate or their apices may be only laterally somewhat expanded (oblong). The "samaras" are similar to those of *S. paralias*, another species in which the dorsal wing is reduced to a small crest. *Stigmaphyllon paralias* is the only other species with a strictly shrubby habit though it is much more robust than *S. harleyi*.

12. Stigmaphyllon cordatum Rose in Smith, Bot. Gaz. 18: 198. 1893.—TYPE: GUATEMALA. Depto. Guatemala, 5000 ft, Mar 1892, *Heyde & Lux 3267* (holotype: US!, photo: MICH!; isotypes: GH! K! NY!).

Vine. Stems and branches sericeous when young, soon becoming glabrate to glabrous. Laminas 6.1–11.5 cm long, 4.5–8.5 cm wide, cordate or narrowly so, apex acuminate or briefly so to mucronate, base auriculate, glabrous adaxially and abaxially, margin eglandular; petioles 2.2–6.3 cm long, sericeous (hairs subsessile) or glabrous in older leaves, not confluent across the node, with a pair of prominent but sessile glands at the apex, each gland 1.1–1.7 mm in diameter; stipules 0.7–1 mm long and wide, free, triangular, eglandular. Flowers 15–20 per umbel or condensed pseudoraceme, these solitary or borne in dichasia (axes glabrate to glabrous). Peduncles 3.3–7 mm long; pedicels 6.5–13.5 mm long, terete; both densely sericeous (hairs subsessile), peduncles (0.3–) 0.5–0.8 times as long as pedicels. Bracts 1.5–2.3 mm long, 0.7–1.3 mm wide, narrowly triangular, apex acute; bracteoles 1.5–1.8 mm long, 0.5–1.2 mm wide, narrowly triangular to sublinear, apex acute, eglandular or with a glandular area in the basal 1/3–1/2; bracts and bracteoles sericeous (hairs subsessile) abaxially. Sepals 2.7–3 mm long, 2.4–2.7 mm wide, glands 1.8–2.5 mm long, 1–1.3 mm wide. All petals with the limb glabrous, yellow, margin erose-dentate, teeth up to 0.3 (–0.4) mm long, limb of lateral petals orbicular; anterior-lateral petals: claw 1.5–2 mm long, limb ca. 12–13 mm long and wide; posterior-lateral petals: claw ca. 1.5 mm long, limb ca. 11–12 mm long and wide; posterior petal: claw 2.2–2.5 mm long, apex not indented, limb ca. 10–11 mm long, ca. 8–9 mm wide, elliptical to obovate. Stamens unequal in size but subequal in shape, those opposite the posterior styles the largest; anthers all loculate, glabrous. Stamen opposite anterior sepal: filament 2.5–3 mm long, anther 1.1–1.3 mm long; stamens opposite anterior-lateral petals: filaments 2.1–2.4 mm long, anthers 1–1.1 mm long; stamens opposite anterior-lateral sepals: filaments 2.8–3.1 mm long, anthers 0.9–1.1 mm long; stamens opposite posterior-lateral petals: filaments 3.5–4.6 mm long, anthers 1.2–1.4 mm long; stamens opposite posterior-lateral sepals: filaments 2.1–2.3 mm long, anthers ca. 0.6 mm long; stamen opposite posterior petal usually slightly shorter than or sometimes subequal to the adjacent two: filament 1.8–2 mm long, anther ca. 0.8 mm long. Anterior style 2.7–3.3 mm long, shorter than the posterior two, terete proximally, the distal 1/4 laterally flattened, glabrous, erect; apex 1.4–1.5 mm long, linear or narrowly lanceolate or expanded proximally and triangular, 0.3–0.6 mm wide, folioles absent. Posterior styles 3.5–4.2 mm long, terete proximally, the distal 1/4 laterally flattened, glabrous, lyrate; apex 1.4–1.8 mm long, linear or expanded into a narrow triangular lip, ca. 0.6 mm wide, folioles absent. Dorsal wing of samara ca. 3 cm long, ca. 1.2 cm wide, upper margin with a blunt or subacute tooth; nut with prominent ribs, ca. 4.5 mm high, ca. 3.5 mm in diameter, without air chambers, areole ca. 2.6 mm long, ca. 2.5 mm wide, convex, carpophore up to ca. 2.2 mm long. Seed not seen. Chromosome number unknown.

Phenology. Collected in flower in December and February, in fruit in December and March.

Distribution (Fig. 61). Guatemala (Huehuetenango, Guatemala); in thickets; ca. 1500–2500 m.

ADDITIONAL SPECIMENS EXAMINED. **Guatemala**. HUEHUETENANGO: Chiantla, *Hunnewell 17152* (GH); Aguacatán, *Skutch 1941* (BM, F, NY, US); vicinity of Aguacatán, near the spring of San Juan, *Standley 83145* (US), *83149* (F).

Stigmaphyllon cordatum is named for its distinctive cordate laminas, which are auriculate and glabrous. It has been confused with the Mexican *S. selerianum*, which has similar leaves but with the margin bearing filiform glands instead of eglandular. The two species can be separated with the following couplet.

1. Lamina with the margin eglandular; peduncles 3.3–7 mm long; bracts and bracteoles narrowly triangular to linear; anthers glabrous; posterior styles efoliolate, the apex linear or expanded into a narrow triangular lip, ca. 0.6 mm wide; Guatemala. 12. *S. cordatum*.
1. Lamina with the margin bearing filiform glands (these often broken off in older leaves but the bases remaining); peduncles 0.5–3 mm long; bracts and bracteoles ovate; anthers pubescent or sometimes glabrous; posterior styles with the apex expanded into a lip or a narrowly triangular or semi-elliptical foliole 0.5–1.3 mm wide; Mexico. 44. *S. selerianum*.

13. Stigmaphyllon romeroi Cuatrecasas, Ciencia (México) 23: 140. 1964.—TYPE: COLOMBIA. Bolívar: entre Sabana Beltrán y Juan Arias, 15 Sep 1963, *Romero-Castañeda 9950* (holotype: US!, photo: MICH!; isotype: COL!).

Vine. Stems and branches sparsely sericeous when young, soon becoming glabrate. Laminas 3.2–7 cm long, 1.7–4.2 cm wide, ovate or narrowly so, apex acuminate or obtuse-mucronate, base truncate to cordate, glabrous adaxially, sparsely and patchily sericeous to glabrate or glabrous abaxially (trabecula 0.3–1 mm long, straight, sessile), margin eglandular; petioles 0.9–2.2 cm long, sparsely sericeous (hairs subsessile), not confluent across the node, with a pair of (or sometimes 3–4) stalked (peg-shaped) glands at the apex, each gland 0.3–0.4 mm in diameter, 0.4–0.8 mm long; stipules 0.3–0.8 mm long, 0.3–0.5 mm wide, free, triangular, eglandular. Flowers ca. (12–) 15–20 per pseudoraceme, these solitary or borne in dichasia (axes to the 2nd order, sparsely sericeous to glabrous). Peduncles 2.7–4 mm long; pedicels 5–7.3 mm long, terete; both sericeous or sparsely so (hairs subsessile), peduncles 0.4–0.7 times as long as the pedicels. Bracts 1.1–1.5 mm long, 0.6–1 mm wide, triangular, apex acute; bracteoles 0.9–1.2 mm long, 0.7–0.9 mm wide, oblong, apex obtuse, each bracteole with an inconspicuous gland centered at the base (each gland 0.2–0.4 mm in diameter); bracts and bracteoles sparsely sericeous (hairs subsessile) abaxially. Sepals 2–2.1 mm long, 1.8–1.9 mm wide, glands 1.7–2 mm long, ca. 1 mm wide. All petals with the limb glabrous, yellow, margin erose; anterior-lateral petals: claw 1.5–2 mm long, limb ca. 8 mm long, ca. 6–8 mm wide, broadly elliptical to orbicular; posterior-lateral petals: claw 1.2–1.5 mm long, limb ca. 7

mm long, ca. 4.5 mm wide, elliptical; posterior petal: claw ca. 2.5 mm long, apex not indented, limb ca. 6 mm long, ca. 4.5 mm wide, elliptical. Stamens unequal in size but subequal in shape, those opposite the anterior-lateral sepals with the longest filaments; anthers all loculate, glabrous or sparsely pubescent. Stamen opposite anterior sepal: filament 1.8–2 mm long, anther ca. 0.9 mm long; stamens opposite anterior-lateral petals: filaments ca. 1.5 mm long, anthers ca. 0.8 mm long; stamens opposite anterior-lateral sepals: filaments 1.9–2 mm long, anthers 0.7–0.8 mm long; stamens opposite posterior-lateral petals: filaments 1.6–1.8 mm long, anthers 0.8–0.9 mm long; stamens opposite posterior-lateral sepals: filaments 1.8–1.9 mm long, anthers 0.7–0.8 mm long; stamen opposite posterior petal always shorter than the adjacent two: filament 1.5–1.6 mm long, anther 0.8–0.9 mm long. Anterior style 1.8–2.2 mm long, slightly shorter than or equaling the posterior two, terete, glabrous, erect or slightly recurved; apex 1–1.1 mm long including a spur ca. 0.1–0.2 mm long, triangular, 0.6–0.9 mm wide, folioles absent but each side with a narrow lip ca. 0.2–0.3 mm at its widest. Posterior styles 1.8–2.2 mm long, canaliculate-complicate, glabrous, lyrate; apex ca. 1 mm long including a spur ca. 0.1 mm long, slightly incurved, folioles absent but with a narrow lateral lip ca. 0.1 mm wide. Mature samara not seen; immature samara with a large dorsal wing, the nut ribbed but without lateral winglets. Chromosome number unknown.

Stigmaphyllon romeroi is known only from two collections from Colombia (Fig. 49), the type and *Romero-Castañeda 9838* (AAU, COL; Colombia. Bolívar: entre Juan Arias y Magangué, 9 Sep 1963). It is readily recognized by its peg-shaped petiole glands. The only other South American species with such glands is *S. jobertii* of Brazil; such glands are also found in the West Indian *S. sagraeanum* and *S. microphyllum*, and occasionally in *S. emarginatum*. *Stigmaphyllon romeroi* is also distinguished by its small flowers borne in pseudoracemes and its homogeneous androecium. The apex of the styles is laterally expanded into a lip but lacks true folioles.

14. Stigmaphyllon lalandianum Adr. Jussieu in St.-Hilaire, Fl. bras. merid. 3: 58. 1833 ["1832"]. *Stigmaphyllon lalandianum* var. *jussieuanum* Niedenzu, Ind. Lect. Lyc. Brunsberg. p. aest. 1900: 6. 1900, nom. superfl.—TYPE: BRAZIL. Rio de Janeiro: "circa Sebastianopolis," *Lalande s.n.* (holotype: P!, photos: F! MICH! US!; isotype: P!).

Vine to 9 m. Stems and branches sericeous when young, soon becoming glabrate. Laminas 5–11.5 cm long, 3–8 cm wide, lanceolate to ovate to elliptical or rarely suborbicular, apex acuminate or acuminate-mucronate, base attenuate or truncate or rarely slightly cordate, glabrous adaxially, sericeous or sometimes sparsely so abaxially (trabecula 0.3–0.6 mm long, straight or wavy, sessile or subsessile), margin eglandular; petioles 1.4–6.5 cm long, sericeous or densely so, not confluent across the node, with a pair of prominent but sessile glands at the apex, each gland 1–2 mm in diameter; stipules 0.5–1.2 mm long, 0.5–1.3 mm wide, free, triangular, eglandular. Flowers ca. 15–40 (–50) per umbel or pseudoraceme, these sometimes solitary but commonly borne in dichasia or compound dichasia or small thyrses (axes to the 4th order, sericeous; in thyrses sometimes only one lateral axis at a node developing and the branching then appearing alternate). Peduncles 3.5–13 mm long; pedicels 4–10 mm long, terete; both sericeous, peduncles 0.5–1.7 times as long as the pedicels. Bracts 1–2.5 mm long, 0.6–1.4 mm wide, triangular or narrowly so, apex acute; bracteoles 0.8–1.6 mm long, 0.6–1 mm wide, triangular or oblong, apex obtuse, eglandular or each bracteole with a pair of inconspicuous glands (each 0.2–0.3 mm in diameter); bracts and bracteoles sericeous abaxially. Sepals 1.8–2.5 mm long, 1.5–2.2 mm wide, glands 1.5–2.1 mm long, 0.8–1.2 mm wide. All petals with the limb glabrous, yellow, margin erose, limb of lateral petals orbicular; anterior-lateral petals: claw 1.2–1.8 (–2.1) mm long, limb 7–10 mm long and wide; posterior-lateral petals: claw 0.8–1.4 mm long, limb 6–8 mm long and wide; posterior petal: claw 2.3–3.1 mm long, apex indented, limb 5.5–7 mm long, 4–5 mm wide, broadly elliptical or sometimes obovate. Stamens unequal in size but subequal in shape, those opposite the posterior-lateral petals the largest but the filaments commonly nearly as long as those opposite the lateral sepals, anthers of those opposite the lateral sepals with the connective slightly enlarged and the locules slightly reduced; anthers all loculate, glabrous. Stamen opposite anterior sepal: filament (1.7–) 2–2.7 mm long, anther 1–1.3 mm long; stamens opposite anterior-lateral petals: filaments (1.3–) 1.5–2 mm long, anthers 0.8–1 mm long; stamens opposite anterior-lateral sepals: filaments 2.1–3.2 mm long, connectives 0.8–1 mm long, locules 0.5–0.9 mm long; stamens opposite posterior-lateral petals: filaments 2.1–3 (–3.3) mm long, anthers 1–1.5 mm long; stamens opposite posterior-lateral sepals: filaments 2–2.5 (–2.8) mm long, connectives 0.7–1 mm long, locules 0.3–0.9 mm long; stamen opposite posterior petal always shorter than the adjacent two: filament 1.5–2.3 mm long, anther 0.7–1 mm long. Anterior style 2.2–3.3 mm long, slightly shorter than or subequal to the posterior two, terete, erect, glabrous or with scattered hairs in the proximal 1/4–1/3; apex (0.9–) 1.2–1.3 mm long including a spur ca. 0.3–0.5 (–0.7) mm long, commonly without folioles but each side with a narrow lip ca. 0.1–0.4 mm wide, or sometimes with small folioles, each foliole 0.3–0.9 mm long, 0.6–0.8 mm wide, triangular to broadly triangular. Posterior styles 2.5–3.5 mm long, terete, lyrate, with scattered hairs in the proximal 1/4–1/2; foliole 0.5–1.1 mm long, 0.6–0.9 mm wide, usually triangular, rarely subsquare, or sometimes each side with only a narrow lip 0.2–0.3 mm wide. Dorsal wing of samara 3.5–4.6 (–5.4) cm long, 1.3–2 (–2.5) cm wide, upper margin with a blunt tooth; nut bearing a pair of rectangular to lunate lateral winglets, these up to 7 mm long, up to 3.5 mm wide, or with 3–6 triangular to rectangular winglets per side, these up 3 mm long, up to 1.5 mm wide, and/or with 2–6 spurs and/or crests; nut 5.3–8 mm high, 3–5 mm in diameter, without air chambers, areole 3–3.7 mm long, 2.7–3 mm wide, concave, carpophore up to 4 mm long. Embryo 5–6.3 mm long, ca. 2 times as long as wide, ovoid, outer cotyledon 4.4–5.8 mm long, 2.4–3.1 mm wide, straight or the distal 1/4 curved but not folded over the inner cotyledon, inner cotyledon 3.8–4.9 mm long, 2–2.6 mm wide, straight. Chromosome number: n = 10 (W. R. Anderson 1993; based on *Anderson 11610, 11666*).

Phenology. Collected in flower mostly from February through May, but also in August, November, and December, in fruit from April through August, and in November and February.

Distribution (Fig. 13). Brazil (Rio de Janeiro and adjacent areas: southern Espírito Santo, eastern Minas Gerais, northeastern São Paulo); common in woods, cerrado, restingas, and roadside thickets; sea level to 670 m.

REPRESENTATIVE SPECIMENS. **Brazil.** ESPÍRITO SANTO: Mpio. Serra, 33 km N of Vitoria on BR-101 (rd to Linhares), *Anderson 11728* (MBM, MICH); between Fundão and Santa Teresa, 17 km from Santa Teresa, *Anderson 11729* (MBM, MICH); Mpio. Itaguassú–Jatiboca, *Brade 18504* (RB, SP); Rod. BR-101, 2 km S of Ibiraçu, *Hatschbach 47702* (MICH).—MINAS GERAIS: Mpio. São Tomé das Letras, ca. 6 km W of São Tomé, *Anderson 11610* (MBM, MICH); Pedro Leopoldo, Fazenda Jaguara-Experiência, *Carauta 1876* (MICH, RB); ca. 1 km S of São Pedro do Suaçui along Hwy 3, *Davidse 11486* (MO); Mpio. Leopoldina, Rod. BR-116, Km 757, *Hatschbach 47668* (MICH); Carangola, caminho para Torre, 20°43′S, 42°01′W, *Leoni 1827* (MICH); Viçosa, Agriculture College lands, *Mexia 4378* (A, BM, CAS, F, G, GB, K, MICH, MO, MT, NY, S, U, US, WIS), *4831*

(A, BM, CAS, F, G, GB, GH, MICH, MO, NY, S, TEX, U, US, WIS); Caldas, *Regnell III308* (F); Teófilo Otoni, *Trinta 745* (R).—RIO DE JANEIRO: Serra dos Orgãos, near Limoeiro, *Anderson 11711* (MBM, MICH); alto de Boa Vista, *Brade 10638* (R); Niterói, Jurujuba, *Brade 11363* (GH, R); cachoeiras de Macacu, Morro do Reservatório, *Carauta 573* (GUA, MICH, NY, RB); Mpio. Magé, cerca de 2 km abaixo do Meio da Serra, *Carauta 2401* (RB); Rio de Janeiro, *Glaziou 715* (BR, C, F, K, P, R); Dois Irmãos, *Occhioni 4519* (MICH); Ilha do Governador, *Trinta 502* (R).—SÃO PAULO: Limeira, *M. Kuhlmann 732* (SP); Rio Claro, *Löfgren 641* (C); Mogymirim, *Mosén 1156* (S); Campinas, *Santoro 641* (RB, US); Pirassenagua, *N. Santos R40548* (R); São José dos Campos, *Usteri SP12030* (GH, SP).

Stigmaphyllon lalandianum is a common species, distinguished by its leaves and small-flowered but abundant inflorescences. The laminas are usually elliptical, attenuate at the base, sericeous abaxially, and lack marginal glands. The anterior style may have tiny folioles but more commonly is only laterally expanded into a narrow lip on each side. This species resembles the much less frequent *S. acuminatum* (no. 15), whose laminas are to-

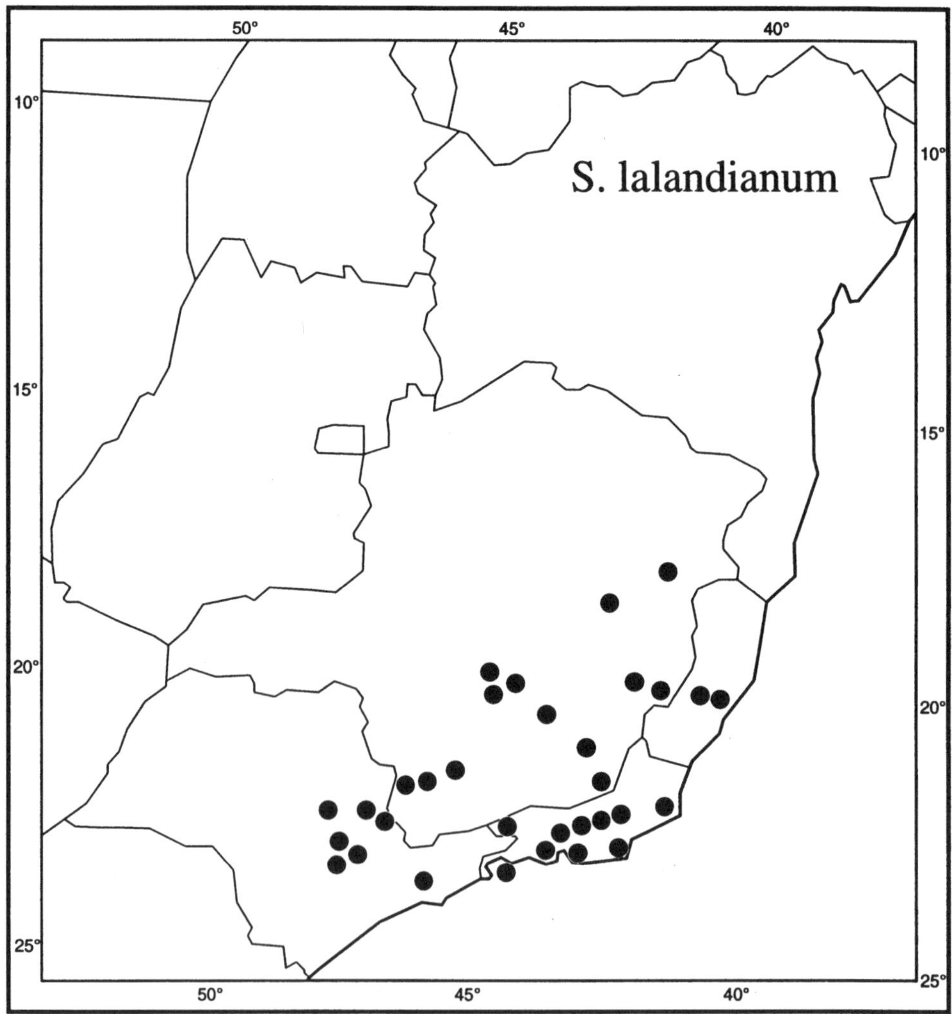

FIG. 13. Distribution of *Stigmaphyllon lalandianum*.

mentose abaxially and often bear irregularly spaced sessile glands adjacent to the margin abaxially. It also differs in that the dorsal wing of the samara, unlike that of *S. lalandianum*, is abruptly narrowed at the nut.

15. Stigmaphyllon acuminatum Adr. Jussieu in St.-Hilaire, Fl. bras. merid. 3: 58. 1833 ["1832"]. *Stigmaphyllon lalandianum* var. *acuminatum* (Adr. Jussieu) Niedenzu, Ind. Lect. Lyc. Brunsberg. p. aest. 1900: 6. 1900.—TYPE: BRAZIL. Rio de Janeiro: "in sylvis caeduis prope villam Uba," *St.-Hilaire C51* (holotype: P!, photo: MICH!; isotypes: P!).

Vine to 2 m. Stems and branches sericeous when young, soon becoming glabrous. Laminas 6–8.7 cm long, 4–6.5 cm wide, broadly lanceolate to ovate to elliptical, apex acuminate, base truncate or attenuate, glabrous adaxially, tomentose abaxially (trabecula 0.4–0.8 mm long, crisped and curled, stalk 0.1–0.2 mm long), margin eglandular or with irregularly spaced sessile glands (ca. 0.3 mm in diameter) borne adjacent to the margin abaxially; petioles 1.3–3.2 cm long, sericeous, not confluent across the node, with a pair of prominent but sessile glands at the apex, each gland 1.2–1.8 mm in diameter; stipules 0.5–1 mm long, 0.8–1.1 mm wide, free, triangular, eglandular. Flowers ca. 15–25 (–30) per condensed or sometimes interrupted pseudoraceme, these borne in dichasia or compound dichasia or small thyrses (axes to the 3rd order, sericeous). Peduncles 5–12 mm long; pedicels 4–8 mm long, terete or distally somewhat distended; both sericeous, peduncles 0.8–2.7 times as long as the pedicels. Bracts 1.4–2.5 mm long, 0.7–1 mm wide, triangular, apex acute; bracteoles 1.1–1.5 mm long, 0.5–0.8 mm wide, oblong, apex obtuse, each bracteole with a pair of inconspicuous glands (each 0.2–0.4 mm in diameter); bracts and bracteoles sericeous abaxially. Sepals 2.2–2.5 mm long and wide, glands 1.9–2.2 (–2.5) mm long, ca. 1 mm wide. All petals with the limb orbicular, glabrous, yellow, margin erose or with fimbriae up to 0.4 mm long; anterior-lateral petals: claw 1.7–2.5 mm long, limb ca. 9 mm long and wide; posterior-lateral petals: claw 1.5–2 mm long, limb ca. 8 mm long and wide; posterior petal: claw 2.5–3 mm long, apex indented, limb ca. 6 mm long and wide. Stamens unequal in size but subequal in shape, those opposite the posterior styles usually the largest; anthers of those opposite the posterior-lateral sepals often with the connective slightly enlarged and the locules slightly reduced; anthers all loculate, glabrous. Stamen opposite anterior sepal: filament ca. 2 mm long, anther 1.1–1.3 mm long; stamens opposite anterior-lateral petals: filaments 1.5–1.7 mm long, anthers 0.9–1 mm long; stamens opposite anterior-lateral sepals: filaments 2.1–2.5 mm long, anther 0.9–1 mm long; stamens opposite posterior-lateral petals: filaments 2.2–2.7 mm long, anthers 1.1–1.3 mm long; stamens opposite posterior-lateral sepals: filaments 2.3–2.5 mm long, connectives 0.9–1 mm long, locules 0.5–1 mm long; stamen opposite posterior petal always shorter than the adjacent two: filament 1.9–2.1 mm long, anther 0.9–1 mm long. Anterior style 2.5–2.7 mm long, shorter than the posterior two, terete, glabrous, erect or slightly recurved; apex ca. 1.2 mm long including a spur ca. 0.3 mm long, linear, folioles absent but each side of apex with a narrow lip ca. 0.8 mm long, 0.1–0.3 mm wide. Posterior styles 2.8–3.2 mm long, terete, with scattered hairs in the proximal 1/5–1/2, lyrate; foliole 0.5–0.7 mm long, 0.5–0.9 mm wide, triangular to subsquare. Dorsal wing of samara 3.6–4.4 cm long, at the nut abruptly narrowed and ca. 0.5 cm wide but flared distally to 1.4–1.8 cm wide, upper margin with a blunt tooth; nut bearing a pair of rectangular or grossly dentate lateral winglets, these 3–4 mm long, 1–1.5 mm wide, and/or spurs and crests; nut 4.5–5 mm high, ca. 3.5 mm in diameter, without air chambers, areole ca. 2.5

mm long, ca. 2 mm wide, convex, carpophore up to 3 mm long. Mature embryo not seen. Chromosome number unknown.

Phenology. Collected in flower from January through May, collected once in fruit in July.

Distribution (Fig. 84). Brazil (southern coastal Bahia, Minas Gerais, and Rio de Janeiro); disturbed mata higrófila, secondary vegetation, and roadsides; 80–700 m.

ADDITIONAL SPECIMENS EXAMINED. **Brazil.** BAHIA: Mpio. São José, estrada São José–Una, ca. de 5 km a partir de São José, *Amorim et al. 1278* (MICH); Mpio. Puerto Seguro, Estação Ecologica, Pau-Brazil, CEPLAC, *de Carvalho et al. 1198* (MICH); Mpio. Santa Cruz Cabralia, Estação Ecologica, Pau-Brazil, 14 km NW of Puerto Seguro, 16°32′S, 39°15′W, *Webster 25064* (K, MICH).—MINAS GERAIS: Belo Horizonte, Campus UFMG, *Domingos 21* (MICH); Monhuaçu, *Heringer 18174* (K, MG, MICH, MO, US).—RIO DE JANEIRO: Serra de Friburgo para Teresópolis, *Duarte 9574* (MICH, SP); près de Petrópolis, sur la vielle route de Minas, *Glaziou 8580* (BR, C, F, G, K, LE, NY, P, S); Petrópolis, Rocio, *Sucre 2227* (RB).

Stigmaphyllon acuminatum is recognized by its numerous though small-flowered inflorescences and elliptical leaves, which are tomentose abaxially and often bear irregularly spaced sessile glands adjacent to the margin abaxially. The apex of the anterior style is laterally expanded into a lip on each side but lacks folioles. The samaras of the only fruiting collection seen (*Webster 25064*) are distinctive in that the large dorsal wing is abruptly narrowed at the nut to only ca. 0.5 cm wide. The two collections from Bahia differ in that the pedicels are very short (ca. 0.5 mm long vs 0.8–1.2 mm long) and appear distally distended.

This species is most likely to be confused with the more common *S. lalandianum* (no. 14), which also has elliptical leaves and small flowers arranged in pseudoracemes. It is readily separated by its laminas, which are sericeous abaxially (the trabecula straight or wavy, and appressed) and lack marginal glands. The dorsal wing of the samara is not abruptly narrowed at the nut.

16. Stigmaphyllon orientale Cuatrecasas, Webbia 13: 535. 1958.—TYPE: COLOMBIA. Meta: Los Llanos, a lo largo del Río Ochoa, 400 m, 19 Dec 1938, *Haught 2471* (holotype: US!, photo: MICH!; isotypes: COL! F! GH! NY!).

Vine. Stems and branches sericeous when young, soon becoming glabrate. Laminas 9–15 cm long, 7–10 cm wide, ovate, apex acuminate, base cordate, glabrous adaxially and abaxially but with abundant appressed hairs along the margin and on the major veins (trabecula 0.5–1.9 mm long, straight, subsessile), margin eglandular or with irregularly spaced sessile glands (ca. 0.4 mm in diameter); petioles 2.7–8.3 cm long, sericeous, not confluent across the node, with a pair of prominent but sessile glands at the apex, each gland 1.5–2.5 mm in diameter; stipules 0.5–1 mm long, 0.7–1 mm wide, free, triangular, eglandular. Flowers ca. 8–12 per umbel, these borne in compound dichasia or small thyrses (axes to the 4th order, sericeous). Peduncles 2–5 mm long; pedicels 4–6.5 mm long; both sericeous (hairs subsessile), peduncles 0.3–0.8 times as long as the pedicels. Bracts 1–1.5 mm long, 1–1.1 mm wide, triangular, apex acute; bracteoles 0.9–1.3 mm long, 0.9–1.1 mm wide, triangular, apex obtuse, eglandular; bracts and bracteoles sericeous abaxially. Sepals 2–2.5 mm long, 2–2.2 mm wide, glands 1.2–1.7 mm long, ca. 1 mm wide. All petals with the limb orbicular, glabrous, yellow or (especially the limb of the posterior petal) red with a yellow margin, margin of the anterior-lateral petals erose, margin of the posterior-lateral and posterior petals digitate-fimbriate, the fimbriae up to 0.2 mm long; anterior-lateral petals: claw 2–2.5 mm long, limb ca. 10 mm long and wide; posterior-lateral petals: claw ca. 1.5 mm long, limb ca. 8 mm long and wide; posterior petal: claw ca. 3 mm long, apex indented, limb 6–6.5 mm long and wide. Stamens unequal in size but subequal in shape; anthers all loculate, glabrous. Stamen opposite anterior sepal: filament 3–3.1 mm long, anther ca. 1.5 mm long; stamens opposite anterior-lateral petals: filaments ca. 2.6 mm long, anthers 1–1.1 mm long; stamens opposite anterior-lateral sepals: filaments 2.5–3.3 mm long, anthers 0.9 mm long; stamens opposite posterior-lateral petals: filaments 2.5–3.3 mm long, anthers ca. 1.1 mm long; stamens opposite posterior-lateral sepals: filaments 2.7–3 mm long, anthers ca. 0.8 mm long; stamen opposite posterior petal always shorter than the adjacent two: filament ca. 2.2 mm long, anther ca. 1.1 mm long. Anterior style 3.5–3.6 mm long, longer than the posterior two, terete, glabrous, erect; apex 1.9–2 mm long, each foliole 1.1–1.2 mm long, 1.6–1.7 mm wide, subtriangular to subrectangular. Posterior styles 3–3.2 mm long, terete, glabrous, lyrate; foliole 1.1–1.3 mm long, 1.4–1.7 mm wide, subrectangular to subsquare. Dorsal wing of samara 3.7–4 cm long, 1.4–1.6 cm wide, upper margin with a blunt tooth; lateral winglets absent; nut 5.2–5.5 mm high, 3 mm in diameter, without air chambers, areole ca. 4 mm long, ca. 3.5 mm wide, slightly convex, carpophore up to ca. 3 mm long. Embryo ca. 5.8 mm long, ca. 2 times as long as wide, ovoid, outer cotyledon 5.4 mm long, 3.4 mm wide, straight, inner cotyledon ca. 5.2 mm long, ca. 2.8 mm wide, straight. Chromosome number unknown.

Stigmaphyllon orientale is known only from two collections from Colombia (Fig. 21), the type (flowers and fruits) and *Shiefer 836* (flowers only) (GH; Colombia. Meta: Mpio. Villavicencio, Road 6 Restrepo, ca. 54 mi from Villavicencio, 23 Jul 1945). It is the only species in which the stamens are unmodified but the styles bear well-developed folioles, and it is one of only five species in which the anterior style is larger than the posterior two. The other four (*S. herbaceum, S. hypargyreum, S. maynense,* and *S. puberum*) all have a heterogeneous androecium. The flowers of *S. orientale*, borne on relatively short peduncles and pedicels, are arranged in 8–12-flowered umbels. The flag petal is red and, like the posterior-lateral petals, has a digitate-fimbriate margin. The laminas are ovate and glabrate abaxially, except for abundant appressed hairs along the margin and on the major veins.

17. Stigmaphyllon puberum (Richard) Adr. Jussieu, Ann. Sci. Nat. Bot., sér. 2, 13: 289. 1840. *Banisteria pubera* Richard, Actes Soc. Hist. Nat. Paris 1: 109. 1792.—TYPE: FRENCH GUIANA. *Leblond 44* (holotype: G!, photos: A! F! GH! MICH!).

Stigmaphyllon puberum β schomburgkianum Bentham, London J. Bot. 7: 129. 1848.—TYPE: GUYANA. *Rob. Schomburgk 2nd. coll. 819* (*1500B*) (holotype: K!, photo: MICH!; isotype: G!).

Vine to 10 m. Stems and branches densely sericeous when young, usually becoming glabrate. Laminas 8.2–20.2 cm long, (2–) 3–12.5 cm wide, commonly lanceolate or narrowly so (rarely linear-lanceolate) to elliptical to ovate to rarely suborbicular, apex acuminate, base attenuate or truncate or sometimes cordate, very sparsely sericeous to glabrate or glabrous adaxially, sericeous to sparsely so or rarely densely so abaxially (trabecula 0.5–1.6 mm long, straight, sessile or with a stalk up to 0.025 mm long), margin with irregularly spaced sessile glands (0.3–0.4 mm in diameter); petioles 1.2–7.2 cm long, densely sericeous, not confluent across the node, with a pair of prominent but sessile glands at the apex, each gland 1–1.8 (–2.2) mm in diameter; stipules 0.5–1 (–1.4) mm long, 0.2–0.8 mm wide, free, triangular to narrowly so to linear, eglandular. Flowers 8–15 per umbel, these borne in dichasia or compound dichasia or small thyrses (axes to the 4th order, densely sericeous), rarely solitary. Peduncles (0.8–) 1.5–4.8 mm long, densely sericeous; pedicels 2.5–7.5 mm long, densely sericeous but just below the flower commonly sparsely pubescent to glabrous, terete; peduncles 0.2–0.8 times as long as the pedicels or rarely subequal. Bracts 1.1–2.2 mm long, 0.7–1.4 mm wide, triangular or broadly so, apex acute; bracteoles 0.8–1.4 mm long, 0.6–1.2 mm wide, broadly triangular to ovate or parabolic, apex obtuse, eglandular or with a glandular region in the basal 1/3–1/2; bracts and bracteoles sericeous abaxially. Sepals (2–) 3–3.2 mm long, 1.6–2.5 mm wide, glands 1.2–1.7 (–2) mm long, 0.8–1.2 mm wide. All petals with the limb orbicular (or limb of the posterior petal broadly obovate or sometimes subsquare), glabrous, yellow with red center or suffused with red or (especially the limb of the posterior petal) mostly red with a yellow margin, margin digitate-fimbriate, fimbriae up to 0.6 (–0.8) mm long; anterior-lateral petals: claw 1.5–2.5 mm long, limb ca. 8–13 mm long and wide; posterior-lateral petals: claw 0.5–1.5 (–2) mm long, limb ca. 7–8.5 mm long and wide; posterior petal: claw 3–3.5 (–4.2) mm long, apex indented, limb ca. 5–7 mm long, 4.5–6.5 mm wide. Stamens unequal, that opposite the anterior style the largest, those opposite the anterior-lateral sepals with the connective enlarged and the locules reduced; anthers all loculate, glabrous or pubescent. Stamen opposite anterior sepal: filament 3.1–4.2 mm long, anther 1.1–1.4 mm long; stamens opposite anterior-lateral petals: filaments 2–2.7 mm long, anthers 1–1.2 (–1.4) mm long; stamens opposite anterior-lateral sepals: filaments 2.5–3.8 mm long, connectives 0.9–1.2 mm long, locules 0.3–0.6 mm long; stamens opposite posterior-lateral petals: filaments 2.1–3 mm long, anthers 0.9–1.2 mm long; stamens opposite posterior-lateral sepals: filaments 2.3–3.1 mm long, anthers 0.5–1 mm long; stamen opposite posterior petal always shorter than the adjacent two: filament 1.7–2.6 mm long, anther 0.6–0.9 mm long. Anterior style 3.5–4.6 mm long, longer than the posterior two, terete, glabrous or sometimes with scattered hairs in the proximal half, incurved; apex (1–) 1.2–1.8 mm long including a spur 0.1–0.3 mm long, each foliole (0.7–) 1–1.8 mm long, (0.6–) 1–1.5 (–1.7) mm wide, commonly narrowly trapezoidal or rectangular to subsquare. Posterior styles 2.5–3.3 (–3.5) mm long, terete, glabrous or sometimes with scattered hairs in the proximal third, lyrate; folioles 0.9–1.5 mm long, 0.6–1.1 mm wide, rectangular to rhombic or sometimes triangular. Dorsal wing of samara 2.5–3.7 (–4.1) cm long, 0.9–1.5 cm wide, usually erect and tapering from the base of the nut; nut smooth or sometimes with 1–5 prominent ribs; nut 7–10 mm high, 5–7 mm in diameter, wall containing spongy tissue, without air chambers, areole 3.5–5.5 mm long, 3.5–5 mm wide, concave, carpophore up to 1.5 mm long. Embryo 7.7–9.4 mm long, 4.5–6 mm wide at the base, ovoid, cotyledons convoluted and folded within each other, outer cotyledon with a cleft in the basal half. Chromosome number unknown. Fig. 14a–m.

Phenology. Collected in flower and in fruit throughout the year.

Distribution (Figs. 15, 16). In the Atlantic lowlands of Central America and also in the Pacific lowlands of Costa Rica (Golfo Dulce area) and Panama; lowlands of northern South America (not recorded from Ecuador but to be expected there): Venezuela (Delta Amacuro, Monagas), Guyana, Suriname, French Guiana, Colombia (northern Chocó and Antioquia, Putumayo), Peru (northern Loreto, Huánuco), Amazonian Brazil (Amapá, Pará, Amazônas, Acre); in the West Indies recorded from Jamaica (very rare, fide C. D. Adams, 1972), the Dominican Republic (not reported from Haiti), Puerto Rico, Guadeloupe, Désiderade, Martinique, Dominica, and St. Vincent; in wet areas: rain forests, gallery forests, riverbanks, and mangrove swamps; sea level to 500 m.

REPRESENTATIVE SPECIMENS. **Jamaica**. [St. Mary, fide Adams, 1972] *McNab s.n.* (GOET). **Dominican Republic**. Santo Domingo, Llano Costero, *Ekman H12508* (F, G, K, MICH, NY, S, US); prov. Samaná, Sánchez,

FIG. 14. *Stigmaphyllon puberum* and *S. hypargyreum*. a–m, *S. puberum*. a. Flowering branch. b. Large leaf. c. Detail of margin and abaxial surface of lamina. d. Posterior petal (the "flag"). e. Androecium; stamen second from left opposes the posterior petal. f. Gynoecium; posterior styles are bent slightly outward to show anterior style (in center). g. Lateral view of anterior style. h. Apex of posterior style. i–l. Four samaras, illustrating variation in size and shape of dorsal wing. m. Embryo. n–p, *S. hypargyreum*. n. Leaf. o. Detail of margin and abaxial surface of lamina. p. Samara. Scale: for a, b, i–l, n, p, bar = 4 cm; for c, o, bar = 2 mm; for d, m, bar = 8 mm; for e, bar = 4 mm; for f–h, bar = 3.3 mm. (Based on: a, c–h, *Chacón G. 1292*; b, *Martínez S. 1696*; i, *Austin & Cavalcante 4144*; j, *Ekman H1508*; k, *Garwood 987*; l, m, *Prance et al. 12015*; n, o, *Croat 7040*; p, *Woodworth & Vestal 621*.)

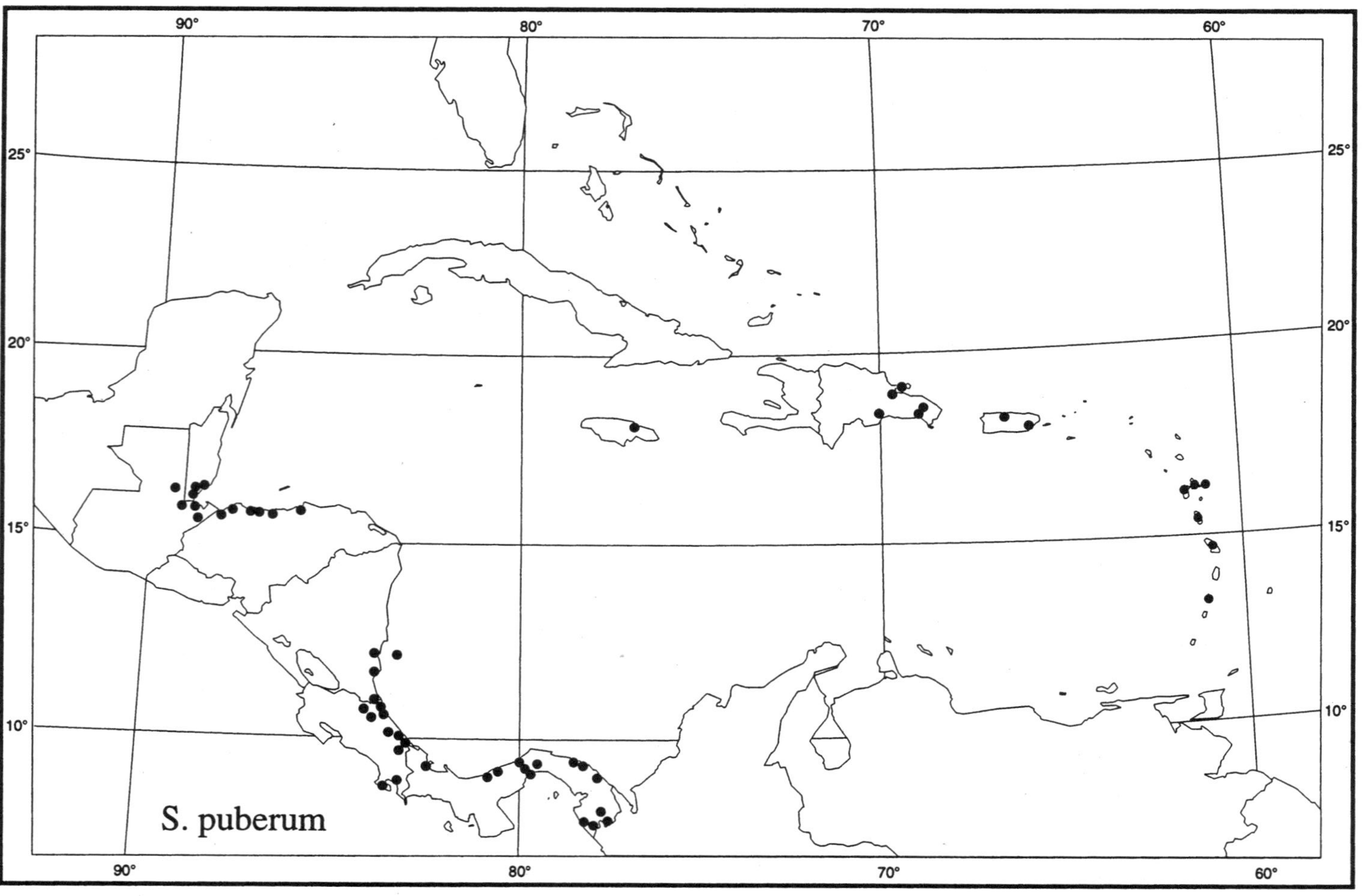

FIG. 15. Distribution of *Stigmaphyllon puberum* in the West Indies and Central America.

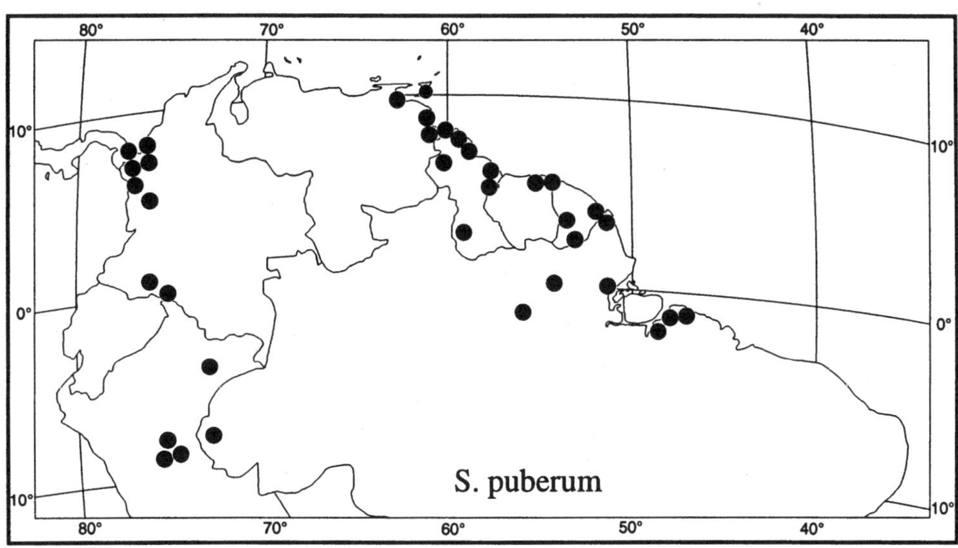

FIG. 16. Distribution of *Stigmaphyllon puberum* in South America.

in the Gran Estero, *Ekman 14796* (S). **Puerto Rico**. Colonía Paraíso, *Liogier et al. 30076* (NY, UPR, US); at Km 28.1 on rte 191 near Florída, *Wagner 1643* (A, U mixed with *S. emarginatum*). **Guadeloupe**. *Duss 2414* (F, MO, NY, US); *Stehlé 410* (P, S, US). **Dominica**. *Eggers 651* (BR, G, GH, GOET, M, P, W); *Hodge 554* (GH, NY); *Howard 11761* (A, BM, NY). **Martinique**. *Duss 1472* (F, GH, MO, US); *Hahn 1132* (BR, G, P, W); *Stehlé & Stehlé 4506* (US). **St. Vincent**. *Smith & Smith 1261* (NY). **Trinidad.** Moruga seashore near Roman Catholic School, *Broadway 7618* (NY, TRIN); Monga, *Broadway 9091* (BM, K).

Belize. TOLEDO: Monkey River, *Gentle 3567* (MICH, MO, NY, U, US, UTD); "El Dorado," Punta Gorda, *Schipp 1009* (A, BM, CAS, F, G, GH, K, MICH, MO, NY, S). **Guatemala**. IZABAL: S of Río Dulce, *LeDoux et al. 2094* (CAS, LL, MEXU, MICH, MO, NY, US, WIS); vic. of Quiriguá, *Standley 24059* (GH, NY, US).—PETÉN: Cadenas, *Contreras 9155* (LL, MICH, UTD). **Honduras**. ATLÁNTIDA: between Tela and Lancetilla, *Yuncker 4577* (A, F, MICH); vic. of La Ceiba, along Río Danto, slopes of Mt. Cangrejal, *Yuncker et al. 8434* (BM, F, G, GH, K, MICH, MO, NY, S, US).—COLÓN: Tanjica Castilla, *Salvoza 798* (F).—CORTÉS: Golfo de Honduras, 2 mi W of Omoa, *Webster et al. 12715* (F, MO). **Nicaragua**. RÍO SAN JUAN: Río Indio, 5 hrs upriver from San Juan del Norte, 11°07′N, 83°50–52′W, *Riviere 242* (MO).—ZELAYA: Isla del Maíz Grande, *Martínez S. 1696* (MICH); Río Chiquito, Caño Dos Oros, a 5–7 km al N de Atlanta, *Téllez 4950* (MEXU, MO). **Costa Rica**. HEREDIA: Finca La Selva, OTS Field Station, Río Viejo just E of its junction with the Río Sarapiquí, *Hammel 9325* (DUKE); near Puerto Viejo de Sarapiquí, *Murray & Johnson 881* (MICH).—LIMÓN: ca. 1 km N of Cahuita, *Almeda 3245* (CAS, MICH); Parque Nacional Tortuguero, Estación Cuatro Esquinas, 10°32′N, 83°30′W, *Chavarría 128* (MICH); banks of the Río Colorado, *Morley 803* (F, GH, MO, US); northern outskirts of Cahuita, ca. 47 km 5 of Limón, *Wilbur et al. 23362* (MICH, DUKE).—PUNTARENAS: Golfito de Osa, *Brenes 12292 (no. 75 of the Porsch Expedition)* (F, W); Corcovado Natl Park, 0–2 km W of park headquarters at Sirena, 08°29′N, 83°36′W, *Liesner 2910* (MO); Santo Domingo de Golfo Dulce, *Tonduz 6985* (BM, F, G, GH, K, MICH, US), *9942* (BR, G, P). **Panama**. BOCAS DEL TORO: Río San Pedro, *Gordon 84C* (MO); Water Valley, *von Wedel 2662* (GH, MO, NY, US); vic. of Chiriquí Lagoon, *von Wedel 2761* (GH, MO, NY, US).—CANAL ZONE: Juan Mina, *Bartlett & Lasser 16510* (MICH, MO); near Gamboa, *Clewell & Tyson 3267* (MO); Barro Colorado Island, *Bangham 496* (A, F, S, US), *Foster 1033* (F, DUKE).—COLÓN: Coclé del Norte, *Hammel 4583* (MICH); Quebrada Santa Marta, on coast rd, 4.5 km SW of PINA, *Nee 11716* (MICH, MO, US); N of Río Guanche, *Davidse 10070* (MO, NY).—DARIÉN: Río Ucunguantí, *Bristan 1154* (MO, US); Río Tuira, between Río Purnusa and Río Mangle, *Duke 14611* (MO, US); Ensenada del Guayabo, 16–19 km SE of Jaqué, *Garwood 987* (MICH).—PANAMÁ: Río Mamoni, below La Caitana, *Pittier 4580* (F, GH, S, US).—SAN BLAS: Mulatuppu, Río Ibadi, *Duke 8474* (MO, US); mainland opposite Playón Chico, *Gentry 6408* (MICH, US); Ailigandí, *Hammel & D'Arcy 4965* (MO).

Colombia. ANTIOQUIA: Mpio. Mutatá, sitio Río Surumbay, 12 km N de Mutatá, 07°20′N, 76°30′W, *Callejas et al. 5745* (MICH); Mpio. Turbo, Bocas del Atrato, Vereda Turbito, en Bahía Colombia (Golfo de Urabá), *Callejas et al. 5027* (MICH); near Río Léon, ca. 15 km W of Chigorodó, ca. 07°45′N, 76°50′W, *Feddema 1894* (MICH, NY, US).—CHOCÓ: Mpio. Quibdó, barrio Yesquita, *Córdoba & García 379* (MO); cerca de Acandí, desembocurada del Río Tolo, *Forero 1035* (COL, MO, NY); hoya del Río Atrato, Mpio. Bojoá, caño de Boyacito, cerca de Bellavista, *Forero 9236* (MO).—PUTUMAYO: márgenes del Río Putumayo en La Concepción, *Cuatrecasas 10803* (COL, US); Umbría, 00°54′N, 76°10′W, *Klug 1797* (A, BM, F, GH, MICH, MO, NY, S, US). **Venezuela**. DELTA AMACURO: Río Acure, between La Margarita and Puerto Miranda, *Steyermark 87775* (MICH, NY, US); 10 km above mouth of Río Cuyubini, *Wurdack 332* (NY).—MONAGAS: Reserva Forestal de Guarapiche (Caño Colorado), *Aristiguieta et al. 7240* (MY, NY, VEN). **Guyana.** Upper Rupununi River, near Dadanawa, 02°45′W, *de la Cruz 1411* (CM, F, GH, MO, NY, US); Pomeroon River, Pomeroon Dist., *de la Cruz 2975* (CM, F, GH, MO, NY, US); Kamakusa, upper Mazaruni River, ca. 59°50′W, *de la Cruz 4087* (CM, F, GH, MO, NY, US); Issarora, Aruka River, 08°10′N, 59°50′W, *Hitchchock 17561* (GH, NY, S, US); margins of Berbice River, S of New Dageraad, 06°N, 57°43′W, *Maas et al. 5574* (MICH). **Suriname**. Pauluskreek ten Zuiden van Paramaribo, *Florschütz & Florschütz 897* (U); Boven Cottica, *Focke 715* (U); *Hostmann 803a* (C, G, GOET, LE, MO, P, S, W); Dombury, fluv. Suriname inferior, *Kramer & Hekking 2345* (LL, U); banks of Marowijne River, from Albins to the N, *Lanjouw & Lindeman 991* (IAN, K, NY, U); La Poule, *Samuels 176* (GH, NY). **French Guiana**. Beira de Rio Aproague, *Black et al. 54-17576,* (IAN), *54-17591* (IAN); Trois-Sauts, *Jacquemin 1613* (CAY), *Lescure 325* (CAY); Karouany, *Sagot 95* (BM, BR, K, P, S, W). **Brazil**. ACRE: Cruzeiro do Sul, Rio Moa, between Igarapés and Ipiranga, *Prance et al. 12015* (INPA, MG, MICH, NY).—AMAPÁ: Mazagão, Rio Vila Nova, *Froés & Black 27456* (IAN); Rio Vila Nova–Macapá, *Rabelo et al. 2068* (MICH).—AMAZÔNAS: Tabatinga, *Schwacke 530* (R).—PARÁ: S of Belém, Rio Mojú, ca. 4 mi N of Mojú, *Austin & Cavalcante 4144* (MG, MICH, MO, NY); Belém, margen de Rio Guamá, *N. T. Silva 57812* (MICH, NY); Lageira, airstrip on Rio Maicuru, 00°55′S, 54°26′W, *Strudwick & Sobel 3353* (K, MICH). **Peru**. HUÁNUCO: Km 11–12, S of Pucallpa–Tingo María rd at Bosque von Humboldt, 08°45′S, 75°01′W, *Gentry 41312* (MICH).—LORETO: Prov. Maynas, Río Afacayu, abajo de Iquitos, *Ayala et al. 2913* (MICH, NY); Prov. Coronel Portillo, quebrada "shesha" affluente del Río Abujao, 08°20′S, 73°45′W, *Díaz et al. 784* (MICH); Iquitos, Lago Quistacocha, *Plowman 2472* (GH, K, NY).

In *S. puberum*, as in *S. orientale*, *S. herbaceum*, *S. hypargyreum*, and *S. maynense*, but unlike the rest of the genus, the anterior style and its opposing stamen are larger than the posterior styles and their opposing stamens. All five species have very short peduncles (up to 5 mm long) and the petals marked with red. *Stigmaphyllon puberum* also differs from most species in its leaves, which are commonly lanceolate and always acuminate. *Stigmaphyllon puberum* occurs in wet lowland habitats, and its distinctive samara appears adapted to dispersal by water; the locule is surrounded by spongy tissue. The dorsal wing is erect and tapers distally; it varies greatly in shape throughout the range of the species (Fig. 14i–l). The embryo is also unusual in that the cotyledons are folded within each other and convoluted, a condition shared with *S. maynense*.

Stigmaphyllon hypargyreum and *S. maynense* both are distinguished by abundant silvery sericeous pubescence on the lower surfaces of the lamina, which is so dense as to obscure the epidermis; in *S. puberum* the laminas are sericeous to sparsely so abaxially, and the epidermis is visible. Although *S. puberum* is readily separated from *S. hypargyreum* and *S. maynense* in fruit (see Figs. 14, 17), flowering specimens with very young, i.e., unusually densely pubescent, leaves may be more difficult to determine in sympatric areas, although the nature of the petal margins is distinctive. In *S. puberum* all petals are digitate-fimbriate [fimbriae up to 0.6 (–0.8) mm long]; those of *S. maynense* are erose or erose-denticulate (teeth up to 0.1 mm long), and those of *S. hypargyreum* are erose or denticulate or sometimes have fimbriae up to 0.2 (–0.3) mm long. Also, the number of flowers per umbel is 8–15 in *S. puberum*, but usually 15–25 in the other two species. See also the discussions of *S. hypargyreum* (no. 19) and *S. maynense* (no. 18). *Stigmaphyllon herbaceum* (no. 20) has the laminas sparsely sericeous to glabrate abaxially, erose petals, and a samara and embryo typical for the genus. *Stigmaphyllon orientale* (no. 16) differs from all four species most notably in its homogeneous androecium.

18. Stigmaphyllon maynense Huber, Bol. Mus. Pará 4: 575. 1906. *Stigmaphyllon fulgens* var. *maynense* (Huber) Macbride, Field Mus. Nat. Hist. Bot. 13: 844. 1950.—TYPE: PERU. Loreto: Pampa de Sacramento, margem da quebrada Chingana, 25 Nov 1898, *Huber 1507* (holotype: MG!, photos: F! MICH!; isotypes: RB! INPA!, photos of RB isotype: F! GH! MICH! NY! US!).

Vine to 10 (–28) m. Stems and branches densely sericeous and also with some T-shaped hairs. Laminas 8.2–19 cm long, 5.5–13.7 cm wide, broadly elliptical to ovate or suborbicular or rarely narrowly elliptical, apex acuminate, base cordate, glabrous adaxially, densely sericeous abaxially (trabecula 0.6–1.2 mm long, straight or slightly wavy, sessile or with a stalk up to ca. 0.05 mm long), margin with irregularly spaced sessile glands (0.2–0.5 mm in diameter); petioles 3.2–12.5 cm long, densely sericeous, not confluent across the node, with a pair of prominent but sessile glands at the apex, each gland 1.5–2.5 mm in diameter; stipules 0.8–2 mm long, 0.6–1.1 mm wide, free, narrowly triangular, eglandular. Flowers ca. (12–) 15–20 per umbel, these solitary or borne in dichasia or compound dichasia or sometimes in small thyrses (axes to the 3rd order, densely sericeous). Peduncles 1–5 mm long; pedicels 2.3–7 mm long, terete; both densely sericeous and also with some T-shaped hairs, peduncles 0.3–0.8 times as long as pedicels. Bracts 1.8–4 mm long, 1–3 mm wide; bracteoles 1.7–3.5 mm long, 1–1.4 mm wide, eglandular; bracts and bracteoles triangular or narrowly so, apex obtuse, densely sericeous abaxially. Sepals 2.5–3 mm long, 2–2.6 mm wide, glands 1.2–2 mm long, 0.9–1 mm wide. All petals with the limb orbicular or sometimes suborbicular, glabrous, yellow with a red center or suffused with red or (especially the limb of the posterior petal) red with a yellow margin, margin erose or erose-denticulate, teeth up to 0.1 mm long; anterior-lateral petals: claw 2.2–2.5 mm long, limb ca. 8.5–10 mm long and wide; posterior-lateral petals: claw (0.8–) 1.5–2 mm long, limb ca. 7–8.2 mm long, ca. 7–7.5 mm wide; posterior petal: claw 3–3.7 mm long, apex indented, limb ca. 5–5.5 mm long and wide. Stamens unequal, that opposite the anterior style the largest, anthers of those opposite the anterior-lateral sepals with the connective enlarged and the locules reduced; anthers all loculate, glabrous. Stamen opposite anterior sepal: filament 3.5–4.1 mm long, anther 1.2–1.7 mm long; stamens opposite anterior-lateral petals: filaments 2.1–2.5 mm long, anthers 1.2–1.6 mm long; stamens opposite anterior-lateral sepals: filaments 2.8–3.5 mm long, connectives 1–1.2 mm long, locules 0.4–0.8 mm long; stamens opposite posterior-lateral petals: filaments 2.5–3.3 mm long, anthers 1–1.3 mm long; stamens opposite posterior-lateral sepals: filaments 2.3–3 mm long, anthers 0.7–0.9 mm long; stamen opposite posterior petal always shorter than the adjacent two: filament 2–2.4 mm long, anther (0.6–) 1–1.1 mm long. Anterior style 4–4.2 mm long, longer than the posterior two, terete, glabrous, erect; apex 1.4–1.6 mm long including a spur 0.1–0.2 mm long, each foliole 1.3–1.6 mm long, 0.8–1.2 (–1.7) mm wide, usually rectangular or narrowly so or rarely subsquare and distally flared, the distal margin shallowly sagittate or coarsely erose. Posterior styles 3–3.6 mm long, terete, glabrous, lyrate, foliole 1–1.5 mm long, 0.6–1.2 mm wide, narrowly to broadly rectangular. Dorsal wing of samara 3.5–4.6 cm long, 1.4–1.8 cm wide, upper margin without a tooth; nut bearing 3–4 semicircular to oblong lateral winglets per side, these 1.5–6.2 mm long, 1.7–4 mm wide; nut 5.5–7 mm high, 4–5 mm in diameter, without air chambers, areole 3.5–4 mm long and wide, concave, carpophore up to 4 mm long. Embryo 7.5 mm long, ca. 2 times as long as wide, ovoid, outer cotyledon ca. 9 mm long, ca. 4 mm wide, inner cotyledon ca. 5 mm long, ca. 3.7 mm wide, both folded and convoluted. Chromosome number unknown. Fig. 17.

Phenology. Collected in flower in June and from September through March, in fruit from November through February and in April and June.

Distribution (Fig. 18). Ecuador (Napo, Pastaza, Zamora-Chinchipe) and Peru (Amazonas, Huánuco, Loreto, San Martín, Pasco, Madre de Dios), one collection from Brazil (southwestern Amazônas); in wet forests and at water's edge; 200–1000 m.

REPRESENTATIVE SPECIMENS. **Brazil**. AMAZÔNAS: Alto [Rio] Purus, Ponta Alegre, *Huber 4729* (MG). **Ecuador**. NAPO: Cantón Lago Agrio, Parroquia Dureno, 00°02′S, 76°42′W, *Cerón M. & Cerón 3081* (MICH); Estación Biológica Jatún Sacha, Río Napo, 8 km al E de Misahualli, *Cerón M. & Iguago 5511* (MICH); Río Napo, between Ahuana and Latas, *Harling 3647* (S).—PASTAZA: Río Curaray, 01°41–40′S, 75°48–55′W, *Brandbyge & Asanza C. 31600* (AAU, MICH).—ZAMORA-CHINCHIPE: Cantón Nangaritza, Río Nangaritza, 2 km N of Miazi, 04°15′S, 78°39′W, *Neill 9565* (MICH). **Peru**. AMAZONAS: Valle del Río Santiago, ca. 65 km N de Pinglo, Quebrada Caterpiza, 2–3 km atras de la comunidad de Caterpiza, *Huashikat 1702* (MICH, MO), *1733* (MO); valle del Río Santiago, Quebrada Caterpiza, 2–3 km atras de la comunidad de Caterpiza, 03°50′S, 77°40′W, *Tunqui 491, 889* (MICH, MO).—HUÁNUCO: Tingo María, *Woytkowski 5301* (BR, C, F, G, MO, NY, P, S).—LORETO: Balsapuerto, *Klug 2895* (BM, F, G, GH, K, MO, NY, S, US).—MADRE DE DIOS: Prov. Manu, Sintuya, *Chávez 839* (MO); Prov. Tambopata, Dtto. Tambopata, Reserva de Tambopata, 12°15′S, 69°17′W, *Cornejo & Rubia 1607* (MO); Parque Nacional del Manu, Río Manu, vicinity of Cocha Cashu Station, *Foster 5164* (F, US), *5255* (F), *6196* (F).—PASCO: Oxapampa, Palcazu valley, 5–6 km W of Iscosacin, 10°12′S, 75°14′W, *D. Smith 3794*

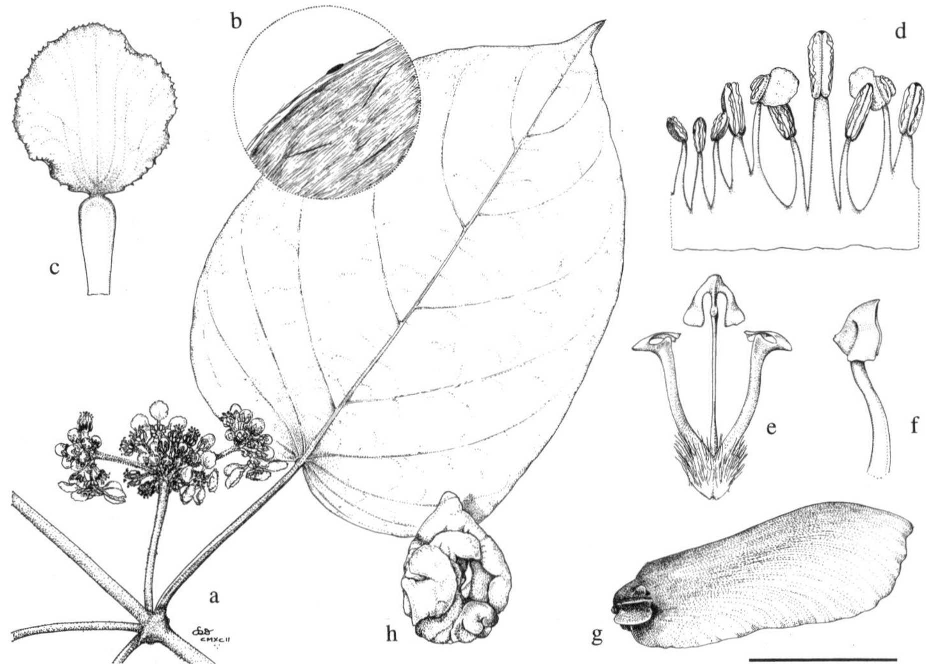

FIG. 17. *Stigmaphyllon maynense*. a. Flowering branch. b. Detail of margin and abaxial surface of lamina. c. Posterior petal (the "flag"). d. Androecium; stamen second from left opposes the posterior petal. e. Gynoecium; posterior styles are bent slightly outward to show anterior style (in center). f. Lateral view of anterior style. g. Samara. h. Embryo. Scale: for a, g, bar = 4 cm; for b, bar = 2 mm; for c, bar = 5 mm; for d, e, f, bar = 4 mm. (Based on: a–f, *Woytkowski 7196*; g, h, *Boeke 1243*.)

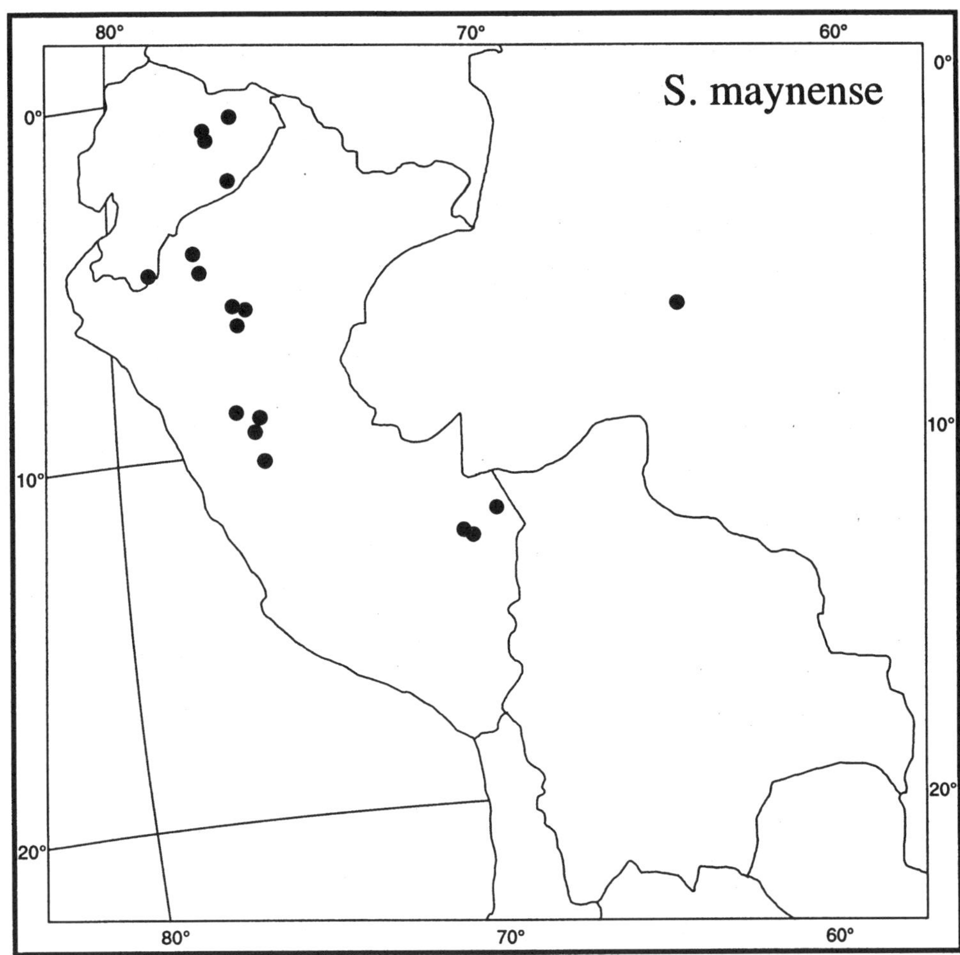

FIG. 18. Distribution of *Stigmaphyllon maynense.*

(MICH).—SAN MARTÍN: Juanjui, *Klug 3918* (CAS, F, GH, MO, NY, S, US, WIS), *Woytkowski 7196* (F, GH, K, MICH, MO, US); Prov. Mariscal Caceres, Dtto. Tocache Nuevo, Challua Yacu, margen izquierda del Río Huallaga, *Schunke V. 4549* (COL, F, GH, MO, NY, US).

Stigmaphyllon maynense, in addition to *S. herbaceum*, *S. hypargyreum*, *S. orientale*, and *S. puberum*, has the anterior style and its opposing stamen larger than the posterior styles and their opposing stamens (for a comparison of these five species, see *S. puberum*, no. 17). Like *S. hypargyreum* (no. 19), *S. maynense* is characterized by a dense silvery sericeous pubescence on the lower surface of the lamina, which obscures the epidermis. *Stigmaphyllon maynense* and *S. puberum* both have distinctive samaras and unusual embryos in which the cotyledons are convoluted and folded within each other (Figs. 14m, 17h). *Stigmaphyllon hypargyreum* is known only from Panama and northern Colombia; however, *S. puberum* also occurs in Peru (but has not been reported from Ecuador) and flowering specimens with young, i.e., unusually densely pubescent leaves, can be confused with *S. maynense*. *Stigmaphyllon maynense* and *S. puberum* can be separated with the following couplet.

1. Petals erose or erose-denticulate, teeth up to 0.1 mm long; bracteoles 1.7–3.5 mm long; anthers glabrous; laminas broadly elliptical to ovate or suborbicular or rarely narrowly elliptical, densely silvery sericeous abaxially, the epidermis hidden; nut of samara with 3–4 semicircular to oblong lateral winglets per side. 18. *S. maynense*.
1. Petals digitate-fimbriate with fimbriae up to 0.6 (–0.8) mm long; bracteoles 0.8–1.4 mm long; anthers glabrous or pubescent; laminas commonly lanceolate or narrowly so (rarely linear-lanceolate) to elliptical to ovate to rarely suborbicular, sericeous to sparsely so abaxially, the epidermis usually visible; nut of samara smooth or ribbed, without lateral winglets or spurs. 17. *S. puberum*.

The only other species in Peru with such densely silvery pubescent leaves is *S. tarapotense* (no. 23); however, its leaf pubescence is not appressed but composed of T-shaped hairs. Also, its anterior style is efoliolate, and its anthers are pubescent.

19. Stigmaphyllon hypargyreum Triana & Planchon, Ann. Sci. Soc. Nat. Bot., sér. 4, 18: 318. 1862.—TYPE: PANAMA. *Duchassaing s.n.* (holotype: P!, photo: MICH!).

Vine to ca. 8 m. Stems and branches densely sericeous when young, usually becoming glabrate. Laminas 7–17.5 cm long, 4.5–12 cm wide, ovate to suborbicular or sometimes elliptical, apex acuminate or acute or rarely obtuse-mucronate, base cordate to truncate or sometimes attenuate, glabrate to glabrous adaxially, densely silvery sericeous abaxially and the epidermis obscured (trabecula 0.5–1.2 mm long, straight, sessile or with a stalk up to 0.05 mm long), margin with irregularly spaced sessile glands (0.3–0.5 mm in diameter) or sometimes eglandular; petioles 2.3–7 cm long, densely sericeous, not confluent across the node, with a pair of prominent but sessile glands at the apex, each gland 1.5–2 mm in diameter; stipules 0.4–1 mm long, 0.3–1 mm wide, free, narrowly triangular to sublinear, eglandular. Flowers (15–) 20–25 per umbel, borne in dichasia or compound dichasia or small thyrses (axes to the 3rd order, densely sericeous). Peduncles 1–4 mm long, densely sericeous; pedicels 4–7 mm long, densely sericeous but just below the flower commonly sparsely pubescent to glabrous, terete; peduncles 0.5–0.7 times as long as pedicels. Bracts 1.1–2 mm long, 1.1–1.5 mm wide, triangular; bracteoles 0.9–1.4 mm long, 0.9–1.3 mm wide, subsquare to parabolic, eglandular; bracts and bracteoles with the apex obtuse, densely sericeous abaxially. Sepals 2.3–2.6 mm long, 1.6–2 mm wide, glands 1.5–1.8 mm long, 0.9–1.1 mm wide. All petals with the limb glabrous, yellow but usually with a red center or suffused with red (especially the limb of the posterior and posterior-lateral petals), margin erose or denticulate or sometimes with fimbriae up to 0.2 (–0.3) mm long, limb of lateral petals orbicular; anterior-lateral petals: claw (1.5–) 2–2.6 mm long, limb ca. 8–9 mm long and wide; posterior-lateral petals: claw 1–1.5 mm long, limb ca. 4.5 mm long and wide; posterior petal: claw 2.9–3.5 mm long, apex indented, limb ca. 4.5 mm long, ca. 3.5 mm wide, broadly elliptical. Stamens unequal, that opposite the anterior style larger than those opposite the posterior styles, those opposite the anterior-lateral sepals usually the longest (sometimes subequal to that opposite the anterior style) and with the connective enlarged and the locules reduced; anthers all loculate, glabrous. Stamen opposite anterior sepal: filament 3–3.5 mm long, anther 1.2–1.3 mm long; stamens opposite anterior-lateral petals: filaments 2.2–2.5 mm long, anthers 0.9–1.2 mm long; stamens opposite anterior-lateral sepals: filaments 3.5–3.8 mm long, connectives 1–1.1 mm long, locules 0.5–0.7 mm long; stamens opposite posterior-lateral petals: filaments 3.1–3.5 mm long, anthers 1–1.1 mm long; stamens opposite posterior-lateral sepals: filaments 3–3.2 mm long, anthers 0.6–0.8 mm long; stamen opposite posterior petal always shorter than the adjacent two: filament 2.2–2.6 mm long, anther 0.7–0.9 mm long. Anterior style 3.4–3.7 mm long, longer than the posterior two, terete, glabrous, erect; apex 1–1.2 mm long, each foliole 0.8–1.1 mm long, ca. 0.6 mm wide, oblong to triangular. Posterior styles 3–3.5 mm long, terete, glabrous, lyrate; folioles 1–1.4 mm long, 0.8–1 mm wide, oblong to obovate. Dorsal wing of samara 3.5–4.2 cm long, at the nut abruptly narrowed and 0.3–0.4 cm wide but flared distally to 1.1–1.4 cm wide, upper margin without a tooth; nut sometimes with a pair of subrectangular lateral winglets, these up to 2.5 mm long, up to 1 mm wide, more commonly bearing only spurs and/or crests, nut 3–5.4 mm high, 2.5–4 mm in diameter, without air chambers, areole 2.3–2.7 mm long, 2.3–3 mm wide, slightly convex, carpophore up to 3 mm long. Embryo ca. 4.5 mm long, ca. 2 times as long as wide, ovoid, outer cotyledon 6.8–7 mm long, ca. 2.5 mm wide, with a longitudinal cleft in the outer surface and the base, the distal 1/3 folded over the inner cotyledon, inner cotyledon ca. 2.3 mm long, ca. 1.2 mm wide, straight. Chromosome number unknown. Fig. 14n–p.

Phenology. Collected in flower from October through July, in fruit from December through April.

Distribution (Fig. 5). Panama (Canal Zone, Darién, Los Santos, Panamá, San Blas) and northern Colombia (Antioquia); in moist tropical forest, at forest edge, river margins, in clearings and scrub, and along roadsides; sea level to 350 m.

REPRESENTATIVE SPECIMENS. **Panama**. CANAL ZONE: Pipeline Rd, 3–5 mi from Gamboa, *Gentry 2415* (F, MO, NY); between Farfan Beach and Palo Seco, *Hunter & Allen 440* (F, G, K, MO, U); ca. 1 mi SW of Cocolí in the Rodman Naval Ammunition Depot, *Wilbur et al. 12871* (F, NY, US); rd from Cocolí to Contractor's Hill, *Tyson & Lazor 6164* (MO); Barro Colorado Island, *Croat 7226* (F, MO, NY), *Wilson 156* (F), *Woodworth & Vestal 621* (A, F, MO).—DARIÉN: Ensenada del Guayabo, 18 km SE of Jaqué, *Garwood et al. 158* (BM, MICH, MO).—LOS SANTOS: 12 km S of Macaracas, *Tyson et al. 2939* (MO).—PANAMÁ: ca. 15 km SW of Cañaza near Río Tortí, 08°52′N, 78°22′W, *Stein 1378* (MICH).—SAN BLAS: along lower Río Ailigandí, *Duke 9328* (MO); Ailigandí, *Hammel & D'Arcy 5029* (MICH).

Colombia. ANTIOQUIA: alrededores de Dabeiba, *Barkeley & Gutiérrez V. 1815* (BM, COL, F); Urabá, en Dabeiba, *Uribe U. 1439* (COL).

Stigmaphyllon hypargyreum is named for the distinctive silvery sericeous pubescence on the lower leaf surfaces, which is so dense that the epidermis is obscured. It is one of five species in the genus in which the anterior style and its opposing stamen are larger than the posterior styles and their opposing stamens (for a comparison of these five species, see *S. puberum*, no. 17). *Stigmaphyllon maynense* (no. 18) is endemic to Peru, and *S. herbaceum* (no. 20) has the laminas sparsely sericeous to glabrate abaxially. *Stigmaphyllon hypargyreum* and *S. puberum* are easily separated by their distinctive samaras (Fig. 14i–l, p), but flowering specimens of *S. puberum* with young, i.e., unusually densely pubescent leaves, may be more difficult to determine in sympatric areas. The species can be separated with the following couplet.

1. Petals erose or denticulate or sometimes with fimbriae up to 0.2 (–0.3) mm long; flowers 8–15 per umbel; anthers glabrous; laminas usually ovate to suborbicular or sometimes elliptical, densely silvery sericeous abaxially, the epidermis not visible. 19. *S. hypargyreum.*
1. Petals digitate-fimbriate, the fimbriae up to 0.6 (–0.8) mm long; flowers (15–) 20–25 per umbel; anthers glabrous or pubescent; laminas commonly lanceolate or narrowly so (rarely linear-lanceolate) to elliptical to ovate to rarely suborbicular, sericeous to sparsely so abaxially, the epidermis visible. 17. *S. puberum.*

20. Stigmaphyllon herbaceum Cuatrecasas, Webbia 13: 532. 1958.—TYPE: COLOMBIA. Boyacá: El Umbo, 130 mi N of Bogotá, 1000 m, 7 Nov 1932, *Lawrance 566* (holotype: A!, photo: MICH!; isotypes: F! G! K! MO! NY! S! U!).

Vine to 20 m. Stems and branches sericeous and also with subsessile T-shaped hairs when young, soon becoming glabrate to glabrous. Laminas 7.2–12.5 cm long, 3.3–7.8 cm wide, elliptical or narrowly so to sometimes ovate, apex acuminate (-mucronate), base cordate to subtruncate, glabrous or glabrate adaxially but the hairs persistent on the midrib, sparsely sericeous to glabrate abaxially (trabecula 0.4–2.3 mm long, straight, sessile to subsessile), margin pubescent and with irregularly spaced glands (0.2–0.5 mm in diameter); petioles 1.3–4.6 cm long, with subsessile T-shaped hairs, not confluent across the node, with a pair of prominent but sessile glands at the apex, each gland 1.2–2 mm in diameter; stipules 0.3–0.7 mm long, 0.3–0.7 mm wide, free, triangular, eglandular. Flowers ca. 10–15 per umbel or pseudoraceme, these borne in dichasia or compound dichasia or small thyrses (axes to the 5th order, with subsessile T-shaped hairs). Peduncles 2.5–5 mm long, sericeous; pedicels 3.2–5 mm long, sparsely sericeous to glabrate; peduncles 0.5–1.3 times as long as the pedicels. Bracts 1.1–2.5 mm long, 0.8–1.2 mm wide, triangular, apex acute; bracteoles 0.7–1.3 mm long, 0.6–1 mm wide, parabolic to triangular, apex obtuse, each bracteole with 0–2 inconspicuous glands (each ca. 0.3 mm in diameter); bracts and bracteoles sericeous or sparsely so abaxially. Sepals 2.4–2.7 mm long, 1.3–2 mm wide, glands 1.4–1.9 mm long, ca. 1 mm wide. All petals with the limb glabrous, yellow or (especially the limb of the posterior petal) red with yellow margin, margin erose; anterior-lateral petals: claw 1.6–2.2 mm long, limb ca. 11.5 mm long and wide, orbicular; posterior-lateral petals: claw 1–1.2 mm long, limb 7–9 mm long and wide, obovate; posterior petal: claw 3.4–3.6 mm long, apex indented, limb ca. 7 mm long, ca. 5 mm wide, broadly obovate. Stamens unequal, that opposite the anterior style the largest, those opposite the anterior-lateral sepals with the connective enlarged and (1–) 2 reduced locules; anthers all loculate, glabrous. Stamen opposite anterior sepal: filament 3.5–4 mm long, anther 1–1.2 mm long; stamens opposite anterior-lateral petals: filaments 2.5–3 mm long, anthers 0.9–1.1 mm long; stamens opposite anterior-lateral sepals: filaments 3.4–3.9 mm long, connectives 0.9–1.2 mm long, locules 0.3–0.5 (–0.6) mm long; stamens opposite posterior-lateral petals: filaments 3–3.5 mm long, anthers 0.8–1 mm long; stamens opposite posterior-lateral sepals: filaments 2.8–3.4 mm long, anthers 0.3–0.5 mm long; stamen opposite posterior petal always shorter than the adjacent two: filament 2.5–2.8 mm long, anther 0.5–0.7 mm long. Anterior style 4–4.1 mm long, longer than the posterior two, terete, glabrous, erect; apex 1.5–1.7 (–2) mm, each foliole 1.3–1.8 mm long, 1.1–1.7 mm wide, subsquare to subrectangular. Posterior styles 3.2–3.7 mm long, terete, glabrous, lyrate; foliole 0.8–1.3 mm long, 1–1.4 mm wide, parabolic to sometimes broadly semi-lunate. Dorsal wing of samara 3–4.4 cm long, 0.5–0.6 mm wide at the nut and flaring to 1–1.8 cm at its widest, upper margin without a tooth; nut without lateral winglets or with 1–2 spurs per side, these up to 4.5 mm long and 1.6 mm wide; nut 4.5–5 mm high, 3.5–4 mm in diameter, without air chambers, areole 3–3.5 mm long and wide, slightly convex, carpophore up to 3 (–3.5) mm long. Embryo 4.5–5.2 mm long, ca. 2 times as long as wide, ovoid, outer cotyledon 4–4.5 mm long, 2.6–3 mm wide, the distal 2/3 folded over the inner cotyledon, inner cotyledon 3.7–4.5 mm long, 1.9–2.3 mm wide, folded at the distal 2/3. Chromosome number unknown.

Phenology. Collected in flower from April through November, in fruit from May through November.

Distribution (Fig. 19). Lowlands of central and western Colombia, one collection from northern Ecuador; tropical wet forest and rain forest, at forest edge, and at roadsides; 150–1000 (–1600) m.

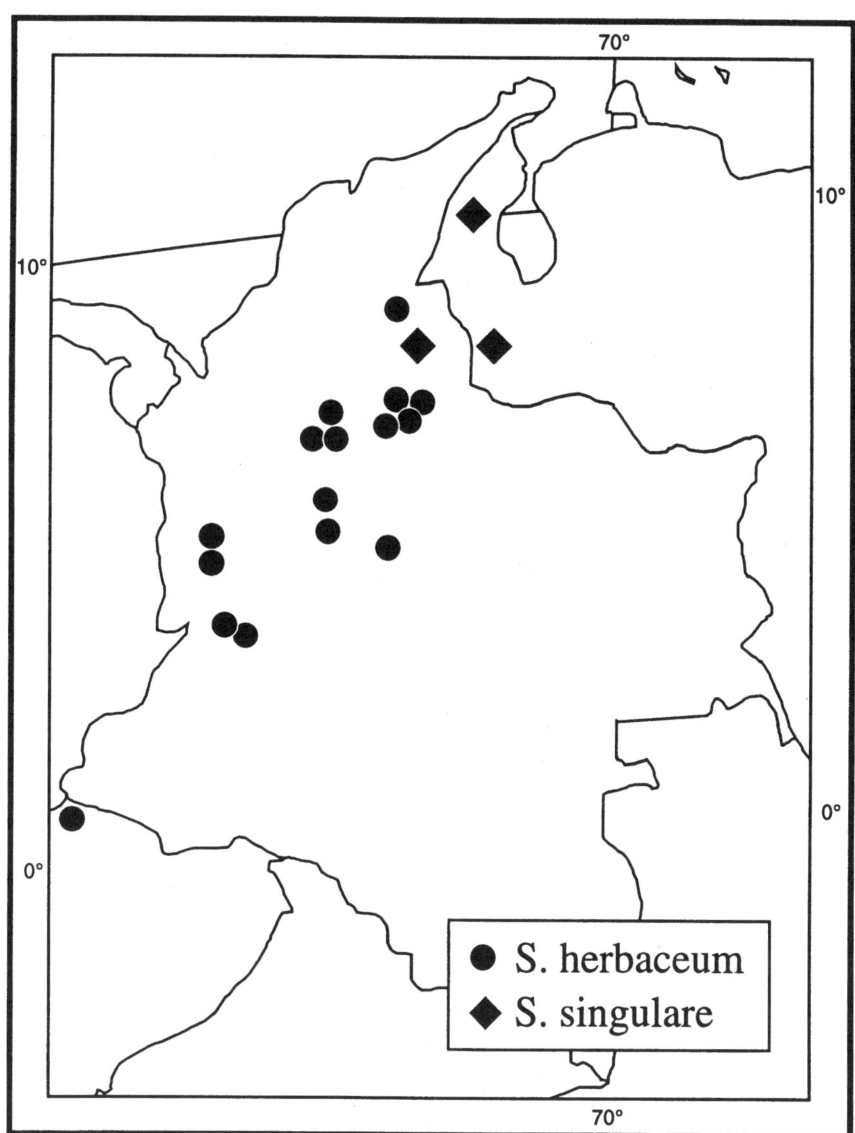

FIG. 19. Distribution of *Stigmaphyllon herbaceum* and *S. singulare*.

REPRESENTATIVE SPECIMENS. **Colombia.** ANTIOQUIA: Mpio. Remedios, Corregimiento El Cerro, Cerra Cabeza, 52 km SO de Zaragoza, 07°08′N, 74°50′W, *Callejas et al. 4690* (MICH); Mpio. San Luis, Cañón del Río Claro, parque ecológico, *Cogollo 710* (MICH); vic. Planta Providencia, 26 km S and 23 km W (air) of Zaragoza, 07°13′N, 75°03′W, *Denslow 2217* (MO, WIS); El Río, Segovia, *Rentería 2330* (NY); selvas de Casabe, *Uribe U. 5885* (COL, US).—BOLÍVAR: Monte Líbano a San Pedro, *Romero Castaneda 1760* (COL).—BOYACÁ: El Humbo, N of Bogotá, *Lawrance 777* (A, F, G, K, MO, US).—CALDAS: Río Samana, *Barkley & Gutiérrez 35365* (US).—CAUCA: "costa del Pacífico, bajo río Micay" [fide Cuatrecasas, 1958], *Lehmann 9059* (K, NY).—CHOCÓ: Mpio. Quibdo, Corregimiento Tutunendo, carretera a Ichó, *Arias 57* (MO); Mpio. Tadó, Corregimiento Play del Oro, en la carretera a Santa Cecilia, *Espina & Arias 1182* (MICH, MO).—SANTANDER: 8–10 km N of Barrancabermeja, on rd to Puerto Wilches, *Gentry 15377* (MICH, MO); Carare Opón, Las Colonias,

Rentería et al. 1546 (MO); Puerto Sogamoso, *Rentería et al. 2008* (MO).—VALLE: hoya del Río Anchicayá, entre Sabaletas y la Quebrada del Tátabro, *Cuatrecasas 22032* (F); Río Cali, Pichindé, *Duque Jaramillo 4044A* (MICH). **Ecuador.** ESMERALDAS. San Lorenzo, new rd to Proyecto NO, Km 8–10, *Sparre 18227* (MICH, S).

Stigmaphyllon herbaceum is one of only five species in the genus in which the anterior style exceeds the posterior ones and the anterior stamen is the largest (for a comparison of these five species, see *S. puberum*, no. 17). It is distinguished by its sparsely sericeous to glabrate laminas borne on relatively short petioles (less than 5 cm long). Of the five species, *S. orientale* (no. 16), known only from Depto. Meta in Colombia, also has glabrous to glabrate leaves, but is readily separated by its homogeneous androecium. The petals of *S. hypargyreum* (no. 19) are erose, and the anthers of the stamens opposing the posterior-lateral sepals are only 0.3–0.5 mm long and lack an enlarged connective. The dorsal wing of the samara 0.5–0.6 mm wide at the nut and flaring to 1–1.8 cm at its widest, similar to that of the samara of *S. hypargyreum*.

21. Stigmaphyllon carautae C. Anderson, sp. nov.—TYPE: BRAZIL. Rio de Janeiro: Bom Jesus de Itabapoana, Fazenda do Dr. Colombino, dentro da floresta, 8 Jun 1982, *Carauta 4309* (holotype: GUA!; isotype: MICH!).

Liana. Laminae 5–7-pinnato-lobatae, margine fimbriato-glanduloso et interdum glandis sessilibus munito; laminae juniores subtus pilos sessiles et pilos T-formes brevissime stipitatos ferentes, vetustiores irregulariter glabrescentes; laminae parvae infloresentiae ciliatae. Inflorescentia dichasialis constata ex umbellis, floribus in quaque umbella ca. 15–20. Pedunculi 10–13 mm longi; pedicelli 6.5–11 mm longi. Petala limbo orbiculari, margine lacero-fimbriato. Stamina heteromorpha, antheris glabris; antherae sepalis anticolateralibus oppositae 2 loculis reductis instructae. Stylus anticus ca. 3.3 mm longi, apice ca. 0.8–0.9 mm longo; styli postici ca. 3.5 mm longi, apice ca. 0.8 mm longo; styli omnes adaxaliter pubescentes.

Vine. Stems and branches sericeous when young, soon becoming glabrate. Laminas 10–13 cm long and wide, pinnately 5–7-lobed, apex of each lobe with a filiform gland (often broken off in older leaves), base auriculate, glabrate adaxially, pubescent with a mixture of sessile to subsessile to short-stalked T-shaped hairs abaxially when young (trabecula 0.4–1 mm long, straight or wavy, stalk up to 0.1 mm long), the hairs sloughed off in patches and older laminas becoming glabrate abaxially, margin shallowly and irregularly dentate, each tooth ending in a filiform gland (up to 5 mm long) or sometimes a stalked capitate glands (0.2–0.3 mm in diameter, ca. 0.2–0.5 mm long) or sometimes the margin also with sessile glands (ca. 0.4 mm in diameter), the laminas of the reduced leaves associated with the inflorescence fringed with filiform glands, petioles of larger leaves 6–7 cm long, pubescent with scalelike T-shaped hairs, not confluent across the node, with a pair of shallowly cupulate glands at the apex, each gland 1.5–2 mm in diameter; stipules 0.8–1.1 mm long, 0.5–0.7 mm wide, free, triangular, eglandular. Flowers ca. 15–20 per umbel, these borne in dichasia (axes to the 2nd order, with scalelike T-shaped hairs) or perhaps also in compound dichasia. Peduncles 10–13 mm long; pedicels 6.5–11 mm long, terete; both densely pubescent with scalelike T-shaped hairs, peduncles 0.6–1 times as long as the pedicels. Bracts 1.8–2.2 mm long, 1.5–2 mm wide, triangular, apex acute; bracteoles 2.5–3 mm long, 2–2.2 mm wide, oblong, apex obtuse, eglandular; bracts and bracteoles with scalelike T-shaped hairs abaxially. Sepals ca. 3.5 mm long, 2–2.3 mm wide, glands 1.8–2 mm long, 1–1.2 mm wide. All petals with the limb orbicular, glabrous, yellow, margin lacerate-fimbriate, fimbriae up to 1 mm long; anterior-lateral petals: claw 2.8–3 mm long, limb ca. 16 (–17) mm long and wide; posterior-lateral petals: claw 2–2.5 mm long, limb ca. 16 mm long and wide; posterior petal: claw ca. 4 mm long, apex indented, limb 14–15 mm long and wide. Stamens unequal, those opposite the posterior styles the largest, those opposite the anterior-lateral sepals with the connective enlarged and the locules reduced; anthers all loculate, glabrous. Stamen opposite anterior sepal: filament ca. 2.8 mm long, anther ca. 1.7 mm long; stamens opposite anterior-lateral petals: filaments ca. 2.2 mm long, anthers ca. 1.5 mm long; stamens opposite anterior-lateral sepals: filaments ca. 3.5 mm long, connective ca. 2 mm long, locules ca. 0.4 mm long; stamens opposite posterior-lateral petals: filaments ca. 3 mm long, anthers ca. 1.6 mm long; stamens opposite posterior-lateral sepals: filaments 1.8–2 mm long, anthers 0.7–0.8 long; stamen opposite posterior petal as long as or slightly longer than the adjacent two: filament 1.8–2 mm long, anther 0.7–0.8 mm long. Anterior style ca. 3.3 mm long, slightly shorter than the posterior two, laterally compressed, the proximal 1/3–1/2 adaxially pubescent, erect; apex 0.8–0.9 mm long, distally blunt, folioles absent. Posterior styles ca. 3.5 mm long, laterally compressed, the proximal 1/3–1/2 adaxially pubescent, lyrate; apex ca. 0.8 mm long, distally blunt, folioles absent. Samara not seen. Chromosome number unknown. Fig. 20.

Stigmaphyllon carautae is known only from the type, a flowering collection from woodlands of northernmost Rio de Janeiro, Brazil (Fig. 9). It is most similar to *S. glabrum* (no. 8) and *S. urenifolium* (no. 7) with which it shares pinnately lobed leaves, showy flow-

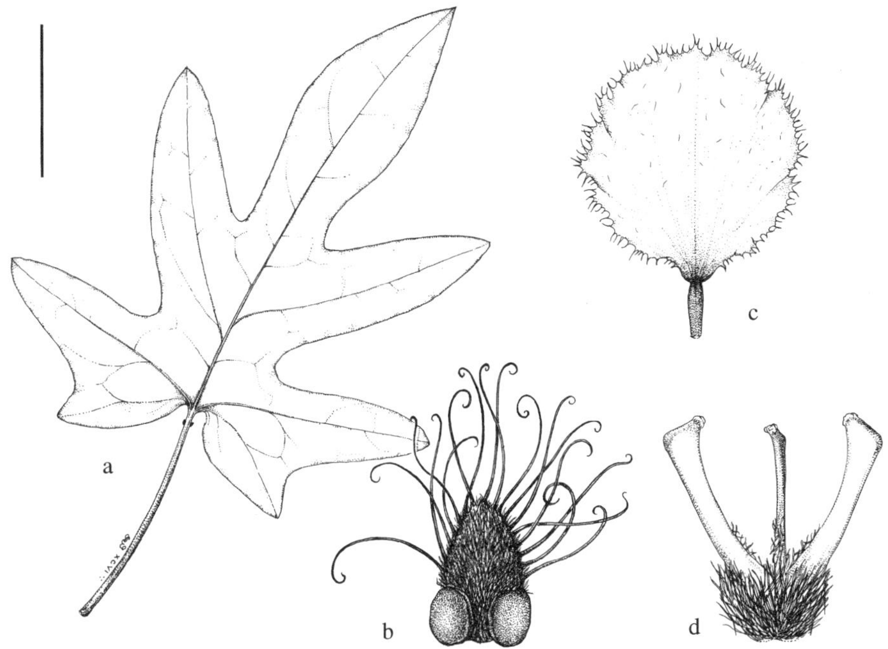

FIG. 20. *Stigmaphyllon carautae*. a. Large leaf. b. Abaxial view of a greatly reduced leaf from a pair subtending an umbel. c. Posterior petal (the "flag"). d. Gynoecium; posterior styles are bent slightly outward to show anterior style (in center). Scale: for a, bar = 4 cm; for b, bar = 4 mm; for c, bar = 1 cm; for d, bar = 2.7 mm.

ers, and efoliolate styles. These three species are readily separated by their leaf vesture. The laminas of *S. carautae* are abaxially pubescent with a mixture of sessile to short-stalked T-shaped hairs; the vesture is appressed and sloughed off in patches, so that larger and older leaves are abaxially glabrescent. The laminas of *S. glabrum* are glabrous, and those of *S. urenifolium* are abaxially persistently pubescent with spreading T-shaped hairs. Also, in *S. carautae* the small leaves subtending an umbel are evenly and abundantly fringed with filiform glands; in the other two species these leaves lack such ornamentation. In *S. glabrum* and *S. urenifolium* the stamens vary only in size; the connective of stamens opposing the anterior-lateral sepals is somewhat enlarged, but the locules are equally long. In *S. carautae*, the connective of stamens opposing the anterior-lateral sepals is greatly enlarged (ca. 2 mm long) and the locules are reduced (ca. 0.4 mm long).

This species is named for the Brazilian botanist J. P. P. Carauta, a specialist in the study of Moraceae, who collected the type.

22. Stigmaphyllon velutinum Triana & Planchon, Ann. Sci. Soc. Nat. Bot., sér. 4, 18: 320. 1862. *Stigmaphyllon incanum* Niedenzu, Ind. Lect. Lyc. Brunsberg. p. aest. 1900: 5. 1900, nom. superfl.—TYPE: COLOMBIA. Norte de Santander: Ocaña, *Schlim 77* (holotype: P!, photo: MICH!; isotypes: BM! BR! F-fragment! G! K!, photos of G isotype: F! GH! MICH!, photo of K isotype: MICH!).

Vine. Stems and branches densely velutinous when young, soon becoming sparsely so. Laminas 6.8–13 cm long, 4.5–8 cm wide, elliptical to ovate to sometimes suborbicular, apex acute to mucronate, base cordate to auriculate, velutinous adaxially, with T-shaped hairs abaxially (trabecula 0.6–1.3 mm long, wavy, stalk 0.1–0.3 mm long), margin with irregularly spaced sessile glands [0.1–0.4 (–0.6) mm in diameter]; petioles 2.2–5 cm long, densely velutinous, not confluent across the node, with a pair of prominent but sessile glands at the apex, each gland 0.6–1.5 mm in diameter; stipules 0.3–0.7 mm long, 0.4–0.8 mm wide, free, triangular, eglandular, obscured by the stem vesture. Flowers ca. 15–60 per pseudoraceme, these borne in small thyrses (axes to the 3rd order, velutinous and pubescent with T-shaped hairs), the pseudoracemes sessile or borne on secondary peduncles up to 11 mm long. Peduncles 5–14 mm long; pedicels 4–7 mm long, terete; both densely sericeous (hairs subsessile), peduncles 0.9–2.3 times as long as the pedicels. Bracts 2–3.5 mm long, 0.5–0.8 mm wide, linear-triangular, apex acute; bracteoles 1.8–2.5 mm long, ca. 0.5 mm wide, linear, apex acute, each bracteole with a pair of inconspicuous to prominent glands (each gland 0.1–0.5 mm in diameter, hidden by the vesture); bracts and bracteoles densely sericeous (hairs subsessile) abaxially. Sepals 2.5–2.7 mm long, 2–2.5 mm wide, glands 1.8–2 mm long, 0.9–1.1 mm wide. All petals with the limb orbicular, glabrous, yellow and often marked or suffused with red, margin digitate-fimbriate and glandular; anterior-lateral petals: claw 2.4–2.5 mm long, limb ca. 10 mm long and wide; posterior-lateral petals: claw 1.8–2.2 mm long, limb ca. 7 mm long and wide; posterior petal: claw ca. 3.5 mm long, apex not indented, limb ca. 6 mm long and wide. Stamens unequal, those opposite the anterior-lateral sepals with the longest filaments and with the connective enlarged and the locules reduced; anthers all loculate, glabrous. Stamen opposite anterior sepal: filament 2.9–3 mm long, anther 1–1.2 mm long; stamens opposite anterior-lateral petals: filaments ca. 2.1 mm long, anthers ca. 0.8 mm long; stamens opposite anterior-lateral sepals: filaments 3.4–3.5 mm long, connectives 1–1.1 mm long, locules 0.2–0.4 mm long; stamens opposite posterior-lateral petals: filaments 3–3.1 mm long, anthers ca. 1 mm long; stamens opposite posterior-lateral sepals: filaments 2.8–3.2

mm long, anthers ca. 0.6 mm long; stamen opposite posterior petal always shorter than the adjacent two: filament ca. 2.4 mm long, anther ca. 0.8 mm long. Anterior style 2.7–3.6 mm long, subequal to the posterior two, terete, glabrous, erect; apex 0.9–1 mm long including a spur ca. 0.2–0.3 mm long, folioles absent but on each side with a narrow linear-triangular lip, 0.1–0.2 mm wide. Posterior styles 2.8–3.7 mm long, terete, glabrous, lyrate; apex 1.1–1.2 mm long, incurved, folioles absent but distally with a narrow or narrowly triangular lip, ca. 0.2 mm wide. Mature samara not seen; immature samaras with the dorsal wing more than 3 cm long, the nut bearing 1–2 large lateral winglets and a second vertical row of ca. 8 spurs and tiny winglets. Chromosome number unknown.

Phenology. Collected in flower in July, August, and November, in young fruit in July.

Distribution (Fig. 21). Colombia (Antioquia, Cundinamarca, Norte de Santander, Quindío, Risaralda); 1000–1600 m.

ADDITIONAL SPECIMENS EXAMINED. **Colombia**. ANTIOQUIA: Salto de Guadeloupe, *Hodge 6947* (F, P, US).—CUNDINAMARCA: route de Guadas a Villeta, *Humbert et al. 27103* (COL).—QUINDÍO: Mpio. La Tebaida, Vrda. El Alambrado, carretera al Valle, ribera Río La Vieja, *Arbeláez et al. 1819* (MO).—RISARALDA: Hacienda Las Colinas, piedmont of W side of Cordillera Central near Río Cauca, ca. 04°51′N, 75°50′W, *Silverstone-Sopkin & Berrío B. 7689* (MICH).

Stigmaphyllon velutinum is an abundantly pubescent species, the only one with velutinous vesture on the stems, petioles, and adaxial surfaces of the laminas; the abaxial surfaces are densely pubescent with T-shaped hairs. The small flowers, clustered in pseudoracemes, are distinctive in their petals, androecium, and styles. The limbs of the petals are glandular-digitate-fimbriate. Of the stamens, only those opposite the anterior-lateral sepals have enlarged connectives and reduced locules. All styles lack folioles but bear a lateral narrow lip up to 0.2 m wide.

23. Stigmaphyllon tarapotense C. Anderson, Contr. Univ. Michigan Herb. 16: 49. 1987.—TYPE: PERU. San Martín: rd between Tarapoto and Juanjui, 21 km SE of Puente Colombia and 4.7 km SE of Juan Guerra, 30 Jun 1984, *Murray & Johnson 1530* (holotype: MICH!).

Vine to ca. 7 m. Stems and branches densely sericeous when young, soon becoming glabrate. Laminas 6.2–16 cm long, 6.5–18 cm wide, ovate to reniform or orbicular, apex mucronate, base cordate to subtruncate or in large leaves auriculate, glabrous adaxially, densely silvery pubescent with T-shaped hairs abaxially (trabecula 1–2 mm long, wavy, stalk 0.1–0.2 mm long), margin eglandular or occasionally with scattered sessile glands; petioles 2–11.5 cm long, sericeous, densely so in older leaves, not confluent across the node, with a pair of prominent but sessile glands at the apex, each gland 1.8–3.2 mm in diameter; stipules 0.4–0.9 mm long, 0.7–1.5 mm wide, free, triangular, eglandular. Flowers ca. 15–30 per umbel or condensed pseudoraceme, these borne in thyrses (axes to the 6th order, sericeous, occasionally with three axes at a node). Peduncles 5.5–9.5 mm long; pedicels 5.5–9 mm long, terete; both densely sericeous, peduncles ca. 0.75 times as long as or subequal or equal to pedicels. Bracts 1.4–2.8 mm long, triangular or narrowly so, apex acute, sericeous or sparsely so abaxially; bracteoles 1.5–2.2 mm long, triangular, apex acute, sericeous abaxially, eglandular. Sepals ca. 1.8–3.2 mm long, ca. 1.8–2.7 mm wide, glands 1.6 mm long, 0.8–1.2 mm wide. All petals with the limb orbicular, glabrous, yellow or (especially the limb of the posterior petal) suffused with red, margin of lateral

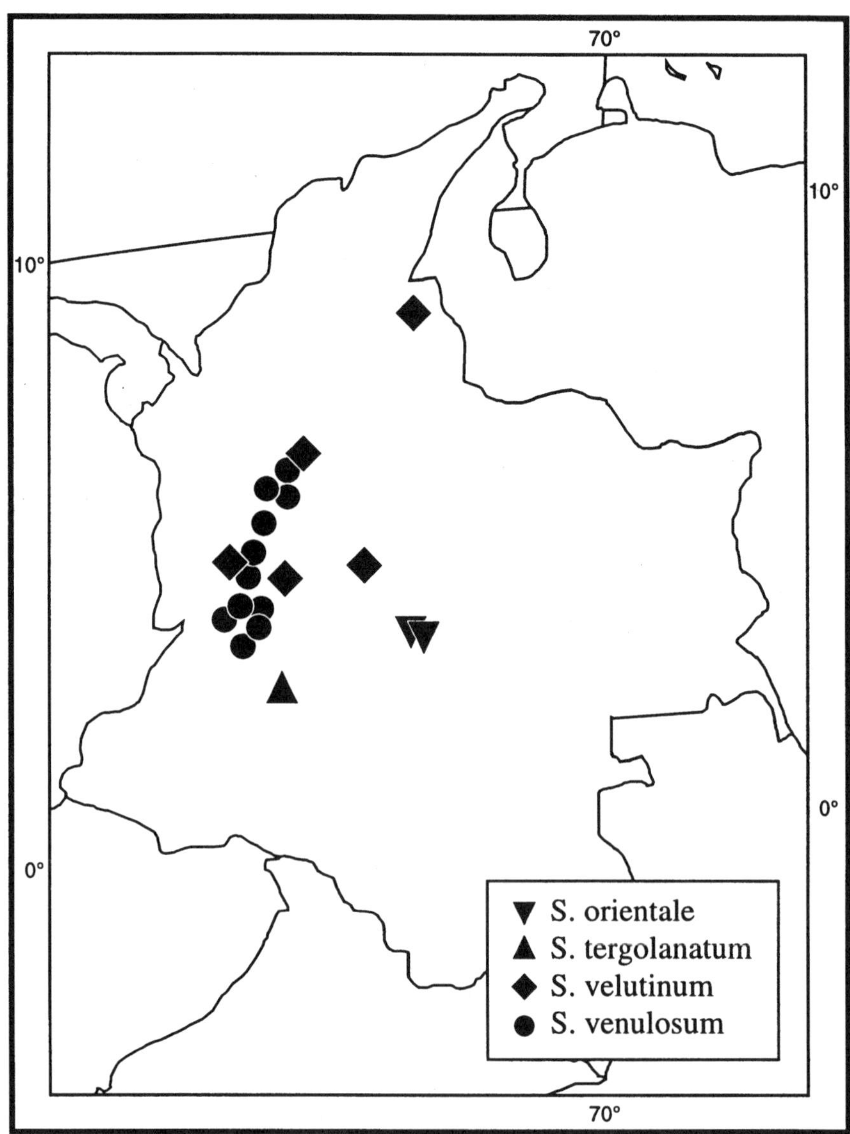

FIG. 21. Distribution of *Stigmaphyllon orientale, S. tergolanatum, S. velutinum,* and *S. venulosum.*

petals erose, anterior-lateral petals: claw 1.8–2 mm long, limb 10.5–11 mm long and wide; posterior-lateral petals: claw ca. 1 mm long, limb ca. 8.5 mm long and wide; posterior petal: claw 2.5–3 mm long, limb ca. 7–8 mm long and wide, margin erose-denticulate or erose-fimbriate, teeth and fimbriae up to 0.2 mm long. Stamens unequal, those opposite the posterior styles the largest, those opposite the anterior-lateral sepals (sometimes also those opposite the posterior-lateral sepals) with the connective enlarged and the locules somewhat reduced and unequal; anthers all loculate, pubescent. Stamen opposite anterior sepal: filament 3–3.3 mm long, anther 2–2.2 mm long; stamens opposite anterior-lateral petals: filaments (1.3–) 1.8–2 mm long, anthers (1.1–) 1.5–1.8 mm long; stamens opposite anterior-lateral sepals: filaments 3–3.5 mm long, connectives 1–1.3 mm long, locules 0.6–0.8 mm long; stamens opposite posterior-lateral petals: filaments 3.8–4 mm long, anthers 1.3–1.6 mm long; stamens opposite posterior-lateral sepals: filaments 2.2–2.5 mm long, anthers 0.9–1.2 mm long, sometimes with reduced locules (ca. 0.9 mm long); stamen opposite posterior petal always shorter than the adjacent two: filament ca. 2 mm long, anther 1–1.3 mm long. Anterior style 2.8–3.5 mm long, shorter than the posterior two, terete proximally, the distal 1/3 laterally flattened, glabrous or with a few scattered hairs near the base, erect; apex 1.1–1.3 mm long including a spur ca. 0.2 mm long, linear, folioles absent. Posterior styles 3.7–4.7 mm long, terete proximally, the distal 1/3–1/2 laterally flattened, glabrous, lyrate; apex 0.7–1.1 mm long, folioles variable, 0.2–0.8 mm long, 0.4–0.8 mm wide, parabolic, oblong, or ligulate, or reduced to a narrow lip. Dorsal wing of samara 3.3–4.5 cm long, 1.5–2 cm wide, upper margin with a blunt tooth; nut with 2–4 lateral winglets per side, usually in 2 rows, these 1.5–4 mm long, 0.3–8 mm wide, linear, triangular, oblong, or subrectangular; nut 5.6–7.5 mm high, 5–5.8 mm in diameter, without air chambers, areole ca. 3–3.5 mm long, ca. 3–4 mm wide, concave, carpophore up to 4 mm long; mature embryo not seen. Chromosome number unknown. Fig. 22.

Phenology. Collected in flower in June and July, in fruit in June and August.

Distribution (Fig. 10). Peru (San Martín); savanna, shrubland, roadside; 300–350 m.

ADDITIONAL SPECIMENS EXAMINED. **Peru.** SAN MARTÍN: Dtto. San Martín, valley of San Martín, 8 km E of Tarapoto, Fundo de San Isidro near Codo Creek, *Belshaw 3239* (F, GH, MICH, MO, NY, U, UTD, WIS); 6 km S of Tarapoto, on rd to Juanjui, *Gentry 37688* (MICH, MO).

Stigmaphyllon tarapotense is distinguished by its ovate to reniform or orbicular leaves, which are densely silvery pubescent abaxially. The androecium is heterogeneous, and all anthers are pubescent. The anterior style is efoliolate. The posterior styles bear variable folioles, which may differ in shape and size within the same umbel or even the same flower; sometimes the folioles are reduced to a narrow lateral lip. In Peru, the only other species with such densely silvery pubescent leaves is *S. maynense* (no. 18), but its vesture is appressed and not composed of T-shaped hairs; its peduncles are very short (1–5 mm long) compared to those of *S. tarapotense* (5.5–9.5 mm long). Also, in *S. maynense* the anterior style and its opposing stamen are longer than the posterior styles and their opposing stamens. All of its styles are foliolate, and the anthers are glabrous.

24. Stigmaphyllon columbicum Niedenzu, Verz. Vorles. Ak. Braunsberg W.-S. 1912–1913: 27. 1912.—TYPE: COLOMBIA. Tolima: Ambalema, *Triana s.n.* (lectotype, designated by Cuatrecasas, 1958: P!, photo: MICH!; isolectotype: G!).

Vine. Stems and branches with scalelike T-shaped hairs, becoming glabrate to glabrous in age. Laminas 5–15.5 cm long, 3.8–14.5 cm wide, cordate or narrowly so to triangular to narrowly ovate, rarely suborbicular, apex acuminate-mucronate, base cordate or sometimes truncate, glabrous adaxially, pubescent with T-shaped hairs abaxially (trabecula 0.4–1.5 mm long, straight to wavy, stalk 0.1–0.4 mm long), margin with irregularly spaced sessile glands (0.2–0.4 mm in diameter) and/or filiform glands up to 1.5 mm long, [in Costa Rica, margin with stipitate (nail-like) glands up to 0.6 mm long]; petioles 1.1–7 cm long, densely beset with scalelike T-shaped hairs, not confluent across the node, with a pair of prominent but sessile glands at the apex, each gland 1–2 mm in diameter; stipules 0.5–0.9 mm long, 0.4–1 mm wide, not fused, triangular to broadly so, glabrous,

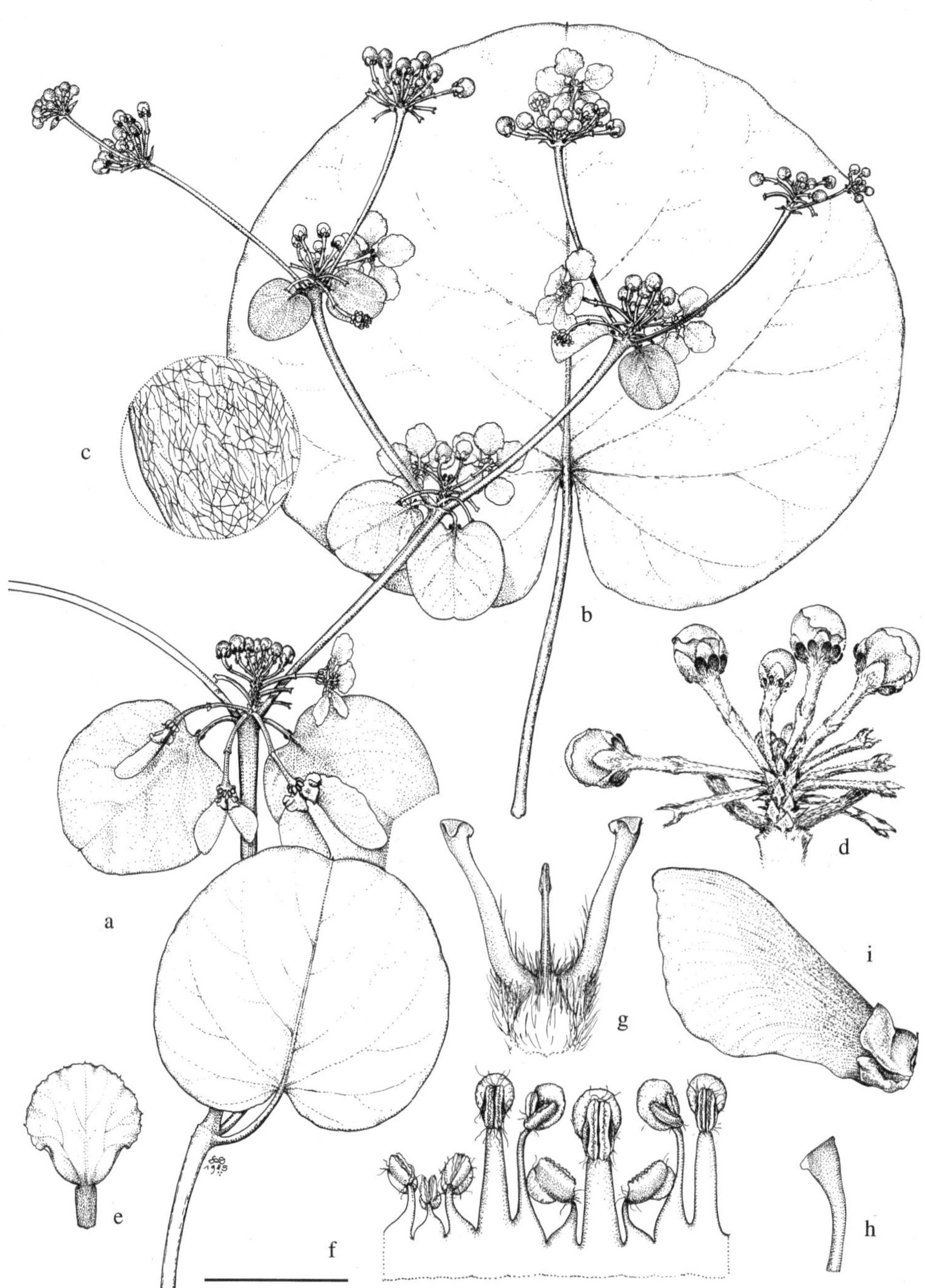

FIG. 22. *Stigmaphyllon tarapotense*. a. Flowering branch. b. Large leaf. c. Detail of margin and abaxial surface of lamina. d. Umbel. e. Posterior petal (the "flag"). f. Androecium; stamen second from left opposes the posterior petal. g. Gynoecium, anterior style in the center. h. Lateral view of anterior style. i. Samara. Scale: for a, b, i, bar = 4 cm; for c, f–h, bar = 4 mm; for d, bar = 1.3 cm; for e, bar = 1 cm. (Based on *Murray & Johnson 1530*.)

eglandular. Flowers (15–) 20–35 (–40) per congested pseudoraceme, these sometimes solitary but usually borne in dichasia, compound dichasia, or small thyrses (axes to the 3rd order, densely beset with scalelike T-shaped hairs). Peduncles (5.5–) 7.5–15 mm long; pedicels 3.5–8.5 mm long, terete; both densely sericeous, peduncles 1.2–3.3 times as long as pedicels. Bracts 1.2–2.8 mm long, 0.5–1.2 mm wide, narrowly triangular, apex acute to acuminate; bracteoles 0.8–1.8 mm long, 0.5–0.9 mm wide, oblong or triangular, apex obtuse, eglandular or each bracteole with two inconspicuous glands (each gland 0.2–0.4 mm in diameter); bracts and bracteoles sericeous abaxially. Sepals 2–2.5 mm long and wide, glands 1.5–2.3 mm long, 0.7–1 mm wide. All petals with the limb orbicular (or that of the posterior petal sometimes broadly obovate), glabrous, yellow, margin erose (in Ecuador with fimbriae up to 0.5 mm long); anterior-lateral petals: claw (1.5–) 1.7–2 (–2.5) mm long, limb 11–12.5 mm long and wide; posterior-lateral petals: claw 1.5–1.8 mm long, limb 9–10 mm long and wide; posterior petal: claw 2.5–3 mm long, apex indented, limb ca. 8–10 mm long and wide. Stamens unequal, those opposite the posterior styles the largest, those opposite the anterior-lateral sepals with the connective enlarged and the locules reduced; anthers all loculate, glabrous (in Ecuador pubescent). Stamen opposite anterior sepal: filament 2.2–3.2 mm long, anther 1–1.4 mm long; stamens opposite anterior-lateral petals: filaments 1.7–2.3 mm long, anthers 0.9–1.1 mm long; stamens opposite anterior-lateral sepals: filaments 3–3.6 mm long, connectives 0.9–1 mm long, locules 0.4–0.8 mm long; stamens opposite posterior-lateral petals: filaments 3.5–4 mm long, anthers 1–1.2 mm long; stamens opposite posterior-lateral sepals: filaments 2.3–3 mm long, anthers 0.7–1 mm long; stamen opposite posterior petal always shorter than the adjacent two: filament 1.8–2.5 mm long, anther 0.7–0.9 mm long. Anterior style 3–3.7 mm long, shorter than the posterior two, terete proximally, flattened distally, glabrous or rarely with a few scattered hairs in the basal 1/2, erect; apex 1.3–1.7 mm long including a spur (0.2–) 0.3–0.5 mm long, apex 0.2–0.3 mm wide, folioles absent. Posterior styles (3.7–) 4–4.6 mm long, terete, glabrous or rarely with a few scattered hairs in the basal 1/2, lyrate; folioles (1.6–) 2–2.6 mm long, (1.2–) 1.7–2.4 mm wide, subrectangular to subsquare. Dorsal wing of samara 3.1–4.6 cm long, 1–2.3 cm wide, upper margin without a tooth; nut with triangular to rectangular and entire to erose lateral winglets and often also spurs, in 1–3 rows per side, winglets 1.2–5 mm long, 1.5–2.7 mm wide, spurs up to 2.7 mm long; nut 4.5–8 mm high, 3–6.3 mm in diameter, woody, the locule surrounded by ca. 7–10 small air chambers, areole 2–3 mm long and wide, convex, carpophore up to 3 mm long. Embryo ca. 6.2 mm long, ca. 2 times as long as wide, ovoid, outer cotyledon ca. 9 mm long, ca. 3.5 mm wide, the distal 2/5 folded over the inner cotyledon, inner cotyledon ca. 4 mm long, ca. 3.2 mm wide, straight. Chromosome number unknown.

Phenology. Collected in Colombia in flower from October through March and from May through July, in fruit from November through March, in Costa Rica and Ecuador in flower in August.

Distribution (Fig. 23, 65). Northern and central Colombia, two collections from Costa Rica (San José), one from Ecuador (Napo); in roadside thickets and matorral, along rivers, at forest edge; 50–1700 m.

REPRESENTATIVE SPECIMENS. **Costa Rica.** SAN JOSÉ: ca. 15.4 km S of Puriscal and 0.3 km S of Salitrales off the road to Quepos, *Almeda et al. 3383* (CAS, F, MICH); ca. 2 km beyond Salitrales towards Parritas or ca. 16 km SE of Puriscal, *Wilbur et al. 23864* (DUKE, F, MICH).

Colombia. ANTIOQUIA: Mpio. Puerto Berrío, vereda Alicante, 1 km S of Finca Penjambo, 06°39′N, 74°33′W, *Callejas et al. 9311* (MICH); Mpio. Sonsón, región de Río Verde, camino hacia Santa Rosa, *Gutiérrez 35657* (F); Malena, *Pennell 3779* (F, GH, K, MO, NY, US); Segovia, *Sandeman 5574* (K, COL); Mpio. Nariño,

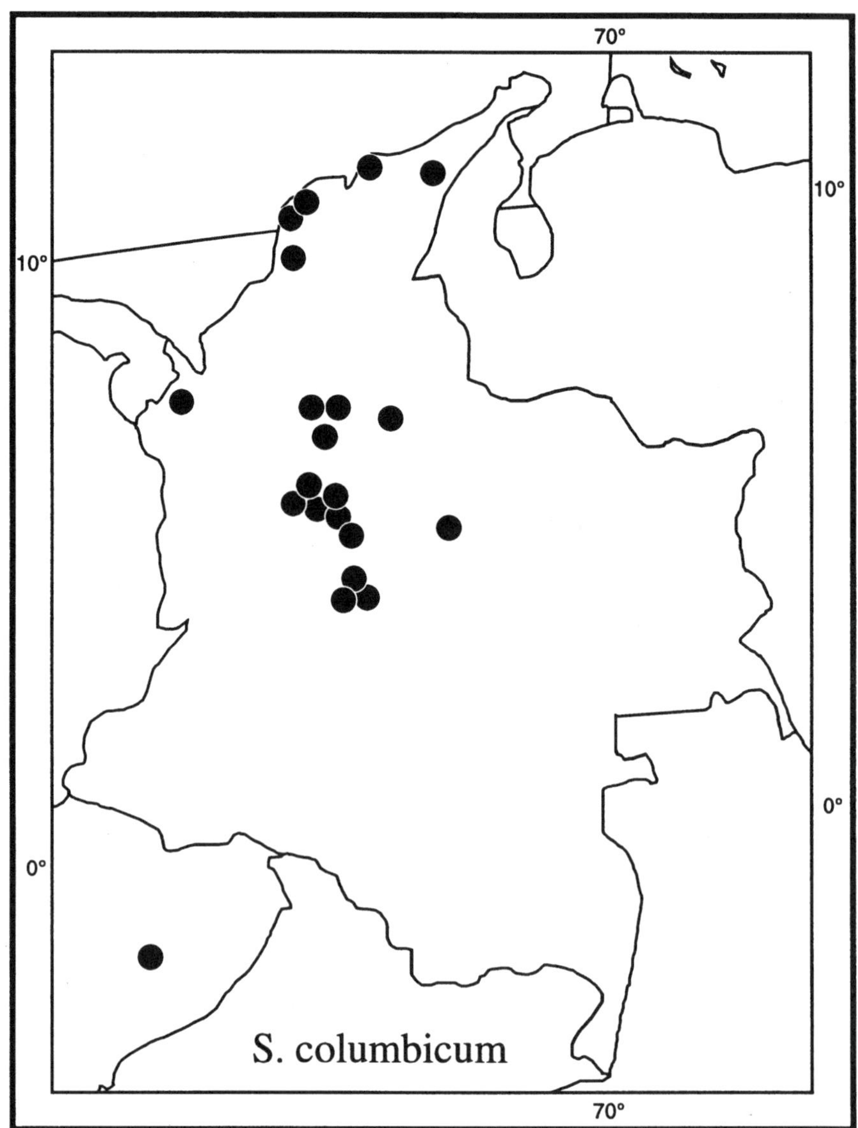

FIG. 23. Distribution of *Stigmaphyllon columbicum* in South America; see Fig. 65 for distribution in Costa Rica.

cerca al Río Samaná, *Uribe U. 1148* (COL, US).—BOLÍVAR: vicinity of Turbaco, *Bro. Heriberto 454* (GH, NY, US), *Killip & Smith 14424* (A, COL, GH, NY, US); N of Arjona, *Killip & Smith 14508* (A, F, NY, US).—BOYACÁ: El Umbo, N of Bogotá, *Lawrance 647* (A, F, G, K, MO, S, US).—CHOCÓ: Mpio. Ríosucio, Parque Natural Nac. Los Katyos, camino Peye–Tilupo, *León 451* (MO).—CUNDINAMARCA: alrededores del puente de San Antonio de Tena (s/el Río Bogotá), *Cuatrecasas 8283* (CM, COL, F, US); Caparrapí, *García B. H7703* (COL, US); Nariño, *Peréz Arbeláez 371* (COL, MA, US); Tocaima, *Peréz Arbeláez 2219* (COL, US); abajo de Santandercito, en Aguaclara, *Uribe U. 4034* (COL, NY).—MAGDALENA: Opía, supra Honda, *Holton 796* (G, K, NY); Mpio. Santa Marta, Don Jaca, *Romero Castañeda 10787* (COL).—SANTANDER: Puerto Nuevo, Río Magdalena, *Pennell 3859* (NY).—SUCRE: trail from Colosó to Reserva de Primatas, 09°30′N, 75°30′W, *Gentry*

34796A (MO).—TOLIMA: cerca al Puente Natural de Icononzo, *Uribe U. 1686* (COL). **Ecuador.** NAPO: carretera Puerto Napo hasta Tena, Km 5, *Dodson et al. 14963* (MICH, MO).

Stigmaphyllon columbicum is distinguished by its flowers, borne in pseudoracemes. The posterior styles bear large folioles, but the apex of the anterior style is efoliolate and extended into a spur/claw up to 0.5 mm long. The stamens opposite the posterior-lateral sepals bear unmodified but very small anthers; only the stamens opposite the anterior-lateral sepals have the connective enlarged and the locules reduced. Although all specimens from Colombia are relatively uniform, the specimens from Costa Rica and Ecuador differ in notable details. The leaves of the two collections (the same population?) from Costa Rica bear stipitate (nail-like) glands up to 0.6 mm long along the margins instead of the sessile and/or filiform glands seen on leaves of Colombian and Ecuadorian material. The single collection from Ecuador has fimbriate instead of erose petals and pubescent instead of glabrous anthers. Additional collections from Central America and Ecuador, especially in fruit, must be examined to determine whether the variants deserve taxonomic recognition.

Niedenzu cited two syntypes, both Triana collections, in the protologue, one from Ambalema (Tolima) and one from Anapoima (Cundinamarca). I overlooked that Cuatrecasas (1958) had lectotypified the name *Stigmaphyllon columbicum* with the collection from Ambalema, and needlessly designated the collection from Anapoima as the lectotype (C. Anderson 1987a).

25. Stigmaphyllon goudotii C. Anderson, Contr. Univ. Michigan Herb. 19: 424. 1993.—TYPE: COLOMBIA. Antioquia: vicinity of Medellín, Los Micos, 18 Dec 1927, *Toro 809* (holotype: NY!, photo: MICH!).

Vine. Stems and branches with scalelike T-shaped hairs when young, soon becoming glabrate. Laminas 8.2–13 cm long, 5.5–10.5 cm wide, ovate, apex acuminate, base truncate to cordate, glabrous adaxially, sparsely pubescent with T-shaped hairs abaxially (trabecula 1.4–1.7 mm long, wavy, stalk 0.1–0.3 mm long), margin eglandular or with irregularly spaced prominent glands (0.2 mm in diameter and long); petioles 3.2–6.5 cm long, with scalelike T-shaped hairs, not confluent across the node, with a pair of shallowly cupulate glands at the apex, each gland 1–1.5 mm in diameter, up to 0.3 mm high; stipules 0.7–0.8 mm long, 0.6–0.8 mm wide, free, triangular, eglandular. Flowers ca. 35–50 per pseudoraceme, these borne in dichasia or compound dichasia or small thyrses (axes to the 3rd order, with scalelike T-shaped hairs). Peduncles 15–21 mm long; pedicels 4–5.5 mm long, terete; both densely beset with scalelike T-shaped hairs, peduncles (1.5–) 3.4–4.2 times as long as the pedicels. Bracts 1.8–2.2 mm long, 0.8–1.1 mm wide, narrowly triangular, apex acute; bracteoles 1.6–2 mm long, 0.8–1 mm wide, triangular, apex obtuse, eglandular or with a glandular area in the basal 1/3–1/2; bracts and bracteoles with scalelike T-shaped hairs abaxially. Sepals ca. 2.2 mm long and wide, glands ca. 2 mm long, ca. 1.2 mm wide. All petals with the limb glabrous, yellow, limb of lateral petals orbicular, margin erose; anterior-lateral petals: claw ca. 1 mm long, limb ca. 12 mm long and wide; posterior-lateral petals: claw ca. 2.3 mm long, limb ca. 12 mm long and wide; posterior petal: claw ca. 3 mm long, apex indented, limb ca. 10 mm long, ca. 8 mm wide, obovate, margin erose or erose-denticulate. Stamens unequal, that those opposite the posterior styles the largest, anthers of those opposite the anterior-lateral sepals with the connective enlarged and the locules reduced; anthers all loculate, glabrous. Stamen opposite anterior sepal: filament ca. 2 mm long, anther ca. 1.5 mm long; stamens opposite anterior-lateral petals: filaments ca. 1.7 mm long, anthers ca. 1.2 mm long; stamens opposite anterior-lateral sepals: filaments ca. 3 mm long, connectives ca. 1.1 mm long, locules 0.7–0.9 mm long; stamens opposite posterior-lateral petals: filaments ca. 3.5 mm long, anthers ca. 1.2 mm long; stamens opposite posterior-lateral sepals: filaments ca. 2 mm long, anthers ca. 1 mm long; stamen opposite posterior petal always shorter than the adjacent two: filament ca. 1.7 mm long, anther ca. 1 mm long. Anterior style ca. 3.5 mm long, shorter than the posterior two, terete, glabrous, erect; apex 1.5 mm long, each foliole ca. 1.5 mm long, ca. 1.2 mm wide, elliptical to parabolic. Posterior styles ca. 4.2 mm long, terete, glabrous, lyrate; foliole ca. 2 mm long and wide, subsquare. Dorsal wing of samara 4.5–5.5 cm long, 1.9–2.2 cm wide, upper margin with a shallow blunt tooth or without a tooth; nut bearing 1–2 rectangular lateral winglets, these 2–5 mm long, 2–3 mm wide, and/or spurs (ca. 1 mm long) and crests; nut to 15 mm high, to ca. 10 mm in diameter, the thick-walled locule surrounded by air chambers, areole ca. 3 mm long and wide, concave, carpophore up to 8 mm long. Embryo ca. 8 mm long, ca. 2 times as long as wide, ovoid, outer cotyledon ca. 13 mm long, ca. 4 mm wide, the distal 2/5 folded over the inner cotyledon, inner cotyledon ca. 5.5 mm long, ca. 2.5 mm wide, straight. Chromosome number unknown. Fig. 24.

Phenology. Collected in flower in September and December, in fruit in February and November.

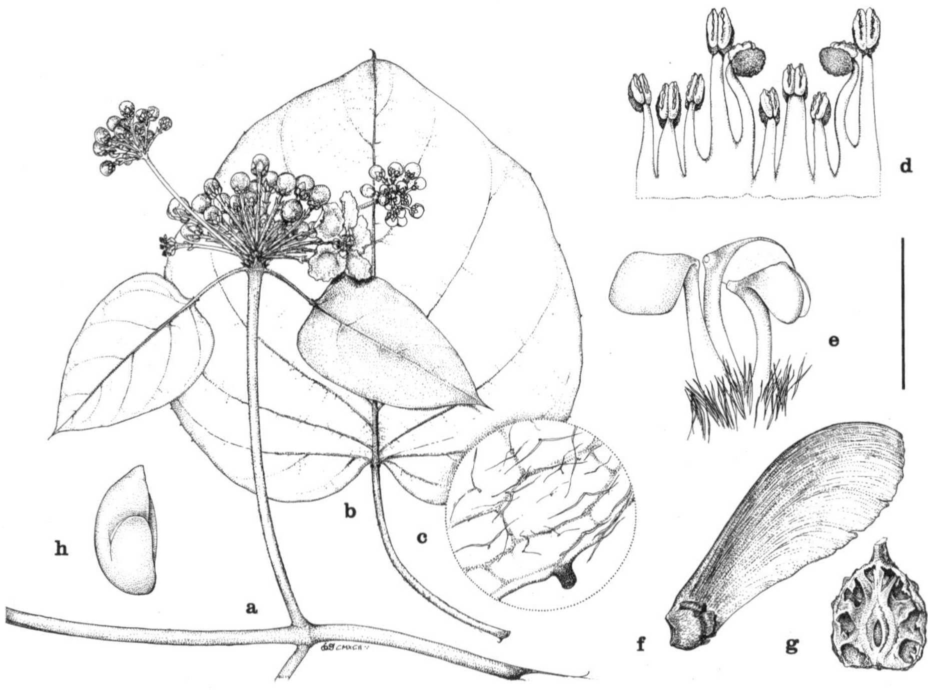

FIG. 24. *Stigmaphyllon goudotii*. a. Flowering branch. b. Large leaf. c. Detail of margin and abaxial surface of lamina. d. Androecium; stamen second from left opposes the posterior petal (the "flag"). e. Gynoecium, anterior style to the right. f. Samara. g. Longitudinal section through nut of samara; note the air chambers surrounding the locule. h. Embryo. Scale: for a, b, bar = 4 cm; for c, bar = 2 mm; for d, e, bar = 4 mm; for f, bar = 2.7 cm; for g, bar = 1.3 cm. (Based on: a, d, e, *Toro 809*; b, c, f, g, *Goudot s.n.*; h, *Holton 795*.)

Distribution (Fig. 5). Colombia (Antioquia, Tolima, Valle); in dry tropical forest and at roadsides; 1300–1350 m.

ADDITIONAL SPECIMENS EXAMINED. **Colombia.** ANTIOQUIA: Mpio. Giraldo, Corregimiento Manglares, 8 km de Manglares, via Santa Fé–Mutatá, 06°40′N, 75°55′W, *Callejas et al. 5608* (MICH); Mpio. Antioquia, Km 34 of rd Cañasgordas–Santa Fé de Antioquia, 06°37N, 75°55W, *Zarucchi et al. 5505* (MO).—TOLIMA: Ibagué, El Tejas, *Goudot s.n.* (P).—VALLE: La Paila, *Holton 795* (K, NY).

Stigmaphyllon goudotii is distinguished by its large compound inflorescences composed of 35–50-flowered pseudoracemes. The branches, petioles, inflorescence axes, peduncles, and pedicels are covered with scalelike T-shaped hairs. The peduncles are usually 3–4 times as long as the pedicels. Only the stamens opposite the anterior-lateral sepals have the connective enlarged and the locules somewhat reduced; all anthers are glabrous. The samara is unusual in that the thick-walled locule is surrounded by air chambers.

26. Stigmaphyllon singulare C. Anderson, Syst. Bot. 11: 126. 1986.—TYPE: VENEZUELA. Táchira: near La Fría, ca. 150 m, 22 Dec 1965, *Breteler 4923* (holotype: NY!, photo: MICH!; isotypes: K! MA! MER! MO! S! U! US! VEN! WAG!).

Vine to more than 12 m. Stems and branches sericeous. Laminas 6.5–15 cm long, 5.3–12.5 cm wide, cordate, apex acuminate, base cordate, glabrate with hairs restricted to veins adaxially, with T-shaped hairs abaxially (trabecula 1.1–1.5 mm long, wavy, stalk 0.1–0.3 mm long), margin with stipitate (nail-like) glands 0.2–0.3 mm long; petioles 2–6.5 cm long, densely sericeous, not confluent across the node, with a pair of prominent but sessile glands at the apex, each gland 1.6–2.1 mm in diameter; stipules 0.3–0.5 mm long, 0.4–0.6 mm wide, free, triangular, eglandular. Flowers (12–) 16–21 per umbel, these borne in small thyrses (axes to the 3rd order, sericeous) or sometimes in dichasia or solitary. Peduncles 5–7.5 mm long; pedicels 4.2–4.5 mm long, distally expanded; both densely sericeous, peduncles and pedicels subequal. Bracts 1.9–2.2 mm long, 1–1.3 mm wide, narrowly triangular; bracteoles 1.3–1.7 mm long, 1–1.3 mm wide, triangular, each bracteole with a pair of glands (each gland 0.5–0.6 mm in diameter); bracts and bracteoles with the apex acute, densely sericeous abaxially. Sepals *deciduous*, ca. 2 mm long and wide, glands 2.3–2.5 mm long, 0.8–1 mm wide. All petals with the limb orbicular or broadly obovate, abaxially sericeous at the center and base, yellow, margin erose; anterior-lateral petals: claw ca. 2 mm long, limb ca. 10 mm long and wide; posterior-lateral petals: claw ca. 1.8 mm long, limb ca. 9 mm long and wide; posterior petal: claw ca. 2.7 mm long, apex indented, limb ca. 8.5–9 mm long and wide. Stamens unequal, those opposite the posterior styles the largest, anthers of those opposite the anterior-lateral sepals with the connective enlarged and the locules (slightly) reduced; anthers all loculate, pubescent but those of stamens opposite the anterior-lateral sepals glabrous. Stamen opposite anterior sepal: filament ca. 3 mm long, anther ca. 1.5 mm long; stamens opposite anterior-lateral petals: filaments ca. 2.1 mm long, anthers ca. 1 mm long; stamens opposite anterior-lateral sepals: filaments ca. 3.3 mm long, connectives 0.8–1 mm long, locules ca. 0.7 mm long; stamens opposite posterior-lateral petals: filaments ca. 4 mm long, anthers ca. 1.5 mm long; stamens opposite posterior-lateral sepals: filaments ca. 2.4 mm long, anthers ca. 1.1 mm long; stamen opposite posterior petal always shorter than the adjacent two: filament ca. 1.8 mm long, anther ca. 1.1 mm long. Anterior style ca. 3.2 mm long, slightly shorter than the posterior two, terete proximally, laterally flattened distally, with scattered hairs in the proximal 1/4–1/3, erect, each foliole 0.9–1 mm long, 0.7–0.8 mm wide, elliptical to parabolic. Posterior styles ca. 3.5 mm long, terete, with scattered hairs in the proximal 1/3–1/2, lyrate; apex ca. 1.2 mm long, foliole ca. 1.5 mm long, ca. 0.9 mm wide, oblong. Dorsal wing of samara ca. 4 cm long, 1.4–1.6 cm wide, upper margin with a blunt tooth; nut bearing two vertical rows of lateral winglets, the larger 6.5–9 mm long, 2–2.5 mm wide, falcate, the smaller 1.6–6.5 mm long, 1–3 mm wide, falcate, rectangular, square, or parabolic; nut 5–5.5 mm high, 3.3–4 mm in diameter, without air chambers, areole ca. 2.7 mm long and wide, concave, carpophore up to 4 mm long; mature seeds not seen. Chromosome number unknown. Fig. 25.

Phenology. Collected in flower in December and February, in fruit in December.

Distribution (Fig. 19). Known only from northeastern Colombia and western Venezuela; in secondary forest (one report); ca. 150–240 (–1000) m.

ADDITIONAL SPECIMENS EXAMINED. **Colombia.** NORTE DE SANTANDER: Ocaña, *Schlim 251* (BM, BR, G, K, P). **Venezuela.** ZULIA: along Río Negro, W Machiques, at base of Sierra Perijá, *Bro. Ginés 69* (US).

Stigmaphyllon singulare differs from all other species of Malpighiaceae in its deciduous sepals and from most species of *Stigmaphyllon* by its pubescent petals. Each bracteole bears a pair of conspicuous glands 0.5–0.6 mm in diameter. All anthers are pubescent, and only those opposite the anterior-lateral sepals consist of an enlarged connective with reduced locules. The nut of the samara bears two winglets on each side. This species might be confused with the sympatric *S. dichotomum* (no. 65), which has glabrous petals, the anthers of all stamens opposing sepals modified, and, of course, persistent sepals; its bracteoles are eglandular or sometimes bear one or two inconspicuous glands up to 0.2 mm in diameter.

27. Stigmaphyllon adenophorum C. Anderson, Syst. Bot. 11: 120. 1986.—TYPE: COSTA RICA. Puntarenas: Telecommunication Hill above the town of Golfito, ca. 500 m, 16 Jul 1977, *Wilbur et al. 22761* (holotype: MICH!; isotype: DUKE!).

Vine. Stems and branches sericeous and with scalelike T-shaped hairs, becoming glabrous. Laminas 8.5–12 cm long, 4.4–7 cm wide, triangular to ovate, apex acuminate-aristate, base truncate or sometimes subattenuate, glabrate to glabrous adaxially, with T-shaped hairs abaxially (trabecula 0.5–0.9 mm long, wavy, stalk ca. 0.1 mm long), margin eglandular; petioles 2–4.3 cm long, not confluent across the node, with a pair of prominent but sessile glands at the apex (at the middle of the petiole in the smaller inflorescence leaves), each gland 1–1.6 mm in diameter; each stipule a prominent, circular gland, ca. 0.8 mm in diameter, with a minute membranous acute tip. Flowers 16–25 per umbel, borne in dichasia or small thyrses (axes to the 3rd order, sericeous and with scalelike T-shaped hairs). Peduncles 4–8 mm long; pedicels 5.2–7.5 mm long, terete; both sericeous and with scalelike T-shaped hairs, peduncles and pedicels subequal or equal. Bracts 1.2–1.6 mm long, 0.9–1.2 mm wide, triangular, apex obtuse; bracteoles 1.2–1.5 mm long, 0.9–1.1 mm wide, ovate, apex acute, each bearing two prominent glands, each gland 0.6–0.8 mm in diameter; bracts and bracteoles densely sericeous abaxially. Sepals ca. 2 mm long, ca. 1.8 mm wide, glands 2.4–2.8 mm long, 1.2–1.3 mm wide. All petals with the limb obovate, glabrous, yellow, margin erose; anterior-lateral petals: claw 2.2–2.5 mm long, limb ca.

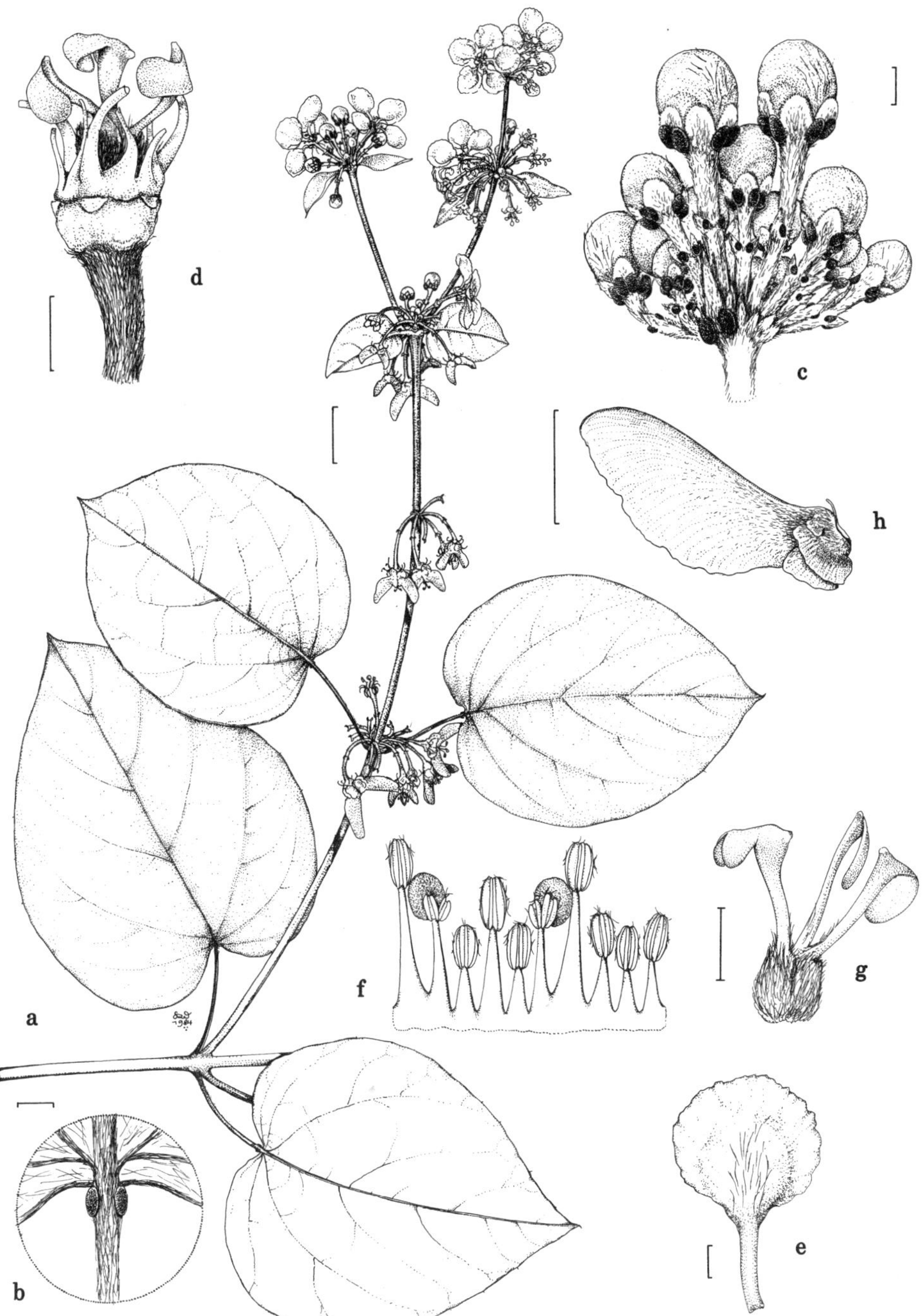

FIG. 25. *Stigmaphyllon singulare*. a. Branch with flowers and young fruits. b. Base of leaf, abaxial view. c. Young umbel. d. Old flower lacking sepals, petals, and anthers. e. Posterior petal (the "flag"); abaxial view, note the sparse pubescence on limb. f. Androecium; stamen second from right opposes the posterior petal. g. Gynoecium, anterior style to the right. h. Samara. Scale: for a, h, bar = 1.5 cm; for b–g, bar = 2 mm. (Based on: a–g, *Breteler 4923*; h, *Ginés 69*.)

11.5 mm long, ca. 10 mm wide; posterior-lateral petals: claw ca. 1.5 mm long, limb ca. 8–9 mm long, ca. 6–7 mm wide; posterior petal: claw 2.5–3 mm long, apex indented, limb ca. 7.5 mm long, ca. 6 mm wide. Stamens unequal, those opposite the posterior styles the largest, anthers of those opposite the anterior-lateral sepals with the connectives enlarged and the locules reduced; anthers all loculate, glabrous. Stamen opposite anterior sepal: filament ca. 2.5 mm long, anther ca. 1.5 mm long; stamens opposite anterior-lateral petals: filaments ca. 2 mm long, anthers ca. 1.3 mm long; stamens opposite anterior-lateral sepals: filaments ca. 3.5 mm long, connectives ca. 1 mm long, locules 0.4–0.5 mm long; stamens opposite posterior-lateral petals: filaments ca. 4.5 mm long, anthers ca. 1.3 mm long; stamens opposite posterior-lateral sepals: filaments ca. 3 mm long, anthers ca. 1–1.2 mm long; stamen opposite posterior petal shorter than the adjacent two: filament ca. 2.5 mm long, anther ca. 1.1 mm long. Anterior style ca. 3.7 mm long, shorter than the posterior two, terete, glabrous, erect or slightly recurved; apex 1.8 mm long, each foliole ca. 1.5 mm long, ca. 0.8 mm wide, oblong. Posterior styles ca. 4.2 mm long, terete, glabrous, lyrate; each foliole 2 mm long, 2.5 mm wide, oblate. Samara not seen. Chromosome number unknown. Fig. 26.

Stigmaphyllon adenophorum is known only from the type from Costa Rica (Fig. 65). It is readily recognized by its stipules, which consist of a large prominent gland with a tiny membranous apex, and by the bracteoles, each bearing a pair of prominent glands 0.6–0.8 mm in diameter.

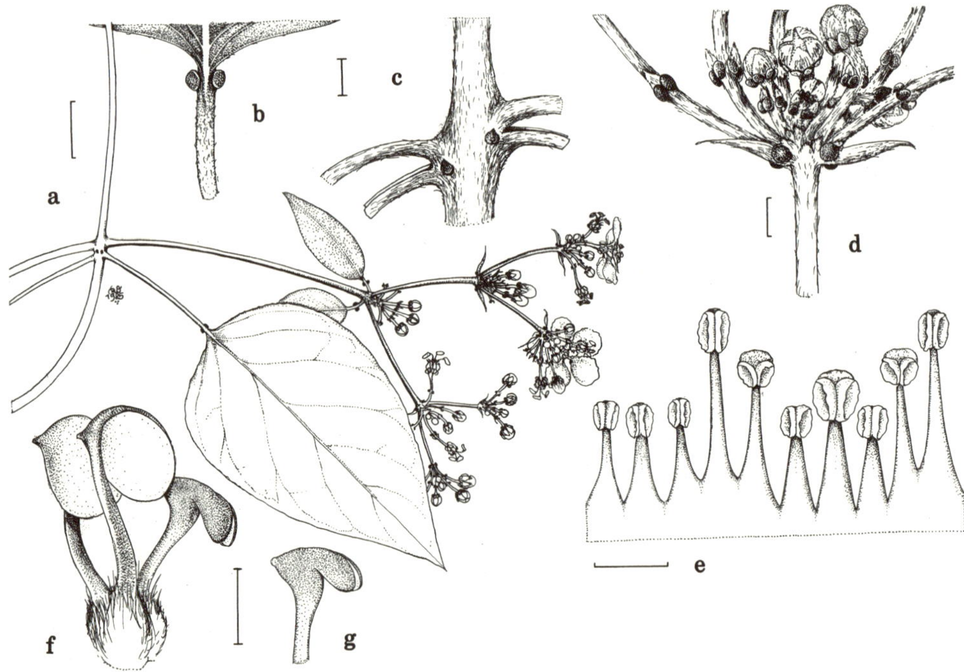

FIG. 26. *Stigmaphyllon adenophorum*. a. Flowering branch. b. Base of leaf, abaxial view. c. Portion of stem with stipules. d. Base of umbel. e. Androecium; stamen second from left opposes the posterior petal (the "flag"). f. Gynoecium, anterior style to the right. g. Distal portion of anterior style. Scale: for a, bar = 1.5 cm; for b–g, bar = 2 mm. (Based on *Wilbur et al. 22761*.)

28. Stigmaphyllon adenodon Adr. Jussieu, Ann. Sci. Nat. Bot., sér. 2, 13: 288. 1840.—TYPE: TRINIDAD. 1824, *de Schach s.n.* (holotype: K!, photo: MICH!).

Vine to 8 m. Stems and branches sericeous and also with T-shaped hairs, soon glabrate to glabrous. Laminas 5–14.5 cm long, 3.5–15 cm wide, cordate or ovate or narrowly so to triangular (especially the smaller laminas), apex mucronate or acuminate-mucronate, base cordate or in smaller leaves truncate, glabrate to glabrous adaxially, with T-shaped hairs abaxially (trabecula 0.4–1.1 mm long, straight, stalk 0.05–0.2 mm long), these often sparsely distributed, margin with stipitate (nail-like) glands 0.2–0.6 mm long and apically disklike to 0.5 mm in diameter; petioles 1.6–11.5 cm long, pubescent with T-shaped to subsessile hairs, sometimes sparsely so, not confluent across the node, with a pair of shallowly cupulate sessile glands at the apex (or to 3 mm below the base of the lamina), each gland 1–2.5 mm in diameter; stipules 0.5–1.3 mm long and wide, free, triangular, eglandular. Flowers 15–30 (–40) per umbel or pseudoraceme, these borne in dichasia or compound dichasia or small thyrses (axes to the 4th order, sericeous and also with T-shaped hairs). Peduncles 3.5–17.5 mm long; pedicels 3.5–8.5 mm long, terete; both densely sericeous and also with T-shaped hairs, peduncles (0.8–) 1–2.75 times as long as pedicels. Bracts 0.8–1.8 (–2) mm long, 0.6–1.4 (–1.8) mm wide, broadly triangular to suboblong, apex acute to subobtuse, sericeous or densely so abaxially; bracteoles 0.6–1.2 (–1.4) mm long, 0.5–1.1 (–1.3) mm wide, ovate to oblong, apex acute to obtuse, sericeous or sparsely so abaxially, eglandular or each bracteole with one or a pair of inconspicuous glands (each gland 0.2–0.4 mm in diameter). Sepals 1.8–2.5 mm long and wide, glands (1.5–) 1.9–2.7 mm long, 1–1.2 mm wide. All petals with the limb orbicular or sometimes broadly ovate, glabrous, yellow, margin erose or erose-denticulate, teeth up to 0.2 mm long; anterior-lateral petals: claw 2–2.5 (–3) mm long, limb 9.5–11 (–13) mm long and wide; posterior-lateral petals: claw 1.5–2.5 mm long, limb ca. 8–9 mm long and wide; posterior petal: claw (2.2–) 2.5–3 (–3.5) mm long, apex indented, limb 8–8.5 (–10) mm long and wide. Stamens unequal, those opposite the posterior styles the largest, anthers of those opposite the lateral sepals with the connective enlarged and the locules reduced or those opposite the posterior-lateral sepals with the locules equaling the connective or only slightly reduced; anthers all loculate, pubescent. Stamen opposite anterior sepal: filament (1.6–) 2.2–2.8 mm long, anther 1–1.4 mm long; stamens opposite anterior-lateral petals: filaments (1.2–) 1.5–2.1 mm long, anthers 0.8–1.2 mm long; stamens opposite anterior-lateral sepals: filaments 2.7–3.5 mm long, connectives 0.8–1.1 mm long, locules 0.4–0.7 mm long; stamens opposite posterior-lateral petals: filaments (2.8–) 3.4–4.3 mm long, anthers (0.8–) 1–1.2 mm long; stamens opposite posterior-lateral sepals: filaments 2–2.7 mm long, connectives 0.7–0.9 mm long, locules 0.5–0.7 mm long; stamen opposite posterior petal always shorter than the adjacent two: filament (1.6–) 1.8–2.2 mm long, anther 0.6–0.9 mm long. Anterior style 2.8–3.7 mm long, shorter than the posterior two, terete, glabrous, erect or slightly recurved; apex 1.8–2.4 mm long, each foliole 1.2–1.5 mm long, (0.5–) 0.8–1.1 (–1.3) mm wide, sometimes sublinear but more commonly narrowly oblong to oblong to parabolic (sometimes one or both folioles somewhat reduced). Posterior styles 3.5–4.7 mm long, terete, glabrous, lyrate; folioles (1.3–) 1.8–2.5 mm long, (1.3–) 1.5–2.1 mm wide, oblong to subsquare. Dorsal wing of samara reduced and erect, 3–4.4 cm high measured from base of nut, 1.1–2.1 cm wide, nut (10–) 17–21 mm high, 12–19 mm in diameter (var. *adenodon*), *or* dorsal wing not reduced but elongate, 5–6.5 cm long, 1.5–2.3 cm wide, nut (10–) 13–15 mm high, (9–) 10–12 mm in diameter (var. *macropterum)*; nut smooth or bearing shallow ridges or crests, these often interconnected, locule surrounded by air chambers, areole (3.5–) 5.5–7 mm long, (4.5–) 5.5–6.5 mm wide, concave, carpophore up to 3.5 mm long. Embryo (8–) 8.8–11 mm long, (4.1–) 4.6–5.5 mm wide, ca. 2 times as long as wide, ovoid, outer cotyledon (12.5–) 14.4–16.5 mm long, (4.1–) 4.7–5.5 mm wide, the distal 2/5–3/5 folded over the inner cotyledon, inner cotyledon (3.5–) 4.5–7.3 mm long, (2.5–) 3–5 mm wide, straight. Chromosome number unknown.

Stigmaphyllon adenodon, a species of wet habitats in the Amazon basin, also occurs in the Delta Amacuro and the Paria Peninsula of Venezuela and on the islands of Grenada, Trinidad, and Tobago. The species is named for the distinctive stipitate and apically flared (nail-like) marginal glands of the leaves. It is also characterized by pubescent anthers.

Throughout most of its range, *S. adenodon* is easily recognized by its distinctive samara. The nut, enlarged by air chambers surrounding the locule, contains a large embryo (to 10.5 mm long) and bears a reduced dorsal wing (Fig. 27g). Four collections from Ecuador and one from adjacent Peru differ from the common representatives of this species only in that the samara's wing is not reduced but elongate, like that of most other species. These differences in size and shape of the dorsal wing are here segregated as var. *adenodon* and var. *macropterum*. Unfortunately, *S. adenodon* has only rarely been collected in fruit in Ecuador, so that the extent of the variation in samara wing cannot be assessed. Because the flowering collections from Ecuador cannot be assigned to either variety with certainty, representative specimens from that country are listed here, following this discussion, rather than under either variety.

The collection from Peru with large-winged samaras (*Berlin 834*, GH, MO) here assigned to var. *macropterum* may represent, perhaps partly, a hybrid or possibly an apomictic population. It consists of fruiting branches as well as of separate inflorescence branches in bud; most likely these were collected from different individuals. The samaras examined contained normal-appearing embryos and the anterior styles still retained have normal-appearing folioles. In the buds examined, the pollen is composed of nearly 100% non-staining, thick-walled, shrunken, misshapen grains. The anterior styles do not bear large folioles; rather, both sides of the apex are variably only somewhat laterally expanded into a lip up to ca. 0.7 mm long and up to 0.5 mm wide. Yet, variability in the ornamentation of the anterior style is also seen in other collections, where the folioles may be somewhat unequal. This is most pronounced in two Peruvian collections [*Asplund 14639* (CAS, G, NY, S, US); *Vásquez & Jaramillo 11736* (MICH); both listed under var. *adenodon*], in which the apex of the anterior style of some flowers bears one large foliole and a lip or two unequal lips. In these specimens, the pollen appears to be normal.

Stigmaphyllon adenodon is most like *S. lacunosum* (no. 29), which has similar samaras (var. *adenodon*) and also pubescent anthers. The leaves of *S. lacunosum* also bear prominent to stipitate marginal glands but are sericeous abaxially instead of beset with T-shaped hairs. *Stigmaphyllon lacunosum* is reported from Amazonian Peru and the Trapecio Amazónico of Colombia. The samaras of the Colombian *S. goudotii* resemble those of *S. adenodon* var. *macropterum*, but that species lacks the distinctive marginal laminar glands and has glabrous anthers.

REPRESENTATIVE SPECIMENS. **Ecuador.** MORONA-SANTIAGO: Río Cuyes, 10 km al W del Río Zamora, 5 km al SW del Gualaquiza, 03°27′S, 78°36′E, *M. A. Baker 6840* (MICH); Sucua, Centro Shuar Yukutais, 8 km SW of Suca, 02°30′S, 78°08′W, *Gómez Andrade 509* (MICH); along Río Palora, 2–3 km upstream from Arapicos, *Lugo S. 6129* (GB, MICH); Cordillera de Cutucú, along trail from Logroño to Yaupi, ca. 02°46′S, 78°06′W, *Madison et al. 3635* (US).—NAPO: Archidona, *Asplund 9502* (G, R, S, US); Cantón Lago Agrio, Parroquia Dureno, 00°02′S, 76°42′W, *Cerón M. & Cerón 3110* (MICH); Las Sachas, rd Coca (Puerto Francisco de Orellana)–Lago Agrio, 30–40 km NE of Coca, *Lugo S. 3366* (GB, MICH); Zatzayacu, *Mexia 7114* (S, US, W).—PASTAZA: upper Río Bobonaza, between Canelos and Iluamayack, *Harling et al. 3339* (S); Mera, Isidro Ayora, *Harling et al. 19703* (GB, MICH); Río Putuime, vicinity of Puyopungu, *Lugo S. 5108* (GB, MICH); Río Conambo near Conambo, *McElroy 363* (QCA); Río Pastaza, riverbanks between outlets of Río Bobonaza and Río Ishpingo, ca. 02°34′S, 76°43′W, *Øllgaard et al. 34985* (AAU, MICH).

KEY TO THE VARIETIES OF STIGMAPHYLLON ADENODON

1. Samaras with a reduced erect wing 3–4.4 cm high measured from base of nut, 1.1–2.1 cm wide (Fig. 27g); Amazon Basin, the Paria peninsula of Venezuela, Trinidad, Tobago, Grenada. 28a. *S. adenodon* var. *adenodon*.
1. Samara with a large elongate dorsal wing 5–6.5 cm long, 1.5–2 (–2.5) cm wide; Amazonian Ecuador and adjacent Peru. 28b. *S. adenodon* var. *macropterum*.

28a. Stigmaphyllon adenodon var. adenodon.

Stigmaphyllon grenadense Niedenzu, Ind. Lect. Lyc. Brunsberg. p. aest. 1900: 26. 1900.—TYPE: TOBAGO. "In convalli fluminis Bacolit ad Cradley versus" (fide Niedenzu), *Eggers 5726* (lectotype, designated by C. Anderson, 1987a: K!; isolectotypes: A! M! P! S! US!).

Stigmaphyllon lalandianum var. *ciliolatum* Niedenzu, Pflanzenreich IV. 141(2): 488. 1928.—TYPE: BRAZIL. Rio de Janeiro: S. Sebastião bei Itabapoana (cultivated, according to label on K sheet), *Glaziou 9674* [excluding the duplicate at P] (holotype: B, destroyed; isotypes: C! K!).

Stigmaphyllon kuhlmannii Pilger, Repert. Spec. Nov. Regni Veg. 42: 178. 1937.—TYPE: BRAZIL. Amazônas: [Rio] Solimões, Yanache, 2 Mar 1924, *J. Kuhlmann 1550* (holotype: RB-26279!, photo: MICH!).

Dorsal wing of samara reduced and erect, 3–4.4 cm high measured from base of nut, 1.1–2.1 cm wide; nut smooth or bearing shallow ridges or crests, these often interconnected, nut 17–21 mm high, 12–19 mm in diameter, locule surrounded by air chambers, areole 5.5–7 mm long, 5.5–6.5 mm wide, concave, carpophore up to 3.5 mm long. Embryo 8.8–11 mm long, 4.6–5.5 mm wide, ca. 2 times as long as wide, ovoid, outer cotyledon 14.4–16.5 mm long, 4.7–5.5 mm wide, the distal 2/5 folded over the inner cotyledon, inner cotyledon 4.5–6.6 mm long, 3.3–5 mm wide, straight. Fig. 27a–i.

Phenology. Collected in flower and in fruit throughout the year.

Distribution (Fig. 28). Amazon Basin, disjunct to the Paria Peninsula of Venezuela, Trinidad, and Tobago, in the Lesser Antilles known only from Grenada; in wet areas, along rivers, and in rain forest and flooded forest; sea level to 1000 m.

REPRESENTATIVE SPECIMENS [flowering specimens from Ecuador listed after species discussion above]. **Grenada.** Tempé, *Broadway s.n.* (Feb 1905: BR; 18 Dec 1904: F, GH; Dec 1905: NY); Gran d'Etang, *Smith 109* (K). **Trinidad.** Guayaguayare, *Baker & Baker 14306* (K); Los Bajos, Erin, *Broadway 7356* (NY); Arima–Blanchisseuse Rd, *Kalloo 1002* (NY, TRIN); Nariva Swamp, along the Large Canal, *Ramcharan 113* (TRIN). **Tobago.** Bacolet, *Hunnewell 19934* (GH).

Venezuela. DELTA AMACURO: Depto. Tucupita, carretera Tucupita–La Horqueta, *Aymard & Stergios 925* (MO); Teima, *Ginés 4908* (US); between La Margarita and Puerto Miranda, Río Arcure, *Steyermark 87783* (MICH, NY).—SUCRE: vic. of Cristóbal Colón, *Broadway 630* (GH, NY, US). **Colombia.** AMAZÔNAS: Río Putomayo, carretera entre Caucayá (Puerto Leguizamo) y La Tagua, *Schultes 3776* (COL, F, GH); Trapecio Amazónico, Loretoyacu River, *Schultes 6302* (COL, GH, NY, US); Puerto Nariño and vicinity, Río Amazônas near mouth of Río Loretoyacu, *Zarucchi & Schultes 1058* (COL, GH, K, MICH).—PUTUMAYO: Río San Miguel en le afluente izquierda Quebrada de la Hormiga, *Cuatrecasas 11102* (COL, F, US); Umbría, 00°54′N, 76°10′W, *Klug 1740* (BM, F, GH, K, MICH, MO, NY, S, US). **Brazil.** AMAPÁ: Reserva Florestal do Governo do Amapá,

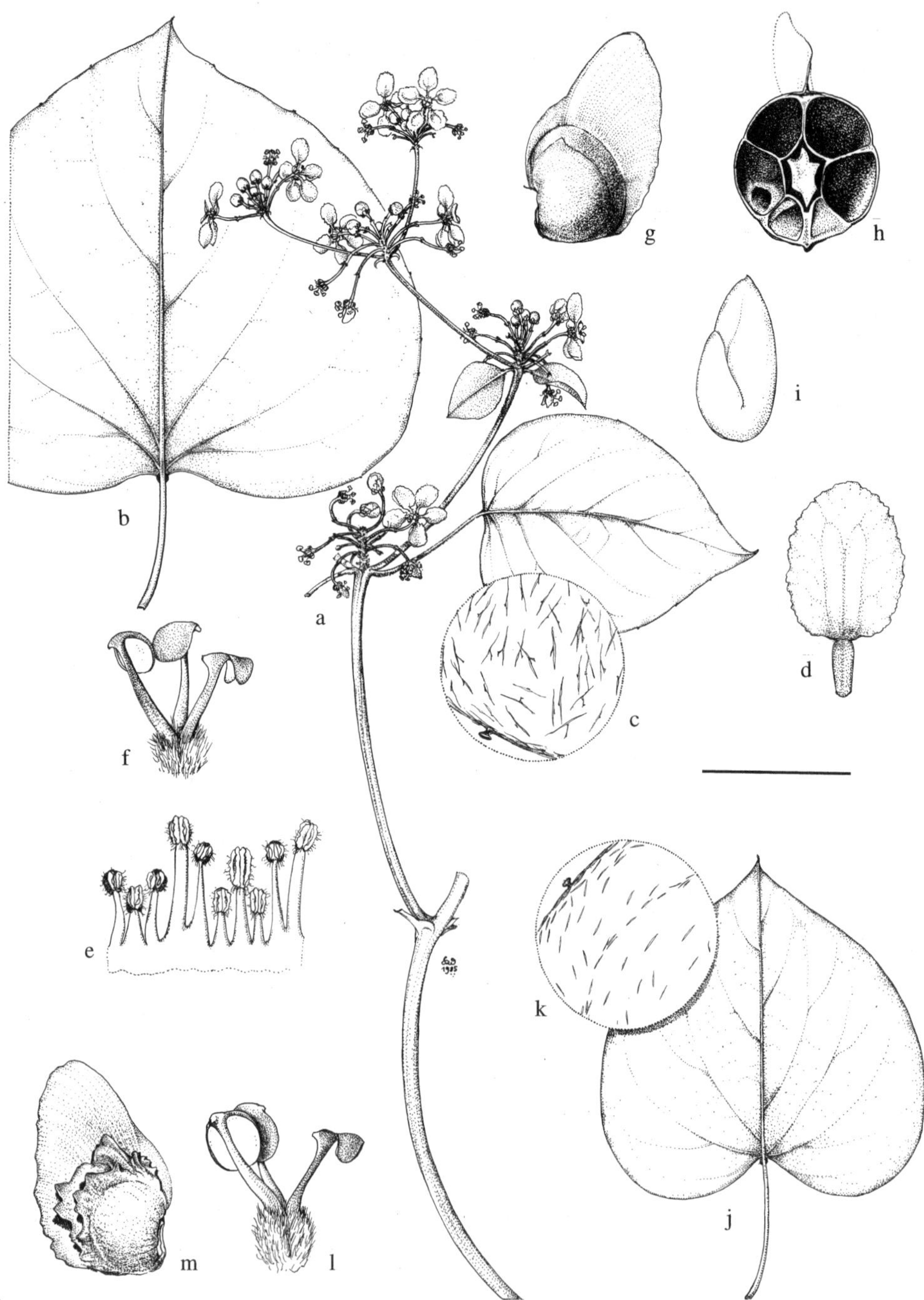

FIG. 27. *Stigmaphyllon adenodon* var. *adenodon* and *S. lacunosum*. a–i, *S. adenodon* var. *adenodon*. a. Flowering branch. b. Large leaf. c. Detail of margin and abaxial surface of lamina. d. Posterior petal (the "flag"). e. Androecium; stamen second from left opposes the posterior petal. f. Gynoecium, anterior style to the right. g. Samara. h. Longitudinal section through nut of samara; note the air chambers surrounding the locule. i. Embryo. j–m, *S. lacunosum*. j. Leaf. k. Detail of margin and abaxial surface of lamina. l. Gynoecium, anterior style to the right. m. Samara. Scale: for a, b, j, bar = 4 cm; for c, k, bar = 4 mm; for d, i, bar = 8 mm; for e, f, l, bar = 5.7 mm; for g, m, bar = 2 cm; for h, bar = 1 cm. (Based on: a, c, *Klug 1740*; b, *Mexia 6329*; d–f, *Wurdack 2239*; g–i, *Prance 24617*; j–l, *Schunke V. 49*; m, *Ayala et al. 2618*.)

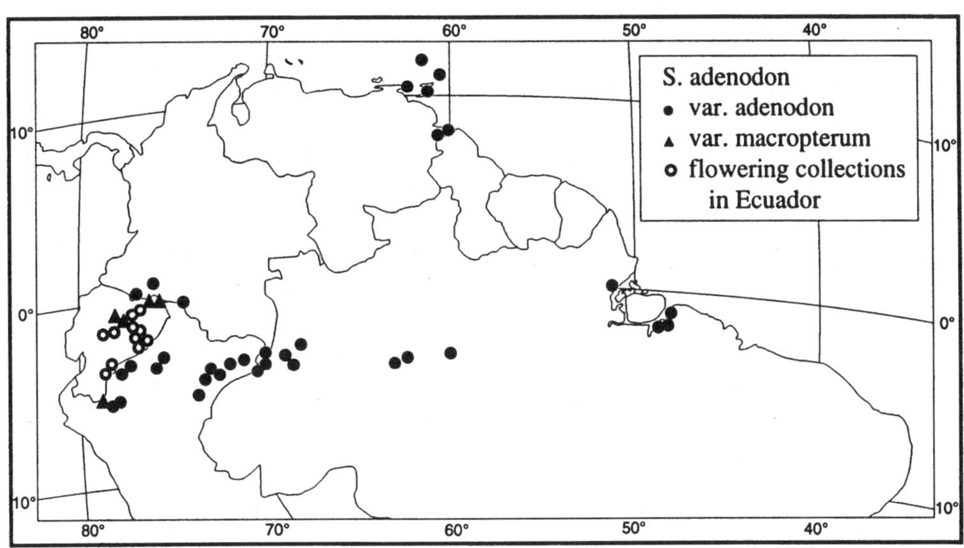

FIG. 28. Distribution of *Stigmaphyllon adenodon.*

entre Vila Amazônas e Fazendinho, *Austin et al. 7016* (IAN, MG, MICH); Rio Matapi, entre a estrada e a foz do rio, Macapá, *Rabelo et al. 1871* (MICH, NY); Rio Vila Nova–Macapá, *Rabelo et al. 2039* (MICH).—AMAZÔNAS: paraná Urariá between Rio Maués-Açu and Rio Apoquitaua, 03°25′S, 57°55′W, *Hill et al. 13180* (MICH); Putumayo, *Jobert 747* (P); Rio Solimões, near Codajás, Mpio. Codajás, *Krukoff 4501* (A, BM, F, G, K, M, MICH, MO, NY, S, U, US); Rio Solimões, Rio Jandiatuba, from mouth to 10 kms upstream, *Mori et al. 9146* (INPA, MICH); Rio Solimões, Igarape Camatía, Saõ Paulo de Olivença, *Prance et al. 24617* (MICH); Rio Solimões, Santo Antonio do Iça, boca do Anati paraná, *M. Silva 2059* (MG).—PARÁ: vicinity of Pará, *C. F. Baker 384* (BM, C, S, U); Rio Moju, between mouth and city of Moju, *Austin 4136 & Cavalcante 2260* (MG, MO); Benevides, estrada p. Mosqueiro, *M. Silva 715* (IAN, MG). **Peru.** AMAZONAS: 10 km from mouth of Quebrada Huampami, Río Cenepa, *Berlin 339* (MO); Río Cenepa, monte orilla de Huampami, *Kayap 163* (MO); Prov. Bagua, 3–5 km above mouth of Río Santiago, *Wurdack 2239* (F, GH, K, NY, P, S, US).—LORETO: Iquitos, above mouth of Río Itaya, *Asplund 14639* (CAS, G, NY, S, US); Dtto. Tigre, Río Corriente, Shiviyacu, camino a Forestales, *Ayala 2320* (AMAZ); Prov. Maynas, lower Río Itaya, near Iquitos, *Davidson & Jones 9835* (F, MICH); Prov. Maynas, Río Manití, *Encarnación 986* (MICH, MO); Prov. Requena, Cocha Iriachua, izquierda del Río Ucayali, abajo de Herrera, *Encarnación 1295* (MO, NY, US); Prov. Requena, Río Tapiche, ca. 1 hr by 40 hp motor above Requena, *Gentry et al. 21261* (AMAZ, MICH); Prov. Alto Amazonas, Andoas, Río Pastaza near Ecuador border, *Gentry & Díaz 28132* (MICH); Río Marañón Valley, *Killip & Smith 29155* (NY, US); Prov. Alto Amazonas, Río Huasaga, Washinta, *Lewis & Vásquez 4044* (NY); Prov. Loreto, Nueva Jerusalem vicinity, Río Macusaré, 02°55′S, 76°15′W, *Lewis et al. 10386* (MO); above Pongo de Manseriche, right bank of Río Santiago, *Mexia 6239* (BM, CAS, F, G, GB, GH, K, MICH, MO, NY, S, TEX, U, US, WIS); Prov. Maynas, Dtto. Iquitos, Río Itaya, *Revilla 1928* (AMAZ, MICH); Prov. Maynas, Río Yavari, Puerto Amelia cerca a Atalaya, boca del Río Atacauai, *Revilla 2166* (MICH); Stromgebiet des Marañón, Santiago-mündung am Pongo de Manseriche, ca. 77°30′W, *Tessmann 4661* (G, S); Prov. Maynas, Quebrada Sucusari, izquierda del Río Napo, Explor. Napo Camp, 03°20′S, 70°55′W, *Vásquez & Jaramillo 11736* (MICH); lower Río Nanay, *Ll. Williams 537* (F, US).

28b. Stigmaphyllon adenodon var. **macropterum** C. Anderson, Contr. Univ. Michigan Herb. 19: 415. 1993.—TYPE: ECUADOR. Napo: Baeza–Tena Rd, ca. 5 km N of Jondachi, ca. 1000 m, *Harling & Andersson 16398* (holotype: MICH!; isotype: GB!).

Dorsal wing of samara elongate (not reduced), 5–6.5 cm long, 1.5–2 (–2.5) cm wide; nut with narrow winglets, up to 11 mm long and 3.5 mm wide or bearing shallow ridges or crests, these often interconnected, nut (10–) 13–15 mm high, (9–) 10–12 mm in diameter, locule surrounded by air chambers, areole 3.5–5.5 mm long, 4.5–6 mm wide, concave, carpophore up to 3.5 mm long. Embryo 8–10 mm long, 4.1–5 mm wide, ca. 2 times as long as wide, ovoid, outer cotyledon 12.5–15.5 mm long, 4.1–5 mm wide, the distal 3/5 folded over the inner cotyledon, inner cotyledon 3.5–7.3 mm long, 2.5–3.5 mm wide, straight.

Phenology. Collected in flower in November and December, and in fruit in January, February, August, November, and December.

Distribution (Fig. 28). Known only from four collections from Ecuador (Napo, Pastaza) and one collection from adjacent Peru (Amazonas); in wet areas, along rivers, and in rain forest and moist forest; 220–1160 m.

REPRESENTATIVE SPECIMENS. **Ecuador.** NAPO: San Pablo Aguarico, Río Shushufindi, *Lescure 2165* (MICH, NY); Yasuní National Park, Río Indillama, a tributary of the Río Napo, at Comuna Pompeya, 00°30′S, 76°40′W, *Neill & Gudiño 10117* (MICH).—PASTAZA: Mera, *Harling 3779* (S). **Peru.** AMAZONAS: Río Cenepa above mouth of Río Huampami and prior to Chávez Valdivia, *Berlin 834* (GH, MO).

29. Stigmaphyllon lacunosum Adr. Jussieu, Ann. Sci. Nat. Bot., sér. 2, 13: 289. 1840.— TYPE: BRAZIL. Unknown collector (holotype: M!, photos: F! GH! MICH! NY! US!; isotype: P-JU!, photo: MICH!).

Vine to 15 m. Stems and branches sericeous when young, soon becoming glabrate to glabrous. Laminas 7–10 cm long, 6–9 cm wide, cordate, rarely 3-lobed, apex mucronate, base cordate, glabrous adaxially, sericeous abaxially (trabecula 0.3–0.6 mm long, straight or wavy, sessile), margin with raised to stipitate (nail-like) glands (0.3–0.5 mm in diameter, up to 0.5 mm long); petioles 3.2–11 cm long, sericeous, not confluent across the node, with a pair of shallowly cupulate sessile glands at the apex, each gland 1.3–2 mm in diameter; stipules 0.5–1.1 mm long and wide, free, triangular, eglandular. Flowers ca. 15–30 per umbel, these borne in compound dichasia or small thyrses (axes to the 4th order, sericeous). Peduncles 3–14.2 mm long; pedicels 3.5–6.5 mm long, terete; both sericeous, peduncles 0.9–2.5 times as long as the pedicels. Bracts 0.9–1.6 mm long, 0.7–1.1 mm wide, narrowly triangular, apex acute; bracteoles 0.7–1.3 mm long, 0.7–1.1 mm wide, ovate, apex obtuse, eglandular or each bracteole with a pair of inconspicuous glands (each gland 0.2–0.3 mm in diameter); bracts and bracteoles sericeous abaxially. Sepals 1.8–2 mm long and wide, glands 1.5–2 mm long, 0.8–1 mm wide. All petals with the limb orbicular, glabrous, yellow, margin erose; anterior-lateral petals: claw ca. 2 mm long, limb ca. 9 mm long and wide; posterior-lateral petals: claw ca. 1.5 mm long, limb ca. 8.5 mm long and wide; posterior petal: claw 2.5–3 mm long, apex indented, limb ca. 7.5 mm long and wide. Stamens unequal, those opposite the posterior styles with the longest filaments, anthers of those opposite the anterior-lateral sepals with the connective enlarged and the locules reduced; anthers all loculate, pubescent. Stamen opposite anterior sepal: filament 1.8–2 mm long, anther ca. 1.2 mm long; stamens opposite anterior-lateral petals: filaments 1.3–1.7 mm long, anthers 0.9–1.1 mm long; stamens opposite anterior-lateral sepals: filaments 2.2–3 mm long, connectives 0.8–0.9 mm long, locules 0.4–0.5 mm long; stamens opposite posterior-lateral petals: filaments 3–3.5 mm long, anthers 1–1.3 mm long; stamens opposite posterior-lateral sepals: filaments 2–2.5 mm long, anthers 0.6–0.8 mm long; stamen opposite posterior petal always shorter than the adjacent two: filament 1.8–2 mm long, anther ca. 0.8 mm long. Anterior style 2.5–3 mm long, shorter than the posterior two, terete, glabrous, erect; apex 1.3–1.8 mm long, each foliole 1–1.2 mm long, 0.4–0.8 mm wide, elliptical to narrowly so, or rarely the folioles reduced to a narrow lip ca. 0.2 mm wide, or the apex distally only rhomboidially expanded, 0.5 mm wide, and the folioles absent. Posterior styles 3.2–4 mm long, terete, glabrous, lyrate; foliole 1.6–1.8 mm long and wide, subsquare. Dorsal wing of samara encircling the nut, (2.5–) 3–4.5 cm high measured from base of nut, 2–2.3 cm wide; nut bearing 2–3 lunate grossly dentate lateral winglets per side, the larger winglet ca. 13.5 mm long, ca. 4.5 mm wide, the smaller ca. 9 mm long, ca. 3 mm wide; nut 16–20 mm high, 13–15 mm in diameter, locule surrounded by air chambers, areole ca. 7 mm long, ca. 8.5 mm wide, convex (?), carpophore absent (?). Mature seed not seen. Chromosome number unknown. Fig. 27j–m.

Phenology. Collected in flower from December through February, in fruit in December and February.

Distribution (Fig. 72). Amazonian Brazil, Colombia, and Peru; along riverbanks and in inundated forest; 100–150 m.

ADDITIONAL SPECIMENS EXAMINED. **Colombia.** AMAZONAS: Trapecio Amazónico, arriba de la desembocadura del Río Loretoyacu, *Duque-Jaramillo 2325* (COL); Trapecio Amazónico, between Río Amazonas and Río Putomayo watersheds, *Schultes 6882* (COL, US). **Brazil**. Without locality, *Martius s.n.* (M). **Peru**. LORETO: Dtto. Tigre, Río Tigre, caserio Viejo Tipishca a 6 horas de Paichi Playa, *Ayala et al. 2618* (MICH); Prov. Maynas, Dtto. Iquitos, Río Nanay, Quebrada de Momón, *Rimachi Y. 7384* (US); Gamitanacocha, Río Mazán, *Schunke 49* (A, F, MICH, NY, US); Prov. Maynas, Nauta, carretera a Iquitos, 04°29′S, 73°35′W, *Vásquez & Jaramillo 8624* (MICH).

Stigmaphyllon lacunosum is easily recognized by its samara with a reduced dorsal wing and enlarged nut in which the locule is surrounded by air chambers. The carpophore is apparently absent; however, the only collection with nearly mature fruits is a poor specimen, and future collections may perhaps show that a small carpophore is present, as in the samaras of the very similar *S. adenodon* var. *adenodon* (no. 28a). *Stigmaphyllon lacunosum* is also distinguished by its cordate leaves bearing stipitate (nail-like) glands along the margin and its pubescent anthers. In one collection (*Duque-Jaramillo 2325*, COL), one of the leaves is 3-lobed. *Stigmaphyllon lacunosum* and *S. adenodon* are readily separated by the pubescence on the lower leaf surface; the laminas of *S. lacunosum* are sericeous abaxially, whereas those of *S. adenodon* are sparsely to moderately beset with T-shaped hairs (the stalk up to 0.2 mm long).

30. Stigmaphyllon cuzcanum C. Anderson, Novon 2: 304. 1992.—TYPE: PERU. Cuzco: below Machu Picchu, 2100 m, *West 6466* (holotype: MO!, photo: MICH!; isotype: GH!).

Vine. Stems and branches densely sericeous and also with scalelike T-shaped hairs when young, soon becoming glabrate. Laminas 7–19 cm long, 5.5–12.7 cm wide, ovate, apex emarginate-mucronate, base cordate to deeply so in larger laminas and to truncate in smaller ones, glabrous adaxially, tomentose abaxially (trabecula 0.9–1.5 mm long, crisped and curled, stalk 0.1–0.3 mm long), margin eglandular or with irregularly spaced sessile glands (0.3–0.4 mm in diameter) and then sometimes shallowly denticulate; petioles 3–7.3 cm long, densely sericeous but glabrous in older leaves, not confluent across the node, with a pair of prominent but sessile glands at the apex, each gland 1.8–2.5 mm in diameter; stipules 1–1.5 mm long, 0.8–1.3 mm wide, free, triangular, eglandular. Flowers ca. 10–35 per umbel, these borne solitary or in small thyrses (axes to the 2nd order, densely sericeous). Peduncles 3.5–15.5 mm long; pedicels 7–13.5 mm long, terete; both densely sericeous; peduncles 0.6–1.4 times as long as the pedicels. Bracts 1.2–2.5 mm long, 0.8–1.3 mm wide, triangular, apex acute, usually only the distal 1/3 sericeous abaxially, sometimes entirely so; bracteoles 1.1–2.3 mm long, 0.7–1.4 mm wide, narrowly triangular, apex obtuse, sericeous abaxially, eglandular or each bracteole with a pair of glands (each 0.4–0.5 mm in diameter). Sepals 2.5–3.5 mm long, 2.5–3.2 mm wide, glands 2–2.5 mm long, 1–1.3 mm wide. All petals with the limb orbicular, glabrous, yellow, margin fimbriate or denticulate-fimbriate, fimbriae/teeth up to 0.5 mm long; anterior-lateral petals: claw ca. 3–4 mm long, limb ca. 16–18 mm long and wide; posterior-lateral petals: claw ca. 2.5–3 mm long, limb ca. 15 mm long and wide; posterior petal: claw 3.5–4.5 mm long, apex not or only very slightly indented, limb ca. 14 mm long and wide, margin sometimes at the base also with a stalked gland (ca. 0.4 mm long). Stamens unequal, those opposite the posterior styles the largest and with the longest filaments, anthers of those opposite the anterior-lateral sepals with the connective enlarged and the locules reduced; anthers all loculate, glabrous or rarely sparsely pubescent. Stamen opposite anterior sepal: filament 3.5–4.2 mm long, anther 1.8–2 mm long; stamens opposite anterior-lateral petals: filaments 2.5–3 mm long, anthers 1.5–1.6 mm long; stamens opposite anterior-lateral sepals: filaments 3.5–4.2 mm long, connectives ca. 1 mm long, locules 0.6–0.8 mm long; stamens opposite posterior-lateral petals: filaments 4–4.7 mm long, anthers ca. 1.5 mm long; stamens opposite posterior-lateral sepals: filaments 3–3.5 mm long, anthers 0.8–1 mm long; stamen opposite posterior petal always shorter than the adjacent two: filament 2.2–3 mm long, anther 0.9–1.1 mm long. Anterior style 4.2–5.1 mm long, shorter than the posterior two, terete, glabrous, erect, apex 2.1–2.6 mm, each foliole 1.4–1.8 mm long, 1–1.5 mm wide, elliptical. Posterior styles 5–6 mm long, terete, glabrous or with a few scattered hairs in the basal 1/3, lyrate; foliole 2–2.5 mm long, ca. 1.8 mm wide, subrectangular. Dorsal wing of samara 3.8–5 cm long, 1.8–2 cm wide, upper margin with a blunt tooth; nut bearing 1–2 grossly dentate rectangular lateral winglets per side, these 3.5–8 mm long, up to 2.5 mm wide, and/or spurs; nut 9.5–10.5 mm high, 5.5–6.5 mm in diameter, without air chambers, areole 4–4.5 mm long, ca. 3.5–5.5 mm wide, concave, carpophore up to 5 mm long. Embryo 8.5 mm long, ca. 2 times as long as wide, ovoid, outer cotyledon ca. 13 mm long, ca. 4.2 mm wide, the distal 1/3 folded over the inner cotyledon, inner cotyledon ca. 7.3 mm long, ca. 3.3 mm wide, straight. Chromosome number unknown. Fig. 29.

Phenology. Collected in flower in February, April, May, June, and August; in fruit in January, February, June, and August.

Distribution (Fig. 10). Peru (Cuzco, Prov. La Convención); in brush forests, matorral, and clearings; 1300–2700 m.

ADDITIONAL SPECIMENS EXAMINED. **Peru.** CUZCO: San Miguel, Urubamba Valley, *Cook & Gilbert 939* (NY, US); Prov. La Convención, Huyro, *Hoogte & Roersch 770* (F); Prov. La Convención, 139 km de Cuzco en Quillomayo, entre Santa Teresa y Chaullay, 13°08′S, 72°36′W, *Núñez V. & Motocanchi 8751* (MICH); Machu Picchu, *Stafford 9* (K); Machu Picchu, Urubamba Valley, *Tutin 1310, 1328* (BM); Prov. La Convención, Machu Picchu, *Vargas C. 814* (F); Prov. La Convención, Montes de Garabito, *Vargas C. 22750* (MO); Prov. La Convención, *Weberbauer 4989* (G).

Stigmaphyllon cuzcanum is notable for its large, abaxially tomentose leaves and its large flowers, borne in umbels aggregated into thyrses. The petals are among the largest

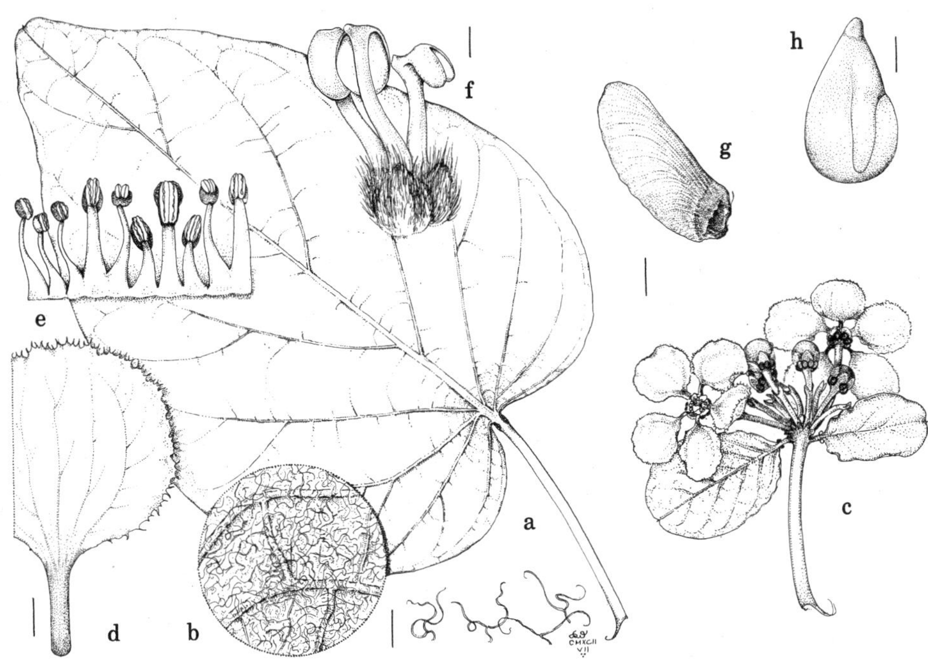

FIG. 29. *Stigmaphyllon cuzcanum.* a. Large leaf. b. Detail of abaxial surface of lamina, and individual hairs. c. Umbel. d. Posterior petal (the "flag"). e. Androecium; stamen second from left opposes the posterior petal. f. Gynoecium, anterior style to the right. g. Samara. h. Embryo. Scale: for a, c, g, bar = 1 cm; for b, bar = 0.5 mm (detail) and 0.3 mm (hairs); for d, h, bar = 2 mm; for e, f, bar = 1 mm. (Based on: a, b, d, *West 6466*; c, e, f, *Stafford 9*; g, *Tutin 1310*; h, *Núñez V. & Motocanchi 8751.*)

in the genus; the limb of the anterior-lateral petals is 16–18 mm in diameter. The stamens opposing the posterior-lateral sepals are not modified, as in most other species. The nut of the samara bears 1–2 grossly dentate rectangular lateral winglets per side. This species is readily separated from the sympatric *S. cardiophyllum* (no. 69), which has very small flowers and sparsely sericeous to glabrous leaves, and *S. strigosum* (no. 73), which has long-fimbriate petals marked with red and modified anthers on the stamens opposing the posterior-lateral sepals. Also, the samaras of *S. cardiophyllum* lack lateral winglets, and those of *S. strigosum* bear 3 to 4 lateral winglets per side.

31. Stigmaphyllon ecuadorense C. Anderson, Contr. Univ. Michigan Herb. 19: 421. 1993.—TYPE: ECUADOR. Manabí: 44 km W of El Empalme on rd from Quevedo to Portoviejo, 1100 ft, 6 Aug 1980, *Wunderlin et al. 8736* (holotype: SEL!; isotype: MO!).

Vine to over 3 m. Stems and branches with scalelike T-shaped hairs. Laminas 10–23 cm long, 8.5–20.5 cm wide, broadly ovate to broadly cordate to suborbicular, apex emarginate-mucronate, base truncate in the smaller laminas to cordate in the larger ones, glabrate to glabrous adaxially, tomentose abaxially (trabecula 0.7–1.2 mm long, crisped and curled, stalk 0.1–0.2 mm long), margin very shallowly dentate and the teeth terminating in filiform glands (up to 4 mm long, broken in large leaves), or margin with irregularly spaced subsessile to stalked glands (0.2–0.4 mm in diameter, 0.1–0.5 mm long), with a pair of prominent but sessile glands at the apex of the petiole, each gland 1.7–2.6 mm in diameter; petioles 3–7.5 cm long, with scalelike T-shaped hairs, not confluent across the node; stipules 0.6–0.8 mm long, 0.8–1 mm wide, free, triangular, eglandular. Flowers ca. 20–40 per pseudoraceme, these borne in compound dichasia (axes to the 4th order, with scalelike T-shaped hairs). Peduncles 7.5–13.5 mm long; pedicels 8–10 mm long, terete; both densely beset with scalelike T-shaped hairs, peduncles 0.9–1.6 times as long as the pedicels. Bracts 1.4–2.2 mm long, 1–1.5 mm wide, triangular, apex acute; bracteoles 1.5–2.2 mm long, 1.1–1.6 mm wide, oblong, apex obtuse, eglandular or each bracteole with 1 or sometimes 2 inconspicuous glands (each 0.2–0.4 mm in diameter); bracts and bracteoles sericeous abaxially. Sepals 2.5–3 mm long and wide, glands 3–3.5 mm long, ca. 1.5 mm wide. All petals with the limb orbicular, glabrous, yellow or also marked with red, margin erose or erose-denticulate, the teeth up to 0.3 mm long; anterior-lateral petals: claw ca. 3 mm long, limb 14–15 mm long and wide; posterior-lateral petals: claw 1.6–2 mm long, limb ca. 13 mm long and wide; posterior petal: claw ca. 3 mm long, apex indented, limb ca. 12 mm long and wide. Stamens unequal, those opposite the posterior styles the largest, anthers of those opposite the anterior-lateral sepals with the connective enlarged and the locules reduced; anthers all loculate, glabrous. Stamen opposite anterior sepal: filament 1.8–2.9 mm long, anther 1.4–1.8 mm long; stamens opposite anterior-lateral petals: filaments 1.5–2 mm long, anthers 1.4–1.8 mm long; stamens opposite anterior-lateral sepals: filaments 4–4.3 mm long, connectives ca. 1.4 mm long, locules 0.4–0.6 mm long; stamens opposite posterior-lateral petals: filaments 4.5–5 mm long, anthers ca. 1.5 mm long; stamens opposite posterior-lateral sepals: filaments 3–3.5 mm long, anthers ca. 1.1 mm long; stamen opposite posterior petal always shorter than the adjacent two: filament 2.5–3 mm long, anther ca. 1.1 mm long. Anterior style ca. 3.5 mm long, shorter than the posterior two, terete, glabrous, erect; apex 1.9–2.2 mm long, each foliole 1.6–2 mm long, ca. 1.6–1.7 mm wide, subsquare to subrectangular. Posterior styles 4.3–4.7 mm long, terete, glabrous, lyrate; foliole ca. 2.5 mm long, 2.3–2.5 mm wide, suborbicular. Immature samara with a dorsal wing up to 4 cm long and 1.5 cm wide, the nut with a pair of lateral winglets up to 8 mm long and 3.5 mm wide. Chromosome number unknown. Fig. 30.

Stigmaphyllon ecuadorense is known from only two collections from the western lowlands of Ecuador (Fig. 43), the type (flowers) and *Steiner 219* (flowers and young fruits) (DAV, MICH; Ecuador. Manabí: 7.5 km W of Río Daule at Pichincha, ca. 800 ft, 16 Jul 1978). It is a showy species notable for its large leaves and large flowers disposed in pseudoracemes. The laminas are tomentose abaxially and have a shallowly dentate margin, the teeth terminating in filiform glands. The limbs of the lateral petals are 13–15 mm in diameter and have an erose or erose-denticulate margin; the anthers of stamens opposite the posterior-lateral sepals have the locules and connective about equally long. The only other species of *Stigmaphyllon* reported from Manabí are *S. eggersii* (no. 47) and *S. ellipticum* (no. 46), which also have large petals but with a fimbriate to lacerate margin; the anthers of stamens opposite all lateral sepals have the connective enlarged and the locules reduced. Their laminas are borne on short petioles (up to 3 cm long) and have eglandular margins. *Stigmaphyllon ecuadorense* might be confused with *S. sinuatum* (no. 33) of the Amazonian region, in which the flowers are also borne in pronounced pseudoracemes and the anthers of stamens opposite the posterior-lateral sepals are unmodified. *Stigmaphyllon sinuatum* is readily separated by the sericeous abaxial pubescence of the laminas.

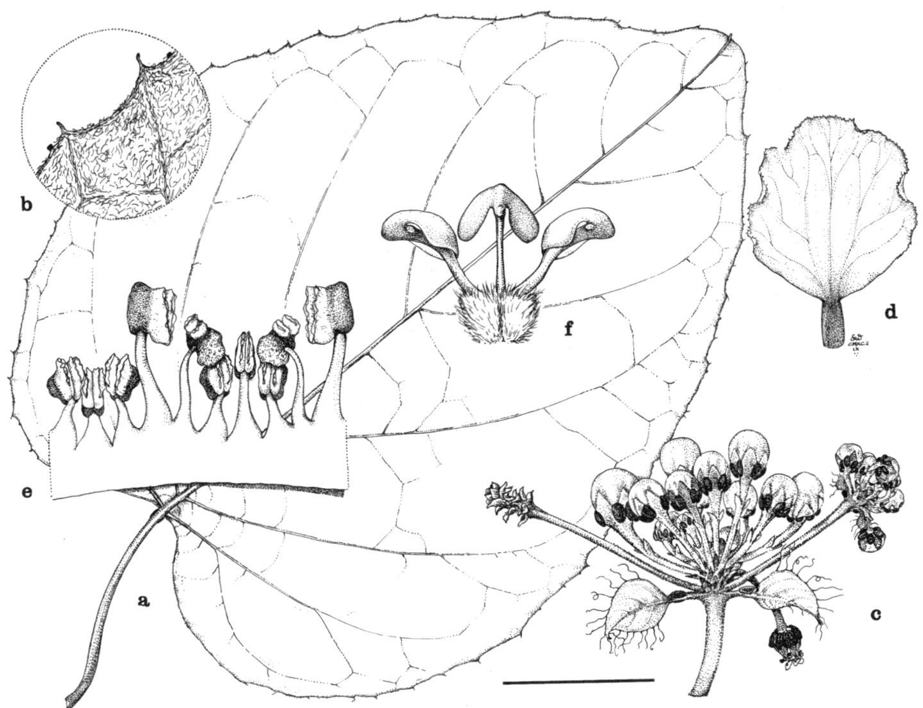

FIG. 30. *Stigmaphyllon ecuadorense*. a. Large leaf. b. Detail of margin and abaxial surface of lamina. c. Portion of inflorescence. d. Posterior petal (the "flag"). e. Androecium; stamen second from left opposes the posterior petal. f. Gynoecium; posterior styles are bent slightly outward to show anterior style (in center). Scale: for a, bar = 4 cm; for b, bar = 8 mm; for c, bar = 2 cm; for d, bar = 1 cm; for e, f, bar = 4 mm. (Based on: a, b, *Steiner 219*; c–f, *Wunderlin et al. 8736*.)

32. Stigmaphyllon nudiflorum Diels, Beitr. Veg. Ecuador [Bibliotheca Botanica 29(116):] 101. 1937.—TYPE: ECUADOR. Guayas: 20 km W von Guayaquil in lichten, regengrünem Walde, ca. 40 m, 29 Sep 1933, *Diels 1231* (holotype: B, destroyed).—ECUADOR. Guayas: Capeira, Km 21 Guayaquil to Daule, *Dodson & Dodson 11199* (neotype, here designated: MO!, photo: MICH!; isoneotype: SEL!).

Vine; flowering and fruiting in leafless conditions (rarely with a few young leaves). Stems and branches sericeous when young, soon becoming glabrate. Laminas 12–18 cm long, 9.8–16 cm wide, cordate to broadly ovate or sometimes 2–3-lobed, apex acuminate, base auriculate, sparsely tomentulose to glabrous adaxially, densely pubescent with T-shaped hairs abaxially (trabecula 1.3–2.3 mm long, wavy, stalk 0.1–0.3 mm long), margin with irregularly spaced sessile glands (0.3–0.5 mm in diameter) and filiform glands (2.5+ mm long); petioles 4.4–5.2 cm long, sericeous, not confluent across the node, with a pair of prominent but sessile glands borne ca. 2.5 mm below the apex, each gland 1.5–2 mm in diameter; stipules 0.6–0.7 mm long, 0.7–0.8 mm wide, free, triangular, eglandular. Flowers ca. 20–25 (–30?) per condensed pseudoraceme, these borne in dichasia or compound dichasia or small thyrses (axes to the 4th order, sericeous). Peduncles 4.5–10 mm long; pedicels 4–10 mm long, terete; both sericeous, peduncles 0.7–1.7 times as long as the pedicels. Bracts 1.2–1.5 mm long, 0.9–1.2 mm wide, triangular, apex acute; bracteoles 1.2–1.6 mm long, 0.9–1.2 mm wide, triangular, apex obtuse, each bracteole usually with a glandular area 0.3–0.5 mm in diameter in the basal 1/3–1/2; bracts and bracteoles sericeous abaxially. Sepals 2.1–2.5 mm long, 2.4–2.5 mm wide, glands 1.4–1.7 mm long, 1–1.1 mm wide. All petals with the limb orbicular, glabrous, yellow, margin denticulate-fimbriate, teeth/fimbriae up to 0.2 mm long; anterior-lateral petals: claw 2.2–2.7 mm long, limb 12–13 mm long and wide; posterior-lateral petals: claw 1.5–1.8 mm long, limb ca. 10 mm long and wide; posterior petal: claw 3.3–4 mm long, apex indented, limb ca. 8–9 mm long and wide. Stamens unequal, those opposite the posterior-lateral petals (and the posterior styles) with the longest filaments, anthers of those opposite the anterior-lateral sepals with the connective enlarged and the locules reduced; anthers all loculate, glabrous. Stamen opposite anterior sepal: filament 2.5–3.5 mm long, anther 1.5–2 mm long; stamens opposite anterior-lateral petals: filaments 1.7–2.3 (–2.9) mm long, anthers 1.1–1.2 (–1.4) mm long; stamens opposite anterior-lateral sepals: filaments 3.1–3.8 mm long, connectives 1–1.3 mm long, locules 0.3–0.7 mm long; stamens opposite posterior-lateral petals: filaments 3.4–4.2 mm long, anthers 1.3–1.5 (–2) mm long; stamens opposite posterior-lateral sepals: filaments 2.5–3.8 mm long, anthers 0.8–0.9 (–1.2) mm long; stamen opposite posterior petal always shorter than the adjacent two: filament 2.1–3 mm long, anther 0.9–1 (–1.3) mm long. Anterior style 3.5–3.8 mm long, shorter than the posterior two, terete, glabrous or with scattered hairs in the proximal 1/5, erect; apex 1.8–2.4 mm long, sometimes including a spur ca. 0.3 mm long, each foliole 1.6–2 mm long, 1.4–1.9 mm wide, parabolic. Posterior styles 4–4.1 mm long, terete, glabrous or with scattered hairs in the proximal 1/7, lyrate; foliole 2.1–2.5 mm long, 2.3–2.5 mm wide, subsquare. Dorsal wing of samara 4.3–4.5 cm long, ca. 1 cm wide, upper margin with a blunt tooth; nut prominently ribbed but without lateral winglets, 1.3–1.5 mm high, ca. 9.5 mm in diameter, woody, without air chambers, areole ca. 8 mm long, 6.5–7 mm wide, deeply concave, carpophore up to 3.5 mm long. Embryo ca. 13 mm long, ca. 2 times as long as wide, ovoid, outer cotyledon ca. 20 mm long, ca. 6.5 mm wide, the distal 3/5 folded over the inner cotyledon, inner cotyledon ca. 7.5 mm long, ca. 4.5 mm wide, straight. Chromosome number unknown.

Phenology. Collected in flower and fruit in leafless condition in July and September, one flowering collection with very young leaves from January, one sterile collection with leaves from February.

Distribution (Fig. 10). Ecuador (southern Guayas); in dry tropical forest and at roadsides; 20–200 m.

ADDITIONAL SPECIMENS EXAMINED. **Ecuador**. GUAYAS: Capeira, Km 21 Guayaquil to Daule, *Dodson & Gentry 12327* (SEL, MO), *12827* (MO), *Gentry 54823* (MICH, MO); Capeira, 22 km N of Guayaquil, on rd to Daule, *Dodson & Dodson 17324* (MO); rd to Miguel Wagner's villa, Km 9 N of Guyaquil on rd to Daule, *Dodson & Thien 712* (WIS); depues de Balzas, aravesando la Cordillera Colonche, en dirección hacia la Prov. Manabí, *Valverde 1857* (SEL); without locality, *Vélez 2818* (VEN).

Stigmaphyllon nudiflorum is known only from the dry tropical forest of southern Guayas, Ecuador, and is the only species in the genus which flowers and fruits in leafless condition during the dry season. The samaras are notable for the large woody nut containing an unusually large embryo (ca. 13 mm long).

33. Stigmaphyllon sinuatum (DC.) Adr. Jussieu, Ann. Sci. Nat. Bot., sér. 2, 13: 288. 1840. *Banisteria sinuata* DC., Prodr. 1: 588. 1824. *Stigmaphyllon hastatum* var. *sinuatum* (DC.) Niedenzu, Ind. Lect. Lyc. Brunsberg. p. aest. 1900: 24. 1900. *Stigmaphyllon sagittatum* var. *sinuatum* (DC.) Niedenzu, Pflanzenreich IV. 141(2): 506. 1928.—TYPE: FRENCH GUIANA. *Perrottet s.n.* (holotype: G-DC!, microfiche: MICH!, photos: F! GH! MICH! NY! US!).

Banisteria rotundifolia Buchoz, Herb. color. Amérique t. 198. 1783, non *Stigmaphyllon rotundifolum* Adr. Juss., 1840.—TYPE: t. 198 in Buchoz, Herb. color. Amérique, 1783.

Banisteria heterophylla Willdenow, Sp. pl. 2: 742. 1799, non *Stigmaphyllon heterophyllum* Hooker, 1843. *Banisteria splendens* DC., Prodr. 1: 588. 1824, nom. superfl. *Stigmaphyllon fulgens* Adr. Jussieu, Ann. Sci. Nat. Bot., sér. 2, 13: 289. 1840, nom. superfl. *Stigmaphyllon splendens* Cuatrecasas, Webbia 13: 531. 1958, nom. superfl.—TYPE: VENEZUELA. "ad Orinocum," *Bredemeyer s.n.* (holotype: B-W 8855, microfiche: MICH!).

Stigmaphyllon martianum Adr. Jussieu, Ann. Sci. Nat. Bot., sér. 2, 13: 289. 1840.—TYPE: BRAZIL. Amazônas: "in sylvis Japurensibus," Dec, *Martius s.n.* (holotype: M!, photos: F! GH! MICH! NY! US!).

Stigmaphyllon richardianum Adr. Jussieu, Ann. Sci. Nat. Bot., sér. 2, 13: 289. 1840.—TYPE: FRENCH GUIANA. Cayenne, July, *Richard s.n.* (holotype: P!, fragment: P-JU!, photos: F! MICH! US!).

Stigmaphyllon hypoleucum Miquel, Linnaea 18: 51. 1844.—TYPE: SURINAME. "ad fluv. Boven Cottica, in sylva," Oct 1842, *Focke 683* (holotype: U!, photo: MICH!).

Stigmaphyllon purpureum Bentham, London J. Bot. 7: 128. 1848.—TYPE: GUYANA. Picarara, *Robt. Schomburgk, 1st coll. 737* (holotype: K!, photo: MICH!; isotypes: G! P! W!).

Stigmaphyllon brachiatum Triana & Planchon, Ann. Sci. Soc. Nat. Bot., sér. 4, 18: 316. 1862.—TYPE: COLOMBIA. Meta: Villavicencio, 450 m, *Triana s.n.* (holotype: P!, photo: MICH!; isotypes: COL! G!, photos of G isotype: F! GH! MICH! K! MO!).

Stigmaphyllon monancistrum Niedenzu, Ind. Lect. Lyc. Brunsberg. p. hiem. 1899–1900: 13. 1899.—TYPE: "Colombia" [VENEZUELA]. Aragua: Maracay, ad rivulis in crepidis, *Moritz 779* (lectotype, designated by C. Anderson, 1993b: LE!, photo: MICH!; fragment of B duplicate: NY!).

Vine to 30 m. Stems and branches sericeous when young, soon becoming glabrate. Laminas 6–21 cm long, 4.5–20 cm wide, triangular to ovate to cordate to elliptical to broadly so to orbicular to oblate to reniform, rarely palmately 3–5-lobed, apex mucronate to emarginate-mucronate to short-acuminate, base acute to truncate to cordate to deeply auriculate, very sparsely sericeous to glabrous adaxially, sparsely sericeous to densely silvery sericeous abaxially [trabecula (0.2–) 0.3–0.5 (–0.7) mm long, straight, mostly sessile but sometimes with a tiny stalk up to 0.01 mm long, especially if vesture is very dense], margin grossly and shallowly crenate to subentire and with irregularly spaced sessile glands (0.4–1.5 mm in diameter) in the sinuses and sometimes also with filiform glands (up to 1.5 mm long); petioles 1.8–13 cm long, sericeous, barely confluent across the node and forming a line bearing the stipules, with a pair of prominent but sessile glands at the apex, each gland 1–3.5 mm in diameter; stipules 0.5–1.3 mm long, 0.5–1.5 mm wide, free or sometimes the opposing stipules united in the basal 1/4–1/3, triangular, eglandular. Flowers ca. 15–35 (–40) per pseudoraceme, these sometimes interrupted, borne in compound dichasia or small thyrses (axes to the 6th order, sericeous). Peduncles 2.5–11 mm long, pedicels 3–9.5 mm long, terete; both sericeous, peduncles 0.4–1.5 times as long as the pedicels. Bracts 0.7–2 mm long, 0.5–1.4 mm wide, triangular or narrowly so, apex acute; bracteoles 0.9–1.6 mm long, 0.7–1.3 mm wide, triangular to ovate to parabolic, apex obtuse, eglandular or more commonly each bracteole with a pair of inconspicuous glands (each gland 0.1–0.5 mm in diameter) or sometimes only with a glandular area in the basal 1/3–1/2; bracts and bracteoles sericeous abaxially. Sepals 1.5–2.5 (–3) mm long, 1.5–2.5 (–3) mm wide, glands (1–) 1.2–2.2 mm long, 0.8–1.3 mm wide. All petals with the limb orbicular or limb of posterior petal broadly elliptical to broadly obovate, glabrous or sometimes pubescent abaxially, yellow or suffused with red, margin erose to erose-denticulate to denticulate, teeth up to 0.3 mm long; anterior-lateral petals: claw (1–) 1.5–2.5 mm long, limb (8–) 10–15 mm long and wide; posterior-lateral petals: claw (0.5–) 0.8–2 mm long, limb (7–) 9–13 mm long and wide; posterior petal: claw 2.6–3.5 (–4) mm long, apex indented, limb (5–) 7–11 mm long and wide. Stamens unequal, those opposite the anterior style and/or the posterior styles the largest, anthers of those opposite the anterior-lateral sepals with the connective enlarged and the locules reduced or sometimes with only one locule or eloculate; anthers glabrous. Stamen opposite anterior sepal: filament (2–) 2.5–4; mm long, anther 1–1.7 mm long; stamens opposite anterior-lateral petals: filaments (1.2–) 1.5–3.5 mm long, anthers 1–1.4 mm long; stamens opposite anterior-lateral sepals: filaments 2.5–3.8 (–4.3) mm long, connectives 0.9–1.5 mm long, locules 0.2–1 mm long or sometimes one or both absent; stamens opposite posterior-lateral petals: filaments 2.7–4 mm long, anthers 0.9–1.8 mm long; stamens opposite posterior-lateral sepals: filaments (2–) 2.4–3.4 mm long, anthers 0.5–0.9 (–1.2) mm long; stamen opposite posterior petal always shorter than the adjacent two: filament 1.9–2.8 (–3.3) mm long, anther 0.5–1 mm long. Anterior style (2.6–) 3.2–4 (–4.8) mm long, shorter than or subequal to the posterior two, terete, glabrous or with a row of hairs adaxially in the proximal 1/5–2/3, erect or slightly recurved; apex (1.1–) 1.4–2.6 mm long, often including a spur 0.1–0.5 mm long, each foliole 0.7–1.9 mm long, 0.6–1.9 (–2.1) mm wide, triangular to parabolic to subrectangular, *or* folioles absent and the apex extended into a claw 0.7–1.3 mm long. Posterior styles (2.8–) 3–4.5 (–5.2) mm long, terete, glabrous or with a row of hairs adaxially in the proximal 1/4–1/2, lyrate; foliole 1–2.2 mm long, (0.8–) 1.2–2.3 mm wide, subsquare to suborbicular or sometimes parabolic or rectangular. Dorsal wing of samara 3.5–5.5 cm long, 1–1.8 cm wide, upper margin with a blunt tooth; nut smooth or more commonly bearing 1–3 subentire to coarsely toothed to lacerate lateral winglets, these up to 6 mm long and 3.5 mm wide, and/or spurs and crests up to ca. 2 mm long and wide; nut 4.5–7 mm high, 2.8–4.4 mm in diameter, without air chambers, areole 2.3–4 mm long, 2,5–4.5 mm wide, concave, carpophore up to 5 mm long. Embryo 5.1–6.8 mm long, ca. 1.5–2 times as long as wide, ovoid, outer cotyledon (4.7–) 5.4–8.5 mm long, 2.5–4.2 mm wide, straight or the distal 1/5–3/5 folded over the inner cotyledon, inner cotyledon 4–6.3 mm long, 2.2–4 mm wide, straight or folded at the distal 1/4–2/3. Chromosome number unknown. Fig. 31.

Phenology. Collected in flower and fruit throughout the year.

Distribution (Fig. 32). Common in the lowlands of Colombia, Venezuela, the Guianas, northern Brazil, Ecuador, northern Peru, and Amazonian Bolivia; in primary and secondary forest, especially wet forest, but also in white sand vegetation, along rivers, at roadsides, in thickets; sea level to 1000 m.

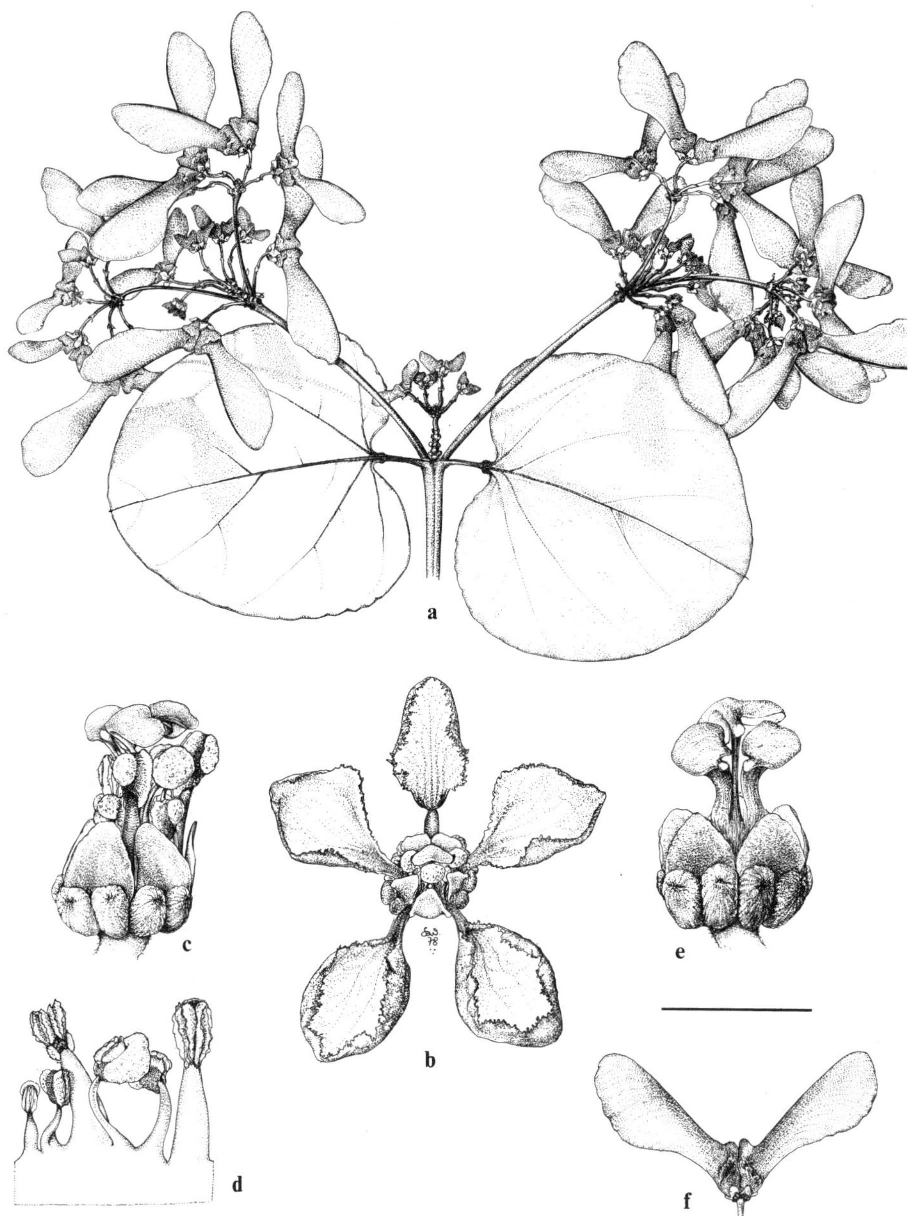

FIG. 31. *Stigmaphyllon sinuatum*. a. Fruiting branch. b. Flower. c. Flower, the petals removed. d. Portion of androecium; posterior stamen on extreme left, anterior stamen on extreme right. e. Flower with petals and androecium removed, anterior style in the center. f. Fruit with two samaras. Scale: for a, bar = 4 cm; for b, bar = 8 mm; for c–e, bar = 4 mm; for f, bar = 2.7 cm. (Based on *Wurdack 34434*.)

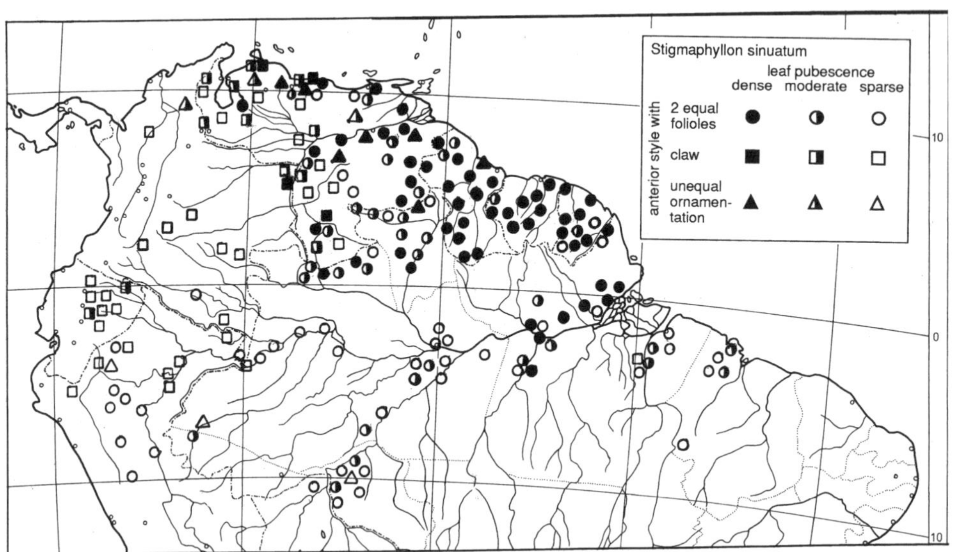

FIG. 32. Distribution of *Stigmaphyllon sinuatum.*

REPRESENTATIVE SPECIMENS (anterior style foliolate). **Colombia.** AMAZONAS: ca. 6 km W of Leticia at Santa Isabella, 04°10′S, 69°58′W, *Gillett & Dickenson 16521* (COL, MO); Araracuara, near Río Caquetá, *Maguire et al. 44142* (MICH, NY); Leticia, ca. 1 km NE of town, *Plowman et al. 2291* (ECON, F, GH, NY); Leticia, *Schultes 8222* (COL, GH, US). **Venezuela.** AMAZONAS: Dept. Río Negro, along Río Mawarinuma, vicinity of Cerro de la Neblina base camp, ca. 00°50′N, 66°09′W, *Liesner 15966* (MICH), *Nee 30880* (MICH); Tamatama, *Ll. Williams 15217* (F, NY, US), *Wurdack & Adderley 43641* (F, MICH, NY, U).—ANZOÁTEGUI: along Río Querecual, SW of Bergantín, *Steyermark 61506* (F, NY, VEN).—ARAGUA: Rancho Grande, *Badillo 1793* (MY).—BARINAS: Allamira, *Curran M-619* (NY).—BOLÍVAR: NE de Canaima, S de Cerro Venado, ca. 06°16′N, 62°46′W, *Agostini 299* (US); Ciudad Bolívar, Maquanta River, ca. 08°10′N, *Bailey & Bailey 1411* (A, NY); Calzeta en la Botella, Río Cuyuní, *Bernardi 6497* (MER, NY); a 48 km NE del caserio Los Rosos, este último a 17 km de Upata, sobre la carretera nueva Upata–San Félix, *Blanco 434* (MER, MO, NY, US); Dtto. Cedeño, 6 km from Manapiure toward Caicara, 06°55′N, 66°30′W, *Boom & Grillo 6487* (MICH); Mpio. Raúl León, Río Paragua, 04°18′N, 62°05′W, *Delgado 104* (MICH); Km 105–112 de la carretera El Dorado–Sta. Elena, *Morillo et al. 2932* (MICH, VEN); Mpio. Piar, camino desde El Plomo a Sta. Barbara, 06°45′30″N, 62°48′W, *Picón Nava 1588* (MICH); Uruyén, Auyantepui, *Schnee 1437* (MY); alrededores de Tumeremo, camino Tumeremo–Bochinche, entre Puesto (GN) Corumo y Caño Matuco del Río Negro, *Stergios et al. 3563* (MICH); Sierra Imataca, along Río Reforma, 1 km above junction with Río Toro, *Steyermark 87917* (MICH, NY, U, US); acercandose a las cabeceras del Río Nichare (affluente del Río Caura) en la dirección de la Sierra Maignalida y Sierra Cervatana, arriba de la desembocadura con el Río Cicuta (Icuta), 06°15′N, 65°05′W, *Steyermark 95736* (NY, US); La Prisión, Medio Caura, *Ll. Williams 11539a* (F); El Palmar, *Ll. Williams 12919* (A, F, K, S); along Fundación rd, *Wurdack 34434* (NY).—DELTA AMACURO: near border (=Río Grande o Toro) between Estado Bolívar and Terr. Delta Amacuro, ca. 08°04′N, 61°44′W, *Breteler 3757* (MER, NY, U, US, VEN); Depto. Tucupita, ca. 13 km by rd ESE of the town of Sierra Imataca, 08°32′N, 62°23′W, *Davidse 16623* (MICH, MO); downstream from San Victor, Río Amacuro, Sierra Imataca, *Steyermark 87299* (G, GH, NY, P).—MIRANDA: Dtto. Páez, Cerro Riberón between Río Guapo and Río Chiquito, 44.5 km (str. line) SE of Caucagua, 10°05′N, 66°01′W, *Davidse 13604* (VEN); between La Cortada and Turumo Bridge, *Pittier 11484* (G, K, NY, US, VEN); Los Mariches, *Pittier 12991* (F, G, M, MO, NY, US, VEN); Cerros del Bachilla, between Quebradas Corozal and Santa Cruz, S of Santa Cruz, 10 km by air W of Cúpira, *Steyermark 116443* (MICH, MO).—MONAGAS: ca. 8 km ESE of Jusepín, *Pursell et al. 9098* (NY, US, VEN).—SUCRE: Río Tatracual, 25 km outside Cumaná on Cumanacoa rd, *Sobel & Strudwick 2274* (MICH). **Guyana.** Near Mazaruni Forest Station, *Archer 2463* (GH, K, US); Pomeroon Dist., Waramuri Mission, Moruka River, *de la*

Cruz 2578 (GH, NY, US); Waini River, NW Dist., 08°20′N, 59°40′W, *de la Cruz 3619* (F, GH, MO, NY, US); upper Demerara–Berbice region, ca. 27 km from Ituni along Ituni–Kwakwanni rd, 05°22′N, 58°07′W, *Gillespie 2991* (MICH); Potaro-Siparuni region, trail from Kato to Paramakatoi, 04°41′N, 59°50′W, *Hahn 5622* (MICH); Akyma, Demerara River above Wismar, 05°09′N, *Hitchcock 17414* (GH, NY, S, US); Kanuku Mts, Rupununi River, near "the farm" of the Captain of Sandcreek, 03°07′N, 59°26′W, *Jansen-Jacobs et al. 206* (MICH, US); Gunn's, Essequibo River, 01°39′N, 56°38′W, *Jansen-Jacobs et al. 1516* (MICH); Potaro-Siparuni region, Chenapou, 50 km upstream from Kaieteur Falls, 05°00′N, 59°34′W, *Kvist 325* (MICH); Cuyuni-Mazaruni region, NW of Conoch Tipu, 05°48′N, 61°03′W, *McDowell 2629* (MICH); Cuyuni River, by portage rd near lower Camaria Landing, *Sandwith 664* (K, NY, U); Essequibo River, near mouth of Orono Creek, ca. 01°35′N, *A. C. Smith 2821* (A, F, G, K, MO, NY, P, S, U, US); W extremity of Kanuku Mts, drainage of Takutu River, *A. C. Smith 3166* (A, F, G, K, MO, NY, P, S, U, US, W); ca. 5 km SW of Mabura Hill towards Essequibo River, 05°19′N, 58°38′W, *Stoffers et al. 35* (MICH); Marudi Mts, Mazoa Hill, near NorMan Mines camp, 02°15′N, 59°10′W, *Stoffers et al. 207* (MICH). **Suriname.** Scotelweg, *Archer 2658* (US); Zandrij, *Archer 2761* (US); Republieck, *van Doesburg, Jr., 70* (U); Dist. Saramacca, Experimental Farm Coebiti, *Everaarts 519* (MICH); in montibus Bakhuis inter flum. Kabalebo et Coppename sinistrum, around Kabalebo airstrip, *Florschütz & Maas 2502* (F, U); Dist. Nickerie, area of Kabalebo Dam project, ca. 22 km SW of Avanavero damsite, *Heyde & Lindeman 103* (MICH, U); without locality, *Hostmann 1029* (BM, G, GH, P, U, W); Wilhelmina Gebergte, Zuid River, 45 km above confluence with Lucie River, 03°10–20′N, 56°29–49′W, *Irwin et al. 57631* (GH, K, MG, MICH, NY, S, U, US); Suriname River near Kabelstation, *Lanjouw 11185* (U); Nickerie Dist., area of Kabalebo Dam project, 04–05°N, 57°30′–58°W, *Lindeman et al. 47* (U); Lely Mts, SW plateaus, *Lindeman et al. 266* (NY, U); Waijombo River, *Linder 89* (GH, NY); vicinity of Sectie O, Km 68, *Maguire & Stahel 25000* (BR, F, G, GH, K, MO, NY, P, RB, U, US); Brownsberg Nature Park, 90 km S of Paramaribo, Mazaroni plateau, trail to Witti-creek, *Mori & Bolten 8397* (MICH, NY); surroundings of Blakawatra, camp 8, 60 km SE of Paramaribo, *den Outer 872* (U); fluv. Saramacca inf. prope Mindrinetti, *Pulle 34* (U); Jodensavanne-Mapanekreek area (Surinam R.), *Schulz 7313* (AAU, COL, MICH, NY, U, US). **French Guiana.** Sinnamary, piste de Ste. Elie, Km 15, *Billiet & Jadin 1101* (BM, BR, CAY, NY); piste Saint Laurent vers Paul Isnard, entre Km 30 et 40, *Billiet & Jadin 1577* (CAY, MICH); piste forestière allant de la route N2 vers Nancibo, *Billiet & Jadin 1845* (CAY); le long de la piste de la route de Cayenne à Régina, ca. 30 km de Régina, *Cremers 5991* (MICH); Bourg d'Apatou, Bassin du Maroni, 05°09′N, 54°20′W, *Fleury 334* (MICH); Haut Oyapock, Trois Sauts, *Garnier 103* (CAY); Haut Oyapock, à 2 km env. en amont de Saut Boko, *de Granville 2465* (NY); Haute Camopi, Mont Belvédère, *de Granville 7010* (CAY, MICH); roche plate Roche Koouton–Bassin Haut-Marounini, 1 km W de la Roche, 02°53′N, 54°04′W, *de Granville 9516* (MICH); St. Georges de l'Oyapock, piste de Maripa, *Grenand 2136* (CAY, MICH); Île de Cayenne, Mont Bachrel, 04°55′N, 52°19′W, *Hoff 5224* (MICH, P); village de Zidockville, Trois Sauts, *Jacquemin 1610* (CAY, MICH); Saül, 03°37′N, 53°12′W, *Marshall & Rombold 169* (CAY, MICH); Maripasoula, *Oldeman 1647* (MICH); rivière Tonégrande, près de port Inini, *Oldeman 1656* (MICH, P); Saül, *Oldeman 1982* (CAY, MICH, P); fleuve Approuague, rivière Arataye, Sauts Pararé, *Poncy 221* (CAY); Acarouani, *Sagot 91* (BM, G, P, S, W); St. Laurent region, ca. 5 km from Rte D9 at Charvein, 05°51′N, 53°51′W, *Skog & Feuillet 7481* (CAY, NY, P, US); main rd through Montagne de Kaw, 04°35′N, 52°15′W, *Weitzman 308* (MICH). **Brazil.** ACRE: Cruzeiro do Sul, Projecto RADAM/BRASIL, aeroporto, *Monteiro & Damião 209* (INPA, MG), *Ramos 116* (INPA), *Ramos & Mota 208* (INPA); near Sta. Lucia, Km 40 on Transamazônica Hwy, E of Cruzeiro do Sul, 07°08′N, 72°33'W, *Pruski 3498* (MICH).—AMAPÁ: 2–10 km N de Ferreira Gomes, BR-156, *Austin 7257* (MICH); Rio Amapari, rd to Porto Terezinha, *Cowan 38466* (K, NY, RB), *Cowan 38473B* (NY); Rio Amapá, Serra do Navio, lower slopes of Fritz Akerman Ore Body, *Cowan & Maguire 38086* (COL, G, GH, MICH, MO, NY, P, S, U, W); Rio Oiapoque, 6 km SE of Clevelandia, 03°48′N, 51°53′W, *Irwin et al. 47382* (IAN, MICH, NY); Mpio. Oiapoque, BR-156, 60 km SSE of Oiapoque, 03°18′N, 51°39′W, *Mori & Cardoso 17134* (MICH); Macapá, Igarapé do Lago, *Rabelo 771* (MG); Matapi, *Ribeiro 1597* (INPA, MICH, MO, NY, RB).—AMAZÔNAS: Mpio. Tefé, Rio Solimões, vila Nogueira, *Amaral et al. 95* (INPA, NY); Manaus–Pôrto Velho hwy, Km 124, *Campbell et al. P20920* (GH, INPA, MICH, MO, NY, S, U); Manaus, Igarapé do Parque 10, *Chagas INPA 3679* (INPA, SP); Rio Uatumã, Mpio. Itapiranga, *Cid et al. 592* (INPA, MICH); BR-172, Manaus–Caracaraí, Km 97, *Cid et al. 945* (INPA, MICH); Mpio. Maraã, Rio Japurá, affluente do Rio Solimões, *Cid & Lima 3434* (INPA, MG, MICH); BR-307, Mpio. Cruzeiro do Sul, 7–8°S, 72–73°W, *Cid Ferreira et al. 5218* (INPA, MICH); Mpio. São Paulo de Olivença, 6 km S of town center, 03°30′S, 68°57′W, *Daly et al. 4442* (MICH); Barcelos, *Duarte 7160* (INPA, RB, SP); Manaus, estrada do Mindú, *Ducke 856* (F, IAN, MG, MO, NY, R, RB, US); Mpio. Maués, ca. 20 km E of Maués, Antarctica Guaraná Plantation, *Hill et al. 13152* (JBSD, MICH, TEX); Manaus, Rua Duque de Caxias, *Maas & Maas 362* (INPA, U); caatinga do Porto Camanaus, *Madison et al. PFE 414* (INPA); basin of Rio Demeni, vicinity of Tototobí, *Prance et al. 10355* (INPA, MG, MICH, NY, U); Manaus, INPA, estrada do Aleixa, Km 3, *Prance & Ramos 20922* (INPA, MICH, MO, NY, U); vic. of Pico Rondon, Perimental Norte, Km 211–220, 01°32′N, 62°48′W, *Prance et al. 28821* (MICH), *Rodrigues et al. 10584* (MICH); Santo Antonio do Içа, p. Vila Militar, *M. Silva 2112* (MG); Fonte Boa, *M. Silva 2183* (MG); Barra [=Manaus], *Spruce 1880* (G, GH, LE, M, MG); Mpio. Humaitá, estrada Humaitá–Lábrea, Km 59, a 3 km ao N, *Teixeira et al. 980* (INPA, MICH); Mpio. Humaitá, estrada Humaitá–Jacarecanga, Km 150, a 63 km ao S, *Teixeira et al. 1340* (INPA, MICH).—MARANHÃO: Mpio. Monção, basin of the Rio Turiaçu, Káapor Indian Reserve, *Balée 886* (NY); ca. 50 km from Santa Luzia on Hwy to Açailândia, 04°05′S, 45°57′W, *Daly et al. D736* (MICH); Rio Alto Turiaçu, Barranquinha, 03°00′S, 45°45′W, *Jangoux & Bahia 161* (MG, NY); margen do Rio Cururupu, *Lisboa 47* (RB, SP); Km 375–380 da rodovia Belém–Brasília, *Oliveira 1072* (IAN, UB).—PARÁ: Belterra, *Black 47-1660* (IAN); Altamira, Km 74 da estrada Transamazônica Itaituba, *Cavalcante & M. Silva 2780* (MG); Mpio. Oriximiná, Rio Trombetas, Lago de Matens, 19 km S de Pôrto Trombetas, *Cid et al. 1793* (INPA, MG, MICH); Jacaracá Island, *da Costa 149* (F); ca. 70 km from Tucuruí, ca. 04°11′S, 49°04′W, *Daly et al. 1435* (INPA, MICH); boca do Lago de Faro, *Ducke 88677* (MG); 4–5 km W of São Francisco do Pará toward Castanhal, *Gentry 13166* (INPA, MICH, MO, NY); Dist. Acará, Thomé Assú, Pau Vermelho, *Mexia 5926* (A, F, G, GB, GH, MICH, MO, NY, S, U, US, WIS); Belém, horta do IAN, *Murça Pires & Black 403* (GH); Mpio. Almeirim, Monte Dourado, estrada MTD, W em direção à mina de bauxita, *Murça Pires et al. 620* (MICH); BR-163, Km 1131, Ciuabá–Santarém hwy, vicinity of Igarapé Natal, *Prance P25427* (MG, MICH); Belém, terreno da EMBRAPA, *Ramos & Rosário 14* (MG); Ourém, *Rodrigues 4024* (MG); Tucuruí, margem direita do Rio Tocantins, *Rosário 93* (MG, NY); Mpio. Almeirim, Monte Dourado, *Santos 437* (NY); Santarém, Km 70 da estrada do Palhão, arredores do Acampamento do Igarapé Guaraná, *M. Silva & R. Souza 2522* (MICH, NY, U), *M. Silva & R. Souza 2522* (CAS, CM, MG, MICH, NY); 7–11 km NW of AMZA camp 3-Alfa on rd to camp 4-Alfa, 05°47′S, 50°34′W, *Sperling 6057* (MICH); Santarém, *Spruce 767* (G, GH, GOET, M, NY, W); Macau airstrip, 1 1/2 hrs upstream from Lageira airstrip, 05°55′S, 54°26′W, *Strudwick & Sobel 3474* (MICH); Taperinha bei Santarém, *Zerny 589* (W).—RONDÔNIA: Pôrto Velho–Cuiabá hwy, 25 km S of Nova Vida, *Forero & Wrigley 7084* (INPA, MG, MICH, NY); basin of the Rio Madeira, cerrado between Jaciparaná and Rio Madeira, *Prance et al. 5180* (INPA, MG, MICH, NY); Km 166–169, Madeira–Mamoré railroad near Mutumparaná, *Prance et al. 5690* (GH, INPA, MG, MICH, NY); foothills of Serra dos Pacaás Novos, 12 km NNE of Guajará–Mirim, *Prance et al. 6658* (F, INPA, MICH, NY); Pôrto Velho–Cuiabá hwy, vicinity of Santa Barbara, 15 km E of Km 117, *Ramos & Prance 6905* (INPA, MICH, NY); Pôrto Velho, Represa Samuel, 08°55′S, 63°16′W, *Thomas et al. 4949, 5034* (MICH).—RORAIMA: Aritumã region, on an azimuth of 011° from Boa Vista at a distance of 210 km, *Coradin & Cordeiro 943* (INPA, MICH, NY); SEMA Ecological Reserve, Ilha de Maracá, 03°21′N, 61°27′W, *Milliken M792* (MICH); Canto Galo, Rio Mucajaí, between Pratinha and Rio Apiaú, *Prance et al. 3964* (INPA, MG, MICH, NY, U); Serra Tepequem, *Prance et al. 4437* (INPA, MG, MICH, NY, U); vicinity of Uaicá airstrip, Rio Uraricoeira, 03°33′N, 63°11′W, *Prance et al. 10909* (INPA, MG, MICH, NY, U); Mpio. Caracaraí, estrada Manaus–Caracaraí, Km 529–550, *dos Santos & Coêlho 700* (INPA, MICH); estrada Manaus–Caracaraí, BR-174, Km 329, army post N of Waimari-Atoari Indian Reserve, *Steward et al. 9* (MICH); Rio Surumu, an einem Bache der Serra do Mel, *Ule 8185* (MG). **Bolivia.** BENI: Vaca Diez, 3 km E of Riberalta on rd to Guayaramerín, 11°00′S, 66°05′W, *Solomon 7682* (MICH). **Peru.** HUÁNUCO: Prov. Leoncio Prado, Moena, cerca a Tingo María, *Woytkowski 1187* (MICH); Santa Tereza, valle del Huallaga, *Woytkowski 1232* (ECON).—LORETO: Prov. Coronel Portillo, Pampa de Sacramento, cerca Pucallpa, *Ferreyra H. 1184* (GH, MICH, US); Prov. Alto Amazonas, entre Yurimaguas y Chambira, *Ferreyra H. 4897* (MICH, US); Prov. Coronel Portillo, on Río Aguaytía, 08°50′S, 75°20′W, *Fosberg 28875* (MO); Balsapuerto, *Klug 3075* (A, BM, F, G, K, MO, NY, S, US); Prov. Ucayali, Canchahuayo (Río Ucayali), 07°05′S, 75°10′W, *Vásquez et al. 6970* (MO); Mariscal Castilla, Caballococha, 03°55′S, 70°30′W, *Vásquez & Jaramillo 9319* (MICH).—SAN MARTÍN: ca. 10 km NE of Tarapoto, *Gentry 37915* (MICH); San Martín, 5–15 km E of Shapaja on rd to Chazuta, 06°36′S, 76°10′W, *Knapp & Mallet 7026* (MICH); Prov. Mariscal Cáceres, Dtto. Uchizo, en la carretera a Río Uchizo 2 km del caserio Nuevo Progreso, *Schunke V. 3219* (COL, F, G, NY, US); Prov. Mariscal Cáceres, Dtto. Tocache Nuevo, Quebrada Luis Sálas (5 km NE de Puerto Pizana), *Schunke V. 6578* (GH, MO).

REPRESENTATIVE SPECIMENS (anterior style clawed). **Colombia.** AMAZONAS: Leticia, Oct 1946, *Black s.n.* (IAN); Río Iagara–Paraná (affl. Río Putumayo), La Chorrera, *Gasche & Desplats 59* (K, MICH).—ANTIOQUIA: Mpio. San Luis Quebrada "La Cristalina," 06°N, 74°45′W, *Ramírez & Cárdenas 474* (MO); Mpio. Caucasia, along rd to Nechí, 24 km from Caucasia–Planeta Rica rd, 08°04′N, 75°05′W, *Zarucchi et al. 4903* (MICH).—CAQUETÁ: 6 km SE of Morelia along rd to Río Pescado (SW of Florencia), *Davidse 5644* (COL, MICH); Río Arteguaza, 9 km S of Florencia, *Plowman & Kennedy 2282* (F, GH, M, NY, P, S, US).—META: Villavicencio, *Killip 34346* (COL, S); Sierra de la Macarena, Río Guapaya, *Philipson et al. 2124* (COL).—VAUPÉS: rd from Mitu to Monfort, *Davis 107* (COL, GH, MICH). **Venezuela.** AMAZONAS: alrededores de San Juan de Manapiare, 05°18′N, 66°03′W, *Agostini 1504* (MICH); Río Orinoco, Isla del Ratón, 05°02′N, 67°45′W, *Breteler 4721* (F, K, MA, MO,

NY, S, US, WAG); Depto. Atures, 26 km SE de Puerto Ayacucho por la carretera Puerto Ayacucho–El Gavilán, 05°32′N, 67°24′W, *Cuello & Fernández 508* (MICH); El Gavilán, 30 km al E de Puerto Ayacucho, *Fernández 2950* (MY); Río Ventuari, La Ceiba, bajo del Salto Tencua, *Foldats 147*A (NY, VEN); Depto. Río Negro, cerca de Shabono Yanomami ubicado a la izquierda del Río Orinoco y a la derecha de la desembocadura del Río Mavaca, 02°30′N, 65°10′W, *Guánchez 656* (MICH); Cerro Camani, *Maguire 31802* (NY); ad flumina Casiquiari, Vasiva, et Pacimoni, *Spruce 3277* (BR, G, K, NY, W); Dtto. Atabapo, Río Cunucunuma, entre el Cerro Duida y Huachamacari, 03°40′N, 65°45′W, *Steyermark 126178* (MICH).—ARAGUA: Pozo del Diablo, cañada del Río Yuare, Maracay, *Baldillo 3782* (F, MY); Rancho Grande, Maracay, *Ferrari 745* (MY), *Vogelsang E10* (MY).—BARINAS: Ticoporo Forest Reserve, 08°15′N, 70°45′W, *Breteler 3678* (G, MER, NY, S, U, US, VEN); entre Km 469–470, carretera Barinas–San Cristóbal, *Cárdenas de Guevara et al. 2565* (MY); 1–2 km NE of Bumbum, ca. 68 km SW of Barinas, *Gentry 11142* (GH, MICH, MO, VEN).—BOLÍVAR: Pica Caicara del Orinoco–San Juan de Manapiare, Río Suapure, 202 km al S de Caicara, ca. 07°N, 67°W, *Delascio & López 2766* (VEN); Santa María de Erebato, Río Erebato, 05°05′N, 64°40′W, *Steyermark 109847* (K, NY).—CARABOBO: Dtto. Valencia, carretera Valencia–El Palotal–El Paito–Los Naranjos, *Bunting 4597* (NY); Guataparo, Valencia, *Saer 865* (VEN).—DISTRITO FEDERAL: Caracas, *Bredemeyer 206* (W); Dtto. Vargas, camino entre Osma y Oritapo, *Benítez de Rojas 588* (MY).—FALCÓN: carretera Yaracal–Araurima, 8 km de la carretera Yaracal–Tucacas, *Cardozo et al. 65* (MICH); El Guanábano, 27 km S de Puerto Cumarebo, *Flora Falcón 343* (MICH, MO, U); Dtto. Silva, ca. 21 km W of Tucacas, *Wingfield 12696* (MICH).—LARA: Dtto. Palavecino, carretera entre Manzanito y El Altar, *Burandt, Jr., & Smith V0075* (MICH).—MÉRIDA: above dam site on Río Caparo, 31 km ESE of Santa Barbara, ca. 07°41′N, 71°28′W, *Liesner & González 9267* (MICH, VEN).—PORTUGUESA: Dtto. Ospino, carretera La Aparición de Ospino–Moroturo, 09°31′N, 69°26′W, *Aymard & Cuello 6577* (MICH, VEN); Dtto. Araure, carretera Hoja Blanca–Guayabal–El Rechazo, *Cuello & Cuello 59* (MICH).—TÁCHIRA: E of San Cristóbal and 40 km W of Santa Barbara, rd W or NW of Abajales, *Sobel & Strudwick 2108* (NY).—TRUJILLO: Boconó, *López-Palacios 395* (MER, VEN).—ZULIA: Dtto. Colón, hacienda El Rosario, 18 km E de la carretera Machiques–La Fría, 12 km N de Río Catatumbo, *Bunting 6475* (MICH, VEN); Dtto. Colón, entre Casigua El Cubo y Km 8 de la vía rumbo al Palmira, *Bunting 7328* (MICH); Dtto. Períja, entre Km 16 de la carretera Machiques–La Fría y Calle Larga y San José, *Bunting 10816* (MICH); Dtto. Períja, between Río Yasa and Río Tucuco along the Machiques and Los Angeles de Tucuco rd, 09°50–56′N, 72°40–44′W, *Davidse 18390* (MICH). **Ecuador.** MORONA-SANTIAGO: Tunantza, Jíbaro settlement near Macuma, ca. 50 km NE of Macas, *Lugo S. 3718* (GB, MICH); El Centro Shuar Kankaim (Cangaime), Río Kankaim (Cangaime), 02°20′S, 77°41′W, *Shiki RBAE219* (NY).—PASTAZA: 31 km N of Puyo on rd to Tena, side rd E of Cajabamba, 00°15′S, 77°50′W, *Boom & Beardsley 8436* (MICH); Curaray (Jesús Pitishka), *Harling & Andersson 17450* (GB, MICH); Río Chullana, ca. 15 km N of Puerto Sarayacu, *Lugo S. 4184* (GB, MICH); Río Curiacu, ca. 8 km W of Puerto Sarayacu, *Lugo S. 4238* (GB, MICH); vicinity of El Porvenir, ca. 5 km W of Puyopunga, *Lugo S. 4942* (GB, MICH); trail to Copataza, 10 km S of Sarayacu, *Lugo S. 5519* (GB, MICH); carretera de Petro–Canada, vía Auca, 115 km S de Coca, 6 km S del Río Tiguino, *Zak & Rubio 4339* (MO).—NAPO: carretera Hollín–Loreto–Coca, entre Avila y Río Pocuno, 00°39′S, 77°22′W, *Cerón et al. 2879* (MICH); Parque Nacional Yasuní, Pozo Petrolero "Cowi" de Conoco, 00°55′S, 76°20′W, *Coello 171* (MICH); Puerto Francisco de Orellana (Coca), ca. 40 km SE of town, Auca oil field, 00°42′S, 76°52′W, *Balslev & Madsen 10595* (AAU, MBM, MO, NY); near end of the Auca oil field rd, 00°44′S, 76°54′W, *Brandbyge & Asanza 30109* (AAU, MICH); rd Coca (Puerto Francisco de Orellana) to Curaray, ca. 40 km SE of Coca, *Harling et al. 14757* (GB, MICH); Dureno on Río Aguarico, *Harling & Andersson 16605* (GB, MICH); Río Auyabeno, near Puerto Montúfar, 00°06′S, 76°01′W, *Holm-Nielsen et al. 21281* (AAU); Río Aguarico, Monte Cristi, 00°18′S, 76°117′W, *Holm-Nielsen et al. 21666* (AAU); Estación Experimental de INIAP, San Carlos, 6 km SE de los Sachas, *Neill et al. 6216* (MICH); a 2 km de Jatún Sacha, en vía a Tena, 5 km de Misahuallí, 01°08′S, 77°30′W, *Palacios 2789* (MICH). **Brazil.** PARÁ: Tucuruí, *Ramos 1115* (INPA). **Peru.** AMAZONAS: valle del Río Santiago, ca. 65 km N de Pinglo, Quebrada Caterpiza, *Huashikat 1110* (MICH, MO); Prov. Bagua, 8 km E of Montenegro at Km 286 E of Olmos on the Mesones–Muro hwy, *Hutchison & Wright 3781* (F, GH, K, M, MICH, MO, NY, P, US); Mirana, *Woytkowski 5649* (G, GH, MO, US); Prov. Bagua, valley of the Río Marañón above Cascadas de Mayasi, Km 276–280 of Marañón rd, *Wurdack 1841* (F, NY, S, US).—LORETO: Maynas, Shusuna, carretera a Zungarococha, *Ayala 437* (AMAZ, MO); Río Yuvineto, affluent du Río Putumayo, Río Putumayo, *Barrier 441* (AMAZ, MICH); cerca de Zúngaro, Cocha, 15 km SW de Iquitos, *Dodson 2809* (MO, SEL, US); Nauta, Río Marañón above mouth of Río Ucayali, 04°30′S, 73°30′W, *Gentry 29968* (AMAZ, MICH); Maynas, Quebrada Sucursari, Río Napo, 03°15′S, 72°55′W, *Gentry 42673* (MICH); Mishuyacu, near Iquitos, *Klug 113* (F, NY, US); Prov. Loreto, Pampa Hermosa and vicinity, Río Corrientes, 1 km S of junction with Río Macusarí, 03°15′S, 72°50′W, *Lewis et al. 10651* (MO); Maynas, Iquitos, Quista Cocha, *McDaniel 10907* (F, MO); Requena, Río Tepiche, Santa Elena, *McDaniel & Marcos 11263* (F, MO); Santa Ana on the Río Nanay, *Ll. Williams 1225* (F, US).

Stigmaphyllon sinuatum is a common and highly variable species of the lowlands of northern South America. Representatives combining the various expressions of characters of the lamina and anterior style have been segregated as species. Although the extremes show some weak geographic correlation, they are linked by intermediates occurring throughout the range and thus are not accorded taxonomic recognition here (see C. Anderson, 1993b, for a detailed discussion of the morphological variability exhibited by this species). The flowers of *S. sinuatum* are borne in pseudoracemes arranged in compound dichasia and thyrses with axes to the sixth order. The petals vary from mostly red to red-yellow to pure yellow. The posterior-lateral stamens bear unmodified anthers; only anthers of the anterior-lateral stamens have the connective enlarged and the locules reduced (sometimes one or both locules are absent). All anthers are glabrous. The styles are usually bearded adaxially but sometimes glabrous. Although the posterior styles are always foliolate, the anterior styles may bear folioles (the common condition in the eastern and central part of the range) or lack them and have the apex extended into a claw (the common condition in the western part of the range); as the specific epithets imply, *S. brachiatum* and *S. monancistrum* are based on variants that lack folioles on the anterior styles. Occasionally the ornamentation of the anterior style is unequal, i.e., the apex may bear only one foliole or two folioles of unequal size and shape. Such variation might suggest hybridization events; however, a survey of pollen showed little correlation between absence of folioles or presence of variable folioles and malformed pollen (C. Anderson 1993b). In most cases, the pollen was 95–99% normal, but in three collections it was 83–86% normal and in one collection 58–72% normal. Only in two collections, one lacking anterior folioles and one with variable anterior folioles, was the pollen composed entirely of misshapen, thick-walled grains. The samara, like that of most species, bears a large, distally flared dorsal wing. The nut is usually ornamented on each side with 1–3 lateral winglets and/or spurs and crests.

In addition to the variation in petal color and the presence of the folioles on the anterior style, *S. sinuatum* is also highly polymorphic in the shape and the pubescence of the laminas. The names *Banisteria heterophylla*, *S. fulgens*, *S. splendens*, *S. hypoleucum*, and *S. purpureum* are based on collections from the northeastern part of the range (along the Rio Amazônas in Pará, the Guianas, and eastern Venezuela). The laminas of plants from this area are usually orbicular to oblate or reniform and very densely white-sericeous abaxially; the epidermis is hidden. The petals have broad red-orange margins or may be entirely red with only a central yellow spot abaxially. All styles are foliolate. Plants from the western part of the range (Colombia, Ecuador, Peru, and Bolivia) also tend to have broad laminas but are moderately to sparsely sericeous abaxially. The anterior style is most commonly efoliolate and has the apex extended into a claw in Colombia, Ecuador, and northern Peru. (This is the variant that was recognized as *S. brachiatum* and *S. monancistrum*.)

The mostly Brazilian form that includes the type of *S. martianum*, based on a collection from Manaus, has triangular to ovate to cordate laminas, which are moderately (the hairs touching or overlapping) to sparsely (the hairs barely or not touching) sericeous abaxially, yellow and slightly larger petals, and foliolate styles. It is the common type encountered in eastern Pará in the region of Belém and along the Rio Toncantins and its tributaries, and west of the Pará border along the Rio Amazônas and the Rio Solimões. It also occurs in Maranhão as well as infrequently in the western Guianas and southeastern Venezuela. The type of *S. richardianum* from French Guiana belongs here.

Stigmaphyllon sinuatum is sometimes confused with the similar *S. convolvulifolium* (no. 34) of the Guianas and northeastern Brazil (Amapá and eastern Pará) and with *S. car-*

diophyllum of the Amazonian lowlands of Ecuador, Peru, Brazil, and Bolivia. *Stigmaphyllon convolvulifolium* is distinguished by its ovate to cordate laminas, which are abaxially very sparsely and minutely sericeous (appearing glabrous to the naked eye) to glabrate to glabrous; the sessile hairs are only 0.1–0.2 mm long. In most of the area of sympatry, *S. sinuatum* has the laminas abundantly pubescent abaxially; even in less densely pubescent forms the vesture, composed of hairs 0.2–0.5 (–0.7) mm long, is never as sparse as in *S. convolvulifolium*. *Stigmaphyllon cardiophyllum* (no. 69) also has very sparsely sericeous to glabrous laminas. The petal limbs of its small flowers are at most 6.5 mm in diameter. The anthers are abundantly pubescent; those of stamens opposite the posterior-lateral sepals consist of an enlarged connective bearing only one tiny locule (very rarely two). The nut of the samara lacks lateral ornamentation.

Jussieu proposed the combination *Stigmaphyllon sinuatum* based on de Candolle's *Banisteria sinuata*. He mistakenly assumed that the Perrottet collection he had on hand was a duplicate of de Candolle's type, but it is actually a specimen of *S. palmatum* (C. Anderson 1993b). As a result, the name *S. sinuatum* has been misapplied since Jussieu's time to *S. palmatum*, and the names *S. fulgens*, *S. splendens*, *S. martianum*, and *S. hypoleucum* have been most commonly used for the species correctly named *S. sinuatum*.

34. Stigmaphyllon convolvulifolium Adr. Jussieu, Ann. Sci. Nat. Bot., sér. 2, 13: 289. 1840.—TYPE: FRENCH GUIANA. Cayenne, *Martin s.n.* (lectotype, designated by C. Anderson, 1987a: P!, photo: MICH!).

Stigmaphyllon latifolium Bentham, London J. Bot. 7: 128. 1848.—TYPE: SURINAME. *Hostmann 146* (holotype: K-herb. Bentham!, photo: MICH!; isotypes: BM! G! K-herb. Hooker! NY-fragment! P! U! W!).

Vine to 15 m. Stems and branches sericeous and sometimes also with scalelike T-shaped hairs when young, soon becoming glabrous. Laminas 5–15 cm long, 4.5–11.5 cm wide, ovate to cordate or narrowly so, apex acuminate-mucronate, base cordate, very sparsely sericeous to soon glabrate or glabrous adaxially, very sparsely and minutely sericeous (appearing glabrous to the naked eye) to glabrate to glabrous abaxially [trabecula 0.1 (–0.2) mm long, straight, sessile], margin with irregularly spaced sessile glands (0.2–0.5 mm in diameter) and filiform glands (up to 1.6 mm long); petioles 1.5–10 cm long, sericeous and/or with scalelike T-shaped hairs, barely confluent across the node and forming a line bearing the stipules, with a pair of prominent but sessile glands at the apex, each gland 1.1–2.6 mm in diameter; stipules 0.3–0.9 mm long, 0.5–1.3 mm wide, free, triangular or broadly so, eglandular. Flowers ca. 15–40 per pseudoraceme, these borne in dichasia or compound dichasia or small thyrses (axes to the 3rd order, sericeous). Peduncles 4–12.5 mm long; pedicels 3–9 mm long, terete; both sericeous or mixed with scalelike T-shaped hairs, peduncles 0.7–2 times as long as or subequal to the pedicels. Bracts 1–1.7 mm long, 0.8–1.2 mm wide, triangular, apex acute; bracteoles 1–1.5 mm long, 0.7–1.3 mm wide, oblong to ovate, apex obtuse, eglandular or each bracteole with a pair of inconspicuous glands (each 0.2–0.4 mm in diameter); bracts and bracteoles sericeous abaxially. Sepals 1.8–2.5 mm long and wide, glands 1.2–1.9 mm long, 0.7–1.2 mm wide. All petals with the limb orbicular, glabrous, yellow suffused with red, margin erose to denticulate-fimbriate, teeth/fimbriae up to 0.2 mm long; anterior-lateral petals: claw (1.5–) 2–2.2 mm long, limb 11–12 mm long and wide; posterior-lateral petals: claw 1–1.5 (–1.7) mm long, limb 8–10 (–11) mm long and wide; posterior petal: claw (3.2–) 2.5–3.5 mm long, apex indented, limb 6–8.5 mm long and wide. Stamens unequal, those opposite the posterior styles the largest, anthers of those opposite the anterior-lateral sepals with the connective enlarged and the locules reduced or rarely with only one reduced locule or eloculate; anthers usually all loculate, glabrous. Stamen opposite anterior sepal: filament 2.5–3 mm long, anther 1.2–1.6 mm long; stamens opposite anterior-lateral petals: filaments 1.8–2.5 mm long, anthers 0.9–1.3 mm long; stamens opposite anterior-lateral sepals: filaments 2.6–3.3 (–3.7) mm long, connectives 1–1.3 (–1.5) mm long, locules 0.3–0.8 mm long; stamens opposite posterior-lateral petals: filaments 2.8–3.9 mm long, anthers 1–1.3 mm long; stamens opposite posterior-lateral sepals: filaments 2–2.7 mm long, anthers 0.7–1 mm long; stamen opposite posterior petal always shorter than the adjacent two: filament 1.7–2 mm long, anther 0.5–1 mm long. Anterior style 2.8–3.3 mm long, shorter than the posterior two, terete, with scattered hairs in the proximal 1/3–1/2, erect or slightly recurved; apex 1.5–1.9 mm long sometimes including a spur ca. 0.2 mm long; folioles variable, the larger folioles 0.9–1.5 mm long, 0.7–1.5 mm wide, parabolic to broadly lunate to subrectangular, sometimes much smaller: ca. 0.6 mm long, ca. 1 mm wide, broadly triangular. Posterior styles 3.1–4 mm long, terete, with scattered hairs in the proximal 1/4–3/4, lyrate; foliole 1.5–1.8 mm long, 1.6–2 mm wide, suborbicular to subsquare to trapezoidal. Dorsal wing of samara 3.4–4.2 cm long, 1.2–2 cm wide, upper margin with a blunt tooth; nut bearing a pair of rectangular to semi-circular to lunate, entire to grossly dentate lateral winglets, these 3.5–6.5 mm long, 1.2–2 mm wide, and often also with a few spurs and crests up to 1.5 mm long and 1 mm wide; nut 4.5–6.7 mm high, 3.5–4.3 mm in diameter, without air chambers, areole 3.3–3.5 mm long, 2.8–3.5 mm wide, concave, carpophore up to 3 mm long. Embryo 6–6.7 mm long, ca. 2 times as long as wide, ovoid, outer cotyledon 9–10.2 mm long, ca. 3.3 mm wide, the distal 1/3 folded over the inner cotyledon, inner cotyledon 6.6–7.3 mm long, 2.8–3.1 mm wide, folded at the distal 1/4–1/3. Chromosome number unknown.

Phenology. Collected in flower and fruit throughout the year.

Distribution (Fig. 33). Guyana, Suriname, French Guiana, and northeastern Brazil (Amapá and eastern Pará), also reported from Martinique and Trinidad (probably introduced); in moist forest, along rivers, and also in secondary growth and along roadsides; sea level to 300 m.

Representative Specimens. **Martinique.** Marigot, Ste. Marie, *Duss 1473* (NY); *Terrasson s.n.* in 1796 (P-JU). **Trinidad.** Maracas, *Broadway 8052* (A, BM, MO, S); NE of Point Fortin, *Davidse 2578* (F, GH, MO, NY); Salybia Bay over bridge, *Dean TRIN13322* (TRIN).

Guyana. Northwest Distr., Waini River, Marabo Creek, *de la Cruz 1267* (NY, US); Pomeroon Distr., Pomeroon River, *de la Cruz 3044* (CM, F, GH, MO, NY, US); Kamakusa, upper Mazaruni River, ca. 59°50′W, *de la Cruz 4149* (CM, F, GH, NY, US); margins of Berbice River, S of New Dageraad, ca. 06°N, 57°43′W, *Maas et al. 5543* (MICH); Essequibo, *Meyer s.n.* (GOET). **Suriname.** Fluv. Coppename, *Boon 1048, 1104* (U); Para, along rd (Meursweg) from Zanderij Hwy to Paranam, 8.3 km from Zanderij Hwy, 05°37′N, 55°08′3″W, *Evans & Lewis 1883* (MICH); Haut Litany, Basin du Litany, 02°31′N, 54°45′W, *de Granville et al. 11945* (MICH); Wilhelmina Gebergte, Lucie River, 03°20′N, 56°49′W, *Irwin et al. 55408* (F, K, MICH, NY, U, US), *Irwin et al. 55463* (C, F, K, MICH, MO, NY, RB, U, US); Paramaribo, *Kramer & Hekking 2338* (U); ad ripas fluv. Marowijne, *Lanjouw & Lindeman 2962* (NY, U); Jandé kreek, boven Suriname rivier, 1 1/2 uur varen beneden Kabel, *Lindeman 4455* (MO, U); Jodensavanne-Mapane kreek area, Suriname River, *Lindeman 5001* (U); Perica River, *Lindeman 5440* (COL, MICH); Saramacca River, rear of village Jacob kondre, *Maguire 23847* (F, GH, K, MO, NY, RB, U, US); fluv. Gonini, *Versteg 47* (U). **French Guiana.** Haut Riv. Mana, amont de Sant Ananas, *Cremers 7531* (CAY, MICH); S de St. Jean du Maroni, *Cremers 7672* (CAY, MICH); route de St. Laurent à Paul Isnard entre les PK 10 et 40, *Cremers 7979* (CAY, MICH); fleuve Tampoc, à 4 km en amont de son confluent avec l'Ouaqui, *de Granville B4834* (MICH, P); RN2, à proximité du pont sur l'Orapu, *de Granville 5036* (CAY, MICH); Crique Cabaret–Bassin de l'Oyapock, entre l'embouchure et la crique Mérignan, 03°55′N, 51°48′W, *de Granville 10235* (MICH, NY); Trois Sauts, Akattis Alasuka, *Haxaire 566* (CAY); Comté, entre Rodre Fondé et

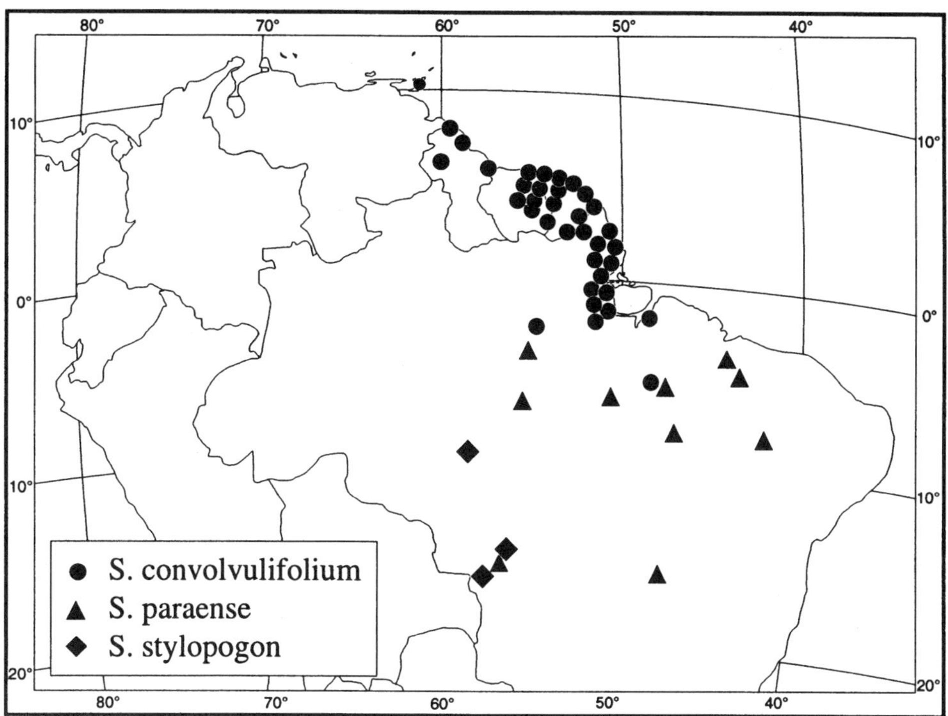

FIG. 33. Distribution of *Stigmaphyllon convolvulifolium, S. paraense,* and *S. stylopogon.*

Belizón, *Oldeman 1449* (MICH); Haut Oyapock, environ 2500 avant l'embouchure de la rivière Eurupoucigne, *Oldeman B3289* (MICH); entre Cabassou et Degrad des Cannes, Île de Cayenne, *Prévost 1257* (CAY, MICH); rivière Inini, affluent du Moyen–Maroni (Lawa), en amont de Maripasoula, *Sastre et al. 3996* (CAY, MICH, P); Sinnamary, route de Ste. Elie, Km 17, *Sastre et al. 4199* (MICH, P, US); fleuve Approuague, près de Régina, *Sastre 4813* (CAY, MICH); montagnes de Kaw, auberge de Brousse des Cascades, 04°35′N, 52°17′W, *Weitzman 272* (MICH). **Brazil.** AMAPÁ: Rio Amapari, Serra do Navio, *Cowan 38206* (MICH, NY); Rio Jari, 0.5–3 km S of Santo Antonio da Cachoeira, 00°55′S, 52°55′W, *Egler & Irwin 46066* (MICH, NY, UB); confluence of Rio Iane with Rio Oiapoque, 02°53′N, 52°22′W, *Egler & Pires 47771* (MICH, NY); Mpio. Mazagão, BR-156, 81 km WSW of Macapá, ca. 11 km SW of Rio Preto, 00°08′S, 51°48′W, *Mori & Cardoso 17432* (MICH); Mpio. Calçoene, BR-156, 53–72 km WSW of Calçoene, ca. 02°33–38′N, 51°16′W, *Rabelo et al. 2964* (MICH); Rio Araguari, between 01°02′N, 51°45′W and 00°57′N, 51°29′W, *Pires et al. 50907* (IAN, MG, MICH, NY); between Rios Cujubim and Flechal, 01°45′N, 50°58′W, *Pires & Cavalcante 52435* (IAN, MG, MICH, NY).— PARÁ: Obidos, Rio Paru de Oeste, *Cavalcante 801* (MG); Gurupá, Rio Amazon, *Killip & Smith 30594* (NY, US); Jari, estrada do Munguba, *N. T. Silva 2138* (IAN, MICH).

Stigmaphyllon convolvulifolium is distinguished by its ovate to cordate leaves, which are abaxially very sparsely and minutely sericeous (appearing glabrous to the naked eye) to glabrate to glabrous; the sessile hairs are only 0.1–0.2 mm long. Only the stamens opposite the anterior-lateral sepals have enlarged connectives and reduced locules, and the styles are pubescent in the basal 1/4–3/4. The petals, especially the flag, are usually suffused with red. In Brazil this species is most likely to be confused with the widely distributed and extremely variable *S. sinuatum* (no. 33) and (in Pará) also with *S. cardiophyllum* (no. 69). In the range of *S. convolvulifolium*, the common form of *S. sinuatum* has the leaves densely white-sericeous abaxially. A variant in French Guiana and in Amazonian Brazil has the leaves less abundantly pubescent abaxially though never as sparsely as in *S. convolvulifolium;* the hairs are 0.2–0.5 (–0.7) mm long. *Stigmaphyllon cardiophyllum* is readily separated by its small pure-yellow petals (the limb at most 6.5 mm in diameter), pubescent anthers, and laterally unornamented samaras.

Stigmaphyllon convolvulifolium has been collected twice in Martinique and a few times in Trinidad; these records probably represent escaped introductions. Niedenzu (1900, 1928) reported *S. convolvulifolium* also from St. Vincent, based on *H. H. Smith & G. W. Smith 418*, presumably a duplicate at B, which is no longer extant. The duplicate at K is a mixture of the South American *S. sinuatum* and one leafy branchlet of *S. finlayanum* (Trinidad and adjacent Venezuela). Neither has been recollected on St. Vincent or elsewhere in the Antilles, and it seems likely that these specimens represent plants escaped from cultivation at the Botanical Garden in St. Vincent.

35. Stigmaphyllon macedoanum C. Anderson, Contr. Univ. Michigan Herb. 17: 10. 1990.—TYPE: BRAZIL. Minas Gerais: Capinópolis, Fazenda Santa Terezinha, as margens do Rio Paranaiba divisa de Goiás e Minas Gerais, 27 Jan 1989, *Macedo 5486* (holotype: RB!; isotypes: G! K! MICH! NY! R! UB! US!).

Vine. Stems and branches sericeous when young, soon becoming glabrate. Laminas 9–18.5 cm long, 6.3–19 cm wide in outline, the larger laminas palmately 3–5 (–7)-lobed to broadly ovate or broadly elliptical, the smaller laminas 2–3-lobed to broadly ovate, apex cuspidate, base deeply cordate to auriculate or in the smaller sometimes sagittate, sparsely pubescent with subsessile T-shaped hairs (trabecula 0.1–0.3 mm long, straight, subsessile) to glabrate adaxially, sparsely sericeous abaxially (trabecula 0.2–0.4 mm long, straight, sessile), margin with filiform glands (up to 1.5 mm long), these terminating the major veins; petioles 4–15 cm long, sericeous, confluent across the node and forming a corky ridge bearing the stipules, with a pair of shallowly cupulate glands at the apex, each gland 0.9–2 mm in diameter, ca. 0.5 mm long; stipules 0.2–0.5 mm long, 0.5–0.6 mm wide, free, triangular, eglandular, caducous. Flowers ca. 15–25 per pseudoraceme, these borne in dichasia or compound dichasia (axes to the 3rd order, sericeous). Peduncles 2.5–4 mm long, sericeous; pedicels 5.5–6 mm long, sericeous but glabrate below the flower; peduncles 0.4–0.7 times as long as the pedicels. Bracts 1.1–1.5 mm long, 0.6–0.9 mm wide, triangular, apex acute; bracteoles 0.8–1 mm long, 0.5–0.8 mm wide, triangular, apex obtuse, eglandular. Sepals ca. 2 mm long and wide, glands 1.8–2 mm long, ca. 1 mm wide. All petals with the limb orbicular, glabrous, yellow, margin of lateral petals with fimbriae up to 0.4 mm long; anterior-lateral petals: claw 1.6–1.7 mm long, limb ca. 9.5 mm long and wide; posterior–lateral petals: claw ca. 1 mm long, limb ca. 7.5 mm long and wide; posterior petal: claw ca. 3 mm long, apex not indented, limb ca. 6.5 mm long and wide, margin with fimbriae up to 1 mm long. Stamens unequal, those opposite the lateral sepals with the longest filaments, anthers of those of stamens opposite anterior-lateral sepals with the connective slightly enlarged and the locules usually obliquely placed; anthers all loculate, glabrous. Stamen opposite anterior sepal: filament 1.8–2 mm long, anther ca. 1.1 mm long; stamens opposite anterior-lateral petals: filaments 1.4–1.5 mm long, anthers ca. 0.9 mm long; stamens opposite anterior-lateral sepals: filaments ca. 2.5 mm long, anthers ca. 0.8 mm long; stamens opposite posterior-lateral petals: filaments 2–2.1 mm long, anthers ca. 1.4 mm long; stamens opposite posterior-lateral sepals: filaments ca. 2.5 mm long, anthers ca. 0.8 mm long; stamen opposite posterior petal always shorter than the adjacent two: filament 1.8–2.1 mm long, anther ca. 0.6 mm long. Anterior style ca. 2.5 mm long, shorter than the posterior two, terete, erect, glabrous; apex ca. 1 mm long including

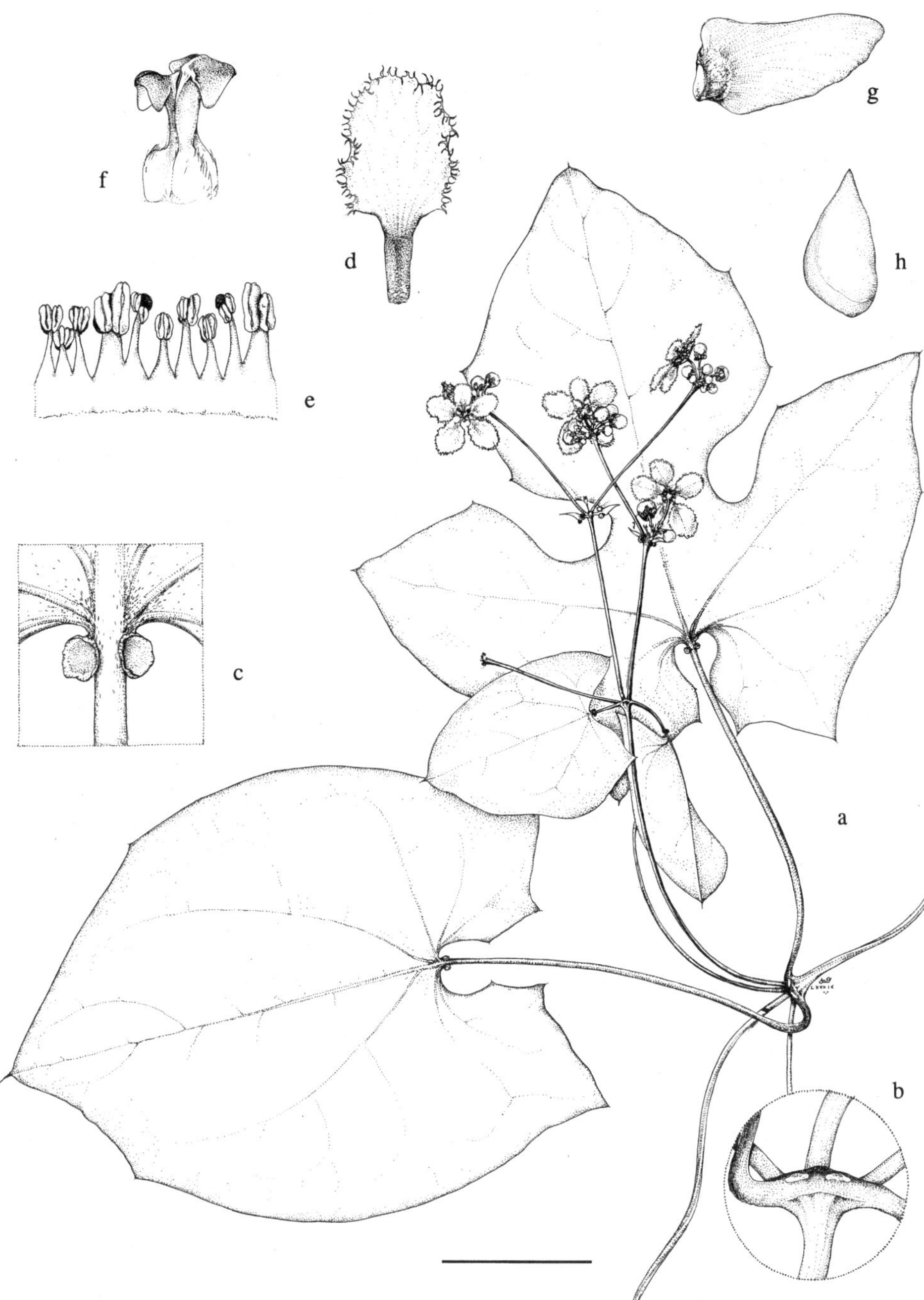

FIG. 34. *Stigmaphyllon macedoanum*. a. Flowering branch. b. Detail of node showing corky ridge formed by the confluent bases of the petioles. c. Base of leaf, abaxial view. d. Posterior petal (the "flag"). e. Androecium; stamen second from left opposes the posterior petal. f. Gynoecium, anterior style to the left. g. Samara. h. Embryo. Scale: for a, bar = 4 cm; for b, bar = 1.3 cm; for c, h, bar = 8 mm; for d, bar = 6.7 mm; for e, f, bar = 4 mm; for g, bar = 2 cm. (Based on *A. Macedo 5486*.)

a spur ca. 0.2 mm long, each foliole ca. 0.8 mm long, 0.6–0.7 mm wide, triangular or broadly triangular. Posterior styles ca. 3 mm long, terete, slightly lyrate, glabrous; apex 1.5 mm long and incurved, foliole ca. 1.2 mm long, ca. 1 mm wide, triangular. Dorsal wing of samara 1.8–2.1 cm long, 1.2 cm wide, upper margin with a blunt tooth; lateral winglets absent; nut ca. 7 mm high, ca. 3.5 mm in diameter, without air chambers, areole ca. 4 mm long and wide, plane, carpophore up to 2.3 mm long. Embryo ca. 7.5 mm long, ca. 2 times as long as wide, ovoid, outer cotyledon ca. 6.6 mm long, ca. 4.2 mm wide, distally curved but not folded over the inner cotyledon, inner cotyledon ca. 6 mm long, ca. 3.8 mm wide, straight. Chromosome number unknown. Fig. 34.

Stigmaphyllon macedoanum is known only from two collections from one area in central Brazil (Fig. 82), the type and *Macedo 5057* (MBM; Brazil. Minas Gerais: Ituiutaba, Capinópolis, Fazenda Santa Terezinha, ramosa de mata em pedreiras, 22 Nov 1971). It is distinguished by it polymorphic, sparsely pubescent leaves, whose major veins terminate in filiform marginal glands. The petioles are conspicuously fused across the node into a corky ridge bearing the caducous stipules. This joining of the petioles occurs in several other species, but is most pronounced in *S. macedoanum*. The anthers of stamens opposite the anterior-lateral sepals are not as greatly modified as in most other species with a heterogeneous androecium; the connective is only slightly enlarged and the locules are usually obliquely placed.

This species may be confused with two species that are not sympatric but also have sparsely pubescent to glabrous lobed laminas, *S. vitifolium* (no. 51), endemic to Rio de Janeiro, and *S. jatrophifolium* (no. 52), found along the Río Paraná and Río Uruguay and their tributaries. Both lack the prominent corky ridge formed by the fusion of the petioles across the node and have a distinctly heterogeneous androecium.

36. Stigmaphyllon angulosum (L.) Adr. Jussieu, Ann. Sci. Nat. Bot., sér. 2, 13: 288. 1840. *Banisteria angulosa* L., Sp. pl. 1: 427. 1753.—TYPE: t. 92 in Plumier, Descr. pl. Amér. 1693.

Banisteria deformis Desvaux ex Hamilton, Prodr. pl. Ind. occ. 40. 1825.—TYPE: herb. Desvaux, fide Jussieu, 1843 (holotype: P?, not located).

Stigmaphyllon angulosum f. *hederifolium* Niedenzu, Ind. Lect. Lyc. Brunsberg. p. aest. 1900: 12. 1900.—SYNTYPES: *Mayerhoff 171*, *Picarda 948*, *1358* (B, destroyed).

Vine to 12 m. Stems and branches densely sericeous and also with scalelike T-shaped hairs when young, becoming sparsely so to glabrate. Laminas 4.2–16 cm long, 4.2–17 cm wide, sinuate-lobate with usually 5–7 (–9) lobes or rarely ovate or suborbicular, apex of each lobe apiculate, base cordate to auriculate, glabrate to glabrous adaxially, sparsely to densely pubescent with T-shaped hairs abaxially (trabecula 0.5–0.8 mm long, wavy, stalk 0.05–0.15 mm long) or the largest and oldest laminas very sparsely pubescent to glabrate abaxially, margin with filiform glands (up to 3 mm long) and/or sessile glands (up to 0.6 mm in diameter); petioles 1.6–7.7 (–9) cm long, densely pubescent with scalelike T-shaped hairs, confluent across the node and forming a band bearing the stipules, with a pair of prominent but sessile glands at the apex, each gland 1.2–1.8 mm in diameter; stipules 0.5–1 mm long, 0.8–1.3 mm wide, broadly triangular, free, eglandular. Flowers 15–35 (–ca. 40) per congested pseudoraceme, sometimes in an umbel, these borne in simple or compound dichasia or sometimes in small thyrses (axes to the 4th order, with scalelike T-shaped hairs, or the more proximal axes of large inflorescences glabrate) or sometimes solitary. Peduncles 3.4–12 mm long; pedicels 5.5–10 mm long, terete; both densely pubescent with scalelike T-shaped hairs, peduncles 0.6–1.3 times as long as the pedicels. Bracts 0.7–1.5 mm long, 0.7–1.2 mm wide, triangular, apex acute; bracteoles 0.8–1.7 mm long, 0.6–1 mm wide, triangular, apex obtuse, eglandular or sometimes each bracteole with a pair of inconspicuous glands (each gland 0.4–0.5 mm in diameter); bracts and bracteoles densely sericeous abaxially. Sepals 2.2–3 mm long, 2.1–2.6 mm wide, glands 1.5–1.8 mm long, 0.8–1.2 mm wide. All petals with the limb orbicular, glabrous, yellow, margin irregularly dentate and/or fimbriate, especially in the distal 2/3, fimbriae up to 0.4 mm long; anterior-lateral petals: claw ca. 2.5 mm long, limb 13–14 mm long and wide; posterior-lateral petals: claw 1.5–2 mm long, limb ca. 11.5–13 mm long and wide; posterior petal: claw 3–3.5 mm long, apex indented, limb ca. 10–11 mm long and wide. Stamens unequal, those opposite the posterior styles the largest, anthers of those opposite the anterior-lateral sepals with the connective enlarged and the locules reduced; anthers all loculate, glabrous. Stamen opposite anterior sepal: filament 2.6–3.2 mm long, anther 1.2–1.5 mm long; stamens opposite anterior-lateral petals: filaments 1.7–2.5 mm long, anthers 1.1–1.2 mm long; stamens opposite anterior-lateral sepals: filaments 2.3–3.2 mm long, connectives 0.8–0.9 mm long, locules 0.5–0.7 mm long; stamens opposite posterior-lateral petals: filaments 3.4–3.7 mm long, anthers 1.2–1.3 mm long; stamens opposite posterior-lateral sepals: filaments 1.8–3.5 mm long, anthers 0.6–0.8 mm long; stamen opposite posterior petal usually shorter but sometimes longer than the adjacent two: filament 2.3–2.8 mm long, anther 0.9–1 mm long. Anterior style 2.6–3.6 mm long, shorter than the posterior two, terete, glabrous or with a few scattered hairs, slightly recurved; apex 1.1–1.6 mm long, each foliole 0.7–1.4 mm long, 0.4–1.2 mm wide, narrowly to broadly parabolic. Posterior styles 3.5–4.2 mm long, terete, glabrous or with scattered hairs, lyrate; folioles 1.3–1.6 mm long and wide, nearly square. Dorsal wing of samara 2.8–4.5 cm long, 1.2–1.8 cm wide, upper margin with a blunt or sometimes acute tooth; nut with a pair of rectangular to triangular lateral winglets, each 5–7.8 mm long, 2.2–4.5 mm wide; nut 4.5–6.5 mm high, 3.5–4.7 mm in diameter, without air chambers, areole 4 mm long, 3.5–4 mm wide, concave, carpophore up to 3.2 mm long. Embryo 5.5–7 mm long, ca. 2 times as long as wide, ovoid, outer cotyledon 7–9.1 mm long, 2.8–3.8 mm wide, the distal 1/4–2/3 folded over the inner cotyledon, inner cotyledon 4.5–5.2 mm long, 2.5–3.8 mm wide, folded at the distal 1/5 to straight. Chromosome number unknown.

Phenology. Collected in flower and in fruit throughout the year.

Distribution (Fig. 35). Endemic to Hispaniola; pine woodlands and mixed hardwood forests, thickets, grassy slopes, and in secondary growth; sea level to 1250 m.

REPRESENTATIVE SPECIMENS. **Dominican Republic**. Prov. Azua, Sierra de Ocoa, San José de Ocoa, *Ekman H11880* (S, US); Barahona, *Fuertes 17* (BM, BR, F, G, GH, K, MO, NY, S, US, W); prov. San Juan, Juan Santiago, *Howard & Howard 9295* (BM, GH, MICH, NY, P, S, US); Monte Cristo, Dist. Sabaneta, *Valeur 66* (G, MICH, S, US); prope Constanza, *von Türckheim 3186* (BM, BR, G, K, M, NY, W); Prov. Santiago, en la ladera oriental de la Loma Diego de Ocampo, 19°35′N, 70°44′W, *Zanoni et al. 42617* (MICH). **Haiti**. Massif du Nord, Port-Margot, *Ekman H2924* (G, K, S, US); Massif du Nord, Port-de-Paix, Haut-Piton, *Ekman H4886* (K, NY, S); trail to Morne Rochelois, Miragoane and vicinity, *Eyerdam 517* (GH, NY, P, US); vicinity of Mission, Fonds Varettes, *Leonard 3630* (NY, US); Nord, vicinity of Dondon, *Leonard 8633* (F, US); vicinity of Bombardopolis, *Leonard & Leonard 13411* (NY, US).

Stigmaphyllon angulosum, endemic to Hispaniola, is easily recognized by its variable but distinctive leaves, which are unique in the family. The lamina is typically shallowly to deeply sinuate-lobate with ca. 5–7 lobes. The only other species on Hispaniola in which

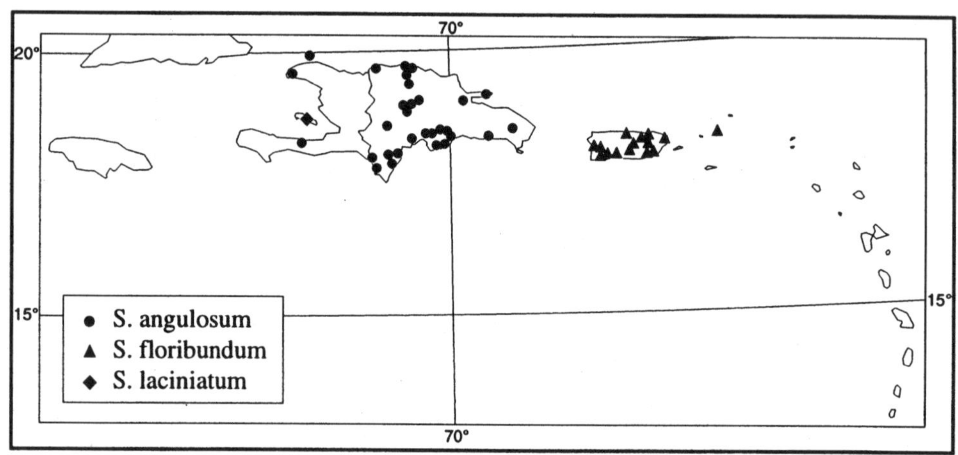

FIG. 35. Distribution of *Stigmaphyllon angulosum, S. floribundum*, and *S. laciniatum.*

all styles are foliolate is *S. puberum* (no. 17), whose laminas are usually elliptical to lanceolate, never lobed. Like *S. laciniatum,* restricted to Gonâve Island, *S. angulosum* resembles more the mainland species of *Stigmaphyllon* than the other West Indian endemics.

37. Stigmaphyllon emarginatum (Cavanilles) Adr. Jussieu, Ann. Sci. Nat. Bot., sér. 2, 13: 290. 1840. *Banisteria emarginata* Cavanilles, Diss. 9: 425. 1790.—TYPE: t. 249 in Cavanilles, Diss. 9. 1790 (lectotype, designated by Niedenzu, 1899).

Banisteria fulgens L., Sp. pl. 1: 427. 1753, non *Stigmaphyllon fulgens* Adr. Jussieu, 1840.—TYPE: specimen in the Clifford herbarium [Hort. Cliff. 169; "Bannisteria foliis ovatis, ramis ramosis, seminibus extrorsum tenuioribus, introrsum lacinulam emittentibus"] (lectotype, here designated: BM!, photo: P!).

Triopterys lingulata Poiret in Lamarck, Encycl. 8: 104. 1808. *Stigmaphyllon lingulatum* (Poiret) Small, N. Amer. fl. 25(2): 140. 1910.—TYPE: DOMINICAN REPUBLIC. "de St. domingue," collector unknown (holotype: P-LAM!, photo: MICH!).

Banisteria umbellulata DC., Prodr. 1: 588. 1824.—TYPE: DOMINICAN REPUBLIC. "in Sancto Domingo," *Bertero s.n.* (holotype: G-DC!, microfiche and photo: MICH!).

Banisteria periplocifolia Desfontaines ex DC., Prodr. 1: 589. 1824. *Stigmaphyllon periplocifolium* (Desfontaines ex DC.) Adr. Jussieu, Ann. Sci. Nat. Bot., sér. 2, 13: 290. 1840.—TYPE: PUERTO RICO. *Bertero s.n.* (holotype: G-DC!, microfiche and photo: MICH!).

Banisteria periplocifolia α *subovata* DC., Prodr. 1: 589. 1824.—TYPE: unknown.

Banisteria periplocifolia β *angustifolia* DC., Prodr. 1: 590. 1824.—TYPE: specimen in herb. Balbis (holotype: TO, fragment: G-DC!).

Banisteria microphylla Hamilton, Prodr. fl. Ind. occ. 40. 1825, nec *Banisteria microphylla* Jacquin, 1768, nec *Stigmaphyllon microphyllum* Grisebach, 1860.—TYPE: specimen in herb. Desvaux, fide Adr. Jussieu, 1843 (holotype: P?, not located).

Stigmaphyllon emarginatum f. *parvifolium* Niedenzu, Ind. Lect. Lyc. Brunsberg. p. hiem. 1899–1900: 7. 1899.—TYPE: JAMAICA. Prope Kingston, *Oersted s.n.* (lectotype, designated by C. Anderson, 1987a: C!, photo: MICH!).

Stigmaphyllon periplocifolium f. *intermedium* Niedenzu, Ind. Lect. Lyc. Brunsberg. p. hiem. 1899–1900: 7. 1899.—TYPE: JAMAICA. *McNab s.n.* (lectotype, designated by C. Anderson, 1987a: GOET!, photo: MICH!).

Stigmaphyllon periplocifolium f. *microphyllum* Niedenzu, Ind. Lect. Lyc. Brunsberg. p. hiem. 1899–1900: 7. 1899.—TYPE: HAITI. *Bertero 28x* (holotype: B, destroyed).

Stigmaphyllon periplocifolium f. *sericans* Niedenzu, Verz. Vorles. Ak. Braunsberg. W.-S. 1912–1913: 24. 1912. *Stigmaphyllon lingulatum* var. *sericans* (Niedenzu) Niedenzu in Urban, Symb. antil. 8: 336. 1920.—TYPE: HAITI: Gonaives, *Buch 616* (holotype: B, destroyed).

Stigmaphyllon haitiense Urban & Niedenzu in Urban, Symb. antil. 7: 243, June 1912. *Stigmaphyllon haitiense* f. *ovatum* Urban & Niedenzu in Urban, Symb. antil. 7: 244, June 1912, nom. superfl.—TYPE: Haiti. *Christ 1772* (holotype: B, destroyed).

Stigmaphyllon haitiense f. *lineare* Niedenzu & Urban in Urban, Symb. antil. 7: 244, June 1912.—TYPE: Haiti. *Christ 1771* (holotype: B, destroyed).

Stigmaphyllon rubrinervum Alain, Mem. New York Bot. Gard. 21: 122. 1971.—TYPE: DOMINICAN REPUBLIC. Cordillera de Yaroa, near trail to Arroyo del Toro, 800–850 m, 28–29 Jun 1968, *Liogier 11863* (holotype: NY!, photo: MICH!; isotypes: GH! P! US!).

Vine to 10 m. Stems and branches sericeous when young, soon becoming glabrous, tuberculate. Laminas 1–13 cm long, 0.5–10.5 cm wide, extremely variable: linear, lanceolate, oblong, elliptical, ovate, or sometimes suborbicular, apex mucronate-emarginate or sometimes mucronate, base truncate to cordate, sometimes oblique, glabrate to glabrous adaxially or in young and smaller leaves sparsely sericeous adaxially, sparsely sericeous but commonly glabrate to glabrous abaxially or sometimes sericeous abaxially (trabecula 0.3–0.9 mm long, straight, stalk absent to 0.02 mm long), margin eglandular; petioles 0.2–2 (–3) cm long, sericeous, commonly glabrate to glabrous in older leaves, not confluent across the node, usually with a pair of glands at the apex, each gland 0.3–1.2 mm in diameter, prominent but sessile or sometimes up to 0.6 mm high, or one or both glands absent; stipules 0.3–0.8 (–1) mm long, 0.2–0.5 mm wide, free, triangular to sublinear, eglandular. Flowers (5–) 15–35 (–45) per open to congested pseudoraceme (if few-flowered then aggregated into an umbel), these usually solitary, sometimes borne in dichasia (axes glabrous or sparsely sericeous). Peduncles 1.3–25 mm long; pedicels 3–23 mm long, terete; both sericeous to sparsely so or glabrous, peduncles (0.2–) 0.3–1 times as long as the pedicels. Bracts 1–2.4 mm long, 0.4–0.9 mm wide, narrowly triangular, apex acute or acuminate, sericeous or sparsely so abaxially; bracteoles 0.6–1.3 mm long, (0.1–) 0.2–0.4 mm wide, linear to narrowly triangular, apex acute or acuminate, sericeous to densely so abaxially, eglandular or with a glandular area in the basal 1/3–1/2. Sepals 2–3.5 mm long, 1.4–2.5 mm wide, glands 1.2–2 mm long, 0.4–1.5 mm wide. All petals with the limb glabrous, yellow, margin erose, limb of lateral petals orbicular; anterior-lateral petals: claw (2–) 2.5–3.5 mm long, limb 7.5–11 mm long and wide; posterior-lateral petals: claw 1–2 mm long, limb 7.5–10.5 mm long and wide; posterior petal: claw 3.2–4.2 mm long, apex indented or slightly so, limb 6.5–9.5 mm long and wide, broadly ovate to orbicular to oblate. Stamens unequal, those opposite the posterior styles the largest, or sometimes stamens opposite the styles subequal, anthers of those opposite the lateral sepals usually with the connective enlarged and the locules reduced (rarely only slightly so); anthers all loculate, glabrous or pubescent. Stamen opposite anterior sepal: filament 2.2–2.6 mm long, anther 0.7–1.1 mm long; stamens opposite anterior-lateral petals: filaments 1.8–2 mm long, anthers 0.6–1 mm long; stamens opposite anterior-lateral sepals: filaments 2.3–3 mm long, connectives (0.5–) 0.6–0.8 mm long, locules 0.4–0.6 mm long; stamens opposite posterior-lateral petals: filaments 2.5–3.4 mm long, anthers 0.8–1.5 mm long; stamens opposite posterior-lateral sepals: filaments 2–2.8 mm long, connectives 0.4–0.8 mm long, locules 0.2–0.6 mm long; stamen opposite posterior petal always shorter than the adjacent two: filament 1.7–2.4 (–2.6) mm long, anther (0.4–) 0.6–0.8 (–1) mm long. Anterior style 2–2.8 mm long, shorter than the posterior two, terete, glabrous, erect but distally somewhat recurved; apex 0.5–0.7 mm long including a spur up to 0.4 mm long or blunt, folioles absent. Posterior styles 2.5–3.6 mm long, canaliculate-complicate, glabrous, erect; apex 0.5–1 mm long, truncate, folioles absent. Dorsal wing of samara 1.6–2.2 cm long, 0.7–0.9 cm wide, upper margin with an acute or obtuse tooth; nut ribbed and commonly bearing spurs and/or crests, 2.8–4 mm high, 2–3 mm in diameter, without air chambers, areole 1.2–2.5 mm long, 1.2–2.1 mm wide, convex, carpophore up to 2 mm long. Embryo 4.2–5.5 mm long, ca. 2 times as long as wide, ovoid, outer cotyledon 4–5.3 mm long, 2.1–2.6 mm wide, inner cotyledon 2.5–4.5 mm long, 1.3–2 mm wide, both straight. Chromosome number unknown. Fig. 36.

Phenology. Collected in flower and in fruit throughout the year.

Distribution (Fig. 37). Jamaica, Hispaniola, Puerto Rico, Virgin Islands, and the Lesser Antilles south to Martinique, not reported from Dominica; on limestone and serpentine outcrops, common in coastal thickets; sea level to 1500 m.

REPRESENTATIVE SPECIMENS. **Jamaica**. ST. ANDREW: Palisadoes, *Yuncker 17293* (BM, F, G, MICH, MO, S); Jack's Hill, *Yuncker 18144* (BM, F, G, MICH, MO, S).—ST. CATHERINE: Spanish Town Rd, *Harris 9232* (F, NY, US); Longville Park to Old Harbour Bay, *Harris 11947* (BM, F, K, MO, NY, S, US).—ST. JAMES: Montego Bay to Round Hill Bluff, *Harris 10350* (BM, F, K, NY, US). **Haiti**. Massif des Matheux, Thomazeau, Morne à Cabrits, *Ekman H930* (F, MICH, S); Massif de la Selle group, Morne des Commissaires, Anses-à-Pitre, *Ekman H6712* (A, G, NY, S, US); Massif du Nord, Vallière, *Ekman H9942* (G, LL, S, US); base of Morne à Cabrits, *Holdridge 1077* (BM, F, GH, MICH, NY, US); Anse Galetter, Gonâve Island, *Leonard 3109* (US); Dept. de l'Artibonite, vic. of Ennery, *Leonard 9713* (BM, GH, S); vic. of Port de Paix, *Leonard & Leonard 12352* (GH, NY, US); vic. of Jean Rabel, E of Bord du Mer, *Leonard & Leonard 13613* (A, GH, US); Tortue Island, vic. of Basse Terre, *Leonard & Leonard 14045* (A, US). **Dominican Republic**. Prov. Samana, Los Bañaderos Prietos, *Ekman H15123* (A, K, NY, S, US); Prov. Barahona, Barahona, *Fuertes 21* (BM, BR, F, G, GH, MO, NY, P, S, U, US, W); Prov. Azua, E of Azua, *Howard & Howard 8649* (BM, GH, MICH, NY, S, US); Prov. Monte Christi, near Puerto Libertador, Manzanilla Bay, *Howard & Howard 9667* (GH, NY, S, US); Prov. Seibo, vicinity of Higüey, *Howard 9752* (A, BM, MICH, NY); prope Constanza, *von Türckheim 3246* (BM, BR, F, G, GH, K, M, MO, NY, S, U, US, W); Prov. Santiago, dist. San José de las Matas, Magua, *Valeur 961* (C, G, K, LL, MO, NY, US). **Puerto Rico**. Culebra Island, *Britton & Wheeler 44* (F, NY, US); vic. of Coamo Springs, *Britton et al. 5960* (G, NY); Condado, *Britton et al. 6632* (G, NY); Guanica, *Liogier 10785* (GH, NY); prope Salinas de Cabo Rojo, *Sintenis 584* (G, GH, GOET, K, M, P, S, US); without locality, *Heller & Heller 474* (F, NY, US). **St. Thomas**. *Britton & Shafer 378* (C, F, K, MO, NY, US); *Eggers 390* (BR, CAS, G, GOET, K, M, P, W); *Wydler 40* (BR, F, G). **St. John**. *Acevedo Rodríguez 4042* (US); *Britton & Shafer 515* (NY, US). **Tortola**. *Britton & Shafer 871* (NY, US); *D'Arcy 793* (A). **Anegada**. *Britton & Fishlock 1045* (NY, US); *D'Arcy 4863* (MO). **Virgin Gorda.** *Fishlock 68* (NY); *Gillis 5825* (A, MSC). **St. Croix**. *Hunnewell 20114* (GH); *Ricksecker 154* (F, GH, MO, NY, US). **Anguilla**. *Howard & Kellogg 19053, 19057* (MICH). **St. Martin**. *Boldingh 2629* (NY, U); 14 Jan 1958, *Hummel s.n.* (GB); *Stoffers 2673* (A, U). **St. Barthélemy**. *Questel 54* (NY, P), *630* (NY, US). **St. Eustatius.** *Boldingh 96, 191* (U). **Barbuda**. *Box 612* (MO, US). **Nevis**. *Proctor 19607* (A). **Antigua**. *Rose et al. 3497* (NY, US); *Box 791* (US). **Guadeloupe**. *Bena 1944* (P); *Stehlé 121* (NY). **Martinique**. *Belanger 553* (P); *Duss 439* (NY).

Typical specimens of this highly variable West Indian endemic have sparsely sericeous to glabrous stems and leaves, the laminas lanceolate to elliptical or ovate, and bear 15–25 showy flowers in solitary pseudoracemes. *Stigmaphyllon emarginatum* is eas-

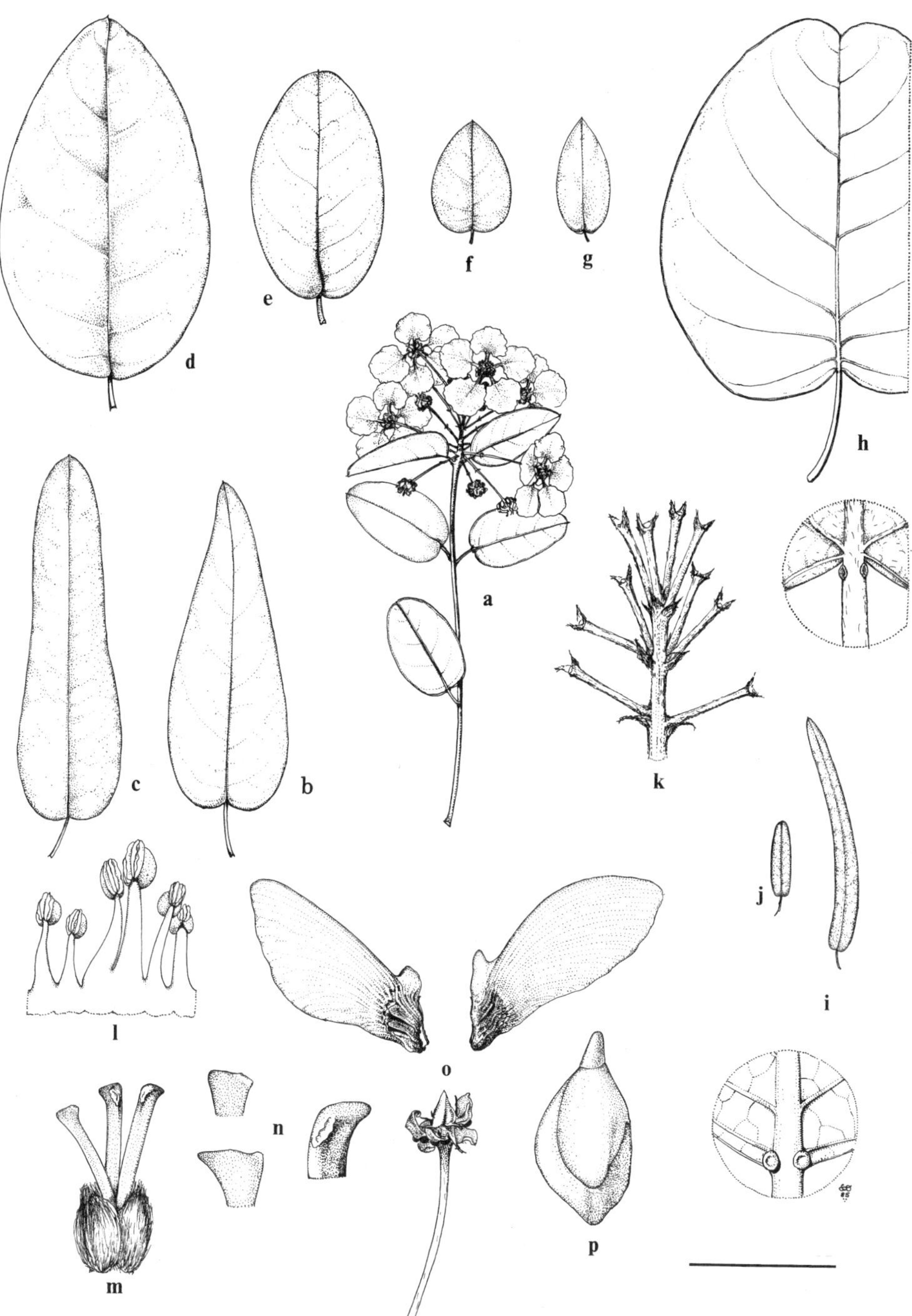

FIG. 36. *Stigmaphyllon emarginatum*. a. Flowering branch. b–j. Leaves illustrating variation of laminar shape, and details of base of leaves (h, i), abaxial view. k. Inflorescence axis with persistent peduncles. l. Portion of androecium; anterior stamen on extreme left, posterior stamen on extreme right. m. Gynoecium, anterior style to the left. n. Apical portion of styles, two anterior style apices on left, posterior style apex on right. o. Two samaras and torus. p. Embryo. Scale: for a–j, bar = 4 cm, for detail of leaf bases, bar = 8 mm (h) and 4 mm (i); for k, bar = 8 mm; for l, m, p, bar = 4 mm; for n, bar = 2 mm; for o, bar = 1.3 cm. (Based on: a, k, *Valeur 416*; b, *Yuncker 17293*; c, *Liogier 15180*; d, *Liogier 17597*; e, *Howard & Howard 8649*; f, *Liogier 14820*; g, *Liogier 12442*; h, *Yuncker 18144*; i, *Liogier & Liogier 26900*; j, *Leonard & Leonard 11787*; l, *Yntema 354A*; m, *Liogier 12442*; n, *Liogier 12442* (anterior style, above, and posterior style), *Yntema 354A* (anterior style, below); o, *Leonard 12352* (left), *Yntema 354A* (right); p, *Yntema 354A*.)

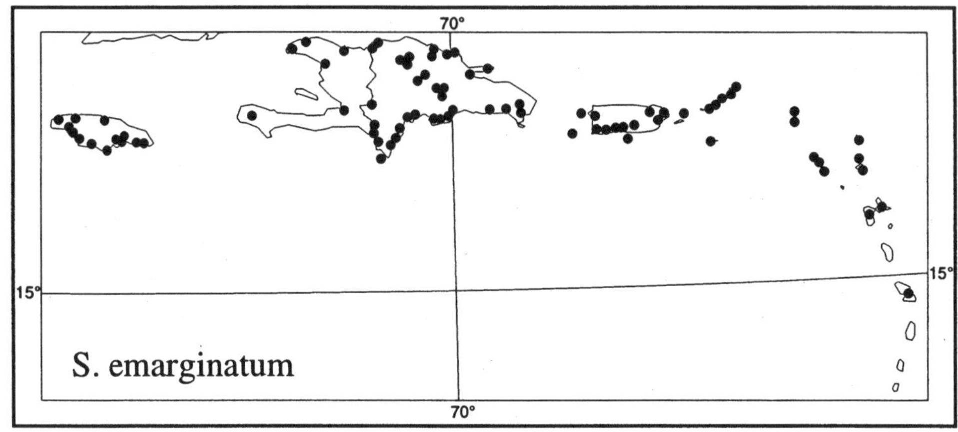

FIG. 37. Distribution of *Stigmaphyllon emarginatum.*

ily distinguished by its unusual posterior styles. All three styles are efoliolate, and the erect, stubby, posterior styles are canaliculate-complicate and apically truncate. Such canaliculate-complicate posterior styles, though foliolate, are also found in the Cuban endemic *S. microphyllum* (no. 38), which resembles small-leaved forms of *S. emarginatum.*

The anthers are sparsely pubescent in nearly all collections from Puerto Rico and the Virgin Islands, whereas in those from Jamaica, Hispaniola, and the Lesser Antilles they are usually glabrous; *Leonard 9713* and *Leonard & Leonard 15209* from Haiti are two exceptional collections with pubescent anthers.

Although conservative in floral characters, *S. emarginatum* is bewilderingly variable in laminar shape and size (Fig. 36a–j). The greatest variety is found on Hispaniola (seven of the ten shapes illustrated here are based on collections from that island), but all shapes may be expected throughout the range. The long list of synonyms reflects attempts to subdivide this species on the basis of leaf morphology; however, all the extremes in variation are linked by intermediate forms. The recognition of *S. periplocifolium* was already questioned by Fawcett and Rendle (1920).

Stigmaphyllon emarginatum is widespread throughout the West Indies, though absent from Cuba and the Bahamas and unknown from Dominica. The Cuban record, *Ekman 8607*, of Niedenzu (1928) and Liogier (1963) is *S. microphyllum. Stigmaphyllon emarginatum* is most frequently confused with *S. diversifolium* (no. 40), another common West Indian endemic with efoliolate styles with which it is sympatric in the Lesser Antilles. These two species are readily separated with the following couplet.

1. Peduncles 1.3–25 mm long, (0.2–) 0.3–1 times as long as the pedicels; anthers all loculate; posterior styles 2.5–3.6 mm long, erect, canaliculate-complicate. 37. *S. emarginatum.*
1. Peduncles absent to 4 mm long, always less than 0.3 times as long as the pedicels; anthers of stamens opposite the lateral sepals sterile (eloculate) or rarely anthers of those opposite the anterior-lateral sepals with one or two greatly reduced locules; posterior styles 2.1–5.8 mm long, lyrate, terete proximally but laterally flattened near the apex and curved around the opposing stamen. 40. *S. diversifolium*

As noted by Fawcett and Rendle (1920), the type of *Banisteria fulgens* L. is a specimen of *S. emarginatum* and not of the South American species *S. sinuatum* (including *S.*

fulgens Adr. Juss., nom. superfl.). Lamarck (1785) first misapplied the Linnaean name and was followed by Cavanilles (1790) and by Jussieu (1840), who made the combination in *Stigmaphyllon* but explicitly excluded Linnaeus's concept (see also C. Anderson, 1993b).

Niedenzu (1899) recognized that Cavanilles (1790), Desfontaines (1829), and Jussieu (1840, 1843) had mixed *S. emarginatum* and the species here called *S. floribundum* [based on *Banisteria tomentosa* Desf. ex DC., non *S. tomentosum* Adr. Juss.], and he lectotypified the name *S. emarginatum* with Cavanilles's plate (C. Anderson 1986).

The names *Stigmaphyllon haitiense*, *S. haitiense* f. *ovatum*, and *S. haitiense* f. *lineare* were proposed by Urban and Niedenzu in 1912; they were first published in the second fascicle (pages 161–304) of volume 7 of Urban's *Symbolae antillanae* in June, 1912, but were also listed by Niedenzu in July, 1912, in his article on *Stigmaphyllon*.

38. Stigmaphyllon microphyllum Grisebach, Pl. Wright. 168. 1860.—TYPE: CUBA. Oriente: Guantánamo, *Wright 93* (holotype: GOET!, photo: MICH!; isotypes: BR! G! GH! K! MO! S! US!).

Vine to ca. 1 m. Stems and branches sericeous when young, soon becoming glabrous, tuberculate or slightly so. Laminas 0.8–3.7 cm long, 0.4–1.4 cm wide, elliptical to oblong or sometimes obovate, apex mucronate-emarginate, sometimes obtuse, base attenuate or truncate to sometimes subcordate, sparsely sericeous to glabrate to glabrous adaxially, sericeous to sparsely so abaxially (trabecula 0.3–0.4 mm long, straight, sessile), margin eglandular; petioles (0.1–) 0.2–0.4 cm long, sericeous or densely so or rarely glabrous, not confluent across the node, with a pair of stipitate glands at the apex, each gland 0.2–0.6 mm long, the disk-shaped apex 0.2–0.3 mm in diameter, or sometimes the glands absent; stipules 0.4–1.1 mm long, 0.2–0.3 mm wide, free, narrowly triangular to sublinear, eglandular. Flowers 4 per solitary umbel. Peduncles 3.5–16 mm long, sericeous or densely so; pedicels (0.7–) 0.9–24.5 mm long, sericeous to sometimes glabrate, terete; peduncles 0.3–1 times as long as the pedicels. Bracts 0.9–14 mm long, 0.4–0.6 (–0.8) mm wide, triangular to broadly so; bracteoles 0.7–13 mm long, 0.3–0.6 mm wide, broadly to narrowly triangular, eglandular; bracts and bracteoles with the apex acute to acuminate, densely sericeous abaxially. Sepals 2–2.2 mm long, ca. 2–2.4 mm wide, glands ca. 1.2–1.7 mm long, 0.7–0.8 mm wide. All petals with the limb orbicular, glabrous, yellow, margin erose; anterior-lateral petals: claw 1.8–2 mm long, limb 8.7–9.3 mm long and wide; posterior-lateral petals: claw 1.1–1.3 mm long, limb ca. 8.5–9 mm long and wide; posterior petal: claw 2.7–3 mm long, apex indented, limb 6.5–7.5 mm long and wide. Stamens unequal, those opposite the posterior styles the largest, anthers of those opposite the lateral sepals with the connective enlarged and the locules reduced; anthers all loculate, pubescent, sometimes sparsely so. Stamen opposite anterior sepal: filament 2–2.2 mm long, anther 0.7–0.9 mm long; stamens opposite anterior-lateral petals: filaments 1.5–1.7 mm long, anthers ca. 0.7 mm long; stamens opposite anterior-lateral sepals: filaments ca. 2.5 mm long, connectives ca. 0.6 mm long, locules ca. 0.2 mm long; stamens opposite posterior-lateral petals: filaments 2.6–3 mm long, anthers 1–1.1 mm long; stamens opposite posterior-lateral sepals: filaments 1.7–2.4 mm long, connectives 0.4–0.5 mm long, locules ca. 0.2 mm long; stamen opposite posterior petal equal to or shorter than the adjacent two: filament 1.7–1.9 mm long, anther 0.4–0.6 mm long. Anterior style 2.3–2.8 mm long, shorter than the posterior two, terete, glabrous, slightly recurved; apex 0.6–0.8 mm long, each foliole 0.4–0.7 mm long, 0.4–0.5 mm wide, obovate or oblong or parabolic, or rarely folioles absent and the apex only laterally expanded, rhombic, ca. 0.5 mm wide. Posterior styles

2.9–3.8 mm long, canaliculate-complicate, glabrous, erect; folioles 0.6–0.7 mm long, 0.4–0.5 mm wide, semi-elliptical or parabolic or obovate. Dorsal wing of samara 1.6–1.7 cm long, 0.6–0.7 cm wide, upper margin with an acute or obtuse tooth; nut commonly bearing spurs and/or crests or smooth, 2.3–3 mm high, 2.5–2.6 mm in diameter, without air chambers, areole ca. 3 mm long, ca. 2.5–2.8 mm wide, convex, carpophore up to 0.6 mm long. Embryo ca. 3.7 mm long, ca. 2 times as long as wide, ovoid, outer cotyledon 3.4–3.5 mm long, 1.9–2 mm wide, inner cotyledon 2.7–2.8 mm long, 1.5–1.6 mm wide, both straight. Chromosome number unknown.

Phenology. Collected in flower in February, March, October, and December, in fruit in May and December.

Distribution (Fig. 38). Cuba; on rocks, in thickets and pastures.

ADDITIONAL SPECIMENS EXAMINED. **Cuba**. ORIENTE: Novaliches, Guantánamo, *Alain 3502* (GH); Guantánamo Bay, *Britton 2252* (NY, US); Ensenada de Mora, *Britton et al. 12947* (NY), *Britton et al. 13028* (F, K, MO, NY, US); Novaliches, Guantánamo, *Hioram 2402* (GH, NY); Naval base, Guantánamo, *Hioram 2339* (US); Estación Naval de Caimanera, *Hioram & Ramsden 2339* (NY); vic. of Manzanillo, *Shafer 12348* (NY).—CAMAGÜEY: Santa Cruz del Sur, *Ekman 8607* (G, NY, S); between Tarafa and Pastilillo, *Ekman 15463* (S).—HAVANA: Batabanó, La Mora, *Ekman 12625* (S).

In this well-named species, the laminas are less than 4 cm long and the petioles less than 0.5 cm long. The petiole glands consist of a stalk flared at the apex into a disk. *Stigmaphyllon microphyllum* is also distinctive in that the flowers are arranged in solitary 4-flowered umbels. As in *S. emarginatum*, the posterior styles are canaliculate-complicate, but differ in bearing folioles. The anterior style is also foliolate; however, in the anterior styles of *Britton et al. 13028* each foliole is reduced to a narrow lip.

Stigmaphyllon microphyllum has been confused with small-leaved forms of *S. emarginatum*, which does not occur in Cuba; see that species (no. 37).

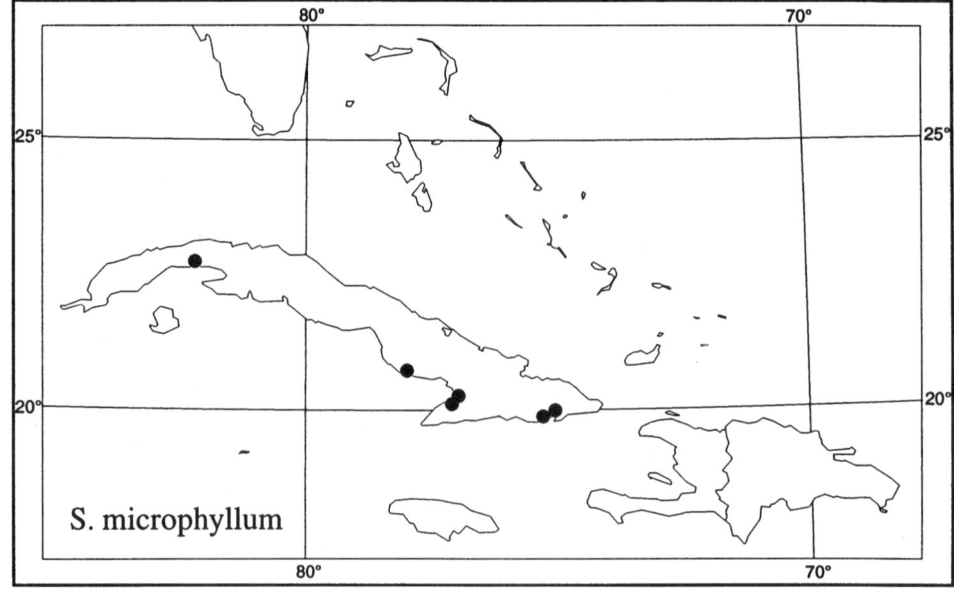

FIG. 38. Distribution of *Stigmaphyllon microphyllum*.

39. Stigmaphyllon sagraeanum Adr. Jussieu, Ann. Sci. Nat. Bot., sér. 2, 13: 290. 1840.—TYPE: CUBA. *de la Sagra s.n.* (holotype: P-JU!; isotype?: P!).

Stigmaphyllon reticulatum Adr. Jussieu, Ann. Sci. Nat. Bot., sér. 2, 13: 290. 1840. *Stigmaphyllon sagraeanum* f. *reticulatum* (Adr. Jussieu) Niedenzu, Ind. Lect. Lyc. Brunsberg. p. hiem. 1899–1900: 11. 1899.—TYPE: CUBA. *de la Sagra s.n.* (lectotype, designated by C. Anderson, 1987a: P-JU!; isolectotype: P!).

Stigmaphyllon obtusum Turczaninow, Bull. Imp. Soc. Naturalistes Moscou 36: 583. 1863.—TYPE: CUBA. *de la Sagra s.n.* (holotype: KW?).

Stigmaphyllon faustinum Wright in Sauvalle, Anales Acad. Ci. Méd. Habana 5: 244. 1868. *Stigmaphyllon sagraeanum* f. *faustinum* (Wright) Niedenzu, Ind. Lect. Lyc. Brunsberg. p. hiem. 1899–1900: 11. 1899.—TYPE: CUBA. *Wright 3522* (holotype: GH!, photo: MICH!; isotypes: K! NY! US!).

Stigmaphyllon sagraeanum f. *wrightianum* Niedenzu, Ind. Lect. Lyc. Brunsberg. p. hiem. 1899–1900: 11. 1899.—TYPE: CUBA. *Wright 2154 p.p.* (holotype: B, destroyed; isotypes: G! GOET!, photo of G isotype: MICH!).

Stigmaphyllon sagraeanum f. *primaevum* Niedenzu, Verz. Vorles. Ak. Braunsberg W.-S. 1912–1913: 26. 1912.—TYPE: CUBA. Pinar del Río, *Baker & Abarca 3699* (Herb. Est. Centr. Agr.) (holotype: B, destroyed; isotype: NY!, photo: MICH!).

Stigmaphyllon coccolobaefolium Alain, Phytologia 8: 369. 1962.—TYPE: CUBA. Oriente: Vía Sur, near Yateritas, on coastal rocks, 10 Jan 1956, *Alain & Morton 4955* (holotype: NY!, photo: MICH!; isotypes: LS, SV, fide Alain).

Stigmaphyllon nipense Alain, Bull. Torrey Bot. Club 90: 188. 1963.—TYPE: CUBA. Oriente: Sierra de Nipe, in charrascales, 300–400 m, 7 Jan 1956, *Morton et al. 8784* (holotype: US!, photo: MICH!).

Vine to 15 m. Stems and branches pubescent with subsessile scalelike T-shaped hairs or sericeous or sparsely so when young, soon becoming glabrous, tuberculate. Laminas (1–) 2.9–13 cm long, 0.2–7 cm wide, linear to oblong to lanceolate to elliptical to suborbicular, apex mucronate or mucronate-emarginate, base truncate or slightly cordate, glabrate to glabrous adaxially, sparsely sericeous (trabecula 0.2–0.4 mm long, straight, sessile) to glabrate to glabrous or rarely pubescent with T-shaped hairs abaxially (trabecula 0.3–0.8 mm long, wavy, stalk ca. 0.1–0.15 mm long), margin eglandular; petioles (0.2–) 0.5–6 cm long, densely sericeous to glabrous, not confluent across the node, with a pair of glands at the apex or sometimes at the base of the lamina, glands usually peg-shaped and up to 1 (–2) mm long or sometimes subsessile, each gland 0.2–1 mm in diameter, or sometimes one or both glands absent; stipules 0.3–0.6 mm long, 0.2–0.4 mm wide, free, narrowly triangular, eglandular. Flowers (8–) 20–25 (–50) per umbel or pseudoraceme, these solitary or borne in dichasia or compound dichasia or sometimes in small thyrses (axes to the 3rd order, sparsely sericeous to glabrate). Peduncles absent to 9 mm long; pedicels (8.5–) 12–27 mm long, terete; both densely to sparsely sericeous, peduncles if present always much shorter than the pedicels. Bracts (0.6–) 1–2 mm long, 0.3–0.7 mm wide; bracteoles 0.6–1 (–1.4) mm long, 0.2–0.3 mm wide, eglandular; bracts and bracteoles narrowly triangular, apex acute, sparsely to densely sericeous abaxially. Sepals 2–2.9 mm long, 1.8–2.3 mm wide, glands 1.1–1.4 mm long, 0.6–1 mm wide. All petals with the limb orbicular or suborbicular, glabrous, yellow, margin erose, limb of lateral petals 8–10.5 mm long and wide; anterior-lateral petals: claw (2–) 2.3–2.6 mm long; posterior-lateral petals: claw 1.2–2.2 mm long; posterior petal: claw 3–4 mm long, apex indented or sometimes only slightly so, limb 6.5–7.5 mm long and wide, glabrous. Stamens unequal, those opposite the posterior styles the largest, anthers of those opposite the lateral sepals with the connective enlarged and the locules reduced, rarely anthers of those opposite the posterior-lateral sepals with only one locule or eloculate; anthers glabrous. Stamen opposite anterior sepal: filament (1.6–) 2.3–3.3 mm long, anther 0.8–1.1 mm long; stamens opposite anterior-lateral petals: filaments 1.8–2.5 mm long, anthers 0.4–0.6 mm long; stamens opposite anterior-lateral sepals: filaments 1.8–2.1 mm long, connectives 0.7–0.9 mm long, locules 0.3–0.6 mm long; stamens opposite posterior-lateral petals: filaments 2.8–3.5 mm long, anthers 0.8–1 mm long; stamens opposite posterior-lateral sepals: filaments 1.6–2.4 mm long, connectives 0.4–0.8 mm long, locules 0.3–0.4 mm long, rarely absent; stamen opposite posterior petal subequal or equal to the adjacent two: filament 1.6–2.7 mm long, anther 0.4–0.7 mm long. Anterior style (2.6–) 3–3.6 mm long, shorter than the posterior two, terete, glabrous, slightly recurved; apex 0.7–0.9 mm long including a spur 0.3–0.5 mm long, linear, folioles absent. Posterior styles 3.3–4 mm long, terete, glabrous, lyrate; folioles 0.9–1.3 mm long, 0.4–0.8 mm wide, oblong to parabolic, or sometimes the foliole reduced to a narrow lip, or rarely absent and the apex linear, ca. 1 mm long including a spur 0.3–0.4 mm long. Dorsal wing of samara 1.8–2.4 cm long, 0.7–1.5 cm wide, upper margin with an acute tooth; nut with prominent veins, 3–4.2 mm high, 1.7–2.8 mm in diameter, without air chambers, areole 2–2.4 mm long, 1.2–1.6 mm wide, convex, carpophore up to 3 mm long; mature seed not seen. Chromosome number unknown.

Phenology. Collected in flower and fruit throughout the year.

Distribution (Fig. 39). Cuba and the Bahamas; on limestone and serpentine outcrops, in coastal thickets, open savannas, and pastures; sea level to 1200 m.

REPRESENTATIVE SPECIMENS. **Bahamas**. ANDROS ISLAND: *Correll & Godfrey 41259* (LL, MO, NY); *Small & Carter 8441* (F, K, NY, US); *Wight 236* (F, GH, NY).—ELEUTHERA: *Correll & Hill 45231* (NY).—LONG ISLAND: *Britton & Millspaugh 6232* (F, NY, US); *Correll 44863, 48173* (NY).—RUM CAY: *Gillis 6240, 6262*

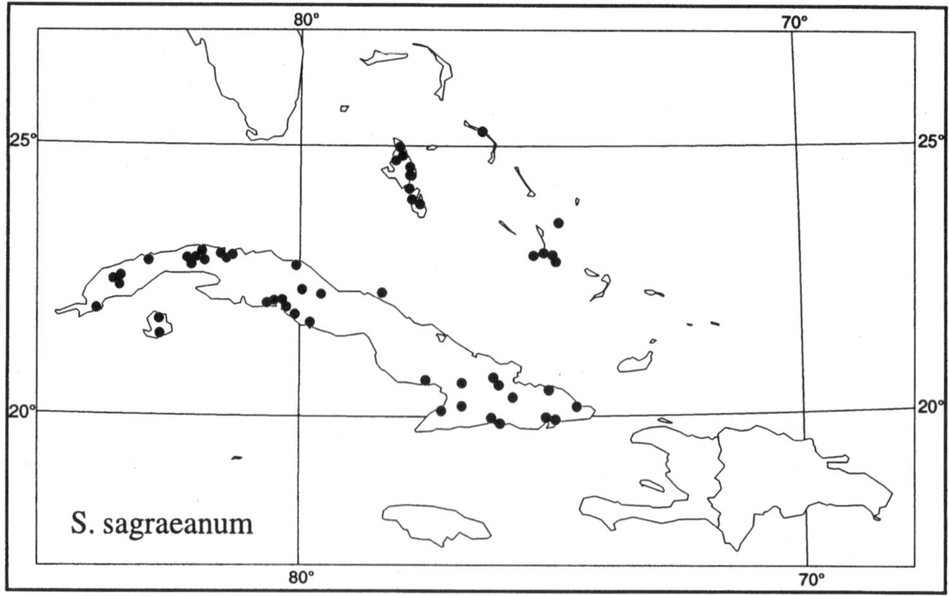

FIG. 39. Distribution of *Stigmaphyllon sagraeanum.*

(MSC). **Cuba.** CAMAGÜEY: 6 mi NW of Cayo Coco, *Shafer 2692* (NY, US); La Gloria, *Shafer 176* (NY, US); Atalaya, *Shafer 979* (NY, US); S of Sierra Cubitas, *Shafer 497* (NY).—HAVANA: vic. of Cojimar, *Britton et al. 6230* (NY); San Antonio, *van Hermann 834* (F); El Morro to Cojimar, *Wilson 9133* (NY); Havana, *Shafer 542* (CM, NY); Isla de Pinos, *Curtiss 213* (A, CM, F, G, M, MO, NY, US), *Jennings 1* (NY), *33* (NY), *520* (CM), *Killip 41660* (GH, US), *43560* (P), *43599* (GH, US), *Millspaugh 1419* (F).—MANTANZAS: vic. of Mantanzas, gorge of the Yumuri, *Britton et al. 496* (CM, F, NY); Mantanzas, *Rugel 157* (BM, GH, K, MO, NY), *Britton et al. 71* (CM, NY), *Ekman 17212* (A, S).—ORIENTE: Río Macaguanigua, *Shafer 3937* (NY, US); Gibara, *Pollard et al. 5* (A, CM, F, GH, MO, NY, US); valley of Río Bayamita, S slope of Sierra Maestra, *Maxon 3912* (F, GH, MO, NY, US); Baracoa, *Underwood & Earle 1353* (NY); Punta Piedra, Nipe Bay, *Britton et al. 12454* (NY, US); Bayate, Sabana Resueña, *Ekman 2819* (NY, S).—PINAR DEL RÍO: Bahía Honda, *Wilson 9418* (NY, U); Sierra de Anafe, *Wilson 11431* (NY, US); Buenaventura to San Juan de Guacamalla, *Wilson 9321* (NY); Laguna Jovero to Las Martinas, *Shafer 11034* (F, MO, NY, US).—LAS VILLAS: 12 km E of Cascajal, *Howard 5586* (GH, NY); Río Toyaba, Trinidad, *Britton et al. 5550* (NY, US); Trinidad, La Viga hill, *Britton & Wilson 5530* (NY); Río San Juan, *Britton et al. 5883* (NY). Without locality: *Wright 97* (BM, BR, G, GH, GOET, K, LE, MO, NY, P, S, US, W).

Stigmaphyllon sagraeanum, a common species of Cuba and the Bahamas, has distinctive coriaceous leaves, which are greatly variable in shape and size. Most of the synonyms apply to segregates described to recognize this diversity. The strongly rugose, sparsely sericeous (glabrous to the naked eye) to glabrous laminas vary from linear to suborbicular but most commonly are linear-oblong to elliptical; they are moderately pubescent abaxially in a few collections (e.g., *Britton 5883, Ekman 17212, Jack 7821, Morton 10479, Rugel 157, Shafer 12352*). This is the only species in the West Indies in which folioles are usually present in the posterior styles but absent in the anterior style; however, as in some other species, the foliole is sometimes much reduced or even absent. Typically, the folioles are broadly oblong to parabolic, but sometimes they are reduced to a narrowly triangular flap or only a lip. Exceptionally, such variation may be seen in a single individual; in *van Hermann 834*, a flower may have one posterior style with a large foliole and the other with only a tiny one. In collections seen from the Sierra de Nipe (Oriente) and in all but two from the Isla de Pinos, all three styles are efoliolate and subequal; the posterior styles of the two exceptions (*Jennings 1, 33*, NY) have large folioles. Niedenzu (1912), who considered the efoliolate condition ancestral, recognized the efoliolate variants as *S. sagraeanum* f. *primaevum*. Because these specimens differ from other representatives of *S. sagraeanum* only in the degree of reduction or absence of the folioles, they are not formally recognized here; however, more detailed work on this variable species may reveal that the efoliolate forms should be accorded taxonomic recognition.

Sometimes *S. sagraeanum* is misidentified as *S. emarginatum* (no. 37), also restricted to the West Indies but unknown from Cuba. *Stigmaphyllon emarginatum* is easily recognized by its distinctive stubby posterior styles (Fig. 36m, n); they are canaliculate-complicate, efoliolate, and truncate. Also, *S. emarginatum* has peduncles 1.3–25 mm long, whereas in *S. sagraeanum* peduncles are absent or, if present, at most 0.9 mm long. In Cuba, *S. sagraeanum* has been confused with *S. diversifolium* (no. 40), another West Indian endemic; they may be separated with the following couplet.

1. Anthers all loculate (rarely anthers of stamens opposite the posterior-lateral sepals with only one locule or eloculate), glabrous; posterior styles foliolate (sometimes the foliole reduced to a narrow lip, or rarely absent), terete, not curved; petiole glands absent or present, peg-shaped and up to 1 (–2) mm long or sometimes subsessile. 39. *S. sagraeanum.*
1. Anthers of stamens opposite the lateral sepals eloculate or rarely anthers of those opposite the anterior-lateral sepals with one or two greatly reduced locules, loculate anthers pubescent (rarely glabrous in Cuba); posterior styles efoliolate, terete proximally but laterally flattened near the apex and curved around the opposing stamen; petiole glands usually present, circular and prominent but sessile. 40. *S. diversifolium.*

40. Stigmaphyllon diversifolium (H. B. K.) Adr. Jussieu, Ann. Sci. Nat. Bot., sér. 2, 13: 290. 1840. *Banisteria diversifolia* H. B. K., Nov. gen. sp. 5: 159. 1822 ["1821"].—TYPE: CUBA. Havana, *Humboldt & Bonpland s.n.* (holotype: P-HBK!).

Banisteria ledifolia H. B. K., Nov. gen. sp. 5: 159. 1822 ["1821"]. *Stigmaphyllon ledifolium* (H. B. K.) Small, N. Amer. fl. 25(2): 141. 1910.—TYPE: CUBA. Havana, *Humboldt & Bonpland s.n.* (holotype: P-HBK!).

Stigmaphyllon lineare Wright ex Grisebach, Catal. pl. cub. 43. 1866.—TYPE: CUBA. Cabo del Rey, *Wright 2156* (holotype: GOET!, photo: MICH!; isotypes: BM! G! GH! K! MO! P!, photos of GH isotype: A! F! NY! S!).

Stigmaphyllon sericeum Wright ex Grisebach, Catal. pl. cub. 43. 1866. *Stigmaphyllon diversifolium* β *sericeum* (Wright ex Grisebach) Gómez, Anales Soc. Esp. Hist. Nat. 19: 232. 1890.—TYPE: CUBA. Río Toro, *Wright 2155* (holotype: GOET!, photo: MICH!; isotypes: G! GH! K! P!).

Stigmaphyllon rhombifolium Wright in Sauvalle, Anales Acad. Ci. Méd. Habana 5: 244. 1868.—TYPE: CUBA. "Protreros de D. Francisco Sauvalle. Bahía Honda y Santa Cruz de los Pinos," *Wright 2153* (holotype: GH!, photos: F! MICH! NY!; isotypes: G! GH! MO! NY! P! US!).

Stigmaphyllon cordifolium Niedenzu, Ind. Lect. Lyc. Brunsberg. p. hiem. 1899–1900: 8. 1899.—TYPE: MARTINIQUE, fide C. D. Adams, pers. comm. (labeled "Fl. trinitatis"), *Sieber 135* (lectotype, designated by C. Anderson, 1987a: G!, photo: MICH!; isolectotypes: F! G! GH! K! M! MO! NY! P! W!, photo of P isolectotype: MICH!).

Vine or scandent shrub to 10 m. Stems and branches sericeous and often mixed with short-stalked T-shaped hairs (stalk up to 0.1 mm long) when young, soon becoming glabrous, slightly tuberculate or smooth. Laminas 1.8–14.7 cm long, 0.2–6.8 cm wide, extremely variable: linear to lanceolate to elliptical to ovate to obovate to rhombic to orbicular, apex mucronate or mucronate-emarginate, base truncate to cordate or acute, glabrous adaxially, commonly glabrous abaxially but also sericeous or with T-shaped hairs to tomentose to sparsely so (trabecula 0.5–1.1 mm long, straight to wavy or crisped or curled, stalk 0.075–0.15 mm long), margin eglandular; petioles 1–13.5 cm long, densely sericeous, not confluent across the node, with a pair of prominent but sessile glands at the apex, each gland 0.3–1 (–1.4) mm in diameter, or sometimes glands absent; stipules 0.2–0.6 mm long, 0.1–0.5 mm wide, free, narrowly triangular to linear, eglandular. Flowers 8–18 (–27) per condensed to interrupted pseudoraceme or umbel, these often solitary, or borne in dichasia or compound dichasia or small thyrses (axes to the 3rd order, sericeous). Peduncles absent to 4 mm long, sericeous; pedicels 7–22 mm long, terete, sparsely to densely sericeous; peduncles always much shorter than pedicels. Bracts 0.9–2.2 mm long, 0.4–1 mm wide; bracteoles 0.5–1.1 mm long, 0.2–0.4 mm wide, eglandular or with a glandular region in the basal 1/3–1/2; bracts and bracteoles narrowly triangular, apex acute to acuminate, sericeous or densely so abaxially. Sepals 1.8–2.6 mm long, 1.8–2.5 mm wide, glands 1.2–2.2 mm long, 0.7–1.5 mm wide. All petals with the limb orbicular, glabrous, yellow, margin erose; anterior-lateral petals: claw 1.6–2.2 (–3.1) mm long, limb 7.5–11 mm long and wide; posterior-lateral petals: claw 1.2–2.2 mm long, limb 7.5–10 mm long and wide; posterior petal: claw 2.2–4 mm long, apex slightly or not indented, limb (6.5–) 7.8–8.5 (–9.5) mm long and wide. Stamens unequal, those opposite the posterior styles the largest, anthers of those opposite the lateral sepals eloculate or rarely anthers of those opposite the anterior-lateral sepals with one or two greatly reduced locules; loculate anthers pubescent or sometimes glabrous. Stamen opposite anterior sepal: filament 1.4–2.2 mm long, anther 0.7–1.3 mm long; stamens opposite anterior-lateral petals: filaments 1–1.8 mm long, anthers 0.6–1.2 mm long; stamens opposite anterior-lateral sepals: filaments 1.1–2.4 mm long, connectives 0.5–0.8 mm long, locules absent (rarely present, ca. 0.1 mm long); stamens opposite posterior-lateral petals: filaments 2.1–4.3 mm long, anthers 0.8–1.2 mm long; stamens opposite posterior-lateral sepals: filaments 1.1–1.2 mm long, connectives 0.3–0.9 mm long, locules absent; stamen opposite posterior petal sometimes shorter or longer than the adjacent two but usually equally long: filament 1.2–2.3 mm long, anther (0.4–) 0.5–0.7 mm long. Anterior style 1.5–3.5 mm long, shorter than the posterior two, terete, glabrous, slightly recurved; apex 0.9–1.7 mm long including a spur 0.6–1.4 mm long, linear and without folioles, or the apex expanded proximally and triangular to rhombic, 0.3–1.2 mm wide, varying from very narrow and with only a lateral lip to sometimes broad and with triangular folioles up to 0.5 mm long. Posterior styles 2.1–5.8 mm long, terete proximally but laterally flattened near the apex, glabrous, lyrate; apex 0.4–0.7 mm long including a spur up to 0.2 mm long or blunt, ca. 0.1 mm wide, folioles absent. Dorsal wing of samara 1.5–2 cm long, 0.5–1 (–1.3) cm wide, upper margin with an acute or obtuse tooth; nut smooth or with prominent ribs, 2.6–4.6 mm high, 2–3.5 mm in diameter, without air chambers, areole 2–3 mm long, 1–2 mm wide, concave, carpophore up to 2.3 mm long. Embryo 4.2–5.5 mm long, ca. 2 times as long as wide, ovoid, outer cotyledon 3.8–5.2 mm long, ca. 2–2.9 mm wide, inner cotyledon 3.3–4.5 mm long, 1–2.2 mm wide, both straight. Chromosome number unknown.

Phenology. Collected in flower and fruit throughout the year.

Distribution (Fig. 40). Cuba and the Lesser Antilles south to Martinique; on limestone and serpentine outcrops, in coastal thickets, pastures, and palm barrens; sea level to 500 m.

REPRESENTATIVE SPECIMENS. **Cuba**. CAMAGÜEY: S of Sierra Cubitas, *Shafer 514* (NY); Camagüey to Santayana, *Britton 2355* (NY).—HAVANA: near Havana, *Shafer 85* (CM, NY); Cojimar, *Killip 13819* (US).—MANTANZAS: near mouth of the Bueyvaca, *Britton & Wilson 60* (NY); vicinity of Mantanzas, gorge of the Yunuri, *Britton et al. 247* (CM, F, NY).—ORIENTE: vicinity of Guantánamo, *Britton 1897* (NY); Río Seboruco to falls of Río Mayari, *Shafer 3689* (NY, US); Ensenada de Mora, *Britton et al. 13032* (NY, US).—PINAR DEL RÍO: Corrientes Bay, *Britton & Cowell 9897* (NY); Sierra de Anafe, *Britton et al. 10340* (NY).—SANTA CLARA: Ciento Viejo Arroyo, 122 km E of Santa Clara, *Howard 5088* (GH, NY); Limones, Soledad Cienfuegos, *Jack 5873* (A, CAS, F, LE, S, US).—ISLA DE PINOS: Vivijagua, *Britton et al. 15025* (NY, US).—Without definite locality: *Wright 2153* (G, GOET, K, LE, MO, NY, P, S, US, W); *Wright 2154 p.p.* (BM, G, GH, GOET, K, MO, P). **Anguilla**. *Boldingh 3513* (U), *Howard & Kellogg 19080* (MICH), *LeGallo 2499* (NY), *Proctor 18581* (A, BM); *Walker 92-038* (A). **St. Martin**. *Arnoldo 3433* (U), *Boldingh 3286* (U). **St. Barthélemy**. *LeGallo 2212* (A). **Barbuda**. *Box 684* (MO, US), *Cowan 1648* (GH, NY, US), *Howard 18518* (A). **St. Kitts**. *Barneby 17782* (MICH, NY), *Britton & Cowell 750* (K, NY, US). **Antigua**. *Rose et al. 3275* (NY, US), *Sauer 2087* (F), *Smith 10494* (A, K, NY, S, US). **Montserrat**. *Proctor 19193* (A, BM), *Shafer 616* (F, NY, US). **Guadeloupe**. *Duss 2413* (F, GH, MO, NY, US), *Questel 1470* (P, US). **Saintes**. *Stehlé 1691* (NY). **Dominica**. *Beard 1459* (GH, NY, S, US), *Stern & Wasshausen 2539* (US), *Wilbur 7641* (F, LL, MICH, MO, TEX, US). **Martinique**. *Duss 437* (F, NY, US), *Duss 438* (F, NY).

Stigmaphyllon diversifolium, well-named for its highly variable leaves, is most reliably distinguished by its stamens and styles. The anthers of stamens opposing the lateral sepals are sterile, i.e., the glandular connectives do not bear locules; rarely, the anthers of stamens opposite the anterior-lateral sepals have one or two rudimentary locules up to 0.1

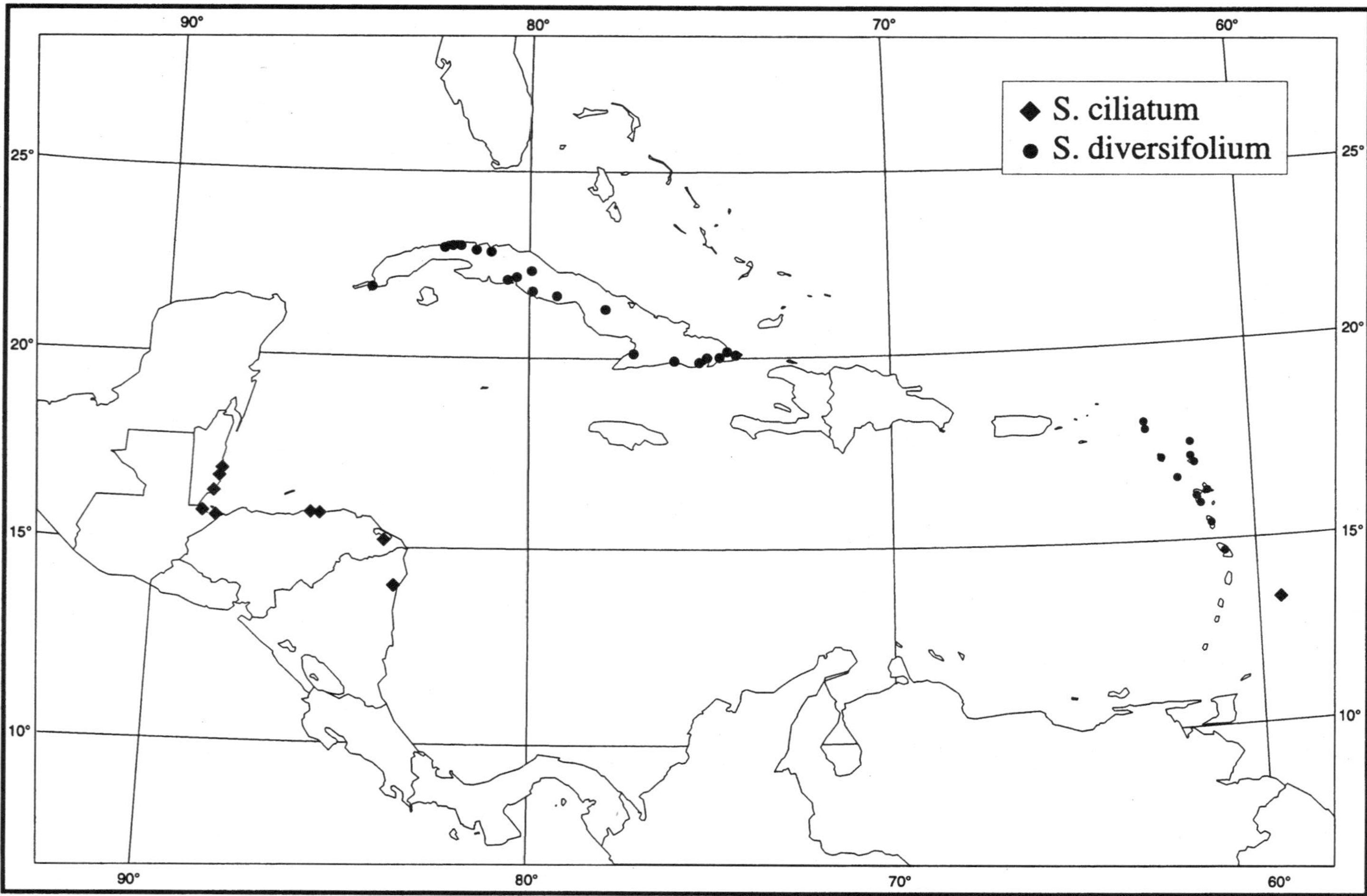

FIG. 40. Distribution of *Stigmaphyllon diversifolium* and of *S. ciliatum* on Barbados and in Central America; see Fig. 54 for distribution of *S. ciliatum* in South America.

mm long. As a rule, the fertile anthers are pubescent in plants from Cuba and glabrous in those from the Lesser Antilles. The efoliolate posterior styles are distally flattened and curve around the opposing stamen; when they are very long, as in most plants from the Lesser Antilles, they are conspicuously curled distally. The apex of the anterior style is extended into a spur and varies from linear to laterally expanded and then triangular to rhombic. When expanded it bears on each side a lip or narrow flap of tissue; occasionally, these lateral extensions are as long as 0.5 mm and form spreading triangular folioles.

The laminas of *S. diversifolium* vary greatly in shape, size, and vesture, which is reflected by the list of synonyms. Because the extremes are linked by intermediates and are conservative in their reproductive characters, they are not recognized taxonomically here. Most of the diversity is found in Cuba, and Niedenzu (1899) segregated the less variable populations from the Lesser Antilles as *S. cordifolium.* These plants usually have ovate laminas pubescent abaxially even at maturity, larger petiole glands, the flowers in umbels, glabrous anthers, and somewhat longer styles. The apex of the anterior style is usually linear and extends into a long claw or less commonly a short spur; the laterally expanded apex often found in Cuban plants is uncommon in those from the Lesser Antilles. Exceptions to this Lesser Antilles form include all collections from Anguilla and St. Martin (linear laminas), *Stern & Wasshausen 2539* from Dominica (broadly elliptical to suborbicular laminas), and *LeGallo 808* from St. Barthélemy (oblong to narrowly elliptical laminas).

In its flowers, *S. diversifolium* is most similar to *S. floribundum* (no. 41), a species restricted to Puerto Rico and the Virgin Islands; however, it is most often confused with two other West Indian endemics: in Cuba with *S. sagraeanum* (no. 39) and in the Lesser Antilles with *S. emarginatum* (no. 37); for separations see those species.

41. Stigmaphyllon floribundum (DC.) C. Anderson, Syst. Bot. 11: 128. 1986. *Banisteria floribunda* DC., Prodr. 1: 589. 1824.—TYPE: PUERTO RICO. *Bertero s.n.* (holotype: TO, photo: MICH!).

Banisteria tomentosa Desfontaines ex DC., Prodr. 1: 589. 1824. *Stigmaphyllon tomentosum* (Desfontaines ex DC.) Niedenzu, Ind. Lect. Lyc. Brunsberg. p. hiem. 1899–1900: 5. 1899, non *Stigmaphyllon tomentosum* Adr. Jussieu, 1832.—TYPE: specimen collected from plants cultivated at the Botanical Garden in Paris (holotype: G-DC!, microfiche: MICH!).

Vine to 6 m or more. Stems and branches pubescent with scalelike T-shaped hairs when young, soon becoming glabrous, commonly tuberculate. Laminas 3.6–18 cm long, 2.5–15.5 cm wide, elliptical or broadly so, oblong, sometimes lanceolate or suborbicular, apex mucronate or mucronate-emarginate, base truncate to slightly cordate or sometimes acute, glabrous adaxially, with (usually golden) T-shaped hairs to tomentose abaxially (trabecula 0.5–1.2 mm long, wavy or sometimes curled, stalk 0.1–0.4 mm long), the vesture often sloughed off in patches and mature laminas then glabrate to glabrous, margin eglandular; petioles (4–) 8–17.5 mm long, densely pubescent with scalelike T-shaped hairs but often glabrate in older leaves, not confluent across the node, with a pair of prominent but sessile glands at the apex, each gland 0.5–0.8 mm in diameter; stipules 0.3–0.7 (–1.2) mm long, 0.2–0.6 mm wide, free, narrowly triangular, eglandular. Flowers (10–) 20–30 (–50) per interrupted or less commonly congested pseudoraceme or sometimes in an umbel, these borne in large thyrses (axes to the 3rd order, densely and usually golden-pubescent with scalelike T-shaped hairs, sometimes with subsidiary axes) or sometimes in dichasia, rarely solitary. Peduncles absent to 4 (–7) mm long, densely pubescent with scalelike T-shaped hairs; pedicels (8–) 10–22 mm long, densely to sparsely pubescent with scalelike T-shaped hairs, terete; peduncles always much shorter than the pedicels. Bracts (0.6–) 1.1–1.7 mm long, (0.4–) 0.5–0.8 mm wide, narrowly triangular; bracteoles (0.2–) 0.7–1.4 mm long, 0.3–0.5 mm wide, narrowly triangular to linear, eglandular or with a glandular region in the basal 0.3–0.5 mm; bracts and bracteoles with the apex acute, densely sericeous abaxially. Sepals 2–2.7 mm long, 2.2–3.2 mm wide, glands 1.3–2.3 mm long, 0.6–1.3 mm wide. All petals with the limb orbicular or suborbicular, glabrous, yellow, margin erose; anterior-lateral petals: claw 2.5–4 mm long, limb (8–) 9–11.5 mm long and wide; posterior-lateral petals: claw 2–3.5 mm long, limb (8–) 9–10.5 mm long and wide; posterior petal: claw 2–3.5 mm long, apex indented or slightly so, limb 6.5–9 mm long and wide. Stamens unequal, those opposite the posterior styles the largest or the one opposite the anterior style the largest or those opposite the styles subequal, anthers of those opposite the lateral sepals eloculate or rarely anthers of stamens opposite the anterior-lateral sepals with one or two reduced locules, sometimes anther of stamen opposite posterior petal with reduced locules; anthers glabrous. Stamen opposite anterior sepal: filament 2.1–3 (–3.3) mm long, anther 0.8–1 mm long; stamens opposite anterior-lateral petals: filaments 2–2.6 mm long, anthers 0.8–1 mm long; stamens opposite anterior-lateral sepals: filaments 1.3–1.7 mm long, connectives 0.3–0.8 mm long, locules rarely present, up 0.4 mm long; stamens opposite posterior-lateral petals: filaments 2.2–2.8 mm long, anthers 0.8–1 mm long; stamens opposite posterior-lateral sepals: filaments 1.3–2.1 mm long, connectives 0.2–0.6 mm long, locules absent; stamen opposite posterior petal always shorter than the adjacent two: filament 1.5–1.8 mm long, connective 0.4–0.7 mm long, locules as long as connective or sometimes reduced, 0.1–0.5 mm long. Anterior style 2.1–3.2 mm long, shorter than the posterior two, sometimes the styles subequal in length, terete, glabrous; apex 0.6–0.7 (–1.2) mm long including a spur 0.2–0.3 (–0.6) mm long, linear, folioles absent. Posterior styles (2–) 2.5–3.6 mm long, terete but laterally flattened toward the apex, glabrous, lyrate; apex 0.5–0.6 mm long including a spur up to 0.2 mm long or blunt, linear, folioles absent. Dorsal wing of samara 1.8–3.2 cm long, 0.6–1.2 cm wide, upper margin with an acute tooth; nut smooth or sometimes bearing spurs and/or crests, 3.5–3.7 mm high, 2.4–2.8 mm in diameter, without air chambers, areole 2.4–3 mm long, 1–2.5 mm wide, concave, carpophore up to 1.2 mm long. Embryo ca. 4.6 mm long, ca. 2 times as long as wide, ovoid, outer cotyledon ca. 4.4 mm long, 2.6 mm wide, inner cotyledon ca. 4.2 mm long, ca. 2.2 mm wide, both straight. Chromosome number unknown. Fig. 41.

Phenology. Collected in flower and in fruit from October through June.

Distribution (Fig. 35). Puerto Rico, Virgin Gorda, and St. John; on limestone and serpentine outcrops, common in coastal thickets, barrens, dunes, pastures, and along roadsides; sea level to 1000 m.

REPRESENTATIVE SPECIMENS. **Puerto Rico**. Las Vilyas, NE of Ponce, *Britton & Britton 7459* (NY); Guayama, *Britton & Britton 9095* (NY); near Dorado, *Britton & Britton 9871* (NY); vic. of San Juan, *Britton & Wheeler* 288 (NY); vic. of Vega Baja, *Britton et al. 5780* (NY, US); Yauco, *Garber 36* (GH, K, NY); Caguas, *Goll 380* (US); near Río Piedras, *Heller & Heller 972* (NY); 5 mi NE of Mayagüez, *Heller 4455* (G, GH, MICH, MO, NY, P, US; F is *S. emarginatum*); along rte 687 near Laguna Tortuguero, *Howard & Nevling 16996* (A, U); Maricao, *Liogier 10753* (GH, NY, US); Cayey, *Liogier et al. 28408, 32547* (UPR); Fajardo, *Martorell & Liogier 28046* (UPR); Mayagüez, Mt. Las Mesas, *Otero 546* (A, CAS, F, MO); Mpio. Maricao, Maricao Insular Forest, *Proctor 39192* (JBSD); inter Sabana Grande et Guanica, *Sintenis 3843* (BM, C, G, GH, K, M, MSC, MO, NY, P, S, US); Manatí, *Sintenis 6716* (F, G, NY, W); 13 km N of Cayey, *Underwood & Griggs 344* (NY, US); near

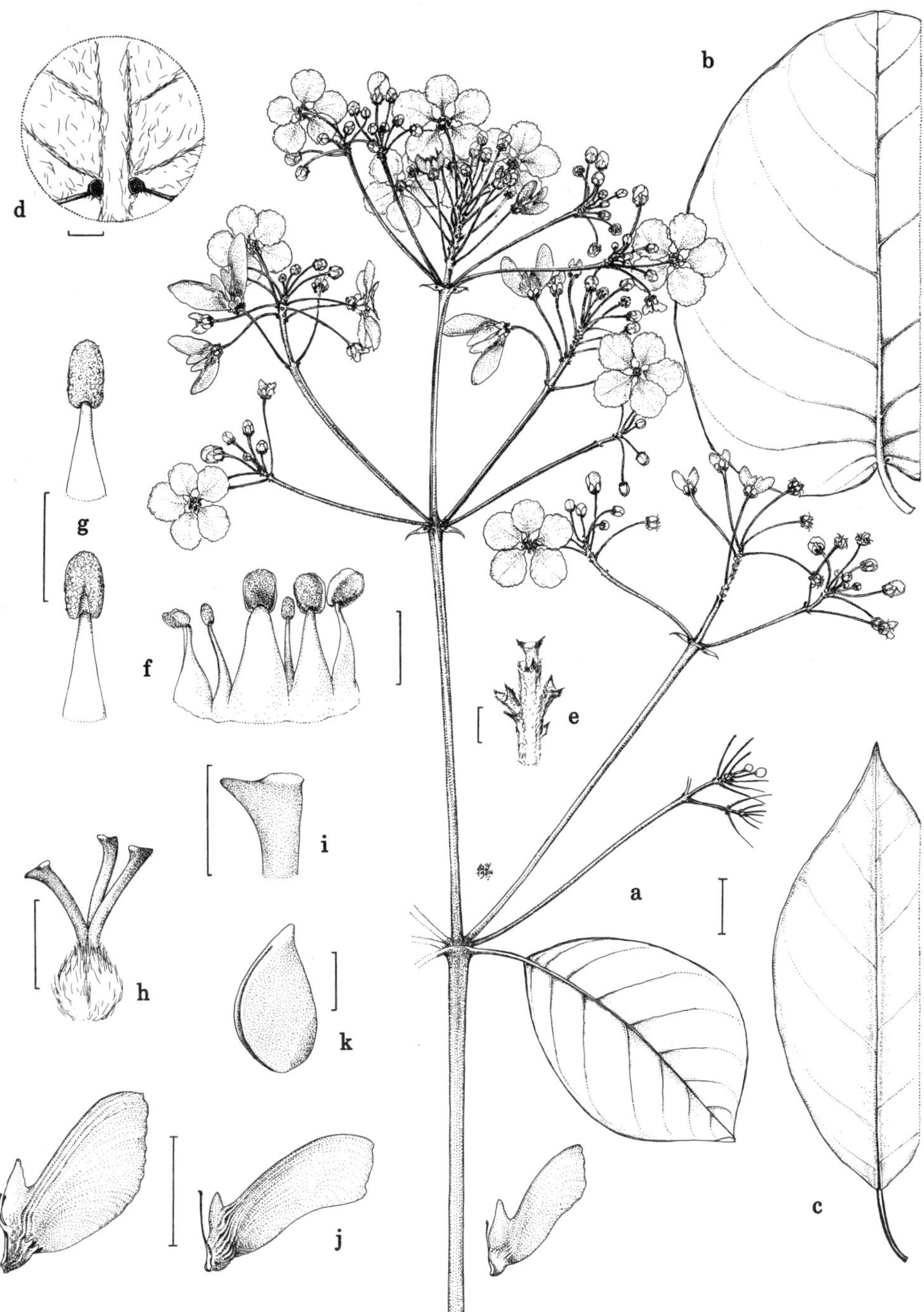

FIG. 41. *Stigmaphyllon floribundum*. a. Flowering branch. b, c. Leaves. d. Base of leaf, abaxial view. e. Portion of inflorescence axis with persistent peduncles. f. Portion of androecium; stamen at extreme left opposes the posterior petal (the "flag"). g. Abaxial (above) and adaxial (below) views of posterior-lateral stamen. h. Gynoecium, anterior style to the left. i. Apex of anterior style. j. Three samaras, illustrating variation in size and shape of dorsal wing and of lateral ornamentation. k. Embryo. Scale: for a–c, j, bar = 1.5 cm; for d–f, h, k, bar = 2 mm; for g, i, bar = 1 mm. (Based on: a, e, *Colwell 577*; b, d, *Britton 9871*; c, *Toro 3*; f, g, *Liogier 33781*; h, i, *Tredwell 751*; j, *Richard s.n.* (left), *Smith 10579* (center), *Sintenis 6716* (right); k, *Sintenis 6716*.)

Coamo Springs, *Underwood & Griggs 458* (NY, US). **St. John**. *Acevedo-Rodríguez 2854, 2855* (US); *Eggers 3259* (C). **Virgin Gorda**. *Fishlock 319* (GH, NY, US); summit and E slope of Virgin Peak, *Smith 10579* (A, K, NY, S, US).

Stigmaphyllon floribundum, a handsome species known only from Puerto Rico and two of the Virgin Islands (St. John and Virgin Gorda), typically has strikingly large, golden-pubescent compound inflorescences of congested or interrupted pseudoracemes of 20–45 flowers each. The large laminas, up to 18 cm long and 15.5 cm wide, are often golden-pubescent abaxially or, less commonly, glabrate to glabrous abaxially, especially in older leaves. Among West Indian species, *S. floribundum* is most similar to *S. diversifolium* (no. 40) of Cuba and the Lesser Antilles, but unknown from Puerto Rico and the Virgin Islands. In both, the peduncles are rudimentary to very short (less than 4 mm long), the stamens opposite the lateral sepals are usually sterile (eloculate), and the styles are efoliolate. *Stigmaphyllon diversifolium* differs in its highly variable smaller leaves (laminas up to 14.7 cm long and 6.8 cm wide), less elaborate inflorescences, and smaller samaras.

Small individuals of *S. floribundum* may be confused with the widespread and sympatric *S. emarginatum* (no. 37), whose styles are also efoliolate. In *S. emarginatum*, the peduncles are (0.2–) 0.3–1 times as long as the pedicels, the anthers are all loculate, and the distinctive stubby posterior styles are erect and canaliculate-complicate.

Collections of *S. floribundum* are often determined as *S. tomentosum* (Desf. ex DC.) Nied. (non Adr. Juss.), an illegitimate name (see C. Anderson, 1986).

42. Stigmaphyllon florosum C. Anderson, Syst. Bot. 11: 123. 1986.—TYPE: PERU. Loreto: Prov. Coronel Portillo, 15 km W of Pucallpa on rd to Tingo María, 175 m, 24 Jul 1964, *Hutchison et al. 6032* (holotype: MICH!; isotypes: F! G! GH! K! M! MO! NY! P! US!).

Vine to 16 m. Stems and branches bearing scalelike T-shaped hairs when young, soon becoming sparsely pubescent. Laminas 6–18 cm long, 6–17 cm wide, cordate to orbicular, rarely palmately 3–5-lobed, apex mucronate, base cordate, glabrous adaxially, with T-shaped hairs abaxially (trabecula 0.3–0.8 mm long, wavy, stalk 0.1–0.2 mm long), margin with scattered sessile glands (0.1–0.2 mm in diameter) and filiform glands (up to 1.5 mm long); petioles 2.5–10 cm long, pubescent with scalelike T-shaped hairs or densely so, not confluent across the node, with a pair of prominent but sessile glands at the apex, each gland 1.6–3.2 mm in diameter; stipules 0.5–1.2 mm long, 0.6–1.1 mm wide, free, triangular, eglandular. Flowers ca. (16–) 20–50 per umbel or congested pseudoraceme, these borne in compound dichasia or thyrses (axes the 4th order, finely beset with scalelike T-shaped hairs). Peduncles 3.3–12.5 mm long; pedicels 3.2–8.5 mm long, terete; both densely sericeous and bearing T-shaped hairs, peduncles up to 2.5 times as long as the pedicels or sometimes subequal or slightly shorter. Bracts 1–1.9 mm long, 0.6–1.2 mm wide, narrowly triangular, apex acute; bracteoles 0.9–1.5 mm long, 0.5–1 mm wide, oblong, apex obtuse, eglandular or each bracteole with a pair of glands (each gland 0.2–0.3 mm in diameter) or with a glandular area in the basal 1/3–1/2; bracts and bracteoles sericeous abaxially. Sepals 1.6–2.4 mm long, 1.5–2.1 mm wide, glands 1.1–1.6 mm long, 0.7–1.2 mm wide. All petals with the limb orbicular or broadly obovate (sometimes limb of posterior petal broadly elliptical), glabrous, yellow, margin denticulate to fimbriate, teeth and fimbriae up to 0.4 mm long; anterior-lateral petals: claw 1.2–2 mm long, limb

7.5–10 mm long and wide; posterior-lateral petals: claw 0.8–1.5 mm long, limb (5–) 6–7.5 mm long and wide; posterior petal: claw (2.3–) 2.5–3.3 mm long, apex indented, limb (5.2–) 5.5–7 mm long, (4.5–) 5–6 mm wide. Stamens unequal, those opposite the posterior styles the largest, those opposite the anterior-lateral sepals sometimes with the longest filaments, anthers of those opposite the lateral sepals with the connective enlarged and the locules greatly or only slightly reduced; anthers all loculate, pubescent. Stamen opposite anterior sepal: filament 2.2–3 mm long, anther 1.2–1.7 mm long; stamens opposite anterior-lateral petals: filaments 1.5–2.1 mm long, anthers 0.8–1.2 mm long; stamens opposite anterior-lateral sepals: filaments 3–3.6 mm long, connectives 1.1–1.3 mm long, locules 0.2–0.6 (–1) mm long; stamens opposite posterior-lateral petals: filaments 2.5–3.2 mm long, anthers 0.8–1 mm long; stamens opposite posterior-lateral sepals: filaments 1.8–2.8 mm long, connectives 0.6–1 mm long, locules (0.2–) 0.4–0.6 (–1) mm long; stamen opposite posterior petal always at least slightly shorter than the adjacent two: filament 2–2.3 mm long, anther 0.7–1 mm long. Anterior style 2.2–2.7 mm long, shorter than the posterior two, terete proximally, laterally flattened in the distal 1/3, with scattered hairs in the proximal 1/2, erect; apex 0.9–1.2 mm long including a spur ca. 0.2–0.3 mm long, linear, folioles absent. Posterior styles 2.8–3.8 mm long, terete proximally, laterally flattened in the distal 1/3–1/2, with scattered hairs in the proximal 1/3–1/2, recurved toward the posterior-lateral sepals; apex 0.5–0.7 mm long including a spur ca. 0.1 mm long or abaxial tip blunt, linear but incurved and with a lateral lip 0.1–0.2 mm wide, folioles absent. Dorsal wing of samara 3.8–4.5 cm long, 1.3–1.9 cm wide, upper margin with or without a blunt tooth; nut with only a vertical ridge per side or with 1–2 spurs or winglets projecting from the ridge, smaller spurs and winglets ca. 1.7 mm long, ca. 0.8 mm wide, the larger 2.5–8 mm long, 1.3–2 mm wide; nut 6.5–9 mm high, 3.5–5.5 mm in diameter, without air chambers, areole 2.5–3.5 mm long and wide, concave, carpophore up to 5.5 mm long. Embryo 5.7–6.5 mm long, ca. 1 1/2–2 times as long as wide, ovoid, outer cotyledon 9.5–10.5 mm long, 3–4.2 mm wide, the distal 2/5 folded over the inner cotyledon, inner cotyledon 3.5–5.7 mm long, 2.5–3 mm wide, folded at the distal 1/3 or straight. Chromosome number unknown. Fig. 42.

Phenology. Collected in flower from July through September (one flowering collection in January from Ecuador), in fruit from July through October.

Distribution (Fig. 43). From eastern Ecuador and Peru to Bolivia and western Brazil (Acre, Rondônia); in forests and secondary growth, and along roadsides; 30–1120 m.

REPRESENTATIVE SPECIMENS. **Ecuador.** NAPO: Auca Oil Field, 60 km S of Coca, *Besse et al. 59* (QCA, SEL); Añaugu, Río Napo, deforested areas around CEPE buildings, 00°31′S, 76°23′W, *Lawesson et al. 39767* (AAU). **Peru.** HUÁNUCO: Prov. Huamalico, zwischen Monzón und Huataya, *Weberbauer 3591* (G); Prov. Huánuco, Dtto. Churubamba, Hacienda San Carlos, trail exit to Derrepente, *Mexia 8123* (BM, F, GB, GH, K, MO, NY, S, U, US).—JUNÍN: Satipo, *Woytkowski 5911* (GH, MO, US); Sanibeni, *Woytkowski 5929* (GH, MO, US); Mazamari, *Woytkowski 5994* (MO).—LORETO: Yurimaguas, lower Río Huallaga, *Killip & Smith 28054* (F, NY, US); Prov. Ucayali, Orellana, Río Ucayali, *McDaniel 14131* (F, MO, NY, RB); Prov. Maynas, Dtto. Pebas, Río Ampiyacu, Estación UNAP, *Revilla 961* (MICH); Prov. Maynas, vicinity of Explorama Lodge opposite Yanamono Island, 03°25′S, 72°48′W, *Steiner 280* (MICH); Prov. Maynas, Sarrangal, margen derecha del Río Itaya, cerca de Yanayaco, 04°10′S, 73°20′W, *Vásquez et al. 366* (G, MICH, MO, NY).—MADRE DE DIOS: Prov. Manu, Manu Park, Cocha Cashu uplands, 11°45′S, 71°00′W, *Núñez 5749* (MO); Prov. Tambopata, Cuzco Amazónico Lodge, 15 km NE of J. Monterroso, Puerto Maldonado, 12°35′S, 69°03′W, *Núñez & Timaná 12172* (MO).—SAN MARTÍN: Prov. Saposoa, Mishquiyacu, cerca de Saposoa, *Ferreyra 4632* (MICH); Prov. Huallaga, entre Tingo de Saposoa y Bellavista, *Ferreyra 4767* (MICH, US). **Brazil.** ACRE: Hwy Abuña to Rio Branco, Km 242–246, vicinity of Campinas, *Forero et al. 6410* (INPA, MG, MICH, NY); Est. Sena Madudeira, Manuel Urbano, Km 38, Rio Caeté, afluente de Rio Jacó, *J. Lima et al. 128* (INPA); 37 km from Rio Branco on Rio

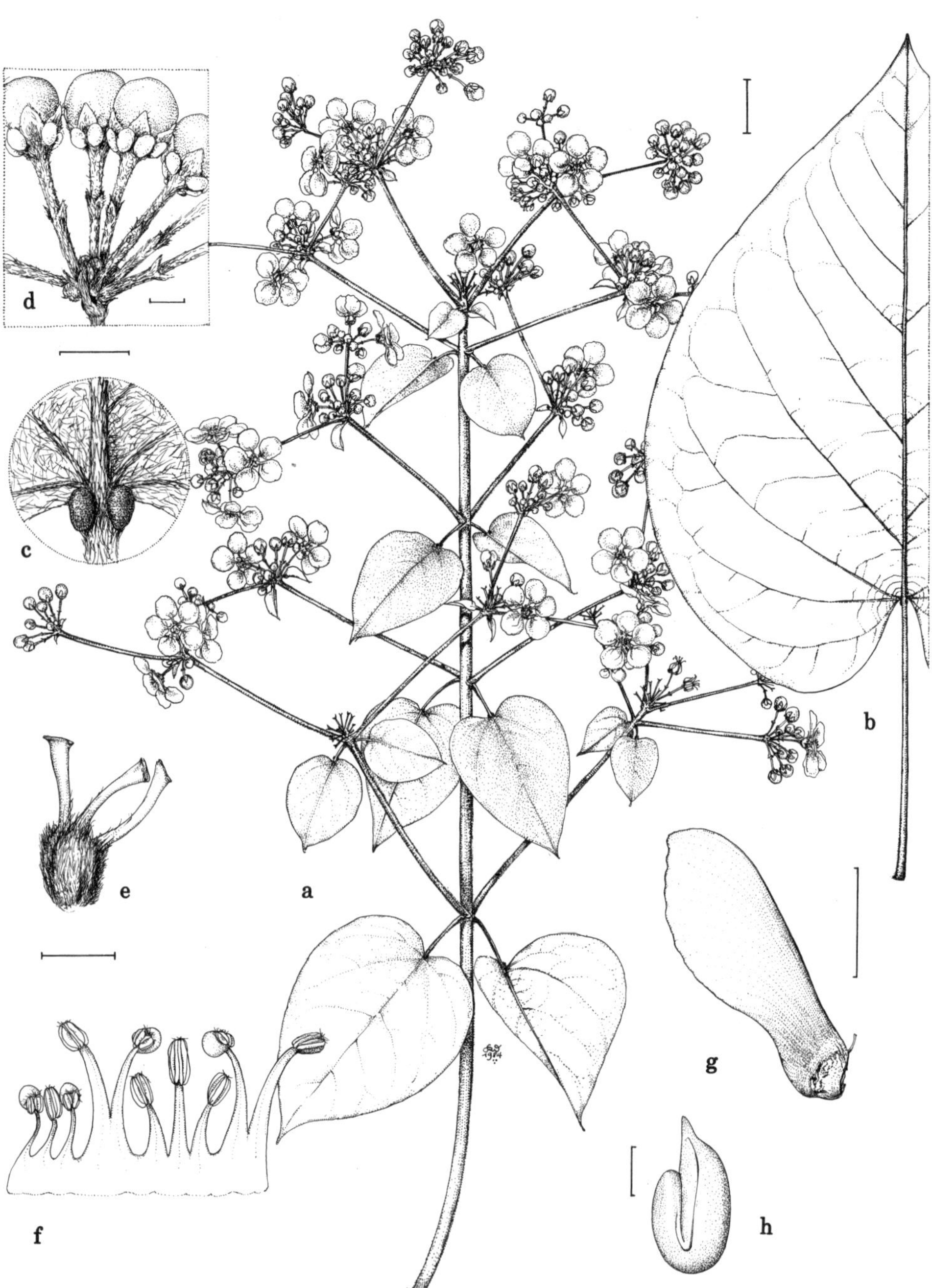

FIG. 42. *Stigmaphyllon florosum*. a. Flowering branch. b. Large leaf. c. Base of leaf, abaxial view. d. Umbel. e. Gynoecium, anterior style to the left. f. Androecium; stamen second from left opposes the posterior petal (the "flag"). g. Samara. h. Embryo. Scale: for a, b, g, bar = 1.5 cm; for c–f, h, bar = 2 mm. (Based on: a, d, e, f, *Hutchison et al. 6032*; b, c, *Cavalcante 3284*; g, h, *Woytkowski 5911*.)

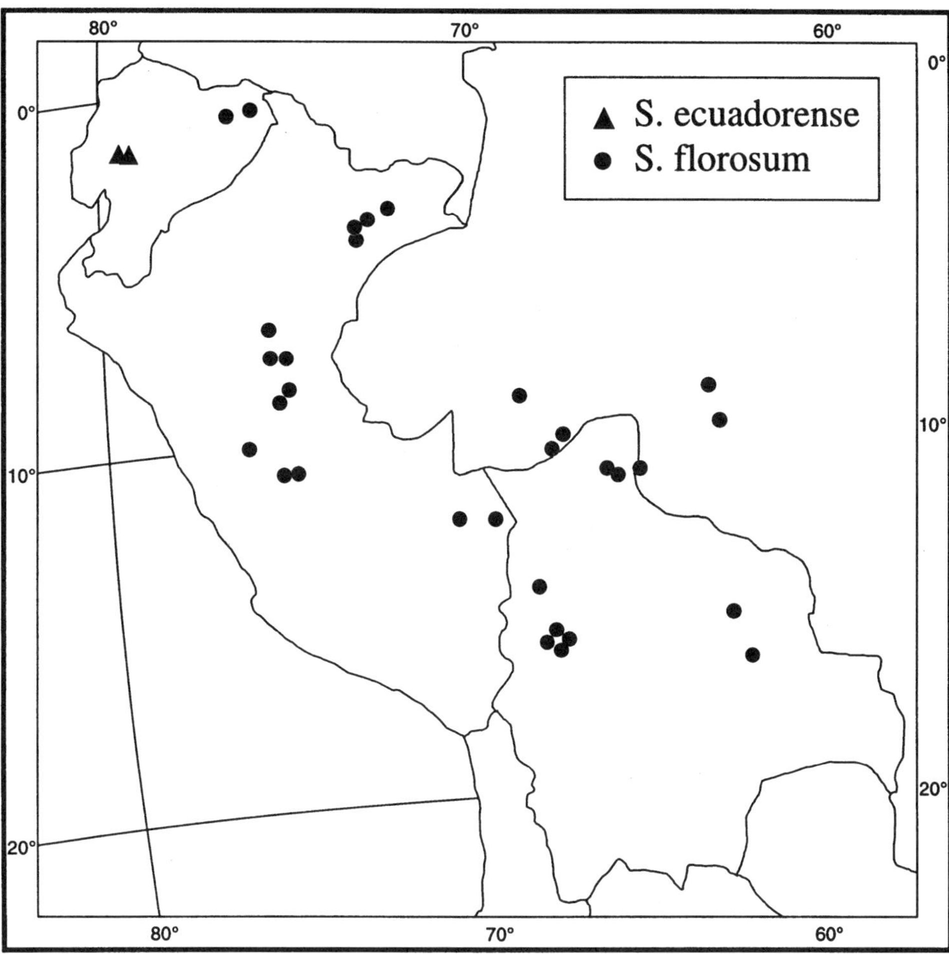

FIG. 43. Distribution of *Stigmaphyllon ecuadorense* and *S. florosum.*

Branco–Santa Rosa rd, *Lowrie et al. 384* (INPA).—RONDÔNIA: Guajará-Mirim, Rio Pacaás Novas, *Cavalcante 3284* (MG); Porto Velho to Cuiabá hwy, 10 km S of Ariquemes, *Forero & Wrigley 7056* (INPA, MG, MICH, NY); Mpio. Porto Velho, UHE de Samuel, antiga BR-364, *Vasques et al. 27* (INPA, NY). **Bolivia.** BENI: Prov. Vaca Diez, 18.4 km E of Riberalta, 11°05′S, 65°50′W, *Solomon 6099* (MO).—LA PAZ: Prov. Nor Yungas, Caranavi, ca. 15 km hacia Puerto Linares, *Beck 4813* (MICH); Prov. Sud Yungas, basin of the Río Bopi, San Bartolomé near Calisaya, *Krukoff 10056* (F, G, K, MICH, MO, NY, S, U, US); Prov. Sud Yungas, concesión de la Cooperativa Sapecho, 15°30′S, 67°20′W, *Seidel 2755* (MICH).—PANDO: Prov. Manuripi, along Río Madre de Dios, 2 km W of Humaita, 12°01′S, 68°17′W, *Nee 31690* (MICH).—SANTA CRUZ: Prov. Nuflo de Chávez, Est. Los Cucis, 45 km NE of Concepción, 15°50′S, 61°45′W, *Killeen 1099* (MICH, NY); Prov. Guarayos, ca. 8 km NW of Perseverancia, E side of Río Negro, 14°40′S, 62°50′W, *Nee 38794* (MICH); Prov. Velasco, Parque Nacional Noel Kempff M., aserradero Moira, 14°35′52″S, 61°21′11″W, *Saldias et al. 2999* (MICH).

Stigmaphyllon florosum is an attractive lowland species, often confused with the superficially similar *S. bogotense*, a widespread species of higher elevations (to 3200 m), because both have efoliolate styles. They are readily separated by characters of the androecium, styles, and embryo, and usually also of the inflorescence and by petal color, as shown in the following couplet.

1. Lowest pair of flowers of an umbel or pseudoraceme not separated a short distance from the rest; petals yellow; anthers pubescent, those of stamens opposing the lateral sepals with the connective enlarged and the locules reduced; apex of posterior styles with a narrow lateral lip; embryo ovoid, the cotyledons not convoluted, the larger outer cotyledon folded over the smaller inner one. 42. *S. florosum.*
1. Lowest pair of flowers of an umbel or pseudoraceme usually separated a short distance from the rest; petals yellow but usually marked or suffused with red; anthers glabrous, all unmodified and subequal; apex of posterior styles without a narrow lateral lip; embryo "brainlike," i.e., spheroid and with the cotyledons convoluted and folded within each other. 2. *S. bogotense.*

43. Stigmaphyllon aberrans C. Anderson, Syst. Bot. 11: 121. 1986.—TYPE: PERU. Junín: Yaupe, 1600 m, 26 Jun 1961, *Woytkowski 6335* (holotype: US!, photo: MICH!; isotypes: CAS! F! GH! K! MO! NY! TEX! US!).

Vine to 8 m. Stems and branches pubescent with scalelike T-shaped hairs when young, soon becoming glabrous. Laminas 8–15.5 cm long, 7.3–16 cm wide, cordate to reniform but smaller laminas ovate or elliptical, apex emarginate-mucronate or mucronate, base cordate or in smaller laminas truncate, glabrous adaxially, with T-shaped hairs abaxially (trabecula 0.6–1.4 mm long, wavy, stalk 0.1–0.4 mm long), margin with irregularly spaced sessile glands (0.2–0.3 mm in diameter); petioles 3.5–11.2 cm long, densely pubescent with scalelike T-shaped hairs, not confluent across the node, with a pair of prominent but usually sessile glands at the apex or up to 7 mm below the base of the lamina, each gland 1.2–2.2 mm in diameter, 0.5–1.2 mm high, in inflorescence leaves the glands usually at the middle or base of the petiole; stipules 0.6–1 mm long, 0.5–0.7 mm wide, free, triangular, eglandular. Flowers ca. 15–35 per condensed pseudoraceme, these solitary or borne in dichasia or small thyrses (axes to the 2nd order, pubescent with scalelike T-shaped hairs). Peduncles 6–19 mm long; pedicels 6.5–10 mm long, terete; pubescent with scalelike T-shaped hairs, peduncles 0.7–3 times as long as pedicels. Bracts 1.2–1.8 mm long, 0.5–0.7 mm wide; bracteoles 1–1.8 mm long, 0.6–0.8 mm wide, one bracteole of each pair with a gland 0.5–0.8 mm in diameter; bracts and bracteoles narrowly triangular, apex acute, densely pubescent with scalelike T-shaped hairs abaxially. Sepals ca. 2 mm long, 1.9–2.3 mm wide, glands 1.8–2 mm long, 0.9–1 mm wide. All petals with the limb orbicular, glabrous, yellow, margin fimbriate or fimbriate-dentate, fimbriae up to 0.3 mm long; anterior-lateral petals: claw 2.3–2.6 mm long, limb ca. 10–11 mm long and wide; posterior-lateral petals: claw 2–2.3 mm long, limb ca. 8–9.5 mm long and wide; posterior petal: claw 3–3.5 mm long, apex slightly indented, limb ca. 6.5–8 mm long and wide. Stamens unequal, those opposite the anterior-lateral sepals with the longest filaments, anthers of those opposite the lateral sepals with the connective enlarged and the locules reduced; anthers all loculate, glabrous. Stamen opposite anterior sepal: filament ca. 2.3 mm long, anther 1.1–1.2 mm long; stamens opposite anterior-lateral petals: filaments 1.5–2 mm long, anthers 0.7–0.8 mm long; stamens opposite anterior-lateral sepals: filaments 3.1–3.5 mm long, connectives ca. 1 mm long, locules 0.6–0.8 mm long; stamens opposite posterior-lateral petals: filaments 2.6–3 mm long, anthers 0.8–1.1 mm long; stamens opposite posterior-lateral sepals: filaments 2.7–3.2 mm long, connectives 0.8–0.9 mm long, locules 0.4–0.6 mm long; stamen opposite posterior petal always shorter than the adjacent two: filament 2.1–2.3 mm long, anther 0.7–0.8 mm long. Anterior style

2.4–2.8 mm long, slightly shorter than the posterior two, terete but distally flattened, glabrous, erect; apex 1.2–1.3 mm long, 0.3–0.4 mm wide, folioles absent. Posterior styles 2.7–3 mm long, terete, glabrous, erect but distally flattened and incurved; apex 1.3–1.4 mm long, 0.3–0.4 mm wide, linear but incurved, folioles absent. Mature samara not seen; immature samaras with a large dorsal wing, up to 5 cm long and 1.7 cm wide, without lateral winglets or spurs. Chromosome number unknown. Fig. 44.

Stigmaphyllon aberrans is known from only two collections in Peru (Fig. 10), the type and *Gentry 40101* (MICH, MO; Peru. Pasco: Oxapampa–San Ramón rd, ca. 15 km S of Oxapampa, 1900 m, 7 Feb 1983). It is the only species in the genus in which one bracteole only of each pair bears a prominent gland. The styles lack folioles. The apex of each is flattened and extends into a spur; in the posterior styles, the apex is incurved. The prominent petiole glands are borne just below or up to 7 mm below the base of the lam-

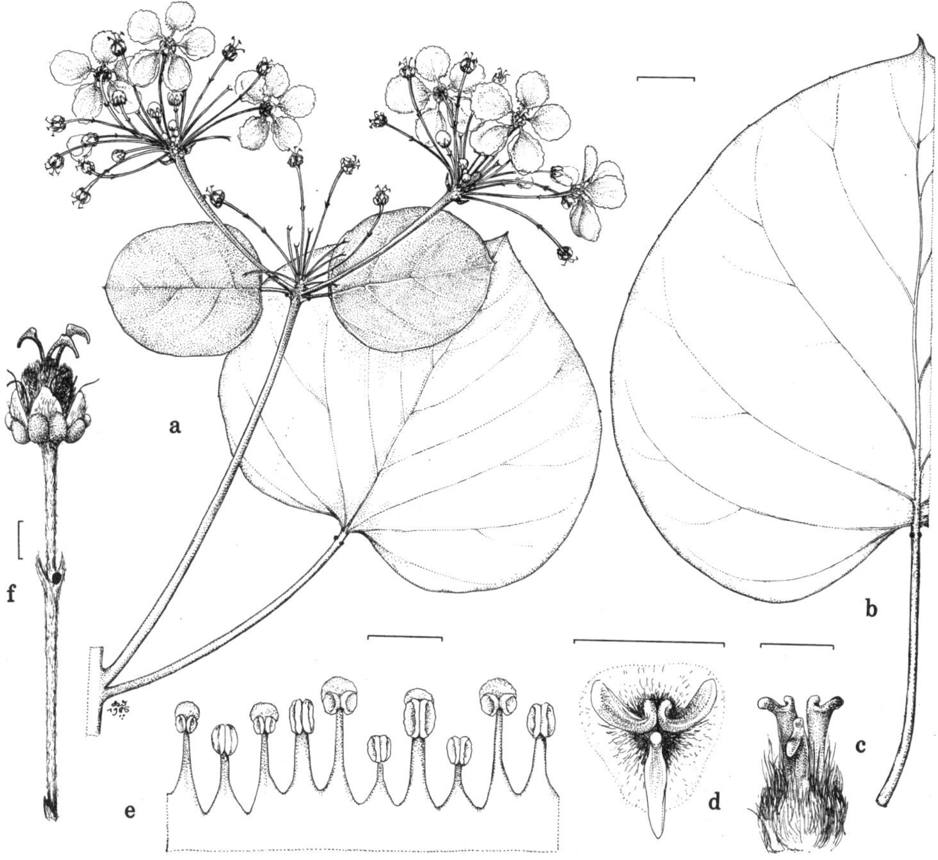

FIG. 44. *Stigmaphyllon aberrans*. a. Flowering branch. b. Large leaf. c. Gynoecium, anterior style in the center. d. Gynoecium seen from above. e. Androecium; stamen second from left opposes the posterior petal (the "flag"). f. Old flower. Scale: for a, b, bar = 1.5 cm; for c–f, bar = 2 mm. (Based on: a, b, f, *Woytkowski 6335*; c–e, *Gentry et al. 40101*).

ina; in the small leaves subtending the pseudoracemes, the glands are often positioned in the middle or at the base of the petiole.

Stigmaphyllon aberrans may be confused with the widespread *S. bogotense* (no. 2), which also has efoliolate styles but lacks the prominent gland on the bracteole and has a homogeneous androecium.

44. Stigmaphyllon selerianum Niedenzu, Ind. Lect. Lyc. Brunsberg. p. aest. 1900: 7. 1900.—TYPE: MEXICO. Oaxaca: Dtto. Nochixtlán, Almoloyas, 15 Nov 1895, *Seler & Seler 1374* (holotype: B, destroyed, fragment: NY!, photos: A! F! MICH! NY! US!).

Vine. Stems and branches pubescent with scalelike T-shaped hairs and also sericeous when young, soon becoming glabrous. Laminas 4–17.5 cm long, 3–15 cm wide, cordate or narrowly so, apex mucronate or emarginate-mucronate, base auriculate, glabrous adaxially, sparsely sericeous to glabrate abaxially (trabecula 0.2–0.4 mm long, straight, subsessile), margin with filiform glands up to 0.8 mm long (these often broken off in the larger leaves but the bases remaining) and sometimes dentate; petioles 1.5–7 cm long, sericeous or glabrous in older leaves, not confluent across the node, with a pair of prominent but sessile glands at the apex, each gland 1–2.8 mm in diameter; stipules 0.5–1.5 mm long, 0.5–1.2 mm wide, free, triangular or sometimes oblong, eglandular. Flowers 8–12 per umbel, solitary or borne in dichasia or rarely in small thyrses (axes to the 4th order, sericeous). Peduncles 0.5–3 mm long; pedicels 6.5–11.5 mm long, terete; both sericeous, peduncles up to 1/3 as long as pedicels. Bracts 1–2 mm long, 0.9–1.8 mm wide, apex acute; bracteoles 0.9–1.9 mm long, 0.8–1.4 mm wide, apex acute or obtuse, eglandular; bracts and bracteoles broadly triangular, sericeous abaxially. Sepals 2.3–3.3 mm long, 2.5–3 mm wide, glands 1.7–2.7 mm long, 0.8–1.5 mm wide. All petals with the limb broadly elliptical or broadly obovate to sometimes orbicular, glabrous, yellow, margin of lateral petals erose, of posterior petal erose or denticulate-fimbriate, teeth/fimbriae up to 0.5 mm long; anterior-lateral petals: claw 2–2.5 mm long, limb ca. 11.5–13 mm long, ca. 10–13 mm wide; posterior-lateral petals: claw 2.5–3 mm long, limb 8.5–11.5 mm long and wide; posterior petal: claw 2.5–3 mm long, apex not indented, limb 8.5–9.5 mm long, 7–8.5 mm wide. Stamens unequal, those opposite the posterior styles the largest, anthers of those opposite the lateral sepals sometimes with the connective enlarged and the locules reduced; anthers all loculate, pubescent or sometimes glabrous. Stamen opposite anterior sepal: filament 2.5–3 mm long, anther 1.1–1.5 mm long; stamens opposite anterior-lateral petals: filaments 2–2.5 mm long, anthers 0.8–1.1 mm long; stamens opposite anterior-lateral sepals: filaments 2.5–3.1 mm long, connectives 1.1–1.3 mm long, locules equaling connective or reduced and 0.7–1 mm long; stamens opposite posterior-lateral petals: filaments 3–4 mm long, anthers 1.2–1.4 mm long; stamens opposite posterior-lateral sepals: filaments 2.2–2.6 mm long, connectives 0.6–1 mm long, locules equaling connective or reduced and 0.5–0.7 mm long; stamen opposite posterior petal always shorter than the adjacent two: filament 1.7–2.5 mm long, anther 0.7–0.8 mm long. Anterior style 2.5–3 mm long, at least slightly shorter than the posterior two, terete proximally, laterally flattened in the distal 1/4, glabrous, erect; apex 1.3–1.6 mm long including a spur 0.4–0.8 mm long, linear or elliptically to obovately expanded distally, folioles absent. Posterior styles 2.8–3.5 mm long, terete proximally, laterally flattened in the distal 1/3, glabrous, lyrate; apex 1.3–1.9 mm long, laterally expanded into a lip or to a narrowly triangular or semi-elliptical foliole 0.5–1.3 mm wide. Dorsal wing of samara ca. 3.5–4 cm long, ca. 1.1–1.5

cm wide, upper margin with an obtuse or subacute tooth; nut smooth or bearing a spur or winglet (ca. 1–2 mm long, ca. 0.7 mm wide); nut ca. 5.5 mm high, ca. 5 mm in diameter, without air chambers, carpophore up to ca. 2.5 mm long; mature seed not seen. Chromosome number unknown.

Phenology. Collected in flower in February, May, and June and from September through March, in immature fruit in December through March and in May.

Distribution (Fig. 64). Mexico (Chiapas, Oaxaca); in evergreen and tropical deciduous forest, and in thickets on limestone hills; 550–1500 m.

ADDITIONAL SPECIMENS EXAMINED. **Mexico.** CHIAPAS: 11 mi from El Sumidero on rd to Tuxtla Gutiérrez, *Anderson & Anderson 5555* (ENCB, MICH, SD); Mpio. Tuxtla Gutiérrez, on rd to El Sumidero, 8 km N of Tuxtla Gutiérrez, *Breedlove 9027* (DS, F, US); Mpio. Jiquipilas, 20 km N of Jiquipilas and Mex Hwy 190, *Breedlove 24135* (MEXU, MICH, MO, NY, UTD); Mpio. La Trinitaria, 18 km S of La Trinitaria on rd to Colonia Morelos and Colonia Chihuahua, *Breedlove 46460* (CAS); Mpio. Villa Corzo, 32 km from Tuxtla along rd to Nueva Concordia, *Breedlove 48714* (CAS); Mpio. Chiapa de Corzo, above El Chorreadero, *Breedlove 50177* (CAS); Tuxtla Gutiérrez, *Goldman 749* (US); Trapichito, Comitán, *Matuda 5686* (F, MEXU, UTD); Mpio. Tzimol, 7 km al SW de Tzimol, *Reyes G. 333* (MICH); Mpio. Mazapa, 10 km al E de Motozintla, sobre la carretera MEX-190, *Reyes G. 1543* (MICH).—OAXACA: de Almoloyas a Santa Catarina, *Conzatti 1680* (NY, US); Mpio. Comitán de Domínguez, 5 km al N de J. Mújica, camino a Tzimol, 16°05′52″N, 92°12′37″W, *Martínez S. 22085* (MICH); 16 km al SE de Dominguillo, 11 km al NW de Tonaltepec, entre Cuicatlán y Telixtlahuaca, *Medrano et al. F-1090, F-1100* (MEXU); Dominguillo, *Miranda 1016* (MEXU); Río de las Vueltas, Dominguillo, *Miranda 4731* (MEXU); Tomellín Canyon, *Pringle 5972* (GH, MICH, US); Nochixtlán, *Quarles van Ufford 348* (U); Mpio. Mazapán, 10 km al E de Motozintla, *Reyes G. 1543* (BM); Jacatlán, *L. C. Smith 531* (US).

Stigmaphyllon selerianum is readily recognized by its cordate, auriculate laminas, which bear filiform glands along the margin. They appear glabrous to the naked eye though they are often sparsely sericeous abaxially. The peduncles are rudimentary (up to 3 mm long). The anterior style is efoliolate, but the posterior ones are apically expanded into at least a lip and commonly into a small, often spreading foliole. The filiform glands are seen best in younger leaves, because they are often broken off in the more mature ones. The laminar tissue is sometimes drawn out into a shallow tooth at the base of the gland; this is most pronounced in *Matuda 5686*, in which the laminas are grossly dentate. Another noteworthy collection is *Medrano et al. F-1100*, which consists of two leafless flowering shoots and a piece of corky-winged stem, ca. 2 cm in diameter.

This species is superficially similar to *S. ciliatum* (no. 55), a species of coastal habitats, which has not been reported from Mexico. It is most likely to be confused with Guatemalan *S. cordatum* (no. 12); for a separation see that species.

45. Stigmaphyllon crenatum C. Anderson, Contr. Univ. Michigan Herb. 19: 419. 1993.—TYPE: BRAZIL. Espírito Santo: Rio Pancas, Aldeiamento dos Indios, *Buena 156* (holotype: R-37607!, photo: MICH!).

Vine. Stems and branches sericeous when young, soon becoming glabrous. Laminas 5.5–13.5 cm long, 4.5–11.3 cm wide, broadly elliptical to suborbicular, apex obtuse or emarginate or emarginate-mucronate or mucronate, base truncate to cordate, in young leaves and those near the inflorescence densely sericeous adaxially and densely pubescent with T-shaped hairs abaxially (trabecula 0.5–1.1 mm long, straight or wavy, stalk 0.1–0.2 mm long), the older laminas glabrate to glabrous adaxially and abaxially, margin deeply crenate, each sinus with a circular flush gland 0.7–1.7 mm in diameter; petioles up to 2 mm long, sericeous or glabrous, not confluent across the node, with a pair of glands borne at the apex, these flush with the epidermis but with a raised margin, each gland 1.3–2.5 mm in diameter; stipules absent (?). Flowers ca. 10 per umbel, these solitary or borne in dichasia or compound dichasia or small thyrses (axes to the 3rd order, sericeous). Peduncles 5–7 mm long (?); pedicels ca. 6 mm long (?), distally expanded; both sericeous, peduncles subequal to the pedicels (?). Bracts 0.8–1.2 mm long and wide, broadly triangular, apex acute; bracteoles 1–1.3 mm long, 0.8–1 mm wide, triangular, apex obtuse, eglandular or each bracteole with a pair of inconspicuous glands (each gland 0.2–0.3 mm in diameter); bracts and bracteoles sericeous abaxially. Sepals ca. 3 mm long, ca. 2 mm wide, glands 2–2.3 mm long, ca. 1 mm wide. Posterior petal not seen; lateral petals with the limb orbicular, glabrous, margin erose-denticulate, anterior-lateral petals: claw ca. 3 mm long, limb 13–14 mm long and wide; posterior-lateral petals: claw ca. 1.5 mm long, limb ca. 12 mm long and wide. Stamens unequal, anthers of those opposite the anterior-lateral sepals with the connective enlarged and the locules reduced; anthers all loculate, glabrous (anthers of posterior-lateral and posterior stamens only seen in young bud). Stamen opposite anterior sepal: filament ca. 2.5 mm long, anther ca. 1 mm long; stamens opposite anterior-lateral petals: filaments ca. 2 mm long, anthers ca. 0.9 mm long; stamens opposite anterior-lateral sepals: filaments 3.2–3.3 mm long, connectives ca. 1 mm long, locules 0.7–0.8 mm long; stamens opposite posterior-lateral petals: filaments ca. 3 mm long, anthers ca. 1.2 mm long; stamens opposite posterior-lateral sepals: filaments ca. 2.6 mm long, mature anther not seen; stamen opposite posterior petal always shorter than the adjacent two: filament ca. 2 mm long, mature anther not seen. Anterior style ca. 2 mm long, shorter than the posterior two, terete proximally, laterally flattened in the distal 1/4, glabrous, erect; apex ca. 2 mm long including a spur ca. 1 mm long, ca. 0.4 mm wide, linear, folioles absent. Posterior styles ca. 3.2 mm long, terete, with scattered hairs in the proximal 1/4, lyrate; foliole ca. 1.2 mm long and wide, parabolic to triangular. Samara not seen. Chromosome number unknown. Fig. 45.

Phenology. Collected in flower in July, November, and December.

Distribution (Fig. 82). Brazil (northern Espírito Santo); in campo rupestre, on rock faces.

ADDITIONAL SPECIMENS EXAMINED. **Brazil.** ESPÍRITO SANTO: Mpio. Venecia, *Duarte 4000* (RB); Colatina, Rio Pancas, *J. Kuhlmann 6651* (RB).

Stigmaphyllon crenatum is the only species with nearly sessile, deeply crenate leaves. Each sinus is marked by a gland up to 1.7 mm in diameter. In many species, leaves of older plants have shed some or sometimes all of the vesture. This phenomenon is apparently particularly strongly expressed in *S. crenatum*. The type, collected in July, consists of three flowering branchlets, one of which bears two pairs of leaves. These laminas are densely sericeous adaxially and densely silvery pubescent with T-shaped hairs abaxially. In *Duarte 4000* and *J. Kuhlmann 6651*, collected in November and December, only the dichasially branched inflorescence axes are retained, and the leaves are glabrous adaxially and abaxially, though on some laminas patches of hairs remain; the immature leaves at branch apices bear the typical vesture. The flowers are distinctive in that the apex of the anterior style is drawn out into a long claw but lacks folioles; the posterior styles are both foliolate. The only mature flower on the type lacked the posterior petal and the anthers of the posterior and posterior-lateral stamens. Examination of a bud indicates that these anthers all have two locules, but it was not apparent whether the connective of anthers of the posterior-lateral stamens would be enlarged at maturity.

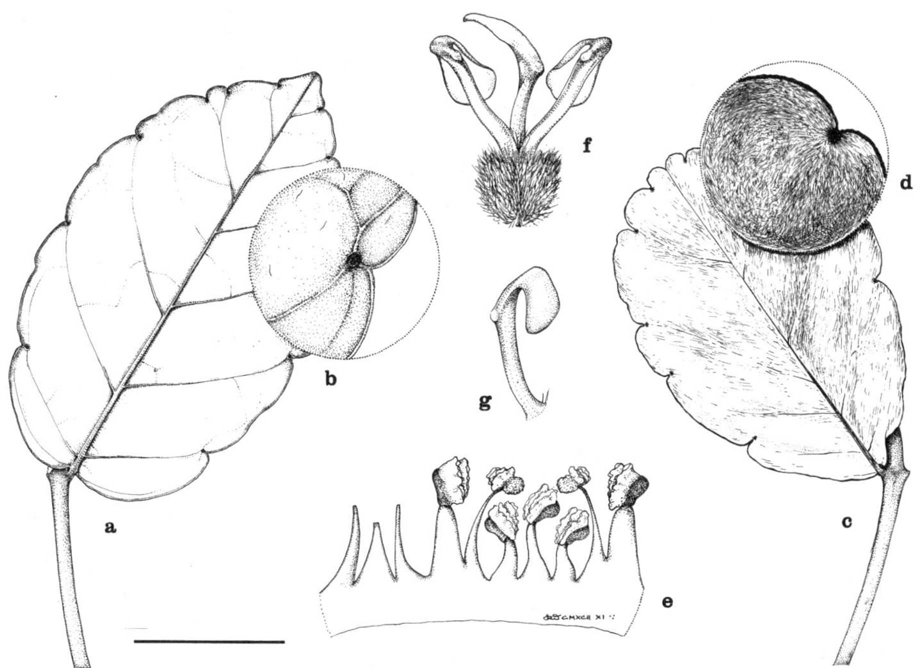

FIG. 45. *Stigmaphyllon crenatum*. a. Glabrous leaf. b. Detail of margin and abaxial surface of glabrous lamina. c. Pubescent leaf. d. Detail of margin and abaxial surface of pubescent lamina. e. Androecium; fourth stamen from right opposes the anterior sepal (anthers of stamens opposing the posterior petal and the posterior-lateral sepals not seen). f. Gynoecium, anterior style in the center. g. Lateral view of a posterior style. Scale: for a, d, bar = 4 cm; for b, c, bar = 1 cm; for e, bar = 4 mm; for f, g, bar = 2.7 mm. (Based on: a, b, *Duarte 400*; c–g, *Buena 156*.)

46. Stigmaphyllon ellipticum (H. B. K.) Adr. Jussieu, Ann. Sci. Nat. Bot., sér. 2, 13: 290. 1840. *Banisteria elliptica* H. B. K., Nov. gen. sp. 5: 161. 1822 ["1821"].—TYPE: ECUADOR. Loja: "Loxa" [Loja], *Humboldt & Bonpland s.n.* (holotype: P-HBK!, photos: F! MICH! US!).

Banisteria mucronata DC., Prodr. 1: 589. 1824. *Stigmaphyllon mucronatum* (DC.) Adr. Jussieu, Arch. Mus. Hist. Nat. Paris 3: 377. 1843.—TYPE: "in Nova-Hispania?," collector unknown (holotype: G-DC!, photo: F! MICH!).

Banisteria ternata DC., Prodr. 1: 591. 1824. *Stigmaphyllon ternatum* (DC.) Adr. Jussieu, Ann. Sci. Nat. Bot., sér. 2, 13: 289. 1840. *Stigmaphyllon mucronatum* var. *ternatum* (DC.) Niedenzu, Ind. Lect. Lyc. Brunsberg. p. aest. 1900: 4. 1900. *Stigmaphyllon ellipticum* var. *ternatum* (DC.) Niedenzu, Pflanzenreich IV. 141(2): 501. 1928.—TYPE: Sessé and Mociño plate, number 6331.1667 in the Torner Collection of Sessé and Mociño Biological Illustrations at the Hunt Institute for Botanical Documentation.

Banisteria billbergiana Beurling, Kongl. Vetensk. Akad. Handl. 1854: 116. 1856.—TYPE: PANAMA. Colón: "Porto Bello [Portobelo], ad litora insulae Manzinella [Manzanillo]," *Billberg 285* (holotype: S!, photos: MICH! NY!; isotypes: S!).

Stigmaphyllon mucronatum var. *nicaraguense* Niedenzu, Ind. Lect. Lyc. Brunsberg. p. aest. 1900: 4. 1900. *Stigmaphyllon ellipticum* var. *nicaraguense* (Niedenzu) Niedenzu, Pflanzenreich IV. 141(2): 500. 1928.—TYPE: NICARAGUA. Matagalpa: Matagalpa, 800 m, *Rothschuh 643* (holotype: B, destroyed).

Stigmaphyllon mucronatum var. *intermedium* Niedenzu, Ind. Lect. Lyc. Brunsberg. p. aest. 1900: 4. 1900. *Stigmaphyllon ellipticum* var. *intermedium* (Niedenzu) Niedenzu, Pflanzenreich IV. 141(2): 500. 1928.—TYPE: PANAMA. Panamá: "Río Bayono [Bayano]," Mar 1858, *Wagner s.n. (Fasc. 11)* (lectotype, designated by C. Anderson, 1987a: M!, photo: MICH!).

Stigmaphyllon felixii Cuatrecasas, Brittonia 14: 55. 1962.—TYPE: PERU. Amazonas: Bagua Chica, 400 m, 3 Mar 1960, *Woytkowski 5610* (holotype: US!, photo: MICH!; isotypes: G! GH! MO! S!).

Vine to 7 m. Stems and branches sericeous when young, soon becoming glabrate or glabrous. Laminas 3.5–17 cm long, 2–9 cm wide, narrowly to broadly elliptical, sometimes lanceolate to ovate, rarely suborbicular, apex mucronate to attenuate, rarely caudate, base truncate to cordate, sometimes attenuate, glabrous adaxially, very sparsely sericeous to glabrous abaxially (trabecula 0.1–0.8 mm long, straight, sessile), margin eglandular; petioles 0.6–2.8 cm long, densely to sparsely sericeous to glabrate or glabrous, not confluent across the node, with a pair of prominent but sessile glands at the apex, each gland 0.5–2 mm in diameter; stipules (0.3–) 0.5–1 mm long and wide, free, triangular, glandular. Flowers 3–9 (–12) per umbel, these solitary or borne in dichasia or small thyrses (axes to the 3rd order, sparsely sericeous). Peduncles 2.5–34 mm long, sericeous or sparsely so; pedicels 2–13 mm long, glabrous, distally expanded; peduncles 0.5–5 times as long as pedicels. Bracts 0.9–2 mm long, 0.7–2 mm wide, triangular, apex acute, sparsely sericeous to glabrous abaxially; bracteoles 0.7–1.8 mm long, 0.6–1.5 mm wide, oblong to ovate, apex obtuse, sericeous or sparsely so abaxially, eglandular or with a glandular area in the basal 1/3–1/2. Sepals 2–3.5 mm long, 1.7–2 mm wide, glands 1.1–2.5 mm long, 0.8–1.5 mm wide. All petals with the limb glabrous, yellow, margin lacerate, lacerate-dentate, fimbriate-lacerate, fimbriate-dentate, or fimbriate, teeth and fimbriae 0.3–1.2 mm long, limb of lateral petals orbicular; anterior-lateral petals: claw 1.8–3.2 mm long, limb (11–) 12–17 mm long, 12–17 mm wide; posterior-lateral petals: claw 1–2 mm long, limb 10–16 mm long, 12–16.5 mm wide; posterior petal: claw (2.8–) 3–5 mm long, apex indented, limb (8–) 11–14.5 mm long, (7–) 10.5–14 mm wide, obovate to broadly elliptical to orbicular. Stamens unequal, those opposite the lateral sepals the longest and their anthers with the connective enlarged and the locules reduced, rarely anthers of those opposite the posterior-lateral sepals eloculate; anthers glabrous. Stamen opposite anterior sepal: filament 2.2–3.1 (–3.4) mm long, anther 1.2–1.6 mm long; stamens opposite anterior-lateral petals: filaments (1.7–) 1.9–2.5 mm long, anthers 0.9–1.3 mm long; stamens opposite anterior-lateral sepals: filaments (3–) 3.2–4.1 mm long, connectives 0.8–1.6 mm long, locules 0.2–0.6 (–0.8) mm long; stamens opposite posterior-lateral petals: filaments long, anthers 1.2–1.6 (–1.8) mm long; stamens opposite posterior-lateral sepals: filaments (2.7–) 3–3.5 (–4) mm long, connectives (0.5–) 0.8–1.4 (–1.6) mm long, locules 0.2–0.4 mm long; stamen opposite posterior petal always shorter than the adjacent two: filament (1.8–) 2–2.8 mm long, anther 0.6–1 mm long. Anterior style (2.3–) 2.7–3.8 mm long, shorter than or sometimes subequal to the posterior two, terete, glabrous, erect; apex 1–1.5 mm long, the distal 1/2–3/4 (–4/5) elliptically to ovately to suborbicularly (rarely triangularly) expanded, (0.3–) 0.4–0.7 (–0.8) mm wide, or sometimes with small folioles, each up to ca. 1 mm long, 0.6 mm wide. Posterior styles (2.3–) 2.6–3.6 mm long, terete, glabrous, erect; folioles 0.9–1.5 mm long, (0.5–) 0.7–1 mm wide, semi-lunate to broadly so to triangular, or only with a lateral lip 0.3–0.5 mm long, 0.4–0.7 mm wide, or the apex only somewhat laterally expanded, 0.3 mm wide. Dorsal wing of samara 2.1–3.5 cm long, (3–) 4–8 cm wide, upper margin usually with an obtuse tooth; nut commonly only bearing lateral spurs and/or crests or only prominently ribbed but sometimes with a pair of rectangular to subsquare lateral winglets, these 3.2–6.5 mm long, 2–3.7 mm wide; nut 5–8 mm high, 4–6 mm in diameter, without air chambers, areole 3.5–5.3 mm long, 2.7–3.8 mm wide, slightly concave to plane to slightly convex, carpophore up to 4 mm long. Embryo 5.5–7 mm long, ca. 2 times as long as wide, ovoid, outer cotyledon 8–10.3 mm long, 2.6–3 mm wide, the distal 1/3–2/5 folded over the inner cotyledon, inner cotyledon 2–4.4 mm long, 2–2.2 mm wide, folded at the distal 1/7–1/10 or straight. Chromosome number unknown.

Phenology. Collected in flower and in fruit throughout the year.

Distribution (Figs. 46, 47, 48). Mexico to northern South America; in evergreen, deciduous, rain, thorn, and pine-oak forest, in second growth, thickets, matorral, and at roadsides and edges of beaches; sea level to 2200 m.

REPRESENTATIVE SPECIMENS. **Mexico**. CAMPECHE: Km 71, camino Escárcega a Candelaria, *Chavelas P. et al. ES-372* (MEXU, MICH); Tuxpeña, *Lundell 975* (A, CAS, F, GH, MICH, MO, NY, US, WIS).—CHIAPAS: 9 mi E of Cintalapa on Mex Hwy 190, *Anderson & Laskowski 4221* (ENCB, GH, MICH, MO, NY); Mpio. Villa Flores, 14 km N of Villa Flores along rd to Tuxtla Gutiérrez, *Breedlove 24605* (DS, MEXU, MICH, MO, NY); Mpio. Solosuchiapa, 2–4 km below Ixhuatán along rd to Pichucalco, *Breedlove 34869* (DS, MEXU, MICH).—OAXACA: gorge of Río Malatengo, *Alexander 150* (MEXU, MICH, NY); Dtto. Cuicatlán, de Palapa a la Raya, *Conzatti 3789* (MEXU, US); Chivela, *Mell 2251* (NY, US).—QUINTANA ROO: Coba, at Lake Coba, *Lundell & Lundell 7668* (MICH, UTD); a 6 km al N de La Unión, *Téllez 1656* (BM, CAS, MEXU, MO).—TABASCO: a 2 km N del camino 25 y a 5.2 km al W de la W-O, Balancán, *Novelo et al. 103* (CAS, K, MEXU, MO).—TAMAULIPAS: irrigation ditch 3 mi S of Ciudad Mante, *Hill, Jr. 39* (TEX).—VERACRUZ: Laguna Encantada, 6 km al N de San Andrés Tuxtla, *Calzada 950* (CAS, F, MEXU, MO); Atoyac, *Matuda 1470* (A, GH, MEXU, MICH, MO, US); Mpio. Dos Ríos, Cerro Gordo, cerca del Salto del Río Grande, *Ventura A. 2624* (CAS, DS, ENCB, F, MICH, NY, P, TEX, US).—YUCATÁN: Izamal, 1888, *Gaumer s.n.* (F, K). **Belize**. BELIZE: Gracie Rock, *Dwyer 12492* (MO).—CAYO: Vaca, *Gentle 2490* (K, MEXU, MICH, TEX).—STANN CREEK: Savannah Forest Station, 16°34′N, 88°20′W, *Hunt 356* (F, US, UTD).—TOLEDO: riverbank of Río Grande, *Gentle 4816* (LL, MICH, UTD). **Guatemala**: ALTA VERAPAZ: along Río Sebol, downstream from Carizal, *Steyermark 45778* (F, LL).—CHIQUIMULA: around Ipala, *Steyermark 30363* (F).—EL PROGRESO: Fiscal, *Deam 6172* (F, GH, MICH, US).—ESCUINTLA: Escuintla, *Harmon & Dwyer 2985* (F, GH, MO).—GUATEMALA: Natl Hwy 5 from Guatemala City to El Chol, 2.9 mi S of boundary with Depto. Baja Verapaz, 14°52′N, 90°37′W, *Croat & Hannon 63553* (MO).—HUEHUETENANGO: near El Reposo, ca. 8 km from Mex. frontier, *Williams et al. 41130* (F).—IZABAL: between Los Amates and Izabal, *Kellerman 7255* (US); near Quiriguá, *Standley 72347* (F).—JUTIAPA: vic. of Jutiapa, *Standley 74973* (F, US).—PETÉN: Guayacán, bordering Laguna Guayacán, *Contreras 7371* (LL, UTD); Parque Nacional de Tikal, Km 59 del camino de El Renate, *Tún Ortíz 373* (F, MO, NY, US).—SANTA ROSA: near Cuilapa, *Standley 78536* (F).—SUCHITEPÉQUEZ: along Río Madre Vieja, above Patutul, *Standley 62209* (F).—ZACAPA: Gualán, *Kellerman 5733* (LL, US); Sierra de las Minas, vic. of Río Hondo, *Steyermark 29379* (F). **Honduras**. ATLÁNTIDA: vaguada del Río Cangrejal, 20 km SE de La Ceiba, *Nelson et al. 3390* (MO).—COMAYAGUA: cerca de La Libertad, matorrales del Río Frío, *Molina R. 7027* (F, GH, US).—COPÁN: 1 mi W of Ruinas Copán, *Molina R. & Molina 30851* (F).—CORTÉS: nacimiento del Río Lindo, *Molina R. 5680* (F).—EL PARAÍSO: matorral del Río Tenpasenti, cerca del pueblo Tenpasenti, *Molina R. 11917* (F, G, NY, US).—GRACIAS A DIOS: arroyada del Río Dursuna, 70 km al O de Puerto Lempira, 15°00′N, 84°13′W, *Nelson 798* (MO).—LA PAZ: inmediaciones de Tepanguare, *Medina 244* (MICH).—MORAZÁN: near Suyapa, along quebrada Suyapa, *Molina R. 702* (F, GH, MEXU).—SANTA BÁRBARA: San Pedro Sula, *Thieme 5168* (GH, US).—YORO: orilla del Río Jacagua, 15 km O de Victoria, *Nelson et al. 7218* (MICH). **El Salvador**: AHUACHAPÁN: vic. of Ahuachapán, *Standley 19847* (GH, NY, US).—LA LIBERTAD: vic. of San Tecla, *Standley 23013* (GH, S, US).—LA UNIÓN: vic. of La Unión, *Standley 20832* (US).—SAN SALVADOR: vic. of Tonacatepeque, *Standley 19468* (GH, MO, S, US).—SANTA ANA: vic. of Metapán, *Standley & Padilla V. 3207* (F).—SONSONATE: vic. of

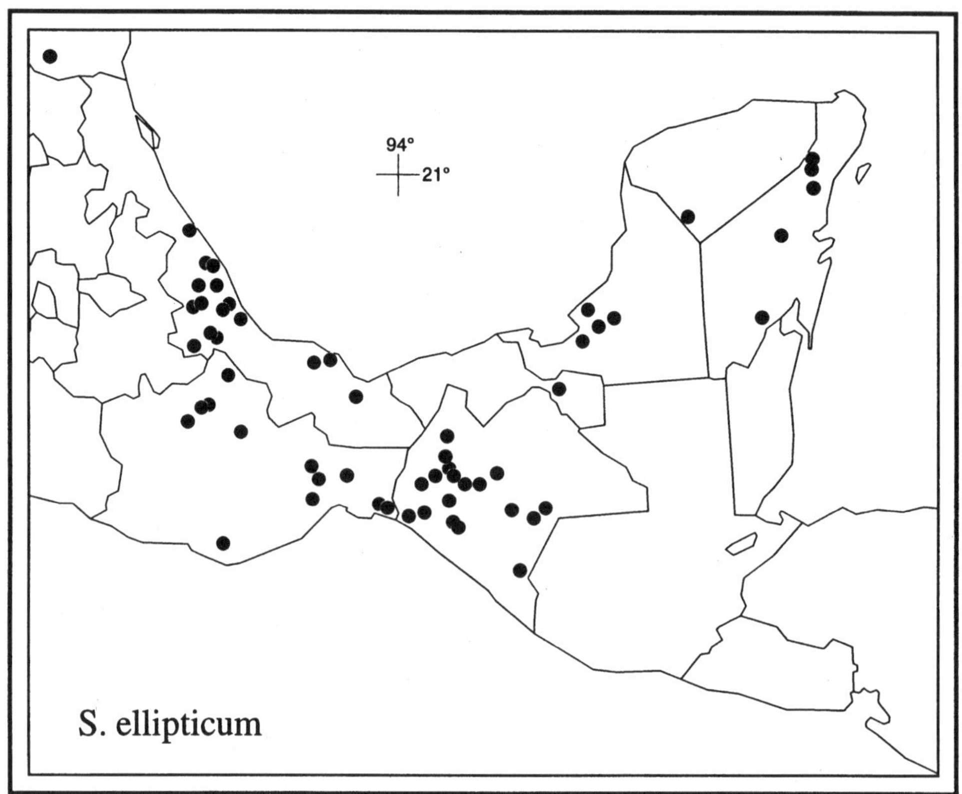

FIG. 46. Distribution of *Stigmaphyllon ellipticum* in Mexico.

Acajutla, *Standley 21973* (GH, US). **Nicaragua**: BOACO: Las Pitas, 12°28′N, 85°35′W, *Moreno 10653* (MO).—CHONTALES: rd from Juigalpa NE toward La Libertad, ca. 17.4 km NE of Río Mayales, ca. 12°12′N, 85°17′W, *Stevens 4188* (MICH).—JINOTEGA: "El Recreo," 4 km al N de Sta. Gertrudis, 13°13′N, 85°53′W, *Moreno 7918* (MICH).—MADRIZ: ca. 5 km SW of San Juan del Río Coco on rd to Telpaneca, ca. 13°31′N, 86°11′W, *Stevens 17672* (MICH).—MASAYA: Parque Nacional Volcán Masaya, on N flank of Volcán Santiago, *Neill 2844* (MICH, MO).—MATAGALPA: E side of Río Tuma between bridge and Río Yasica, ca. 13°03′N, 85°44′W, *Stevens 19188* (MO).—NUEVA SEGOVIA: a 6 km del Jícaro, carretera a Murra, 13°44′N, 86°0S′W, *Moreno 8308* (MICH).—RÍO SAN JUAN: 10 km al SSE de San Miguelito, sobre la carretera a San Carlos, *Sandino 5038* (MO).—ZELAYA: Monkey Point, 11°36′N, 83°40′W, *Moreno 12359* (MO); ca. 2.5 km NW of Rama, ca. 12°10′N, 84°14′W, *Stevens 17465* (MO). **Costa Rica**. ALAJUELA: Canton Atenas, Angeles de Atenas, barranca del Río Grande, *Smith 2446* (GH, US).—GUANACASTE: ca. 10–20 km NE of Liberia on Camino Santa María, *Utley & Utley 3114* (DUKE, MA, MICH).—LIMÓN: vic. of Westfalia, S of Limón, *Almeda et al. 3225* (CAS, MICH).—PUNTARENAS: Cabo Blanco Nature Reserve, S tip of Nicoya peninsula, 9°35′N, 85°06′W, *Burger & Liesner 6649* (F); Cantón Buenos Aires, Cañón del Río Grande de Térraba, 08°57′N, 83°20′W, *Kennedy 4659* (MICH); near Rincón de Osa, *Liesner 2214* (MO).—SAN JOSÉ: from Palmital to San Ignacio, *Khan et al. 193* (BM, MICH). **Panama**: BOCAS DEL TORO: Isla Colón, vic. of Chiriquí Lagoon, *von Wedel 2828* (GH, MICH, MO, NY).—CANAL ZONE: near Gatún Lake, *Croat 4707* (F, MO, NY); near Salamanca Hydrographic Station, *Dodge et al. 16966* (G, K, MO, S, U); Barro Colorado Island, *Foster 1320* (F, GH, MICH).—CHIRIQUÍ: Volcán, dist. near Las Lagunas, *D'Arcy 10054* (MICH, MO, NY, U); Burica peninsula, N of San Felix, *Mori & Kallunki 6024* (MICH, MO, NY, US).—COCLÉ: vic. of El Valle, *Allen 1777* (MO).—COLÓN: María Chiquita, *Ebinger 434* (F, GH, MO).—DARIÉN: Río Chucunaque, above Yaviza, *Gentry 13478* (COL, F, MICH, MO).—HERRERA: above Chapo de las Minas, *Folsom et al. 7005* (MICH, MO).—LOS SANTOS: Los Toretos, *Dwyer 2438* (MO, US).—PANAMÁ: Cerro Jefe, *D'Arcy 9733* (LL, MO, P, TEX); Isla Taboga, *Pittier 3579*

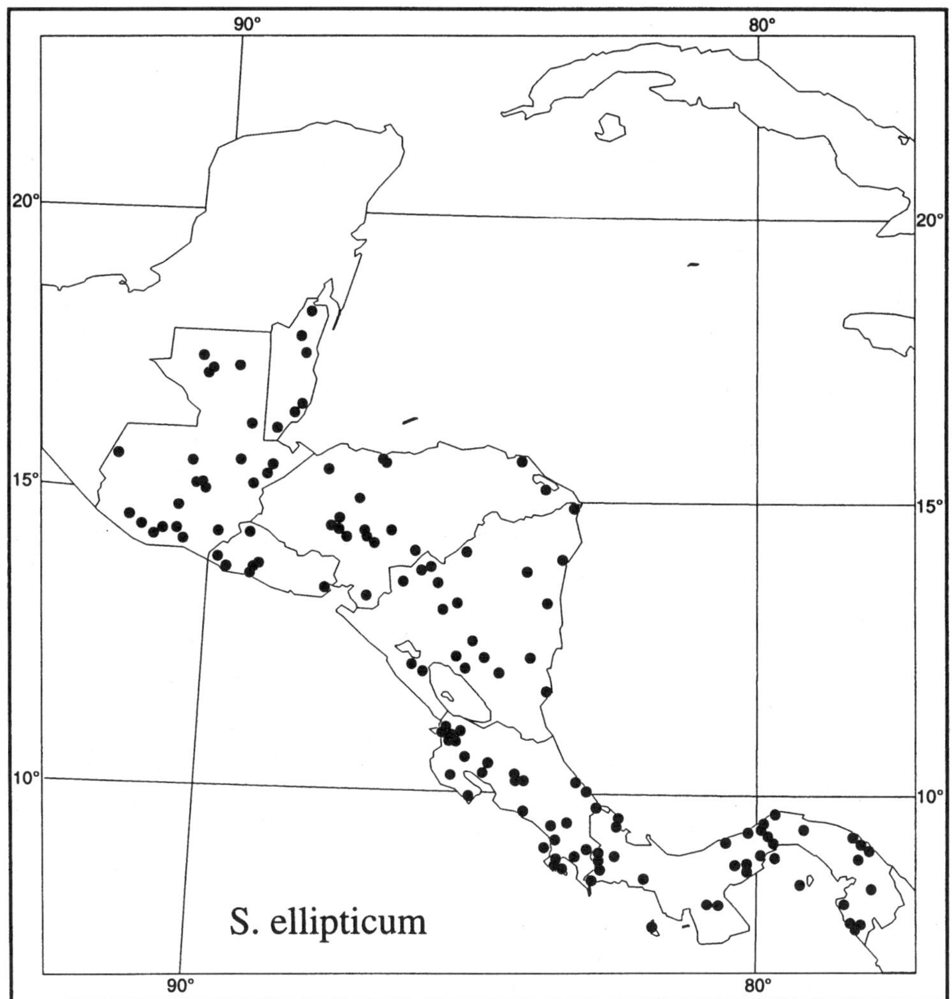

FIG. 47. Distribution of *Stigmaphyllon ellipticum* in Central America.

(US).—SAN BLAS: Ailigandí, *Dwyer 6858* (MO).—VERAGUAS: between Santa Fe and Escuela Agrícola Alto Piedras, *Croat & Folsom 33836* (CAS, MO, TEX).

Colombia. ANTIOQUIA: Mpio. Andes, 6 km de Andes hacia Vereda Mourblau, *Fonnegra R. et al. 2385* (MICH).—ATLÁNTICO: entre Palmar de Varela y Ponedera, orillas del Río Magdalena, *Dugand 4717* (F, COL, US).—CHOCÓ: Bahia Solano, *Haught 5568* (COL, US); camino entre Nuquí y Panguí, *Fernández 241* (COL, US); Cupica, *Fernández 361* (COL, US); a lo largo del Río Bandó, 15 km desde Pizarro al La Porquera, *Fuchs et al. 21712* (COL); hoya del Río San Juan, abajo de Docordó, ca. 04°15′N, 77°22′W, *Forero et al. 4415* (COL, MO).—GUAJIRA: via Mingueo–San Antonio, *Cuadros V. 2407* (MO).—MAGDALENA: Santa Marta, *H. H. Smith 341* (A, BM, BR, CM, COL, F, G, GH, K, MA, MICH, MO, MT, NY, P, S, TEX, U, US, UTD); Parque Nacional Tayrona, ca. 11°19′N, 73°58′W, *Kirkbride 2553* (COL, F, K, NY, US).—NARIÑO: E side of Gorgona Island, *Killip & García 33223* (COL, US).—VALLE: north shore of Buenaventura Bay, Playa Basaú, *Killip & Cuatrecasas* (F, US); costa del Pacífico, Río Yurumangú, *Cuatrecasas 15940* (F, US); costa del Pacífico, Río Cajambre, *Cuatrecasas 17663* (F, US). **Ecuador**. AZUAY: Pasaje–Girón Rd, ca. 10 km E of Uzhcurrumi, *Harling & Andersson 18729* (GB, MICH).—BOLÍVAR: Km 20 Babahoya–Montalvo rd, *Dodson 6121* (AAU, MO, SEL).—COTOPAXI: Tenefuerste, Río Pilalo, Km 52–53, Quevedo–Latacunga, *Dodson & Dodson 11872* (MICH, SEL).—EL ORO: Km 25, Santa Rosa–Pinas, *Dodson 9166* MO, SEL).—ESMERALDAS: Limones–Borbón, 5 km before Bor-

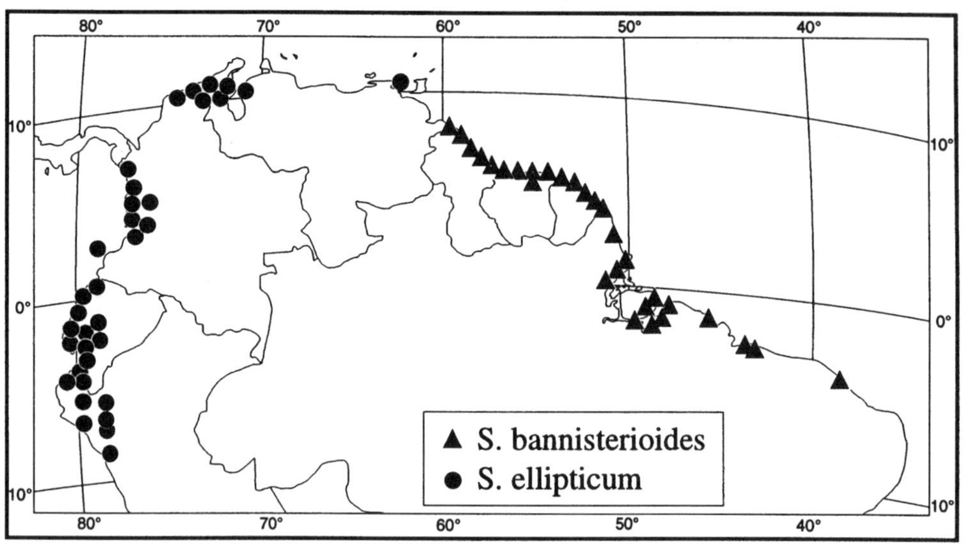

FIG. 48. Distribution of *Stigmaphyllon bannisterioides* in the Guianas and Brazil and of *S. ellipticum* in South America.

bón, 01°07′N, 79°00′W, *Holm-Nielsen et al. 26044* (AAU).—GUAYAS: Guayaquil, *Asplund 5203* (NY, R, S, US); Capeira, Km 21, Guayaquil to Daule, *Dodson & Gentry 12655* (MO, SEL);1 km S of Recinto Olon, 3 km N of Montañito, *Gentry 10028* (MICH, MO, US).—LOJA: Celica–Zapotillo Rd, ca. 5–6 km S of Sabanilla, *Harling & Andersson 18271* (GB, MICH); 1–4 km E of Macará along old rd to Cariamanga, *Harling & Andersson 22538* (S).—LOS RÍOS: Hacienda Clementina on Río Pita, *Asplund 5447* (CAS, G, K, NY, P, S, US).—MANABÍ: slopes of Montecristi, *Besse et al. 143* (QCA, SEL); Pedro Pablo Gómez area, Km 5, *Dodson & Thien 1698* (GB, NY, WIS); ad Portoviejo, *Mille 1969* (F).—PICHINCHA: [Sto. Domingo de los Colorados]: Sto. Domingo–Quevedo rd, Km 35, Hacienda Margarita, *Werling & Leth-Nissen 590* (AAU, NY). **Peru**. CAJAMARCA: Prov. Santa Cruz, bosque de Monteseco, *Sagástegui 12372* (MICH); Prov. Cutervo, Izco, *Stork & Horton 10209* (F, G, K).—LA LIBERTAD: Prov. Trujillo, alrededores de Quirihuac (Laredo), *López M. & Sagástegui 7878* (MO, NY, US).—LAMBAYEQUE: Mesone–Muro Hwy, 28 km E of Omos, *Hutchison & Wright 6714* (F, K, MO, NY, US).—PIURA: Morropón, Palo Blanco, *López M. 8757* (MICH).—TUMBES: Parque Nacional "Cerros de Amotape," Quebrada del Plátano, *Díaz & Peña 4074* (MICH); Cerro de Amotape, Quebrada Los Conejos, ca. 25 km SE of Cherratique, 04°09′S, 80°37′W, *Gentry 58206* (MICH). **Venezuela**. ZULIA: Dtto. Bolívar, cerca de Piedras Blancas, sitio 8 km al NE de Campo Lara, *Bunting 8009* (MICH, VEN); Dtto. Mara, entre Carraquero y Playa Bonita, *Bunting 9968* (MICH); Sierra de Perijá, Río Palmar, *Delascio Chitty & Benkowski 3094* (MICH, VEN).

Stigmaphyllon ellipticum is an easily recognized lowland species. The laminas are borne on short petioles (less than 3 cm long) and are most commonly elliptical and usually so sparsely sericeous abaxially that they appear glabrous to the naked eye. In three variant collections (*Díaz & Peña 4074*, *Gentry 58206*, both from Tumbes, Peru; *Harling & Andersson 18729*, Azuay, Ecuador), the appressed hairs on the lower surface of the lamina are more abundant and the trabecula is terete rather than flattened. The large flowers have fimbriate petals, are borne on glabrous and distally expanded pedicels, and are arranged in 3–9 (–12)-flowered umbels. The pedicels are usually shorter than the peduncles in plants from Mexico to Costa Rica, whereas in Panama and South America, they are commonly subequal or longer; however, exceptions to these generalizations are not infrequent. The nature of the folioles on all styles is variable throughout the range. In most collections from Mexico and Central America, the anterior style lacks folioles although the apex is distally elliptically to obovately to suborbicularly expanded and then bears a narrow lip on each side; the posterior styles bear semi-lunate folioles. In South America, in some plants the apex of the anterior style may be more broadly expanded and may even bear small folioles (the posterior styles have semi-lunate folioles), whereas in others the apex of the anterior style is only slightly expanded, and the folioles of the posterior styles are quite small or reduced to a lateral lip. Sometimes the stylar appendages are so greatly reduced that the styles are essentially efoliolate; this extreme reduction has been recognized as *S. felixii*.

The only species with which *S. ellipticum* might be confused is *S. eggersii* (no. 47) of coastal Ecuador. The species are very similar in many aspects but readily separated by the abaxial vesture of the laminas. Those of *S. eggersii* are not sparsely sericeous to glabrous abaxially but bear T-shaped hairs.

47. Stigmaphyllon eggersii C. Anderson, Contr. Univ. Michigan Herb. 19: 422. 1993.—TYPE: ECUADOR. Manabí: Estero Perro Muerto, Madralilla National Park, below San Sebastián, 01°36′N, 80°42′W, 23 Jan 1991, *Gentry & Josse 72667* (holotype: MICH!; isotypes: F! MO!).

Vine. Stems and branches sericeous when young, becoming glabrous. Laminas 5.3–18 cm long, 3–10.7 cm wide, ovate to broadly elliptical, apex mucronate or emarginate-mucronate to acuminate, base cordate to truncate, glabrous adaxially, with T-shaped hairs abaxially (trabecula 0.3–1.4 mm long, straight to wavy, stalk 0.1–0.4 mm long), margin eglandular; petioles 1.2–3 cm long, sericeous, not confluent across the node, with a pair of prominent but sessile glands at the apex, each gland 1–1.7 mm in diameter; stipules 0.5–1.2 mm long, 0.5–1 mm wide, free, triangular, eglandular. Flowers ca. 8–15 per condensed pseudoraceme, these solitary or borne in dichasia or compound dichasia (axes to the 2nd order, sericeous). Peduncles 1.8–9 mm long, sericeous or sparsely so; pedicels 6–9 mm long, distally expanded, glabrous; peduncles 0.3–1.3 times as long as the pedicels. Bracts 1.2–1.8 mm long, 0.7–1.1 mm wide, triangular, apex acute; bracteoles 1–1.6 mm long, 0.7–1 mm wide, ovate, apex obtuse, eglandular or with a glandular area in the basal 1/3–1/2; bracts and bracteoles sparsely sericeous abaxially. Sepals 2.5–3 mm long, 2.5–2.8 mm wide, glands 1.8–2.5 mm long, 0.8–1.3 mm wide. All petals with the limb orbicular, glabrous, yellow, margin fimbriate to fimbriate-lacerate, fimbriae and teeth up to 1.6 mm long; anterior-lateral petals: claw 2.5–2.8 mm long, limb 14–15 mm long and wide; posterior-lateral petals: claw ca. 2 mm long, limb 13–14 mm long and wide; posterior petal: claw 3.6–4.5 mm long, apex indented, limb 10–12 mm long and wide, sometimes at the base each side with a pair of stout gland-tipped fimbriae (0.5–0.6 mm long, 0.3–0.4 mm in diameter). Stamens unequal, those opposite the anterior-lateral sepals with the longest filaments, anthers of those opposite the lateral sepals with the connective enlarged and the locules reduced; anthers all loculate, glabrous. Stamen opposite anterior sepal: filament 2–2.8 mm long, anther 1–1.3 mm long; stamens opposite anterior-lateral petals: filaments 2–2.5 mm long, anthers ca. 1 mm long; stamens opposite anterior-lateral sepals: filaments 2.9–3.4 mm long, connectives ca. 1 mm long, locules 0.4–0.6 mm long; stamens opposite posterior-lateral petals: filaments ca. 2.7–3 mm long, anthers 1.3–1.5 mm long; stamens opposite posterior-lateral sepals: filaments 2.4–3 mm long, connectives ca. 1 mm long, locules 0.4–0.5 mm long; stamen opposite posterior petal always shorter than the adjacent two: filament 1.8–2 mm long, anther ca. 1 mm long. Anterior style 3–3.3 mm long, subequal to the posterior two, terete, glabrous, erect; apex 1.1–1.5 mm long, folioles variable in size, each foliole 0.3–0.7 mm long, 0.4–0.5 mm wide, parabolic. Posterior styles ca. 3.2 mm long, terete, glabrous, slightly lyrate; foliole 0.5–1 mm long, 0.4–0.8 mm wide, parabolic. Mature samara not seen; dorsal wing of nearly mature samara ca. 2.3 cm long, ca. 1.2 cm wide, upper margin with a shallow blunt tooth, lateral winglets absent; nut ca. 4.5 mm high, ca. 3.5 mm in diameter, without air chambers, areole ca. 3.5 mm long, ca. 2.3 mm wide, concave but convex at center, carpophore up to ca. 1 mm long. Embryo not seen. Chromosome number unknown.

Phenology. Collected in flower in January, June, and October, in immature fruit in January and June.

Distribution (Fig. 72). Coastal lowlands of Ecuador (Guayas, Manabí); in wet forest and matorral; sea level to 470 m.

ADDITIONAL SPECIMENS EXAMINED. **Ecuador**. GUYAS: Balao, *Eggers 14335* (A, M, US); Cantón Naranjal, camino de Santa Rosa de Flandres a Puerto Baquerizo, *Valverde 880* (MO).—MANABÍ: Rd Santa Elena–Jipijapa, Río Pital at Puerto López, 01°34′S, 80°48′W, *Holm-Nielsen et al. 27837* (AAU).

Stigmaphyllon eggersii is most similar to the widespread and sympatric *S. ellipticum* (no. 46), with which it shares large flowers borne on distally expanded pedicels, fimbriate-lacerate petals, and short-petioled leaves. They are readily separated by their leaf pubescence. The laminas of *S. eggersii* are pubescent with T-shaped hairs abaxially, whereas those of *S. ellipticum* are sparsely sericeous to glabrous abaxially.

48. Stigmaphyllon venulosum Cuatrecasas, Webbia 13: 536. 1958.—TYPE: COLOMBIA. Valle: Cordillera Occidental, La Popala, entre Ansama y Las Brisas, 1200 m, 28 Oct 1946, *Cuatrecasas 22702* (holotype: US!, photo: MICH!; isotypes: F! GH!).

Vine. Stems and branches sericeous when young, soon becoming glabrate to glabrous. Laminas 9–17.5 cm long, 5.4–11.8 cm wide, cordate to narrowly ovate or triangular, apex mucronate, base truncate or cordate, glabrous adaxially, sericeous to glabrous abaxially (trabecula 0.3–1 mm long, straight, sessile to subsessile), margin eglandular; petioles 2.5–5.3 cm long, sericeous to glabrate, confluent across the node and forming a band bearing the stipules, with a pair of prominent but sessile glands at the apex, each gland 1.6–2.5 mm in diameter; stipules 0.5–0.7 mm long, 0.7–1.2 mm wide, triangular, eglandular, the opposing stipules commonly fused into a bifid structure. Flowers ca. 10–14 per umbel, these solitary or borne in dichasia or small thyrses (axes to the 3rd order, sericeous or sparsely so). Peduncles 3.2–5.5 mm long, sericeous; pedicels 4–6 mm long, glabrous except for a row of hairs extending from the base of the pedicel to the base the posterior petal, distally expanded; peduncles 0.6–1 times as long as the pedicels. Bracts 1.1–1.5 mm long, 0.8–1.5 mm wide, triangular, apex acute; bracteoles 1–1.5 mm long and wide, ovate to subsquare, apex obtuse, eglandular; bracts and bracteoles sericeous abaxially. Sepals 2–2.8 mm long, 2.3–2.5 mm wide, glands 1.3–1.8 mm long, 0.9–1.2 mm wide. All petals with the limb orbicular (or sometimes limb of posterior petal broadly elliptical), glabrous, yellow and often marked with red, margin fimbriate, fimbriae up to 2 mm long; anterior-lateral petals: claw 3–3.4 mm long, limb ca. 15 mm long and wide; posterior-lateral petals: claw 1.2–2 mm long, limb ca. 15 mm long and wide; posterior petal: claw 3.5–4 mm long, apex indented, limb ca. 13 mm long and wide. Stamens unequal, those opposite the lateral sepals with the longest filaments and with the connective enlarged and the locules reduced; anthers all loculate, glabrous. Stamen opposite anterior sepal: filament 2.5–2.8 mm long, anther 1.1–1.4 mm long; stamens opposite anterior-lateral petals: filaments 2–2.3 mm long, anthers 0.9–1.2 mm long; stamens opposite anterior-lateral sepals: filaments 3.3–3.6 mm long, connectives 0.7–1 mm long, locules 0.3–0.5 mm long; stamens opposite posterior-lateral petals: filaments 2.9–3 mm long, anthers 1.2–1.3 mm long; stamens opposite posterior-lateral sepals: filaments 3–3.7 mm long, connectives 0.7–1 mm long, locules 0.3–0.5 mm long; stamen opposite posterior petal always shorter than the adjacent two: filament 2.3–3.4 mm long, anther 0.9–1 mm long. Anterior style 2.6–3.2 mm long, shorter than the posterior two or styles subequal, terete, glabrous, erect; apex 1.2–1.3 mm long, each foliole 0.3–0.7 mm long, 0.5–0.8 mm wide, triangular or narrowly so. Posterior styles 2.7–3.2 mm long, terete, glabrous, erect; foliole 1–1.1 mm long, 0.6–0.7 mm wide at insertion, broadly semi-lunate. Dorsal wing of samara 3.6–4.2 cm long, 1.3–1.9 cm wide, upper margin with a blunt tooth; nut rugose, ca. 10 mm high, ca. 5 mm in diameter, the locule surrounded by narrow air chambers, areole ca. 5.5 mm long, ca. 4 mm wide, concave, carpophore up to ca. 4 mm long. Embryo 8 mm long, ca. 2 times as long as wide, ovoid, outer cotyledon ca. 7.5 mm long, ca. 4.5 mm wide, the distal 1/3 folded over the inner cotyledon, inner cotyledon ca. 4.8 mm long, 4 mm wide, folded at the distal 1/5. Chromosome number unknown.

Phenology. Collected in flower and fruit from January through March, May, and July through November.

Distribution (Fig. 21). West-central Colombia; in secondary growth, "matorrales," "matorrales xerofiticos," "zona cafetera"; 600–1800 m.

REPRESENTATIVE SPECIMENS. **Colombia**. ANTIOQUIA: Mpio. Betania, Veredas La Cascajosa–Santa Ana, 05°45′N, 75°45′W, *Callejas & Fonnegra G. 7317* (NY); Fredonia, *Toro 317* (NY); Mpio. Salgar, Km 4 of rd Salgar–Hcda El Dauro (Chocó), 05°58′N, 76°01′W, *Zarucchi & Echeverry 4732* (K, MICH).—CALDAS: Mpio. La Merced, vereda La Primera Quiebra, 05°30′N, 75°35′W, *Arias 64* (MO).—RISARALDA: entre La Felisa y Filadelfia, *Albert de Escobar & Brand 2067* (MICH).—VALLE: Lobo Guerrero, *Cuatrecasas 17791* (F, US); hoya del Río Cali, Pichindé, El Abismo, *Cuatrecasas 18671* (F, US); La Cumbre, *Cuatrecasas 19559* (F, US); Mpio. Buga, Yocoto, *Cucalón S. & Estévez G. 3* (U); La Cumbre, *Hazen 11846* (GH, NY, US); 1 km above and E of Bitaco, *Hutchison & Wright 3266* (COL, F, US).

Stigmaphyllon venulosum is most similar to *S. echitoides* (no. 49) and *S. tergolanatum* (no. 50), from which is it is readily separated by the vesture on the lower surfaces of the leaves. The laminas of *S. venulosum* are sericeous abaxially (the trabecula straight and appressed), whereas those of *S. echitoides* and *S. tergolanatum* are pubescent with T-shaped hairs to tomentose abaxially (the trabecula wavy to crisped and curled, and stalked). With these species *S. venulosum* shares large flowers with long-fimbriate petals, borne on short peduncles and pedicels, and disposed in umbellate clusters. As also in *S. echitoides*, the nut of the samara is somewhat enlarged, because the locule is surrounded by air chambers; the samaras of *S. tergolanatum* are unknown. See also the discussion of *S. echitoides*.

49. Stigmaphyllon echitoides Triana & Planchon, Ann. Sci. Soc. Nat. Bot., sér. 4, 18: 318. 1862.—TYPE: COLOMBIA. Norte de Santander: "La Enllanada près d'Ocaña," 1300 m, *Triana s.n.* (lectotype, designated by Cuatrecasas, 1958: P!, photo: MICH!; isolectotype?: BM!, photo: MICH!).

Stigmaphyllon ipomoeoides Triana & Planchon, Ann. Sci. Soc. Nat. Bot., sér. 4, 18: 317. 1862.—TYPE: COLOMBIA. Antioquia: Medellín, 1500 m, *Triana s.n.* (holotype: P!, photo: MICH!; isotype?: BM!, photo: MICH!).

Vine. Stems and branches sericeous when young, soon becoming glabrate to glabrous. Laminas 8.5–16.5 cm long, 5–11.5 cm wide, cordate or lanceolate to broadly elliptical or sometimes suborbicular, apex acuminate or mucronate, base cordate, glabrate to glabrous adaxially, with T-shaped hairs to tomentose abaxially (trabecula 0.3–1.1 mm long, wavy to crisped and curled, stalk 0.1–0.3 mm long), margin eglandular; petioles 1.4–4.7 cm long, sericeous to glabrate, confluent across the node and forming a band bearing the stipules, with a pair of prominent but sessile glands at the apex, each gland 1–2.3 mm in diameter; stipules 1–1.8 mm long, 0.6–0.8 mm wide, triangular, eglandular, the opposing stipules commonly fused into a bifid structure. Flowers ca. 10–14 per umbel, these solitary or borne in dichasia or sometimes in small thyrses (axes to the 3rd order, sericeous or sparsely so). Peduncles 1.5–6.5 mm long, densely sericeous; pedicels 3.3–7 mm long, glabrous except for a row of hairs extending from the base of the pedicel to the base the posterior petal, distally expanded; peduncles 0.3–0.75 (–1) times as long as the pedicels. Bracts 1–1.8 mm long, 0.6–1.4 mm wide, triangular, apex acute; bracteoles 0.9–1.6 mm long, 0.7–1.3 mm wide, oblong to subsquare, apex obtuse, eglandular; bracts and bracteoles sericeous or sparsely so abaxially. Sepals 2–3 mm long, 1.8–2.9 mm wide, glands 1.4–1.9 mm long, 0.9–1 mm wide. All petals with the limb orbicular (or that of the posterior petal sometimes broadly obovate), glabrous, yellow and marked or suffused with red or the limb of the posterior petal mostly red, margin fimbriate, fimbriae up to 2.2 mm long; anterior-lateral petals: claw 2.5–3 mm long, limb ca. 15 mm long and wide; posterior-lateral petals: claw 1.3–2 mm long, limb ca. 15 mm long and wide; posterior petal: claw 3.7–4 mm long, apex indented, limb ca. 13 mm long and wide. Stamens unequal, those opposite the lateral sepals with the longest filaments and with the connective enlarged and the locules reduced; anthers all loculate, glabrous. Stamen opposite anterior sepal: filament 2.5–2.7 mm long, anther 1.4–1.5 mm long; stamens opposite anterior-lateral petals: filaments 2–2.1 mm long, anthers ca. 1.2 mm long; stamens opposite anterior-lateral sepals: filaments 3.3–3.5 mm long, connectives ca. 1.5 mm long, locules 0.4–0.5 mm long; stamens opposite posterior-lateral petals: filaments 2.5–3 mm long, anthers 1.3–1.5 mm long; stamens opposite posterior-lateral sepals: filaments 3–3.2 mm long, connectives 0.9–1.1 mm long, locules 0.4–0.5 mm long; stamen opposite posterior petal always shorter than the adjacent two: filament 2.2–2.8 mm long, anther 0.6–0.9 mm long. Anterior style 2.8–3.3 mm long, shorter than the posterior two or styles sometimes subequal, terete, glabrous, erect; apex 1.3–1.5 mm long, each foliole ca. 0.5 mm long, 0.7–1 mm wide, triangular, or folioles absent but the distal 2/3 of the apex elliptically to suborbicularly expanded and each side with a lip 0.3–0.5 mm wide. Posterior styles 2.6–3.5 mm long, terete, glabrous, erect; foliole 1–1.5 mm long, 0.8–1 mm wide at insertion, broadly semi-lunate. Dorsal wing of samara 3.5–4 cm long, 1.3–1.5 cm wide, upper margin with a blunt tooth; nut bearing rectangular to lunate lateral winglets, these 6.5–7.5 mm long, 2–2.7 mm wide or only 2–4 large teeth or only rugose to prominently ribbed; nut 8–9 mm high, 5–5.5 mm in diameter, wall containing spongy tissue, locule surrounded by air chambers, areole ca. 5 mm long, 3–3.5 mm wide, concave, carpophore up to 3.5 mm long. Embryo 9 mm long, ca. 2 times as long as wide, ovoid, outer cotyledon ca. 10.2 mm long, ca. 3.8 mm wide, the distal 1/3 folded over the inner cotyledon, inner cotyledon ca. 9 mm long, 3.3 mm wide, folded at the distal 1/4. Chromosome number unknown.

Phenology. Collected in flower and fruit throughout the year.

Distribution (Fig. 49). Colombia; in thickets and forests; 350–2000 m.

REPRESENTATIVE SPECIMENS. **Colombia**. ANTIOQUIA: alrededores de Dabeiba, *Barkley & Gutiérrez 1771* (COL, F, S); Medellín, quebrada La Hueso, *Uribe U. 1181* (COL, US); Dabeiba, *Uribe U. 1440* (COL).—

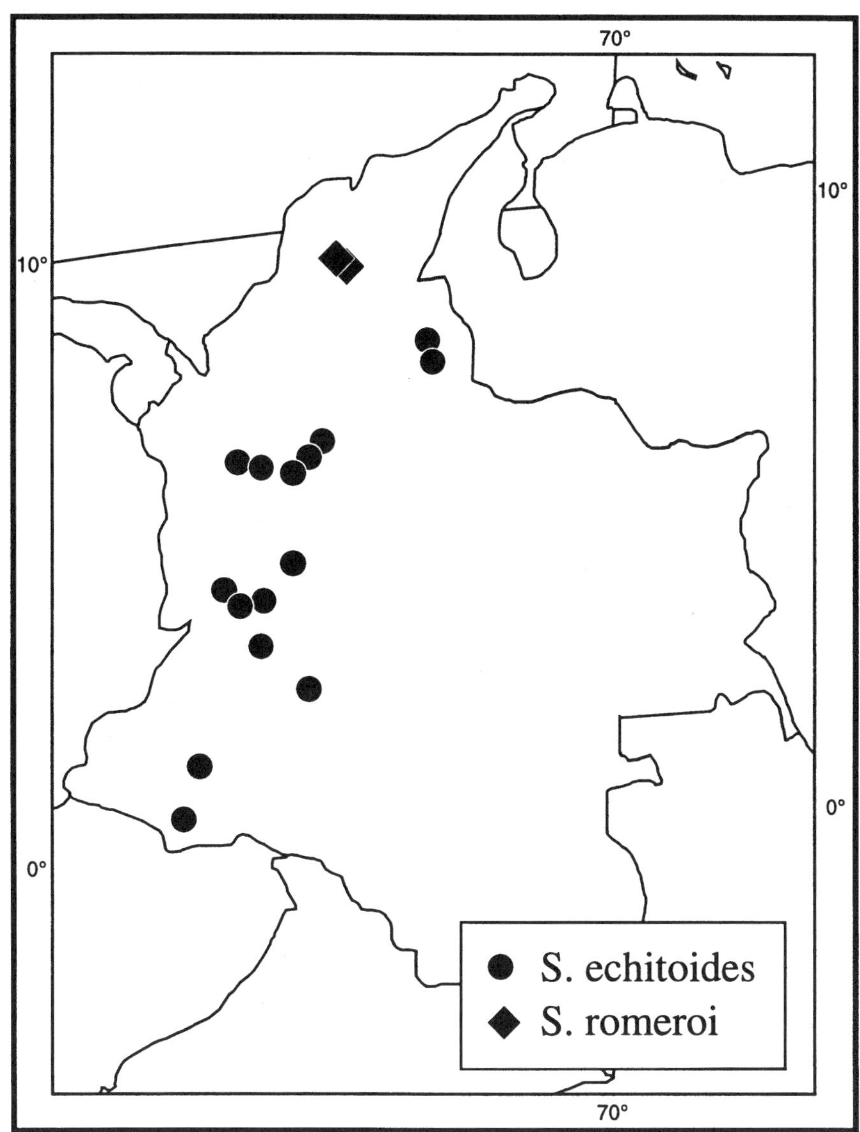

FIG. 49. Distribution of *Stigmaphyllon echitoides* and *S. romeroi*.

CAUCA: 13 km below Dagua and 1 km above Lobo Guerrero, *Hutchison & Idrobo 3122* (COL, F, K, MO, NY, US); rd between Patía and El Estrecho, *Plowman & Vaughan 5359* (GH, MICH).—CESAR: Quebrada del Gobernador, carretera de San Martín a Ocaña, Km 21–22, *Cuatrecasas & Rodríguez 27967* (COL, MA, US).—HUILA: E of Neiva, *Ruby & Pennell 447* (NY).—NARIÑO: Cerro de las Piñas, *Ewan 15930* (MO, P, S, US).—NORTE DE SANTANDER: around Ocaña, *Kalbreyer 636* (K).—VALLE: Cartago, Hacienda La Silvia, *Duque Jaramillo 4106-B* (COL); bosque de Ema, ca. 60 km N of Cali, 03°38′N, 76°33′W, *Gentry 59522* (MICH).

Stigmaphyllon echitoides is most similar to *S. venulosum* and *S. tergolanatum*, both also from Colombia. They all have large flowers with long-fimbriate petals, which are borne on short peduncles and pedicels, and are arranged in umbellate clusters. The pedicels are glabrous except for a row of hairs extending from the base of the pedicel to the base the posterior petal. The petioles, less than 5.5 cm long, are confluent across the node and form a band bearing the stipules. The nut of the samaras of *S. echitoides* and *S. venulosum* is somewhat enlarged, because the locule is surrounded by air chambers; the samaras of *S. tergolanatum* are unknown. The three species may be separated with the following key.

1. Laminas abaxially sericeous. 48. *S. venulosum.*
1. Laminas abaxially pubescent with T-shaped hairs to densely tomentose.
    2. Laminas abaxially pubescent with T-shaped hairs to moderately tomentose (epidermis always visible); foliole of posterior styles broadly semi-lunate. 49. *S. echitoides.*
    2. Laminas abaxially densely woolly-tomentose (epidermis hidden); foliole of posterior styles triangular. 50. *S. tergolanatum.*

50. Stigmaphyllon tergolanatum Cuatrecasas, Webbia 13: 538. 1958.—TYPE: COLOMBIA. Huíla: Cordillera oriental, La Junta, 6 km W of Algeciras (San Juanito), 38 km S of Neiva, 850 m, 3 Dec 1942, *Fosberg 19230* (holotype: US!, photo: MICH!; isotype: NY!).

Vine. Stems and branches pubescent with scalelike T-shaped hairs when young, soon becoming glabrate. Laminas 5.3–10 cm long, 4.5–7 cm wide, ovate or elliptical, apex emarginate-mucronate or obtuse-mucronate, base cordate, glabrous adaxially, very densely tomentose abaxially (trabecula 0.6–1.2 mm long, crisped and curled, stalk 0.1–0.2 mm long), the vesture obscuring the epidermis, margin eglandular; petioles 1.8–2.5 cm long, sericeous or densely so, confluent across the node and forming a band bearing the stipules, with a pair of prominent but sessile glands at the apex, each gland 2–2.8 mm in diameter; stipules 1–1.2 mm long, 1.6–2.5 mm wide, opposing stipules basally fused into a triangular bifid structure, eglandular. Flowers ca. 8–12 per umbel, these borne in compound dichasia (axes to the 3rd order, densely sericeous). Peduncles 2.5–4 mm long, densely sericeous; pedicels 7–8 mm long, glabrous except for a row of hairs extending from the base of the pedicel to the base the posterior petal, distally expanded; peduncles 0.3–0.5 times as long as the pedicels. Bracts 1.4–1.7 mm long, 1.1–1.2 mm wide, triangular, apex acute, sparsely sericeous; bracteoles 1.5–2.2 mm long, 1–1.3 mm wide, triangular, apex acute to obtuse, each bracteole with a basal gland (each gland 0.3–0.4 mm in diameter), glabrous to sparsely sericeous near the base. Sepals ca. 2.5 mm long and wide, glands 2–2.3 mm long, 1–1.2 mm wide. All petals with the limb orbicular, glabrous, yellow (and marked with red?), margin fimbriate, the fimbriae up to 1 mm long; anterior-lateral petals: claw ca. 3.5 mm long, limb ca. 19 mm long and wide; posterior-lateral petals: claw ca. 2 mm long, limb ca. 16 mm long and wide; posterior petal: claw 4.5–4.8 mm long, apex indented, limb ca. 15 mm long and wide. Stamens unequal, those opposite the anterior-lateral sepals the largest, anthers of those opposite the lateral sepals with the connective enlarged and the locules reduced; anthers all loculate, glabrous. Stamen opposite anterior sepal: filament ca. 3 mm long, anther ca. 1.6 mm long; stamens opposite anterior-lateral petals: filaments ca. 2 mm long, anthers ca. 1.3 mm long; stamens opposite anterior-lateral sepals: filaments ca. 3.7 mm long, connectives ca. 1.5 mm long, locules ca. 0.4 mm long; stamens opposite posterior-lateral petals: filaments ca. 3 mm long, anthers ca. 1.3 mm long; stamens opposite posterior-lateral sepals: filaments ca. 3.5 mm long, connectives ca. 1.1 mm long, locules ca. 0.4 mm long; stamen opposite posterior petal always shorter than the adjacent two: filament ca. 2.8 mm long, anther ca. 1.2 mm long. Anterior style ca. 3.5 mm long, subequal to the posterior two, terete, with scattered hairs in the proximal 1/4, erect; apex 1.1 mm long including a spur ca. 0.3 mm long, each foliole ca. 0.8 mm long, ca. 0.7 mm wide, triangular. Posterior styles ca. 3.6 mm long, terete, with scattered hairs in the proximal 1/4, lyrate; foliole ca. 0.7 mm long, ca. 1.1 mm wide, triangular. Samara not seen. Chromosome number unknown.

Stigmaphyllon tergolanatum is known only from the type collection from Colombia (Fig. 21). It is distinguished by its laminas, which are so densely tomentose abaxially that the epidermis is hidden. It is similar to *S. echitoides* and S. *venulosum* in that its flowers, with long-fimbriate petals, are borne on short pedicels and peduncles, and arranged in umbellate clusters. In all three, the pedicels are glabrous except for a row of hairs extending from the base of the pedicel to the base the posterior petal. For a separation of these three species, see *S. echitoides* (no. 49).

51. Stigmaphyllon vitifolium Adr. Jussieu in St.-Hilaire, Fl. bras. mer. 3: 50. 1833 ["1832"].—TYPE: BRAZIL. Rio de Janeiro: "prope Sebastianopolim," Dec, *Gaudichaud s.n.* (holotype: P!, photos: F! MICH! US!; isotype: P!).

Stigmaphyllon gaudichaudianum Adr. Jussieu in St.-Hilaire, Fl. bras. mer. 3: 50. 1833 ["1832"].—TYPE: BRAZIL. Rio de Janeiro: Sebastianopolis, *Leandro do Sacramento s.n.* (lectotype, here designated: P!, photo: MICH!; isolectotypes: BR! M! P!).

Vine. Stems and branches pubescent with scalelike T-shaped hairs when young, soon becoming glabrate. Laminas 9.5–16 cm long, 8–15 cm wide, commonly shallowly to deeply palmately 3–5-lobed or sometimes triangular to ovate, apex of lamina and of each lobe mucronate, base deeply auriculate, glabrous adaxially and abaxially but often very sparsely sericeous on the midrib abaxially, margin grossly dentate, each tooth ending in a filiform gland (up to 3 mm long), sometimes also with stalked capitate glands (0.2–0.3 mm in diameter, ca. 0.3–0.4 mm long) in the sinuses; petioles 3–10.5 cm long, pubescent with scalelike T-shaped hairs to glabrate, not confluent across the node, bearing a pair of shallowly cupulate glands 4–20 mm below the base of the lamina, each gland 1–1.9 mm in diameter, 0.8–2 mm high; stipules 0.6–1.3 mm long, 0.8–1.7 mm wide, free, triangular, glabrous abaxially, eglandular. Flowers ca. 8–10 per umbel, these solitary or borne in dichasia (axes to the 2nd order, densely sericeous). Peduncles 2.5–8 mm long; pedicels 3.5–7 mm long, distally expanded; both sericeous (hairs subsessile) or the pedicels glabrous in the distal 1/2–1/3, peduncles 0.5–1.3 times as long as the pedicels. Bracts 1.5–2.5 mm long, 1–1.8 mm wide, oblong, apex obtuse, glabrate to glabrous abaxially; bracteoles 0.7–1.8 mm long, 0.9–1.5 mm wide, ovate to semicircular, apex obtuse, sparsely sericeous (hairs subsessile) to glabrous, eglandular or with a glandular area in the basal 1/3–1/2. Sepals 2–2.5 mm long, 2.3–2.6 mm wide, glands 2.3–3 mm long, 1.2–1.6 mm wide. All petals with the limb orbicular, glabrous, yellow or (especially the limb of the posterior petal) suffused with red, margin erose; anterior-lateral petals: claw 2–2.5 mm long, limb 8–9 mm long and wide; posterior-lateral petals: claw 1–1.5 mm long, limb 7–8 mm long and wide; posterior petal: claw 2.5–3 mm long, apex indented, limb 6–7 mm long and wide. Stamens unequal, those opposite the posterior styles the largest but filaments of those opposite the anterior-lateral sepals equal or subequal, anthers of those opposite the lateral sepals commonly with the connective enlarged and the locules reduced, or anthers of stamens opposite the anterior-lateral sepals unmodified; anthers all loculate, glabrous. Stamen opposite anterior sepal: filament 1.5–2.4 mm long, anther 1–1.1 mm long; stamens opposite anterior-lateral petals: filaments 1–1.7 mm long, anthers 0.8–0.9 mm long; stamens opposite anterior-lateral sepals: filaments 1.6–2.5 mm long, connectives 0.7–0.9 mm long, locules 0.3–0.8 mm long; stamens opposite posterior-lateral petals: filaments 1.8–2.5 mm long, anthers 1–1.3 mm long; stamens opposite posterior-lateral sepals: filaments 1.8–2.2 mm long, connectives 0.8–0.9 mm long, locules 0.3–0.6 mm long; stamen opposite posterior petal always shorter than the adjacent two: filament 1.5–1.8 mm long, anther 0.6–0.7 mm long. Anterior style 2.1–2.7 mm long, shorter than the posterior two, terete, glabrous, erect; apex 1.4–1.5 mm long, in the distal half laterally expanded into small triangular folioles, each 0.2–0.3 mm long, 0.3–0.4 mm wide. Posterior styles 2.6–3.5 mm long, terete, glabrous, lyrate; foliole 1.4–2 mm long, 1.5–1.8 mm wide, subsquare. Dorsal wing of samara 3.4–3.5 cm long, 1.3–1.4 cm wide, upper margin with a blunt tooth; nut smooth, 6.5–7 mm high, 5.3–5.5 mm in diameter, without air chambers, areole 3–3.5 mm long and wide, concave, carpophore up to 3.5 mm long. Mature seed not seen; immature embryo ca. 2 times as long as wide, the outer cotyledon folded at the distal 2/5 over the inner cotyledon. Chromosome number unknown.

Phenology. Collected in flower from August through April; date of the single fruiting collection unknown.

Distribution (Fig. 50). Brazil (Rio de Janeiro); in open woods and roadside thickets; sea level to 200 m.

ADDITIONAL SPECIMENS EXAMINED. **Brazil.** RIO DE JANEIRO: Itaipú, Morro das Andorinhas, *Araujo 3955* (GUA); Botafogo, mata atrás do Colégio Santa Ursula, *Braga 2438* (MICH); Jacarepaguá, Pau-da-Fome, *Casari 376* (GUA); Rio Comprido, *Goés RB73296* (RB), *Schwacke 5072* (NY, RB); Lagoa de Freitas, *Glaziou 1146* (BR, C, F, P); Guaratiba, *Lutz 1721* (R); Niteroi, *Brade R78328* (R), *Brade 12819* (RB), *Holway & Holway 1050* (US), *L. B. Smith & Brade 2335* (GH), *Ule 776* (NY); próximo a Lagoa de Araruama, *da Rocha 379* (GUA); Organ Mts, *Wilkes Expedition s.n.* (US). Without locality, *Burchell 1267* (GH, K, P), *Burchell 1341* (GOET, K),

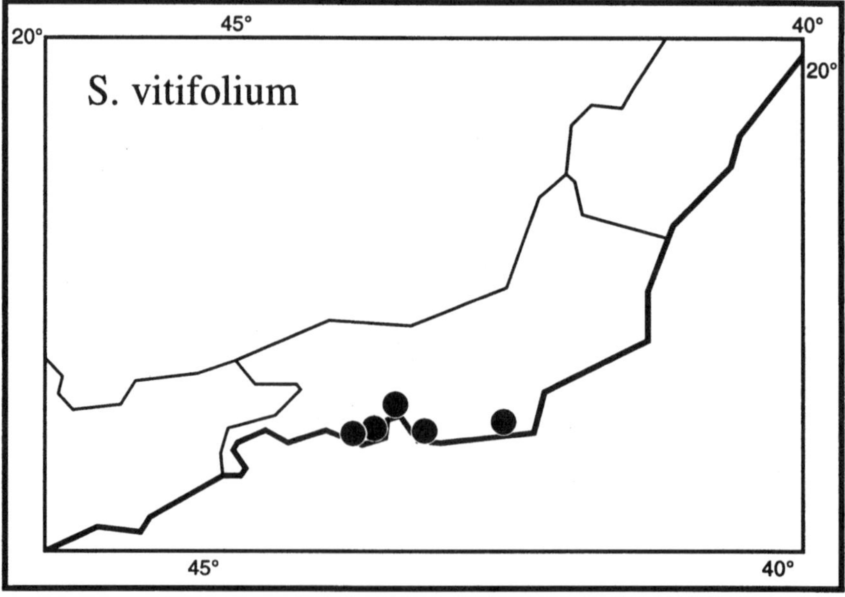

FIG. 50. Distribution of *Stigmaphyllon vitifolium.*

Burchell 1505 (GOET, K, P), *Glaziou 10372* (K, P, R), *Riedel & Luschnath 1209* (A, LE, S, US), *Schüch s.n.* (BR), *Sello 1285 S118* (NY), *Weddell 540* (G, P), *Widgren 508* (S), *Wilkes Expedition s.n.* (GH, NY).

Stigmaphyllon vitifolium is readily recognized by its unique leaves. The glabrous, grossly dentate laminas commonly are shallowly to deeply palmately 3–5-lobed but also vary from triangular to ovate; each tooth terminates in a filiform gland. The large basal glands are borne on the petiole, up to 2 cm below the base of the lamina. Jussieu based *S. gaudichaudianum* on specimens with unlobed leaves and *S. vitifolium* on a collection with deeply lobed leaves, but later collections show that leaf shape is variable in this species; e.g., the duplicate of *Glaziou 1146* at BR consists of a sprig with three large, attached leaves, of which one is unlobed and two are deeply lobed.

This species is also distinguished by its umbellate clusters of 8–10 flowers, borne on small peduncles and pedicels. It shares this inflorescence type and vesture, red flag petal, stem vesture, petiole glands, and laterally unornamented samaras with the Bolivian *S. boliviense* (no. 59; samaras unknown) and *S. coloratum* (no. 60) and the Brazilian *S. puberulum* (no. 61; Bahia, Pernambuco) and *S. bradei* (no. 58; São Paulo; samaras unknown), but differs from these species, in addition to laminar shape and vesture, in its erose rather than fimbriate petals and very small folioles of the anterior style (in *S. boliviense* only the posterior petal is fimbriate). *Stigmaphyllon vitifolium* is also similar to *S. macedoanum* (no. 35) of Minas Gerais, distinguished by the prominent corky ridge formed by the confluence of the petioles.

52. Stigmaphyllon jatrophifolium Adr. Jussieu in St.-Hil., Fl. bras. merid. 3: 51. 1833 ["1832"].—TYPE: BRAZIL. Rio Grande do Sul: "in lapidosis juxta castrum de Salto non procul a ripis Uruguay," *St.-Hilaire 2525* (holotype: P!, photo: MICH!; isotypes: P!, photo: MICH!).

Stigmaphyllon jatrophifolium f. *gracile*, Niedenzu, Pflanzenreich IV. 141(2): 503. 1928.—TYPE: BRAZIL. Rio Grande do Sul: "bords de l'Uruguay, ancienne mission de S. Borja," *Isabelle s.n.* (lectotype, here designated: G!, photo: MICH!).

Vine to 8 m. Stems and branches sericeous when very young, soon becoming glabrous. Laminas in outline 4.2–18.2 cm long and 3.5–17 cm wide, sometimes triangular to elliptical to suborbicular but usually palmately to pedately lobed or smaller laminas sometimes hastate, lobes (2–) 5–7 (–9), apex (of each lobe) acuminate or acuminate-mucronate, base auriculate or deeply cordate, glabrous adaxially, very sparsely sericeous (especially near the major veins) to glabrous abaxially (trabecula 0.3–0.5 mm long, straight, sessile), margin commonly coarsely dentate, with filiform glands (up to 4 mm long) and sometimes also with irregularly spaced sessile glands (0.2–0.4 mm in diameter); petioles 1.6–5 cm long, sericeous to glabrous, confluent across the node and forming a band or ridge bearing the stipules, with a pair of prominent but sessile glands at the apex, each gland 0.9–2.3 mm in diameter; stipules 0.4–1 mm long, 0.5–1.2 mm wide, free, triangular, eglandular, caducous. Flowers 8–15 per umbel, these solitary or borne in dichasia (axes sericeous or sparsely so). Peduncles 3.5–16 mm long, sericeous or sparsely so; pedicels 4–7.5 mm long, sparsely sericeous to glabrous, distally expanded; peduncles 0.75–2 times as long as pedicels. Bracts 0.8–1.9 mm long, 0.8–1.5 mm wide, triangular, apex acute; bracteoles 0.7–1.5 mm long, 0.7–1.2 mm wide, triangular to parabolic, apex obtuse, each bracteole with a pair of inconspicuous glands (each 0.2–0.4 mm in diameter) or sometimes eglandular; bracts and bracteoles sericeous to glabrous abaxially. Sepals 2.3–3.5 mm long,

2.5–2.8 mm wide, glands 1.7–2.2 mm long, 1–1.3 mm wide. All petals with the limb orbicular, glabrous, yellow, margin with fimbriae up to 1 mm long; anterior-lateral petals: claw 2–3 mm long, limb 10.5–12 mm long and wide; posterior–lateral petals: claw 1–2 mm long, limb (7.5–) 9–11 mm long and wide; posterior petal: claw 3–4.2 mm long, apex indented, limb 7–8.5 mm long and wide. Stamens unequal, those opposite the lateral sepals with the longest filaments (sometimes stamens opposite the posterior-lateral petals with the filaments equally long) and usually with the connective enlarged and the locules reduced; anthers all loculate, glabrous. Stamen opposite anterior sepal: filament 2.5–3.2 mm long, anther 1.3–1.5 mm long; stamens opposite anterior-lateral petals: filaments 1.8–2.1 mm long, anthers 0.8–1.2 mm long; stamens opposite anterior-lateral sepals: filaments 3–3.8 mm long, connectives 0.8–1.1 mm long, locules 0.4–0.8 mm long; stamens opposite posterior-lateral petals: filaments 2.8–3.5 mm long, anthers 1.4–1.6 mm long; stamens opposite posterior-lateral sepals: filaments 2.4–3 mm long, connectives 0.6–0.9 mm long, locules 0.4–0.9 mm long; stamen opposite posterior petal always shorter than the adjacent two: filament 2–2.5 mm long, anther 0.6–0.9 mm long. Anterior style 3.6–4.1 mm long, slightly shorter than the posterior two, terete, glabrous, slightly incurved; apex 1.3–1.8 mm long, each foliole 1.7–2.3 mm long, 1.2–1.6 mm wide, rectangular to square to rhombic. Posterior styles 4.1–4.4 mm long, terete, glabrous, lyrate; foliole 1.5–2 mm long, 1.6–2 mm wide, square to sometimes narrowly trapezoidal. Dorsal wing of samara 3–4 cm long, 1.2–2 cm wide, upper margin without a tooth; nut smooth or with a few prominent veins, 9–11 mm high, 6.5–7.2 mm in diameter, wall containing spongy tissue, without air chambers, areole 5–8 mm long, 4.5–6.5 mm wide, strongly concave, carpophore up to 3.5 mm long. Embryo 9.5–9.7 mm long, ca. 2 times as long as wide, ovoid, outer cotyledon 11–12.5 mm long, 5.3–6 mm wide, the distal 1/3 folded over the inner cotyledon, inner cotyledon 8–9.7 mm long, 4.5–5.1 mm wide, folded at the distal 1/3. Chromosome number: n = 10 (W. R. Anderson 1993; based on *Anderson 12371*). Frontispiece.

Phenology. Collected in flower and fruit from September through June.

Distribution (Fig. 51). Mostly along the Río Paraná and Río Uruguay and their tributaries in eastern Argentina, western Uruguay, southern Paraguay, and southeastern Brazil; in wet places, along streams, at forest edge, and at roadsides; sea level to 280 m.

Representative Specimens. **Argentina**. Corrientes: Depto. San Martín, Yapeyú, *Krapovickas & Cristóbal 28978* (CTES, MICH); Depto. Alvear, 30 km NE de Alvear, *Schinini et al. 16491* (MICH); Depto. San Tomé, Garruchos, *Schinini 19954* (CTES).—Entre Ríos: Depto. Gualeauaychú, *Burkart & Troncoso 27861* (MICH); Depto. Colón, Parque Nac. El Pamar, camino a La Virgen, *Cusato 969* (CTES); Depto. Federación, Col. Ensanche, *Troncoso et al. 1341* (MICH).—Misiones: inter Posadas et La Granja, *Ekman 1527* (G, MO, NY, S, US); Depto. San Javier, Balneario 4 Bocas, 11 km NE de San Javier, *Krapovickas & Cristóbal 28877* (CTES, MICH, MO); Cataratas del Iguazú, *Meyer 5352* (A, U). **Brazil**. Paraná: Rio Perdido, Mpio. Laranjeiras do Sul, *Hatschbach 19361*; Carimá, Mpio. Foz do Iguazú, *Hatschbach 23143* (MICH, P).—Rio Grande do Sul: *Fox 16* (K); Tristeza, pr. Pôrto Alegre, *Malme 770* (G, R, S); Roque Gonzales, *Pedersen 11990* (C, CTES, MICH); Pôrto Alegre, Sant'Ana, *Rambo 220* (SP, W).—Santa Catarina: Descanso, *Klein 5118* (P); Itapiranga (Peperí–Misiones), *Rambo 60330* (P). **Paraguay**. Alto Paraná: cerca del pueblo de Hernandarias, *Fernández & Molero 5724* (G, MO, NY).—Caapaza: Tavai, 26°10′S, 55°17′W, *Soria 3392* (CTES, MA, MO).—Caaguazú: N edge of Caaguazú, *Anderson 12389* (MICH).—Central: Yaguarón, monte Oratorio, *Krapovickas et al. 12252* (CTES, MICH, P).—Concepción: Horequeta, *Lurvey 473* (CTES).—Cordillera: Cordilleras de Altos, *Fiebrig 536* (F, G, GH, K, M, NY), *Hassler 3501* (BM, G, GH, K, NY, P, W); Cordillera de Altos, Cerro de Tobatí, *Schinini 24023* (MICH).—Guiará: Villa Rica, *Jörgensen 3827* (A, C, DS, F, MO, NY, S, US); Cerro Neville, 5 km E de Mbocayaty, 25°42′S, 56°25′W, *Schinini et al. 27889* (MICH).—Itapúa: circa Colonia Naranjito, *Bernardi 20503* (G, MO).—Misiones: Arroyo Concepción, 16 km NW of Asara on Ruta 2, *Anderson 12371* (MICH).—Paraguarí: 15 km N of Paraguarí, *Anderson 12383* (MICH); Cerro Palacios, 5 km N of Paraguarí, 25°25′S, 57°10′W, *Zardini 9842* (MICH).—San Pedro: alrededores de Lima,

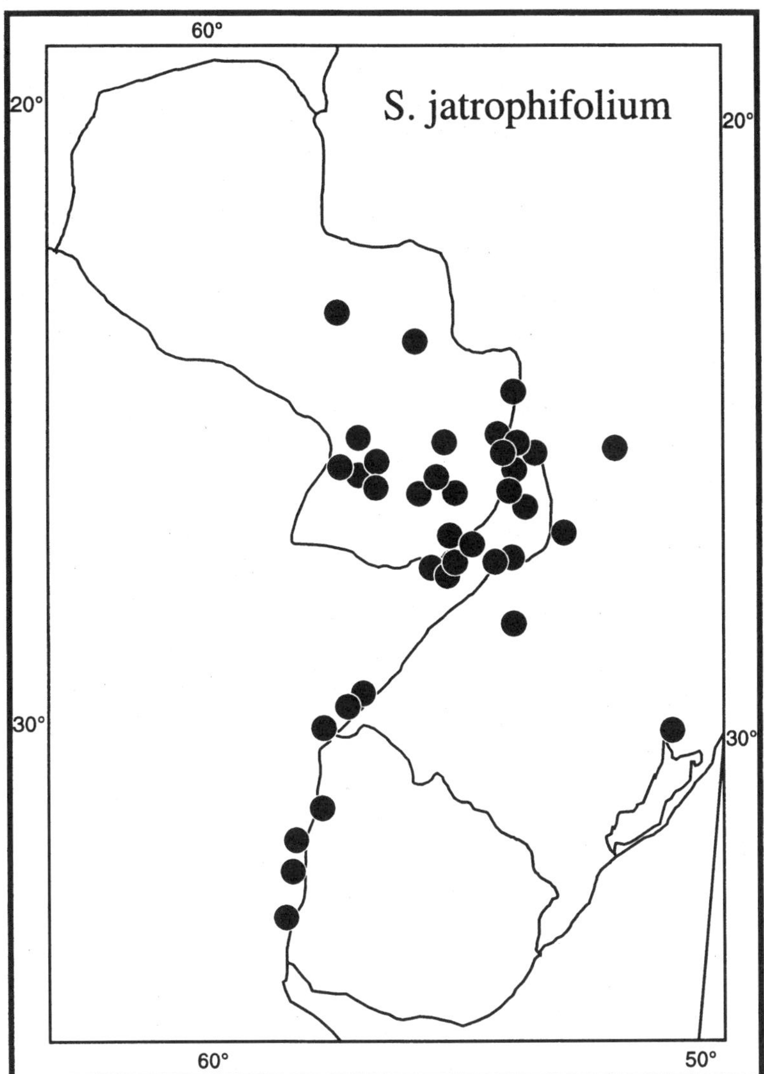

FIG. 51. Distribution of *Stigmaphyllon jatrophifolium.*

Krapovickas et al. 14262 (C, CTES, LL, P). **Uruguay**. Depto. Salto, Isla Belén, *Berro 3445* (P); Montevideo, *Fruchard s.n.* (P); Concepción del Uruguay, Feb 1877, *Lorentz s.n.* (BM, K); Depto. Salto, Salto Grande, *Osten 5421* (US); *Tweedie s.n.* (K).

Stigmaphyllon jatrophifolium is named for its distinctively palmately to pedately lobed leaves. The laminas vary from shallowly lobed to nearly divided to the base, although the smaller leaves near the inflorescence are often unlobed and triangular to elliptical. Exceptionally, as in *Hatschbach 19361*, even the large leaves are not lobed but triangular to suborbicular and grossly dentate. The vegetative parts are essentially glabrous or at most very sparsely sericeous.

53. Stigmaphyllon auriculatum (Cavanilles) Adr. Jussieu in St.-Hilaire, Fl. bras. mer. 3: 48. 1833 ["1832"]. *Banisteria auriculata* Cavanilles, Diss. 9: 428, t. 255. 1790.—TYPE: BRAZIL. Rio de Janeiro, *Commerson s.n.* (holotype: P!, photo: MICH!; isotype?: MA!, photos: F! GH!).

Banisteria angulata Vellozo, Fl. flum. 4: 191, t. 160. 1829. *Stigmaphyllon hastatum* Grisebach, Linnaea 13: 208. 1839, nom. superfl.—TYPE: unknown.

Stigmaphyllon aristatum Lindley, Bot. Reg. 20: t. 1659. 1834.—TYPE: based on living material "obtained from Mrs. Marryatt of Wimbledon" (holotype, fragments in herb. Bentham: K!, photo: MICH!).

Stigmaphyllon repandum Grisebach, Linnaea 13: 202. 1839.—TYPE: BRAZIL. [fide Urban, 1893:] São Paulo: between Santos and São Paulo, Oct 1827, *Sello V. it. e 14* (holotype: B, destroyed, photos: F! MICH! NY! US!).

Vine to 5 m. Stems and branches sparsely sericeous when very young, soon becoming glabrous. Laminas 3.2–10.5 cm long, 2–8 cm wide, triangular, ovate, elliptical, hastate, or sagittate, apex mucronate and with 2–4 filiform glands on the margin along each side near the apex, base truncate to auriculate, glabrous adaxially and abaxially or rarely glabrate with a few sessile hairs especially on the major veins, margin eglandular or with irregularly spaced filiform glands (up to 2 mm long), especially (or sometimes only) on the basal lobes of hastate or sagittate laminas, the margin of laminas of the reduced leaves associated with the inflorescence bearing filiform glands or sometimes also with irregularly spaced sessile glands (0.4–0.5 mm in diameter); petioles 1.2–6 cm long, glabrous or sometimes sparsely sericeous, confluent across the node and forming a band bearing the stipules, or forming only a very narrow band, or barely confluent and forming a line, with a pair of prominent but sessile glands at the apex (or borne in the middle of the petiole of the leaves immediately below the inflorescence), each gland 0.8–1.7 mm in diameter; stipules 0.4–0.6 mm long, 0.4–0.8 mm wide, free, triangular, eglandular, caducous. Flowers ca. 8–15 per umbel, these solitary or borne in dichasia, axes glabrous. Peduncles 2.7–11 mm long, very sparsely sericeous to glabrous, rarely sericeous; pedicels 3–8 mm long, glabrous or rarely glabrate, distally expanded; peduncles 0.3–1.6 times as long as the pedicels. Bracts 1–1.6 mm long, 0.8–1.2 mm wide, triangular, apex acute; bracteoles 0.9–1.4 mm long, 0.7–1.2 mm wide, broadly triangular to oblong, apex obtuse, each bracteole with a pair of glands (each gland ca. 5 mm in diameter) or with a glandular region in the basal 1/3–1/2; bracts and bracteoles glabrous abaxially. Sepals 2–2.7 mm long, 2.4–2.8 mm wide, glands 1.7–2.8 mm long, 1–1.4 mm wide. All petals with the limb orbicular, glabrous, yellow, margin with fimbriae up to 1.4 mm long; anterior-lateral petals: claw 1.8–2.8 mm long, limb 12.5–14 mm long and wide; posterior-lateral petals: claw 1–1.5 mm long, limb 11–12 mm long and wide; posterior petal: claw 3–4 mm long, apex indented, limb ca. 10 mm long and wide. Stamens unequal, those opposite the posterior styles the largest or those opposite the anterior-lateral sepals with the longest filaments, anthers of those opposite the lateral sepals with the connective enlarged and the locules reduced, anthers of those opposite the posterior-lateral sepals eloculate or with one or two locules; anthers all glabrous. Stamen opposite anterior sepal: filament 2–3.2 mm long, anther 1.3–1.6 mm long; stamens opposite anterior-lateral petals: filaments 1.5–2.5 mm long, anthers 0.8–1 mm long; stamens opposite anterior-lateral sepals: filaments 2.2–4 mm long, connectives 0.9–1 mm long, locules 0.3–0.8 mm long; stamens opposite posterior-lateral petals: filaments 2.6–3.5 mm long, anthers 1.2–1.5 mm long; stamens opposite posterior-lateral sepals: filaments 2–2.4 mm long, connectives 0.7–1.1 mm long, locules

0.3–0.6 mm long or absent; stamen opposite posterior petal always shorter than the adjacent two: filament 1.6–3 mm long, anther 0.6–0.8 mm long. Anterior style 4–4.5 mm long, at least slightly shorter than the posterior two, terete, glabrous, erect; apex 1.5–1.7 mm long, each foliole 1.8–2.3 mm long, 1.4–2 mm wide, subsquare to subrectangular. Posterior styles 4.3–5.1 mm long, terete, glabrous, lyrate, apex incurved; foliole 1.4–2.2 mm long, 1.3–2 mm wide, parabolic to subrectangular. Dorsal wing of samara 4–4.5 cm long, 1.6–1.8 cm wide, upper margin without a tooth; nut smooth, 9–15 mm high, 6–11 mm in diameter, wall containing spongy tissue, without air chambers, areole 6–13 mm long, 6–10 mm wide, strongly concave, carpophore up to 7 mm long. Embryo ca. 11.5 mm long, ca. 2 times as long as wide, ovoid, outer cotyledon ca. 15.3 mm long, ca. 6 mm wide, folded at the distal 1/4 but not enfolding the inner cotyledon, inner cotyledon ca. 7.1 mm long, ca. 5.1 mm wide, straight. Chromosome number unknown.

Phenology. Collected in flower from February through December, in fruit in February, March, June, August, and October through December.

Distribution (Fig. 52). Eastern Brazil (Ceará to Rio de Janeiro, São Paulo?); in sandy habitats: in Rio de Janeiro in restingas, in woods, and at forest margin, in Ceará to Bahia in caatingas; sea level to 700 m.

REPRESENTATIVE SPECIMENS. **Brazil**. BAHIA: 5 km E of Conceição do Coité on rd from Serrinha, 11°35′S, 39°15′W, *Anderson 11738* (MICH); ca. 3 km NW de Irecé, *Bastos 74* (CEPEC, MICH); 14 km SW of Cancanção on rd to Queimadas, 10°47′S, 39°34′W, *Harley 16479* (K, MICH); Mpio. Bom Jesus de Lapa, Rio das Rãs, *Hatschbach 55168* (MICH); Mpio. Jequié, BR-330, trecho Jequié/Ipiau, 13°52′S, 40°03′W, *Mori & King 12196* (US); Serra de São José, F. de Santana, *Noblick 3118* (MICH); Mpio. Morpará, Fazenda São Domingos, *Salgado & Bautista 314* (GUA, MG, MICH); Curaça, Fazenda Caldeirão de Serra, 09°46′S, 39°39′W, *da Silva & Pinto 306* (K, NY).—CEARÁ: Baturité, *Ducke 1218* (MG); Cedro, Serra de Estava, *Löfgren 1077* (R).—PARAÍBA: Mpio. Areia, prox. a Alagôa de Remigio, *Coêlho de Moraes 736* (RB); região de Agrestá, Arára, *Coêlho de Moraes 1990* (IAN).—PERNAMBUCO: Afrânio, *Heringer et al. 300* (R, RB, UB).—RIO DE JANEIRO: Restinga de Grumari, *Araujo 212* (RB); Restinga de Tijuca, au Porto do Marisco, *Glaziou 10373* (BR, C, G, K, LE, NY, P, R); Mpio. Niteroi, Itaipú–Açú, Morro do Alto Moirão, ca. 22°58′S, 43°02′W, *Plowman 12859* (F, MICH, NY); Camorim, *Costa 228* (GUA); Serra do Mendanha, *Sucre et al. 6433* (RB); Recreio dos Bandeirantes, *Sucre 7935* (MICH, SP).

Stigmaphyllon auriculatum, a large-flowered vine of sandy places in eastern Brazil, is usually entirely glabrous. Two forms, not worthy of formal recognition, may be discerned. Plants from Rio de Janeiro occur in coastal areas, mostly in restingas, and usually have triangular to hastate or sagittate laminas, whereas plants from the northern part of the range, Ceará to Bahia, grow more inland, in caatingas, and mostly have ovate to elliptical laminas. Also, in the more northern plants, the nodal band formed by the petiole bases is often not as pronounced as in most plants from the Rio area. Exceptions to these generalizations are uncommon but do occur, e.g., the laminas of *Anderson 11738* from caatinga in Bahia are ovate/elliptical as well as hastate/sagittate.

Stigmaphyllon auriculatum is sometimes confused with *S. ciliatum* (no. 55), also a vine of coastal habitats with small, basally lobed leaves. Both have showy flowers with fimbriate petals borne on distally expanded pedicels, though *S. auriculatum* usually has more flowers per umbel (8–15) than *S. ciliatum* (3–8). *Stigmaphyllon ciliatum* is named for its laminas, which are always fringed with long filiform glands (up to 5.5 mm long), and which are commonly so deeply auriculate that the basal lobes overlap; it lacks the nodal petiolar band. In *S. auriculatum*, the leaf margins are eglandular or have a few widely spaced filiform glands (up to 2 mm long), especially (or sometimes only) on the basal lobes of hastate or sagittate laminas; sometimes the margin also bears irregularly

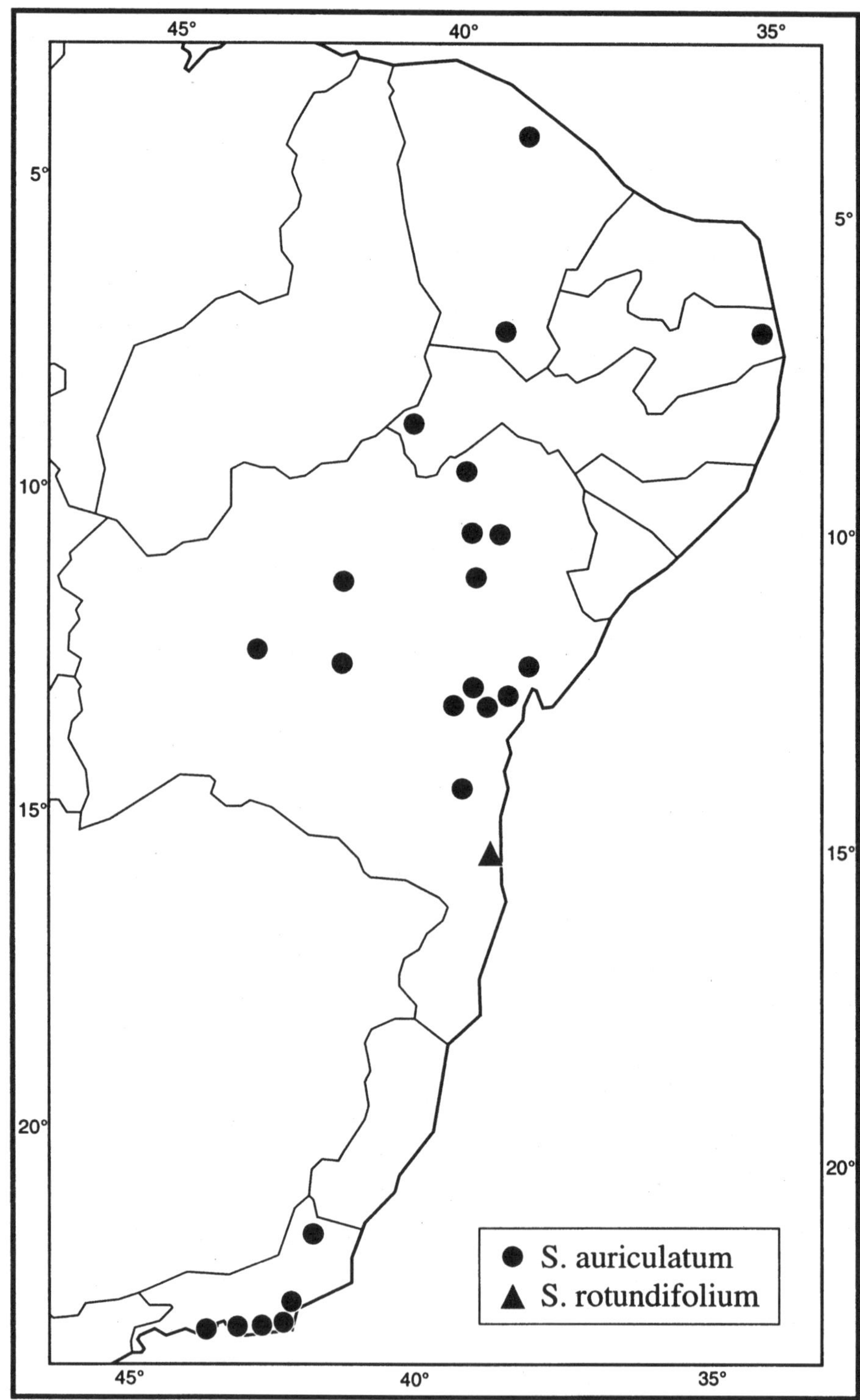

FIG. 52. Distribution of *Stigmaphyllon auriculatum* and *S. rotundifolium.*

spaced sessile glands. The basal lobes of the lamina never overlap, and the petioles are confluent at the node. The samaras of both species lack lateral winglets but otherwise differ greatly. Those of *S. auriculatum* are of the type common in the genus, i.e., an ovoid to spherical nut with a large dorsal wing, whereas those of *S. ciliatum* are lenticular, i.e., the nut laterally flattened and encircled by the dorsal wing; the embryo is also laterally flattened.

Niedenzu (1900, 1928) listed two collections of *S. auriculatum* from São Paulo, the type of *S. repandum* (*Sellow V. it. e 14*) and *Lund 16* (C!). In citing the Sellow collection from São Paulo, Niedenzu followed Urban (1893). Lund collected in São Paulo as well as in Rio de Janeiro (Urban 1906), but the specimens I have seen lack locality data. The holotype of *S. repandum* was at B, now destroyed, but photos show the specimen to represent *S. auriculatum*. The range of this species may extend into coastal São Paulo, but I have not seen any collections from that state.

54. Stigmaphyllon jobertii C. Anderson, Contr. Univ. Michigan Herb. 19: 426. 1993.—TYPE: BRAZIL. Piauí: Nazareth, Rio Piauí, 1877–78, *Jobert 1114* (holotype: P!, photo: MICH!).

Vine. Stems and branches sparsely sericeous when young, soon becoming glabrous. Laminas 6.5–10.3 cm long, 5.5–7 cm wide, cordate, apex apiculate, base cordate to auriculate in the larger laminas, glabrous adaxially, very sparsely sericeous but soon becoming glabrate to glabrous abaxially, margin eglandular; petioles from 3 cm long (complete petioles of larger leaves not seen), glabrous, not confluent across the node, with a pair of stipitate glands up to 2 mm below the base of the lamina, each gland 0.8–1.5 mm long, 0.6–0.9 mm in diameter; stipules 0.6–1 mm long, ca. 0.8 mm wide, free, triangular, eglandular. Flowers ca. 10–15 per pseudoraceme, these borne in dichasia or small thyrses (axes to the 2nd order, very sparsely sericeous to glabrous). Peduncles 2–2.5 mm long; pedicels 5.6–6.5 mm long, terete; both sericeous, peduncles ca. 0.4 times as long as the pedicels. Bracts 1.8–2.3 mm long, 1.8–2 mm wide, apex acute or obtuse; bracteoles 1.5–1.7 mm long and wide, apex obtuse, eglandular; bracts and bracteoles broadly triangular, sericeous abaxially. Sepals ca. 2.5 mm long, 2.5–3 mm wide, glands 1.8–2 mm long, ca. 1 mm wide. Petals not seen. Stamens unequal in size, those opposite the anterior-lateral sepals and posterior-lateral petals with the longest filaments, anthers of those opposite the lateral sepals not seen, other anthers all loculate, glabrous. Stamen opposite anterior sepal: filament ca. 2.5 mm long, anther ca. 1.3 mm long; stamens opposite anterior-lateral petals: filaments ca. 1.8 mm long, anthers ca. 1.1 mm long; stamens opposite anterior-lateral sepals: filaments ca. 2.7 mm long, anthers not seen; stamens opposite posterior-lateral petals: filaments ca. 2.7 mm long, anthers ca. 1.2 mm long; stamens opposite posterior-lateral sepals: filaments ca. 2.5 mm long, anthers not seen; stamen opposite posterior petal always shorter than the adjacent two: filament ca. 2.2 mm long, anther ca. 0.8 mm long. Anterior style ca. 2.6 mm long, shorter than the posterior two, terete, glabrous, erect; apex ca. 1.3 mm long, each foliole ca. 1.2 mm long, ca. 0.8 mm wide, triangular. Posterior styles ca. 3 mm long, canaliculate-complicate, glabrous, slightly lyrate; each foliole ca. 1.3 mm long, ca. 0.6 mm wide, parabolic. Dorsal wing of samara encircling the nut, 2.6 cm high measured from base of nut, 1.4 cm wide; nut smooth, locule embedded in spongy tissue, without air chambers, areole ca. 3.5 mm long, ca. 4 mm wide, concave, carpophore up to 5 mm long. Seed not seen. Chromosome number unknown. Fig. 53.

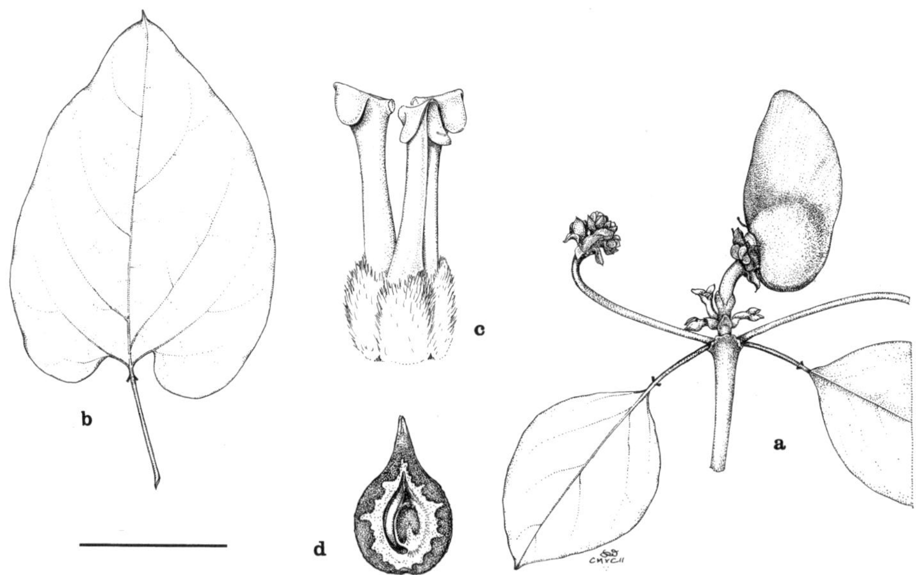

FIG. 53. *Stigmaphyllon jobertii*. a. Portion of inflorescence, the center umbel with one samara. b. Leaf. c. Gynoecium, anterior style in the center. d. Longitudinal section through nut of samara; note the spongy tissue surrounding the locule. Scale: for a, d, bar = 2 cm; for b, bar = 4 cm; for c, bar = 2.7 mm. (Based on *Jobert 1114*.)

This distinctive species is known from only one incomplete specimen from Piauí, Brazil (Fig. 90). It is easily recognized by its distinctive petiole glands and unique samaras. The cordate laminas are nearly glabrous, and the basal glands are peg-shaped and borne on the petiole, often obliquely, up to 2 mm below the base of the lamina. In South America, such glands are found only in the Colombian *S. romeroi*; they are also present in *S. sagraeanum* and *S. microphyllum*, and occasionally in *S. emarginatum*, all of the West Indies. The flowers are borne on very short peduncles and pedicels, and are aggregated into umbellate clusters. The samara, similar to that of *S. adenodon* var. *adenodon* and *S. lacunosum*, has a reduced dorsal wing surrounding the enlarged nut but lacks lateral winglets. In *S. adenodon* and *S. lacunosum*, as well as several other species, the nut is enlarged by air chambers surrounding the locule. *Stigmaphyllon jobertii* differs in that the locule is surrounded by a thick layer of spongy tissue, as in *S. puberum*. The type includes two samaras, but in each the embryo was destroyed by insects.

55. Stigmaphyllon ciliatum (Lamarck) Adr. Jussieu in St.-Hil., Fl. bras. merid. 3: 49. 1833 ["1832"]. *Banisteria ciliata* Lamarck, Encycl. 1: 369. 1785 ["1783"].—TYPE: BRAZIL. *Commerson s.n.* (holotype: P!, photo: MICH!; isotype?: C!).

Banisteria glauca Desfontaines, Tabl. école bot., ed. 3, 406. 1829.—TYPE: based on living material at the Botanical Garden at Paris.

Banisteria nitida Vellozo, Fl. flum. 4: 188, t. 148. 1829, non *Banisteria nitida* Lamarck, 1785.—TYPE: unknown.

Vine to 8 m. Stems and branches sericeous to sparsely so when young, soon becoming glabrous. Laminas 4.3–9.5 cm long, 3.5–7.3 cm wide, broadly ovate or cordate, apex mucronate, base deeply auriculate and in the larger laminas the lobes overlapping, glabrous adaxially and abaxially, margin fringed with filiform glands up to 5.5 mm long; petioles 1.6–5.1 cm long, sericeous to sparsely so or glabrate, not confluent across the node, with a pair of prominent but sessile glands at the apex, each gland 0.8–1.3 mm in diameter; stipules 0.1–0.5 mm long, 0.1–0.3 mm wide, free, triangular, eglandular; the pair of leaves subtending an umbel abruptly smaller and commonly reduced to bracts fringed with filiform glands. Flowers 3–8 per solitary umbel. Peduncles absent to 5.3 mm long, sericeous or sparsely so; pedicels 6–13 mm long, glabrous, distally expanded; peduncles up to 0.5 times as long as pedicels. Bracts 1–2 mm long, 0.9–2 mm wide, ovate to broadly so, apex acute or acuminate or obtuse; bracteoles 0.9–1.3 mm long, 0.8–1.3 mm wide, subsquare to broadly obovate, apex obtuse; bracts and bracteoles sericeous or sparsely so to glabrous abaxially, eglandular. Sepals 2.5–3.6 mm long, 2.5–3.3 mm wide, glands 2.3–3 mm long, 1.2–2.4 mm wide. All petals with the limb orbicular, glabrous, yellow, margin fimbriate or sometimes denticulate-fimbriate, fimbriae up to 0.5 (–0.9) mm long; anterior-lateral petals: claw 3.1–4 mm long, limb 13.5–18 mm long and wide; posterior-lateral petals: claw 1–3 mm long, limb 11.5–16 mm long and wide; posterior petal: claw 3.5–4.8 mm long, apex not indented, limb 8–11 mm long and wide. Stamens unequal, those opposite the posterior styles the largest, anthers of those opposite the lateral sepals with the connective enlarged and the locules reduced or with only one reduced locule or eloculate; anthers glabrous. Stamen opposite anterior sepal: filament 3.5–4.1 mm long, anther 1–1.3 mm long; stamens opposite anterior-lateral petals: filaments 2–2.5 mm long, anthers 0.7–1 mm long; stamens opposite anterior-lateral sepals: filaments 3.5–4 mm long, connectives 1.2–1.4 mm long, locules 0.1–0.3 (–0.5) mm long or absent; stamens opposite posterior-lateral petals: filaments 4.1–4.8 mm long, anthers (1–) 1.2–1.4 mm long; stamens opposite posterior-lateral sepals: filaments 2–3.5 mm long, connectives 0.6–1 mm long, locules 0.1–0.2 mm long or absent; stamen opposite posterior petal always shorter than the adjacent two: filament 1.6–2.6 mm long, anther (0.3–) 0.5–0.9 mm long. Anterior style 3.4–4.2 mm long, shorter than the posterior two, terete, glabrous, erect; apex 1.4–1.5 mm long, each foliole (0.9–) 1.4–1.5 mm long, 0.9–1.2 (–1.5) mm wide, parabolic or oblong. Posterior styles 4.1–5.6 mm long, terete, glabrous, lyrate; folioles (1.3–) 1.8–2.3 mm long, 1.9–2.4 mm wide, square to parabolic. Samara lenticular, dorsal wing 2–2.5 cm long, 1.6–1.8 cm wide, encircling the nut, upper margin with a blunt tooth; nut smooth, 7.5–9 mm high, 3.4–4 mm in diameter, without air chambers, areole ca. 5 mm long, 2.8–3.5 mm wide, concave, carpophore up to ca. 1.5 mm long. Embryo ca. 8.5 mm long, ca. 3 times as long as wide, flattened, outer cotyledon 7.8–8 mm long, 4.1–4.3 mm wide, inner cotyledon 4.1–5 mm long, 2.2–2.5 mm wide, both straight. Chromosome number unknown.

Phenology. Collected in flower and in fruit throughout the year.

Distribution (Figs. 40, 54). Atlantic lowlands of Belize, Guatemala, Honduras, Nicaragua, Colombia, Venezuela, Brazil, and Uruguay, also in Trinidad, and naturalized in Barbados; not reported from Costa Rica, Panama, and the Guianas but should be expected there; most commonly in wet localities: along rivers, in mangrove or freshwater swamps, on or near beaches, and also at forest edge and roadsides; sea level to 50 m.

REPRESENTATIVE SPECIMENS. **Barbados**. Bioser Hill, St. Joseph, *Gooding 335* (NY); Chemin de Bridgetown à Bathsheba, *Stehlé 2946* (NY); Chemin du Turner's Hall Wood, *Stehlé 2979* (NY). **Trinidad**. Erin Beach, *R. E. D. Baker TRIN14207* (TRIN); Manzanilla, *Britton & Britton 2171* (NY), *Broadway 2623* (F, G); *Finlay 195* (P).

Belize. STANN CREEK: Gragra Creek, Commerce Bight, *Gentle 8019* (LL, MICH, UTD); Dangriga, *Proctor 36604* (MO); Stann Creek, *Schipp S59* (F, G), *880* (F, G, GH, MICH, MO, NY, S, WIS).—TOLEDO: Cow-

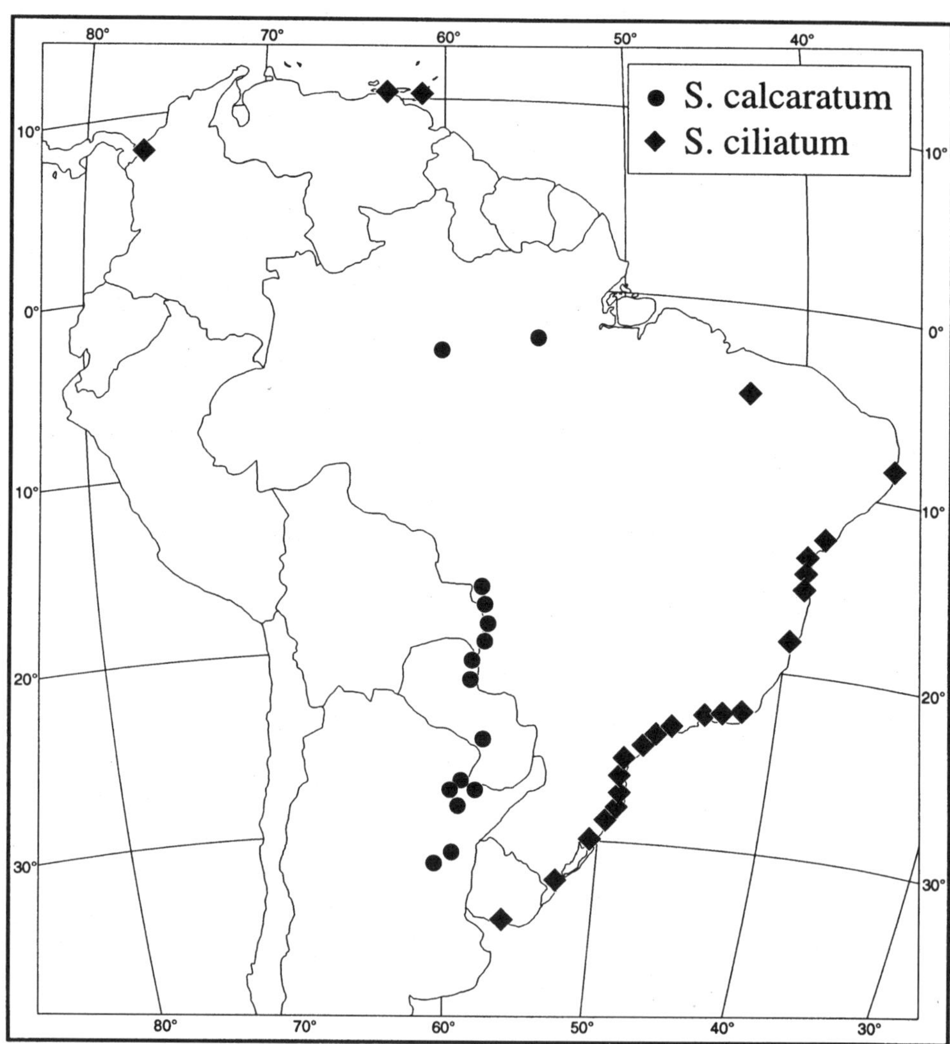

FIG. 54. Distribution of *Stigmaphyllon calcaratum* and of *S. ciliatum* in South America; see Fig. 40 for distribution of *S. ciliatum* in Central America and the West Indies.

pen, Swasey Branch, Monkey River, *Gentle 4018* (A, F, MICH, MO, NY, U, UTD). **Guatemala**. IZABAL: Puerto Barrios, *Deam 6018* (GH, MICH, US); Livingston, *Donnell Smith 1805* (US), *Lewton 430* (F, GH, MEXU, US), *von Türckheim II-1356* (US). **Honduras**. COLÓN: 4.5 mi NE of Trujillo on old road to Castilla, 15°57′N, 85°54′W, *Saunders 399* (MO, TEX).—GRACIAS A DIOS: alrededores de Puerto Lempira, *Clare 153* (MICH). **Nicaragua**. ZELAYA: vicinity of Awastara, ca. 14°19′N, 83°12–13′W, *Stevens 17741* (MO).

Colombia. ANTIOQUIA: ca. 1 km W of Turbo, ca. 08°05′N, 76°43′W, *Feddema 2032* (MICH, NY, S, US); Turbo, *Haught 4838* (COL, US). **Venezuela**. SUCRE: vic. of Cristóbal Colón, Aricana beach, *Broadway 602* (NY, US). **Brazil**. BAHIA: estrada Ilhéus–Olivença, *Hage 12* (CEPEC); Belmonte, ca. 15°52′S, 38°53′W, *Harley 17469* (K, M, MICH, MO, NY, P, U); between Alcobaça and Caravelas, 23 km S of Alcobaça, 17°44′S, 39°15′W, *Harley 18061* (K, MICH, NY, RB, U); Mpio. Itacaré, a 3 km S de Itacaré, *Mori et al. 13082* (MICH, NY).—ESPÍRITO SANTO: Conceição da Barra, *Hatschbach 46990* (MICH, US); *J. Kuhlmann RB102183*; 10 km W de Santana, via Naque–Santana, *Mattos 111008* (SP).—MARANHÃO: ad Caxias, *Martius s.n.* (M).—PARANÁ: Guaratuba, *Dusén 13744* (F, G, S); Mpio. Paranaguá, Baia de Paranaguá, Piaçaguera, *Hatschbach 21395* (P); Mpio. Antonina, Saivá, *Hatschbach 37981* (MICH, MO); Mpio. Morretes, Ilha do Malha, *Kummrow 989* (C, MICH).—PER-

NAMBUCO: Rest. Rio Dolce, Olinda, *Leal & da Silva 46* (RB).—RIO DE JANEIRO: Paquetá, *Dusén 4* (S, US, W); Mpio. Magé, Guanabara Bay, *Eiten & Eiten 7858* (MICH, NY); Rio de Janeiro, *Glaziou 717* (BR, C, F, P); Gruta da Impresa, *Pabst 4789* (M, NY); base of Dois Irmaos, *Rose & Russell 20244* (NY, US).—RIO GRANDE DO SUL: Pôrto Alegre, *Fox 295* (K); Quinta pr. Rio Grande, *Malme 400* (S).—SANTA CATARINA: Mpio. Sombrio, *Dobereiner & Tokarnia 449* (RB); Laguna, *Krapovickas & Cristóbal 37737* (G, MICH); Rio Ravares, Ilha Santa Catarina, *Reitz 5093* (US); Barra do Saí-guaçú, São Francisco do Sul, *Reitz & Klein 1428* (US).—SÃO PAULO: Mpio. Cananeia, ca. 5 km N of Cananeina, 24°59′S, 47°57′W, *Eiten & Clayton 6113* (NY, US); Mpio. Iguape, E point of Iguape Island, 24°39′S, 47°24′W, *Eiten & Clayton 6223* (K, MO, NY, UB, US); Piruibe, *Löfgren 1663* (C, SP); Santos, *Mosén 2796* (C, S). **Uruguay**. Montevideo, 1816, *Marsaud s.n.* (DS); *Serre s.n.* (P).

Stigmaphyllon ciliatum is named for its leaf margins, which are evenly fringed with persistent filiform glands. In this easily recognized species, the ovate to cordate laminas are usually so deeply auriculate that the basal lobes overlap. The large flowers, borne on distally expanded pedicels, are arranged in 3–8-flowered solitary umbels, which are subtended by a pair of greatly reduced leaves (commonly a pair of ciliate bracts). The distinctive samaras consist of a lenticular, i.e., laterally flattened, nut encircled by the dorsal wing; the embryo is also laterally flattened.

In eastern Brazil, *S. ciliatum* is sometimes confused with *S. auriculatum* (no. 53), a showy species of sandy habitats that has small, often hastate to sagittate laminas, but these are never fringed with filiform glands. For a comparison, see that species.

56. Stigmaphyllon calcaratum N. E. Brown, Trans. & Proc. Bot. Soc. Edinburgh 20: 48. 1894.—TYPE: BRAZIL. Mato Grosso do Sul: marsh near Corumbá, Jan 1892, *Moore 1012* (lectotype, here designated: BM!, photo: MICH!; isolectotype: K!).

Stigmaphyllon hasslerianum Niedenzu, Bull Herb. Boiss., 2. sér., 7: 291. 1907.—TYPE: PARAGUAY. "Gran Chaco, ad rimpam occidentalem flum. Paraguay, 23°20′–23°30′S," Jan 1903, *Hassler 2881* (holotype: G!, photos: F! GH! MICH!; isotype: G!).

Stigmaphyllon hagmannii Markgraf, Notizbl. Bot. Gart. Berlin 15: 219. 1940.—TYPE: BRAZIL. Pará: Santarem, Taperinha, Westufer des Ayayá, 20 m, 26 Dec 1938, *Markgraf & Hagmann 3884* (holotype: B, destroyed; isotype: RB-39924!, photo: MICH!).

Vine to 5 m. Stems and branches sericeous when young, soon becoming glabrate. Laminas 4.8–12.3 cm long, 1–4.2 cm wide, linear to lanceolate to oblong to elliptical or narrowly so, apex obtuse to acute-mucronate, base cordate to auriculate or sometimes each gland-bearing lobe drawn out into a tooth and the laminar base then hastate, sparsely sericeous to glabrate adaxially, sparsely sericeous to glabrate abaxially (trabecula 0.3–0.5 mm long, straight, sessile), margin eglandular or with a sessile gland (0.2–0.3 mm in diameter and up to 1.3 cm from the costa) or a filiform gland on one or both basal lobes, or sometimes also with irregularly spaced sessile glands (each gland 0.3–0.4 mm in diameter); petioles 0.7–2.1 cm long, sericeous or sparsely so, not confluent across the node, with a pair of prominent but sessile glands at the apex, each gland 0.8–1.5 mm in diameter; stipules 0.5–1.2 mm long, 0.5–0.8 mm wide, free, triangular or narrowly so, the proximal 1/3–1/2 glandular, the distal portion herbaceous. Flowers 4–6 (–8) per umbel, these solitary or borne in dichasia (axes sericeous). Peduncles absent to 10 mm long; pedicels 2.5–10.3 mm long, distally expanded; both densely sericeous, peduncles up to 1.1 times as long as the pedicels. Bracts 0.9–2 mm long, 0.8–1.4 mm wide; bracteoles 0.9–1.5 mm long, 1–1.3 mm wide, eglandular or with a glandular area in the basal 1/3–1/2; bracts and bracteoles triangular, apex acute to obtuse, sericeous to sparsely so abaxially. Sepals 2–3 mm long, 1.7–2.8 mm wide, glands 1.8–2 mm long, 0.8–1.3 mm wide. All petals with the limb glabrous, yellow, limb of the lateral petals orbicular, margin fimbriate or erose-fimbriate, fimbriae up to 0.6 (–0.7) mm long; anterior-lateral petals: claw 2–3 mm long, limb ca. 9–14 mm long and wide; posterior-lateral petals: claw 1.5–2 mm long, limb ca. 9–12 mm long and wide; posterior petal: claw 2.7–3 mm long, apex very slightly or not indented, limb ca. 7–10.5 mm long, ca. 6–7 mm wide, obovate or broadly so, margin erose to erose-fimbriate to fimbriate, fimbriae up to 0.8 mm long. Stamens unequal, those opposite the posterior styles the largest, anthers of those opposite the lateral sepals with the connective enlarged and the locules reduced, or anthers of those opposite the anterior-lateral sepals eloculate, or anthers of those opposite the posterior-lateral sepals with the locules only slightly or not reduced; anthers all glabrous. Stamen opposite anterior sepal: filament 2.2–2.4 mm long, anther 1.2–1.5 mm long; stamens opposite anterior-lateral petals: filaments 1.3–2 mm long, anthers 0.8–1 mm long; stamens opposite anterior-lateral sepals: filaments 2.3–2.7 mm long, connectives 1–1.2 mm long, locules 0.2–0.6 mm long or one or both locules absent; stamens opposite posterior-lateral petals: filaments 4–5.2 mm long, anthers 0.8–1.3 mm long; stamens opposite posterior-lateral sepals: filaments 2–2.5 mm long, connectives 0.7–1 mm long, locules 0.1–1 mm long; stamen opposite posterior petal always shorter than the adjacent two: filament 1.7–2.2 mm long, anther 0.7–0.8 mm long. Anterior style 2.8–3.7 mm long, shorter than the posterior two, terete, glabrous, erect or slightly recurved; apex 1.2–2 mm long, each foliole (0.9–) 1.5–1.7 mm long, (0.8–) 1.1–1.6 mm wide, usually subsquare or subrectangular, sometimes parabolic. Posterior styles 4.6–5.2 mm long, terete, glabrous, lyrate; foliole 2.3–2.5 (–3) mm long and wide, subsquare. Samara 1.5–1.8 cm high, 0.7–0.9 cm wide, dorsal wing reduced to an erect crest, lateral winglets absent, the nut covered with numerous bulbous and warty excrescences composed of spongy tissue; nut 3.5–5.5 mm high, ca. 3.5–5.5 mm in diameter, without air chambers, areole 3–4 mm long, 2–2.2 mm wide, convex, carpophore absent. Embryo 4.7–5.9 mm long, ca. 2 times as long as wide, ovoid, outer cotyledon ca. 8 mm long, 3.2–3.5 mm wide, the distal 1/2 folded over the inner cotyledon, inner cotyledon up to ca. 5 mm long or sometimes reduced to a narrow lip ca. 0.5 mm long, ca. 3 mm wide, folded at the distal 3/4 or straight. Chromosome number unknown. Fig. 55.

Phenology. Collected in flower and fruit from September through June.

Distribution (Fig. 54). In northern Brazil along the Rio Amazônas near Manaus and Santarem, and in northeastern Argentina, Paraguay, the Pantanal of Brazil (Mato Grosso), and adjacent Bolivia; in marshy areas and along rivers; sea level to 50 m.

REPRESENTATIVE SPECIMENS. **Argentina**. CHACO: Depto. Gral. Donovan, 2 km N de Makallé, *Schinini 20026* (CTES, MICH); Depto. Bermejo, Isla del Cerrito, confluencia de los ríos Paraguay y Paraná; Colonia Benítez, borde del Río Tragadero, *Schulz 90* (CTES, SI), *120* (CTES), *2338* (CTES), *4057* (CTES), *9938* (CTES), *17920* (CTES, G, P).—CORRIENTES: Depto. Esquina, Isla Correntina, *Burkart et al. 26988* (MICH).—ENTRE RÍOS: Depto. La Paz, Dtto. Tacuaras, Isla Curuzú–Chali, *Burkart & Bacigalupo 21245*, *Burkart et al. 26986* (MICH).—SANTA FE: Depto. Gral. Obligado, Islas, frente a Los Laureles, *Blanchoud 2291, 2296* (CTES). **Bolivia.** Río Pilcomayo, *Kerr s.n.* (K). **Brazil.** AMAZÔNAS: Lago de Pesqueiro próximo ao Lago do Rei, Igapó, *Rodriguez & Coêlho 5888* (INPA, NY).—PARÁ: Igarapé Maica, en aval de Santarem, 02°35′S, 54°30′W, *Bamps 5294* (BR, MICH); Cacanal Grande, beira do Canal Maroja Neto, *Black 52-15612* (IAN).—MATO GROSSO: Fazenda Acurizal, Rio Paraguay, 17°52′S, 57°32′W, *Prance et al. 26076* (CEN, MICH).—MATO GROSSO DO SUL: Mpio. Corumbá, inv. Volta Grande, Nabileque, 19°42′S, 57°03′W, *Pott 4488* (CTES); Mpio. Poconé, Tranpantaneira, Km 110, *Chagas & Silva 1171* (MICH). **Paraguay**. Chaco septentrionalis, Puerto Tularara, *Fiebrig 1327, 1327a* (G); Depto. Presidente Hayes, Río Confuso, *Mereles 1592* (G); Bahia Negra–Chaco Paraguayo, *Rojas 13776* (W); Central, Tavarory, Río Paraguay, 25°30′S, 57°30′W, *Zardini 24132* (MO).

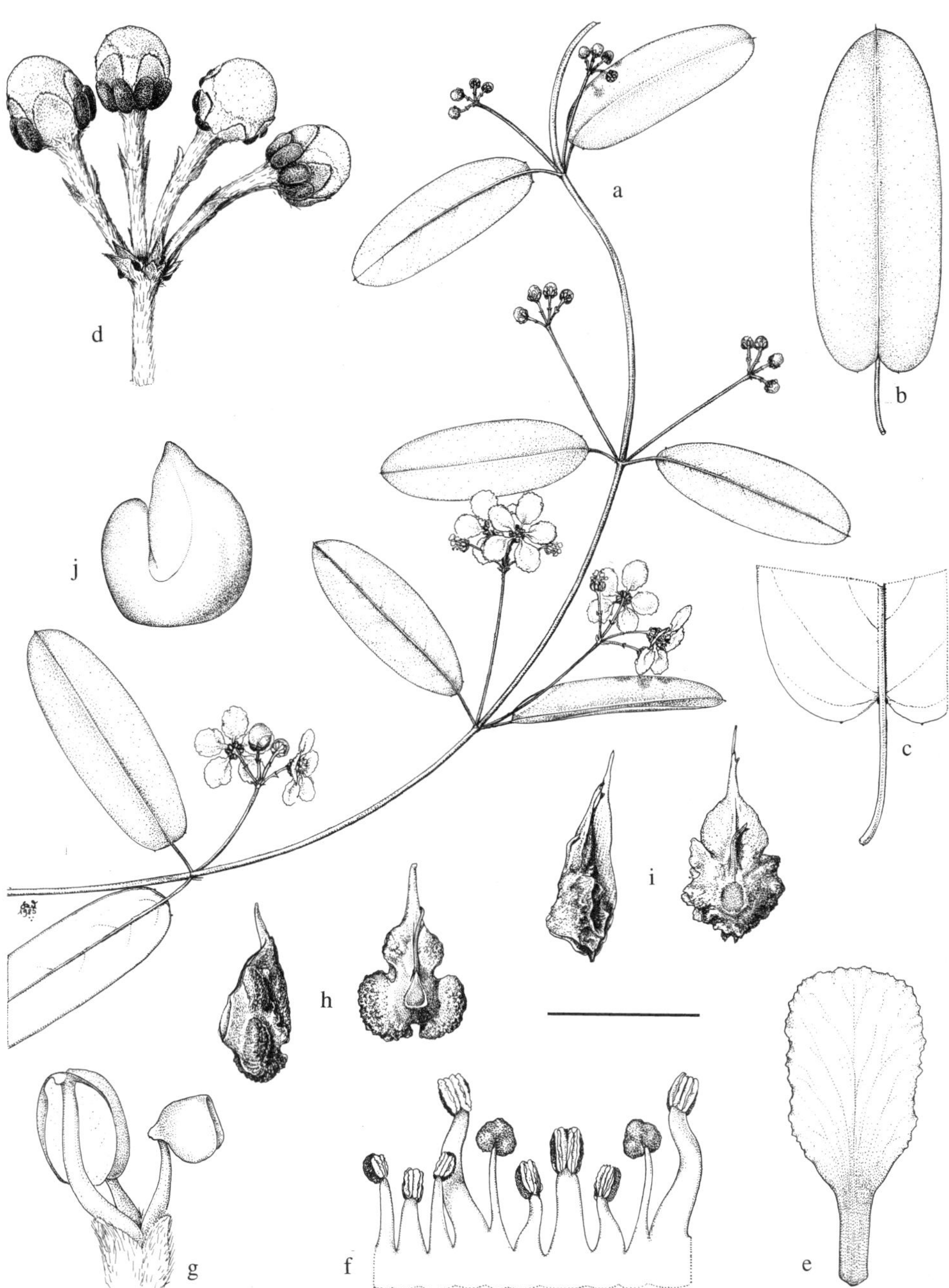

FIG. 55. *Stigmaphyllon calcaratum.* a. Flowering branch. b. Leaf. c. Base of leaf, abaxial view. d. Umbel. e. Posterior petal (the "flag"). f. Androecium; stamen second from left opposes the posterior petal. g. Gynoecium, anterior style to the right. h, i. Lateral (left) and adaxial (right) views of two samaras. j. Embryo. Scale: for a, b, bar = 4 cm; for c, bar = 2 cm; for d, bar = 8 mm; for e, j, bar = 5 mm; for f, g, bar = 4 mm; for h, i, bar = 1.3 cm. (Based on: a, d, *Schaller 253*; b, c, *Prance et al. 26076*; e, f, g, h, *Hassler 2881*; i, *Rodriguez & Coêlho 5888*; j, *Fiebrig 1327*.)

Stigmaphyllon calcaratum, a most distinctive species of marshy habitats, is easily recognized by its greatly modified "samara." The dorsal wing is reduced to an erect crest, and, instead of lateral winglets, the nut bears numerous bulbous and warty excrescences composed of spongy tissue; there is no carpophore. Unlike in other species of wet habitats with modified samaras, the embryo is not unusually large. The laminas vary from linear to oblong, less commonly elliptical, and are borne on short petioles (up to 2.1 cm long). The laminar base varies from cordate to auriculate, or to hastate and then each lobe is drawn out into a tooth and often terminates in a filiform gland. The stipules are unusual in that the basal portion is glandular, as in *S. adenophorum* and *S. bradei*. The 4–6 (–8)-flowered umbels are borne in the leaf axils, either solitary or in dichasia. The only other South American species with such narrow leaves is the poorly known *S. stenophyllum* (no. 5) from the uplands of Chocó, Colombia. It is readily separated by its efoliolate styles and leaf pubescence composed of T-shaped hairs; the samaras are unknown.

At present, *S. calcaratum* is known from the marshy regions of northern Argentina and Paraguay, the Pantanal of Mato Grosso, Brazil, and adjacent Bolivia, as well as from several localities along the Rio Amazônas. The gap between the disjunct ranges may reflect absence of suitable habitats or only paucity of collections.

57. Stigmaphyllon laciniatum (Ekman ex Niedenzu) C. Anderson, Contr. Univ. Mich. Herb. 16: 29. 1987. *Stigmaphyllon angulosum* f. *laciniatum* Ekman ex Niedenzu in Urban, Arkiv Bot. 22A(17): 19. 1929.—TYPE: HAITI. Île de la Gonâve, *Ekman 8820* (holotype: B, destroyed; fragment of holotype: NY!; isotypes: NY! S! US!, photo of US isotype: MICH!).

Vine. Stems and branches densely pubescent with scalelike T-shaped hairs when young, soon becoming sparsely so to glabrate. Laminas 3–12 cm long, 4–11 cm wide in outline, laciniate, apex of each division mucronate, base cordate, glabrate adaxially, sparsely pubescent with subsessile to T-shaped hairs to glabrate abaxially (trabecula 0.4–0.8 mm long, wavy, stalk 0.03–0.1 mm long), margin with sessile glands and filiform glands up to 2.5 mm long; petioles 1–5 cm long, sparsely to densely pubescent with scalelike T-shaped hairs, confluent across the node and forming a band bearing the stipules, with a pair of prominent but sessile glands at the apex, each gland 1–1.7 mm in diameter; stipules 0.7–1 mm long, 0.6–1.2 mm wide, free, triangular, eglandular. Flowers 13–22 per umbel or open to congested or sometimes interrupted pseudoraceme, these borne in dichasia or compound dichasia (axes to the 3rd order, densely pubescent with scalelike T-shaped hairs) or sometimes solitary. Peduncles 2.3–7.5 mm long; pedicels 4–7 mm long, terete; both densely pubescent with scalelike T-shaped hairs, peduncles 0.8–1.3 times as long as the pedicels. Bracts 1–1.5 mm long, 0.6–1 mm wide; bracteoles 1.1–1.7 mm long, 0.7–1 mm wide, with a glandular region in the basal 0.3–0.5 mm; bracts and bracteoles triangular, apex acute to obtuse, sparsely sericeous abaxially. Sepals 2–2.4 mm long, 2.2–2.3 mm wide, glands 1.2–1.5 mm long, 0.7–0.8 mm wide. All petals with the limb orbicular, glabrous, yellow, margin denticulate or entire near the claw, distally with fimbriae up to 0.4 mm long; anterior-lateral petals: claw 1–1.5 mm long, limb 11–12 mm long and wide; posterior-lateral petals: claw ca. 1.5 mm long, limb ca. 10 mm long and wide; posterior petal: claw 3.3–3.5 mm long, apex indented, limb ca. 9 mm long and wide. Stamens unequal, those opposite the posterior styles the largest, anthers of those opposite the anterior-lateral sepals eloculate, those opposite the posterior-lateral sepals with the connective enlarged and the locules reduced; anthers glabrous. Stamen opposite anterior sepal: filament ca. 2.5 mm long, anther 1.1–1.2 mm long; stamens opposite anterior-lateral petals: filaments ca. 1.8 mm long, anthers 0.8–1 mm long; stamens opposite anterior-lateral sepals: filaments 2.2–2.8 mm long, connectives 0.8–0.9 mm long, locules absent; stamens opposite posterior-lateral petals: filaments 3.3–3.4 mm long, anthers 1.1–1.3 mm long; stamens opposite posterior-lateral sepals: filaments 2.7–3.2 mm long, connectives 0.7–0.8 mm long, locules 0.4–0.7 mm long; stamen opposite posterior petal always shorter than the adjacent two: filament 2.5–2.8 mm long, anther 0.8–0.9 mm long. Anterior style 3–3.2 mm long, shorter than the posterior two, terete, glabrous, slightly recurved; apex ca. 0.7 mm long, each foliole 1.3–1.4 mm long, 1.2–1.3 mm wide, parabolic. Posterior styles 3.6–3.7 mm long, terete, glabrous, lyrate; folioles 1.7–2 mm long, ca. 1.7 mm wide, subsquare. Dorsal wing of samara 1.5–1.8 cm long, 0.5–0.7 cm wide, upper margin with an obtuse tooth; nut with a pair of triangular lateral winglets commonly extending only along the basal 1/2 of the nut, these 1.4–3.2 mm long, 0.8–1.1 mm wide, or sometimes only with a narrow ridge on each side; nut 3.5–4 mm high, 2.5–3 mm in diameter, wall containing spongy tissue, without air chambers, areole ca. 3.7 mm long, ca. 3.2 mm wide, concave, carpophore up to 2.5 mm long. Embryo ca. 4.6 mm long, ca. 2 times as long as wide, ovoid, outer cotyledon ca. 4.1 mm long, ca. 2.7 mm wide, inner cotyledon ca. 3.3 mm long, ca. 2.4 mm wide, both straight. Chromosome number unknown.

Phenology. Collected in flower and in fruit in July and August.

Distribution (Fig. 35). Endemic to Île de la Gonâve, west of Haiti.

ADDITIONAL SPECIMENS EXAMINED. **Haiti**. Île de la Gonâve: *Ekman 8670* (K, NY-fragment, S), *Eyerdam 63* (F, GH, US), *219* (A, F, GH, MO, NY, US).

Stigmaphyllon laciniatum is easily recognized by its laciniate leaves. Although a number of species have lobed leaves, only in this one is the lamina so greatly and finely divided. It was first recognized as a form of *S. angulosum* (no. 36), endemic to Hispaniola and characterized by sinuate-lobate leaves. *Stigmaphyllon laciniatum* differs, in addition to its unique leaves, in its stamens and smaller samaras. In *S. laciniatum*, the anthers of stamens opposite the anterior-lateral sepals are eloculate, and the dorsal wing of the samara is 1.5–1.8 cm long and 0.5–0.7 cm wide; in *S. angulosum*, all anthers are loculate, and the samaras bear a dorsal wing 2.8–4.5 cm long and 1.2–1.8 cm wide.

58. Stigmaphyllon bradei C. Anderson, Contr. Univ. Michigan Herb. 17: 7. 1990.— TYPE: BRAZIL. São Paulo: Registro, 21 Jun 1963, *Moura s.n. [SP123438]* (holotype: SP!, photo: MICH!).

Vine. Stems and branches bearing scalelike T-shaped hairs when young, soon becoming glabrate. Laminas 6.5–14.3 cm long, 6.8–12.5 cm wide, cordate to suborbicular, apex mucronate in larger laminas to caudate in smaller ones, base auriculate in larger laminas or cordate in smaller ones, glabrous adaxially, glabrous or sparsely pubescent with T-shaped hairs only on major veins abaxially (trabecula 0.3–1.1 mm long, wavy, stalk 0.06–0.13), margin with irregularly spaced stalked glands (0.2–0.3 mm in diameter, 0.3–0.6 mm long) and/or filiform glands (up to ca. 2 mm long); petioles 2.8–7 cm long, pubescent with scalelike T-shaped hairs, sparsely so in older leaves, not confluent across the node, with a pair of shallowly cupulate glands borne 1–10 mm below the base of the lamina, each gland 1.5–1.8 mm in diameter, up to 1 mm high; stipules 1.3–2.5 mm long, 0.9–1.6 mm wide, free, the proximal 1/3 a circular gland, the distal 2/3 herbaceous. Flowers ca. 15–25 per umbel, these solitary (and borne in dichasia?). Peduncles rudimentary to 1.5 mm long, densely sericeous (hairs subsessile); pedicels 8–13.5 mm long, densely sericeous (hairs subsessile) but glabrous just below the flower, terete; peduncles up to 0.1 times as long as the pedicels. Bracts ca. 1.5–2.5 mm long, ca. 1.5–2 mm wide, broadly triangular, apex acute; bracteoles ca. 1.3–2 mm long, 1.1–1.3 mm wide, broadly triangular, apex obtuse, eglandular; bracts and bracteoles sericeous abaxially. Sepals ca. 2.2–2.4 mm long, ca. 2.5–2.7 mm wide, glands ca. 2.5–2.8 mm long, 1.3–1.5 mm wide. All petals with the limb orbicular, glabrous, yellow and suffused with red, limb of the anterior-lateral petals with the margin erose or irregularly denticulate-erose, teeth up to 0.3 mm long, limb of the posterior-lateral and posterior petals with the margin fimbriate, fimbriae up to 0.6 mm long; anterior-lateral petals: claw 3–3.2 mm long, limb ca. 11 mm long and wide; posterior-lateral petals: claw 2.5–3 mm long, limb ca. 10 mm long and wide; posterior petal: claw 3.2–4 mm long, apex slightly indented, limb ca. 7.5–8 mm long and wide. Stamens unequal, those opposite the posterior styles the largest, anthers of those opposite the lateral sepals eloculate or with the connective enlarged and the locules reduced; anthers glabrous. Stamen opposite anterior sepal: filament 2.3–2.5 mm long, anther ca. 1.3 mm long; stamens opposite anterior-lateral petals: filaments 1.3–1.8 mm long, anthers ca. 1.1 mm long; stamens opposite anterior-lateral sepals: filaments 2.5–2.8 mm long, connectives 1–1.1 mm long, locules 0.3–0.4 mm long or absent; stamens opposite posterior-lateral petals: filaments 2.8–3.1 mm long, anthers 1.2–1.3 mm long; stamens opposite posterior-lateral sepals: filaments ca. 2.6–3 mm long, connectives 1–1.2 mm long, locules ca. 0.3 mm long or absent; stamen opposite posterior petal always shorter than the adjacent two: filament ca. 2.5–2.7 mm long, anther ca. 0.9 mm long. Anterior style 2.6–3.2 mm long, shorter than the posterior two, terete, with scattered hairs in the proximal 1/4, erect; apex 1.4–1.9 mm long, each foliole 1.2–1.4 mm long, 1.1–1.2 mm wide, triangular to parabolic. Posterior styles 3.3–4 mm long, terete, glabrous or with scattered hairs in the proximal 1/3, lyrate; foliole 1.8–2 mm long and wide, subsquare to suborbicular. Samara not seen. Chromosome number unknown. Fig. 56.

Stigmaphyllon bradei is known from only two collections from São Paulo, Brazil (Fig. 84), the type and *Brade 7987* (R; Brazil, São Paulo: Iguape, Morro das Pedras, 1921). It is distinguished by its glandular stipules, broadly ovate to suborbicular auriculate laminas, and large shallowly cupulate petiole glands. The inflorescences are compact owing to the rudimentary to very short peduncles. All petals are suffused with red. The anthers of stamens opposing the anterior-lateral sepals are eloculate but those opposing the posterior-lateral sepals bear two reduced locules. This species resembles most closely the Bolivian *S. boliviense* (no. 59) and *S. coloratum* (no. 60) and the Brazilian *S. puberulum* (no. 61; Bahia, Pernambuco). They differ in their eglandular stipules. Also, *S. boliviense* and *S. coloratum* bear glands on all five sepals; the laminas of *S. puberulum* have regularly spaced nail-like marginal glands.

59. Stigmaphyllon boliviense C. Anderson, sp. nov.—TYPE: BOLIVIA. Santa Cruz: Prov. Nuflo de Chávez, Las Trancas, 16°32′45.4″S, 61°50′21.5″W, 29 Mar 1995, *Jardim 1901* (holotype: MO!; isotype: MICH!).

Liana. Laminae 9.5–15 cm longae, 7.3–10.8 cm latae, cordatae, supra glabrae, subtus dense pubescentes pilis T-formibus ferentes, margine glandulis filiformibus et breve stipitatis munito. Inflorescentia solitaria constata ex umbellis, floribus in quaque umbella ca. 10. Pedunculi rudimentalis vel usque ad 1 mm longi; pedicelli 8–12 mm longi. Petala

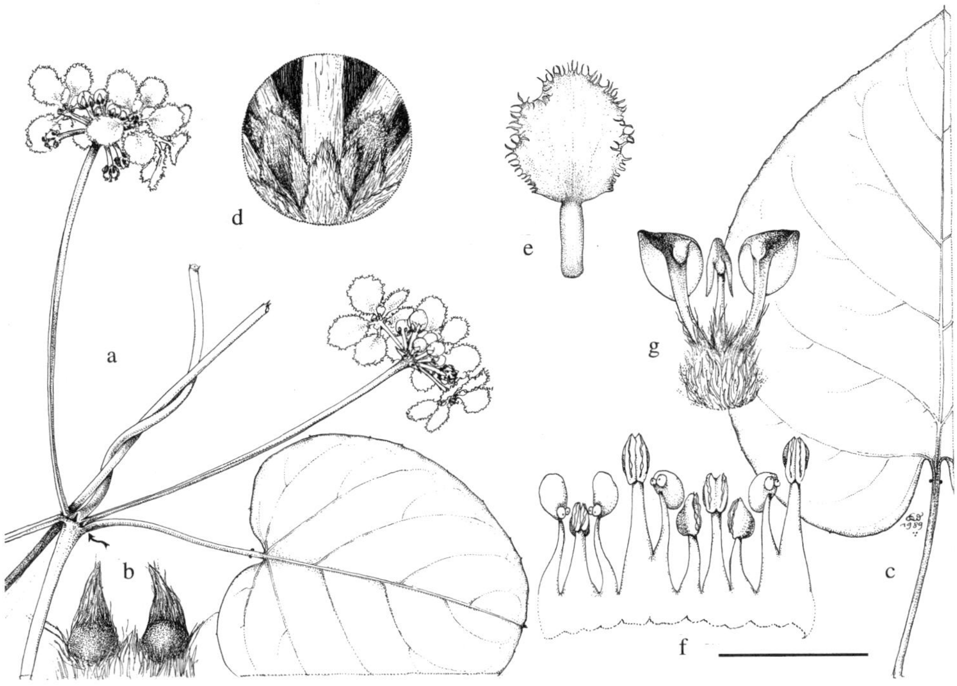

FIG. 56. *Stigmaphyllon bradei*. a. Flowering branch. b. Enlargement of stipules. c. Large leaf. d. Detail of base of umbel; the peduncle is rudimentary and thus the base of each pedicel is hidden by two bracteoles and one bract. e. Posterior petal (the "flag"). f. Androecium; stamen second from left opposes the posterior petal. g. Gynoecium; posterior styles are bent slightly outward to show anterior style (in center). Scale: for a, c, bar = 4 cm; for b, d, f, g, bar = 4 mm; for e, bar = 8 mm. (Based on: a, b, d–g, *Moura s.n. SP123438*; c, *Brade 7987*.)

limbo orbiculari; limbus petalorum antico-lateralium margine eroso, limbus petalorum postico-lateralium margine denticulato-eroso, limbus petali postici margine fimbriato. Stamina heteromorpha; antherae sepalis antico-lateralibus oppositae steriles, antherae sepalis postico-lateralibus oppositae loculis reductis instructae, antherae ceterae fertiles. Stylus anticus ca. 3.5 mm longus, glaber, utroque foliolo ca. 2 mm longo, 1.8 mm lato, subquadrato; styli postici ca. 4.5 mm longi, lyrati, glabri, foliolo ca. 2.5 mm longo, ca. 2.8 mm lato, suborbiculari vel subquadrato. Samara ignota.

Vine to 5 m. Stems and branches pubescent with scalelike T-shaped hairs when young, becoming glabrate. Laminas 9.5–15 cm long, 7.3–10.8 cm wide, ovate to cordate, apex apiculate, base auriculate, glabrous adaxially, densely pubescent with T-shaped hairs abaxially (trabecula 0.8–2.3 mm long, wavy, stalk 0.1–0.4 mm long), along the margin also with scalelike T-shaped hairs, margin with filiform glands (up to 6 mm long) and a few irregularly spaced slightly raised glands (0.3–0.4 mm in diameter, ca. 0.2 mm long); petioles 3.4–6 cm long, pubescent with scalelike T-shaped hairs, sparsely so in older leaves, not confluent across the node, with a pair of shallowly cupulate glands borne 5–13 mm below the base of the lamina, each gland 1.2–1.5 mm in diameter, up to 1 mm high; stipules 0.5–0.6 mm long, ca. 0.4 mm wide, free, triangular or narrowly so, eglandular. Flowers ca. 10 per solitary umbel. Peduncles rudimentary, up to 1 mm long, sericeous (hairs subsessile); pedicels 8–12 mm long, sericeous (hairs subsessile) but the distal 1/4–1/3 sparsely so, distally slightly expanded; peduncles up to 0.1 times as long as the pedicels. Bracts 1–1.2 mm long and wide, ovate to semicircular, apex obtuse; bracteoles 0.8–1.3 mm long and wide, semicircular, apex obtuse, eglandular; bracts and bracteoles sericeous or sparsely so or glabrate abaxially. Sepals 2–2.5 mm long and wide, *all* bearing a pair of glands, glands 2–2.5 mm long, 1–1.2 mm wide. All petals with the limb orbicular, glabrous, yellow but that of the posterior petal streaked with red; anterior-lateral petals: claw ca. 3 mm long, limb ca. 12 mm long and wide, erose; posterior-lateral petals: claw ca. 2 mm long, limb ca. 10–11 mm long and wide, erose-denticulate, the teeth up to 0.3 mm long; posterior petal: claw 3.2–3.5 mm long, apex indented, limb ca. 9.5–10 mm long and wide, margin with fimbriae up to 0.5 mm long. Stamens unequal, those opposite the posterior-lateral petals (and the posterior styles) with the longest filaments, anthers of those opposite the anterior-lateral sepals eloculate, anthers of those opposite the posterior-lateral sepals with the connective enlarged and the locules reduced; anthers glabrous. Stamen opposite anterior sepal: filament ca. 2.5 mm long, anther ca. 1.5 mm long; stamens opposite anterior-lateral petals: filaments ca. 1.5 mm long, anthers ca. 1.4 mm long; stamens opposite anterior-lateral sepals: filaments ca. 2.5 mm long, connectives ca. 1.2 mm long, locules absent; stamens opposite posterior-lateral petals: filaments ca. 4 mm long, anthers ca. 1.3 mm long; stamens opposite posterior-lateral sepals: filaments ca. 2.3 mm long, connectives ca. 0.8 mm long, locules ca. 0.4 mm long; stamen opposite posterior petal always shorter than the adjacent two: filament ca. 2.3 mm long, anther ca. 0.8 mm long. Anterior style ca. 3.5 mm long, shorter than the posterior two, terete, glabrous, erect; apex ca. 1.7 mm long including a tiny spur up to 0.2 mm long, each foliole ca. 2 mm long, 1.8 mm wide, subsquare. Posterior styles ca. 4.5 mm long, terete, glabrous or with some scattered hairs in the proximal 1/4, lyrate; foliole ca. 2.5 mm long, ca. 2.8 mm wide, suborbicular to subsquare. Samara not seen. Chromosome number unknown. Fig. 57.

Stigmaphyllon boliviense is known only from the type, a flowering collection from Santa Cruz, Bolivia (Fig. 8), and is most similar to *S. coloratum* (no. 60), also known only from Bolivia. These two differ from all other species in *Stigmaphyllon* in that all five sepals bear glands instead of having the anterior sepal eglandular. They are readily separated by the laminar pubescence. In *S. boliviense* the laminas are abaxially densely pubescent with T-shaped hairs, whereas in *S. coloratum* they are abaxially very sparsely pubescent to glabrous. In *S. boliviense* anthers of stamens opposing the anterior-lateral sepals consist of only a glandular connective lacking locules, but the anthers of stamens opposing the posterior-lateral sepals have the connective enlarged and bear two reduced locules. In *S. coloratum* the anthers of stamens opposing all lateral sepals commonly lack locules, though occasionally those of stamens opposing the posterior-lateral sepals have two tiny locules; additional collections of *S. boliviense* may reveal similar variation its androecium. *Stigmaphyllon boliviense* and *S. coloratum* also resemble *S. bradei* (no. 58) and *S. puberulum* (no. 61) of eastern Brazil, but are readily separated by the calyx. All four have leaves with large shallowly cupulate glands on the petiole, inflorescences composed of umbellate clusters of flowers borne on short peduncles and pedicels, and petals marked with red.

60. Stigmaphyllon coloratum Rusby, Mem. Torrey Bot. Club 6: 14. 1896.—TYPE: BOLIVIA. La Paz: between Guanai [Huanay] and Tipuani, Apr–Jun 1892, *Bang 1366* (holotype: NY!, photo: MICH!; isotypes: BM! F! G! GH! K! LE! M! MICH! MO! NY! US! W!).

Vine. Stems and branches pubescent with scalelike T-shaped hairs when young, soon becoming glabrate. Laminas 6.8–16.5 cm long, 6.2–13.8 cm wide, ovate to cordate, apex

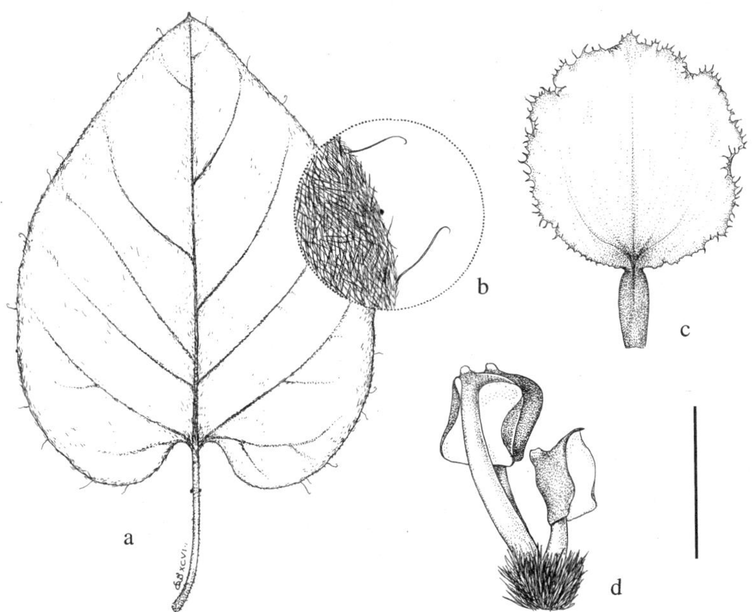

FIG. 57. *Stigmaphyllon boliviense*. a. Large leaf. b. Detail of margin and abaxial surface of lamina. c. Posterior petal (the "flag"). d. Gynoecium, anterior style to the right. Scale: for a, bar = 4 cm; for b, bar = 8 mm; for c, bar = 6.7 mm; for d, bar = 3.3 mm. (Based on *Jardim 1901*.)

mucronate, base cordate to auriculate, glabrous adaxially, very sparsely pubescent with T-shaped hairs abaxially to glabrous (trabecula 0.8–2.3 mm long, wavy, stalk 0.1–0.4 mm long), margin with irregularly spaced stipitate (nail-like) glands (0.2–0.3 mm in diameter, 0.1–0.5 mm long) and/or filiform glands (up to 5 mm long); petioles 1.8–9 cm long, pubescent with scalelike T-shaped hairs, sparsely so in older leaves, not confluent across the node, with a pair of shallowly cupulate glands borne 1.5–18 mm below the base of the lamina, each gland 1.5–2.3 mm in diameter, up to 1 mm high; stipules 0.5–1.8 mm long, 0.4–1.1 mm wide, free, triangular or narrowly so, eglandular. Flowers ca. 10–20 per umbel, these solitary or rarely borne in dichasia. Peduncles 1–4 mm long, densely sericeous (hairs subsessile); pedicels 3.3–10 mm long, sericeous (hairs subsessile) but the distal 1/4–1/3 glabrate to glabrous, terete or distally slightly expanded; peduncles 0.1–0.6 times as long as the pedicels. Bracts 1–1.7 mm long, 0.8–1.3 mm wide, ovate to semicircular, apex obtuse; bracteoles 1–1.3 mm long and wide, semicircular, apex obtuse, eglandular; bracts and bracteoles sericeous or sparsely so abaxially. Sepals 1.5–2 mm long, 2–2.5 mm wide, *all* bearing a pair of glands, glands 2.5–3.3 mm long, 1.2–1.5 mm wide. All petals with the limb orbicular, glabrous, yellow or (especially the limb of the posterior petal) suffused with red, margin digitate-fimbriate, fimbriae up to 0.6 mm long; anterior-lateral petals: claw 2.2–3 mm long, limb ca. 10–12 mm long and wide; posterior-lateral petals: claw 1–2 mm long, limb ca. 8.5–9.5 mm long and wide; posterior petal: claw 2.5–2.7 mm long, apex indented, limb ca. 8–9 mm long and wide. Stamens unequal, those opposite the posterior-lateral petals (and the posterior styles) with the longest filaments, anthers of those opposite the lateral sepals eloculate or rarely anthers of those opposite the posterior-lateral sepals with the connective enlarged and the locules reduced; anthers glabrous. Stamen opposite anterior sepal: filament 2.3–3 mm long, anther ca. 1.5 mm long; stamens opposite anterior-lateral petals: filaments 1.5–2 mm long, anthers 1.2–1.4 mm long; stamens opposite anterior-lateral sepals: filaments ca. 3 mm long, connectives 0.9–1.1 mm long, locules absent; stamens opposite posterior-lateral petals: filaments 4–4.5 mm long, anthers 1.2–1.5 mm long; stamens opposite posterior-lateral sepals: filaments 2.5–2.8 mm long, connectives 0.7–0.9 mm long, locules ca. 0.2 mm long but usually absent; stamen opposite posterior petal always shorter than the adjacent two: filament 1.9–2.3 mm long, anther 0.6–0.9 mm long. Anterior style 2.7–3 mm long, shorter than the posterior two, terete, glabrous, erect; apex 1.6–1.8 mm long including a tiny spur up to 0.2 mm long, each foliole 2.2–2.5 mm long, 1.7–2 mm wide, rectangular. Posterior styles 4.3–4.8 mm long, terete, glabrous or with some scattered hairs in the proximal 1/4, lyrate; foliole 2.3–3 mm long, 2.1–2.7 mm wide, suborbicular to subrectangular. Dorsal wing of samara 3–3.1 cm long, 1.1–1.4 cm wide, upper margin with a blunt tooth; lateral winglets absent; nut 8–10 mm high, 6–7.2 mm in diameter, without air chambers, areole 4.5–5 mm long, ca. 4 mm wide, deeply concave, carpophore up to 4.5 mm long. Embryo 9–9.3 mm long, ca. 2 times as long as wide, ovoid, outer cotyledon 12–13 mm long, 4.5–5 mm wide, the distal 1/3 folded over the inner cotyledon, inner cotyledon 5.6–7.3 mm long, 2.5–3.1 mm wide, straight. Chromosome number unknown.

Phenology. Collected in flower in January, from March through June, and in November, in fruit in January and from April through June.

Distribution (Fig. 8). Bolivia (La Paz, Santa Cruz); forest and roadsides; 340–820 m.

ADDITIONAL SPECIMENS EXAMINED. **Bolivia**. LA PAZ: Prov. Larecaja, along rd between Caranavi and Canay, 28.1 km N of Caranavi, 15°27′S, 67°50′W, *Croat 51688* (MO); Prov. Sud Yungas, camino de Palos Blancos hacia la cumbre del Cerro Marimono, *Vargas & Seidel 2036* (MICH).—SANTA CRUZ: Prov. Ichilo, 2 km WSW of Buena Vista on rd to El Cairo, 17°27′S, 63°14′W, *Nee & Coimbra S. 36091* (MICH, NY); Prov. Andres Ibánez, 2 km W of center of La Belgica, 17°24′S, 63°14′W, *Nee 33768* (NY); Prov. Ichilo, Buena Vista, *Steinbach 7931* (A, BM, F, G, K, MO, S).

Stigmaphyllon coloratum and *S. boliviense* (no. 59), both known only from Bolivia, are the only species in the genus in which each of the five sepals bears a pair of glands; in all other species the anterior sepal is eglandular. They differ most notably in that the abaxial surface of the lamina is sparsely pubescent to glabrous in *S. coloratum*, whereas in *S. boliviense* it is densely pubescent with T-shaped hairs. Both also resemble to *S. bradei* (no. 58) and *S. puberulum* (no. 61) of eastern Brazil, but are readily separated by the calyx. For further discussion, see *S. boliviense*.

61. Stigmaphyllon puberulum Grisebach, Linnaea 13: 205. 1839.—TYPE: BRAZIL. Bahia or Espírito Santo: "inter Vittoria et Bahia" (fide Niedenzu, 1900, 1928), *Sello s.n.* (holotype: B, destroyed, fragment: NY!, photos: A! F! MICH! NY! US!; isotype: K!).

Vine. Stems and branches pubescent with scalelike T-shaped hairs when young, becoming glabrate. Laminas 6.8–17.5 cm long, 6.8–13.5 cm wide, cordate to orbicular, apex mucronate or emarginate-mucronate, base cordate to auriculate, glabrate to glabrous adaxially, with T-shaped hairs abaxially (trabecula 0.4–1.3 mm long, wavy to crisped, stalk 0.1–0.15 mm long), margin with irregularly spaced sessile and/or stipitate (nail-like) glands (0.2–0.4 mm in diameter, up to 0.4 mm long) and/or filiform glands (up to 1 mm long); petioles 2.5–6.3 cm long, pubescent with scalelike T-shaped hairs to glabrous, not confluent across the node, with a pair of shallowly cupulate glands at the apex or more commonly up to 5 mm below the base of the lamina, each gland 1–2.5 mm in diameter, up to 1 mm high; stipules 0.7–1 mm long, 0.7–1.2 mm wide, free, triangular, eglandular. Flowers ca. (10–) 15–20 per umbel, these solitary or borne in dichasia or small thyrses (axes to the 3rd order, densely pubescent with scalelike T-shaped hairs). Peduncles rudimentary to 6 mm long, sericeous (hairs subsessile); pedicels 4.5–10.5 mm long, sericeous (hairs subsessile) but usually glabrous in the distal 1/4–1/3, distally expanded; peduncles to equally as long as the pedicels. Bracts 1.5–2.2 mm long, 1.3–1.7 mm wide, triangular or sometimes lanceolate, apex acute or sometimes obtuse; bracteoles 1.1–1.5 mm long, 1–1.5 mm wide, broadly triangular, apex obtuse, eglandular or with a glandular area in the basal 1/3–1/2; bracts and bracteoles sericeous (hairs subsessile) abaxially. Sepals 2.2–2.5 mm long, 2.1–3 mm wide, glands 2.5–2.7 mm long, 1.3–1.5 mm wide. All petals with the limb orbicular, glabrous, yellow or (especially the limb of the posterior petal) suffused with red, margin fimbriate, the fimbriae up to 0.5 (–0.6) mm long; anterior-lateral petals: claw 2–2.6 mm long, limb ca. 10 mm long and wide; posterior-lateral petals: claw 1.5–2 mm long, limb ca. 8 mm long and wide; posterior petal: claw 2.5–2.8 mm long, apex indented, limb ca. 7–7.5 mm long and wide. Stamens unequal, those opposite the posterior styles with the longest filaments, anthers of those opposite the lateral sepals eloculate; anthers glabrous. Stamen opposite anterior sepal: filament 2.1–2.4 mm long, anther 1–1.4 mm long; stamens opposite anterior-lateral petals: filaments 1.3–1.9 mm long, anthers 0.8–1 mm long; stamens opposite anterior-lateral sepals: filaments 2.7–2.8 mm long, connectives 1.1–1.2 mm long, locules absent; stamens opposite posterior-lateral petals: filaments ca. 3 mm long, anthers 1–1.3 mm long; stamens opposite posterior-lateral sepals: filaments 2.5–2.6 mm long, connectives 0.8–1.1 mm long, locules absent; stamen opposite posterior petal always shorter than the adjacent two: filament 2–2.2 mm long, anther 0.7–0.8 mm long. Anterior style 2.6–3.1 mm long, shorter than the posterior two, terete, glabrous, erect; apex 1.7–1.9 mm long, each foliole 0.7–1 mm long, 1–1.3 mm wide, broadly triangular to trapezoidal. Posterior styles 3.5–4.1 mm long, terete, glabrous, lyrate; foliole 1.8–2.2 mm long and wide, subsquare to suborbicular. Mature samara not seen; immature samara with a large dorsal wing, the nut prominently ribbed but without lateral winglets (*Blanchet 7*). Chromosome number unknown.

Phenology. Collected in flower in September, November, and December, date of fruiting collection unknown.

Distribution (Fig. 9). Brazil (Bahia, Pernambuco, the type perhaps from Espírito Santo); lowlands, at river margins.

ADDITIONAL SPECIMENS EXAMINED. **Brazil**. BAHIA: Itabuna, margem do Rio da Cachoeira, *Belém 1794* (CEPEC, IAN, NY, UB); Ilhéus, *Blanchet 7* (G); Mpio. Ilhéus, área do CEPEC, *dos Santos 3396* (MICH).—PERNAMBUCO: Tapera, *Pickel 131* (BM, F), *1196* (DS, F, GH, IAN, MICH, NY, US, WIS).

Stigmaphyllon puberulum is characterized by its leaves, inflorescences, and flowers. The margin of the lamina bears irregularly spaced sessile and/or stipitate (nail-like) glands (0.2–0.4 mm in diameter, up to 0.4 mm long) and/or filiform glands (up to 1 mm long); the basal glands are shallowly cupulate and commonly borne on the petiole up to 5 mm below the base of the lamina (each gland 1–2.5 mm in diameter, up to 1 mm high). The peduncles are rudimentary to 6 mm long. The petals are fimbriate and especially the posterior one is suffused with red; the stamens opposing the sepals are eloculate. As also noted by Grisebach (1839), the samaras lack lateral winglets. This species is most similar to *S. bradei* (no. 58) from São Paulo, Brazil, which has glandular stipules, and the Bolivian *S. boliviense* (no. 59) and *S. coloratum* (no. 60), the only species in the genus bearing glands on all five sepals.

62. Stigmaphyllon paralias Adr. Jussieu in St.-Hilaire, Fl. bras. mer. 3: 59. 1833 ["1832"]. *Brachypterys australis* Adr. Jussieu in Delessert, Icones sel. pl. 3: 20, t. 34. 1838 ["1837"], nom. superfl. *Brachypterys paralias* (Adr. Jussieu) Hutchinson, Gen. fl. pl. 2: 589. 1967.—TYPE: BRAZIL. Rio de Janeiro: a laco Pacuarema usque ad civitatem Cabo Frio, Dec, *St.-Hilaire B2/214* (holotype: P!, photo: MICH!; isotype: P!).

Shrub to 2 m. Stems and branches sericeous when young, soon becoming glabrate to glabrous. Laminas 3–16 cm long, 1.5–8 cm wide, narrowly to broadly elliptical to lanceolate to broadly ovate to rarely suborbicular to orbicular, apex commonly obtuse or sometimes acute to acuminate, base attenuate to truncate or rarely cordate, glabrous or rarely sparsely sericeous to glabrate adaxially, glaucous and usually sparsely sericeous but sometimes more densely sericeous abaxially (trabecula 0.3–0.4 mm long, straight, sessile), margin eglandular; petioles 0.4–1.5 cm long, sericeous or densely so, not confluent across the node, with a pair of glands at the apex, these flush with the epidermis but with a raised margin, each gland 0.8–2 mm in diameter; stipules 0.4–2.2 mm long, each 0.3–0.8 mm wide and triangular but commonly fused into a bifid structure 0.5–1.5 mm wide, eglandular. Flowers (3–) 4–15 per umbel or rarely in a pseudoraceme, these commonly solitary or sometimes borne in dichasia or compound dichasia (axes to the 4th order, densely sericeous). Peduncles 0.2–2.8 (–4.5) mm long, sericeous; pedicels 8–30 mm long, densely sericeous, terete; peduncles 0.02–0.2 times as long as the pedicels. Bracts 1–2 mm long, 0.8–1.7 mm wide, triangular, apex acute; bracteoles 0.7–1.8 mm long, 0.6–1.9 mm wide, triangular or broadly so, apex obtuse or sometimes acute, eglandular or with a glandular region in the basal 1/3–1/2; bracts and bracteoles sericeous abaxially. Sepals 2–3.5 mm long, 2–3 mm wide, glands 1.5–2 mm long, 0.9–1.3 mm wide. All petals with the limb orbicular, glabrous, yellow, margin of the lateral petals erose or erose-dentate or erose-fimbriate, teeth/fimbriae to 0.5 (–0.7) mm long or sometimes erose-lacerate and the teeth up to 1 mm long; anterior-lateral petals: claw 2.5–4 mm long, limb 14–17 mm long and wide; posterior-lateral petals: claw 1.5–2.5 mm long, limb ca. 12–16 mm long and wide; posterior petal: claw 3.5–5 mm long, apex not indented, limb ca. 10–12 mm long and wide, margin erose or fimbriate-dentate or fimbriate-lacerate or lacerate, or the proximal 1/4–1/2 erose or erose-dentate and the distal portion fimbriate or lacerate, teeth/fimbriae up to 1 (–1.5) mm long, also proximally with 1–3 knob-shaped glands per side (each gland 0.2–0.5 mm in diameter) or sometimes with a narrow band of glandular tissue 1.2–1.9 long per side, glands rarely absent. Stamens unequal, those opposite the lateral petals usually with the longest filaments, anthers of those opposite the posterior petal and the posterior-lateral sepals eloculate or with the connective enlarged and the locules reduced, anthers of those opposite the anterior-lateral sepals usually with the connective enlarged and the locules reduced; anthers all glabrous. Stamen opposite anterior sepal: filament (2.4–) 3–4 mm long, anther 1.2–1.7 mm long; stamens opposite anterior-lateral petals: filaments 1.8–3.3 mm long, anthers 0.9–1.1 mm long; stamens opposite anterior-lateral sepals: filaments 3–4.8 mm long, connectives 1–2 (–2.5) mm long, the distal 1/4–1/3 commonly recurved, locules 0.4–1 mm long; stamens opposite posterior-lateral petals: filaments 3.5–5 mm long, anthers (1–) 1.2–1.5 mm long; stamens opposite posterior-lateral sepals: filaments 2.2–3.5 mm long, connectives 0.6–1.2 mm long, locules usually absent, rarely present, 0.5–0.7 mm long; stamen opposite posterior petal always shorter than the adjacent two: filament 2–3.1 mm long, connective 0.6–1 mm long, locules absent or 0.1–0.2 (–0.4) mm long. Anterior style 4–6 mm long, shorter than or subequal to the posterior two, terete, glabrous, erect; apex 1.5–2 mm long, each foliole 1.2–2 mm long, 1.1–1.6 mm wide, subsquare to parabolic to subrectangular. Posterior styles 4.5–5.3 mm long, terete, glabrous, lyrate; foliole 1.5–2.3 mm long, 2.3–3.3 mm wide, triangular to semi-elliptical. Dorsal wing of samara reduced to a triangular crest 1.3–2 mm long, 2.5–3.5 mm wide at the base; samara from base of nut to apex of crest 7–8.8 mm high, 4.8–5.5 mm wide; nut 5.5–7 mm high, 5–6.7 mm in diameter, surface rugose to reticulate, without air chambers, areole 3.8–5 mm long, 3.8–4.5 mm wide, deeply concave, carpophore absent. Embryo 6.3–6.6 mm long, circular to horseshoe-shaped, ca. 2 times as long as wide, outer cotyledon ca. 9.5 mm long, 3.2–3.4 mm wide, the distal 2/5 folded over the inner cotyledon, inner cotyledon 3.7–4.1 mm long, 2.3–4.5 mm wide, straight. Chromosome number: $n = 10$ (Ormond et al. 1981; based on *Ormond 650*). Fig. 58a–h.

Phenology. Collected in flower throughout the year, in fruit in January, March, May, and June through October.

Distribution (Fig. 59). Eastern Brazil from Maranhão to Rio de Janeiro, also reported from Tucuruí, Pará, and once from Posse, Goiás; in sand: in restingas, caatingas, near beaches, in taboleiros, cerrado, and dry woodlands and grasslands; sea level to 1100 m.

REPRESENTATIVE SPECIMENS. **Brazil**. ALAGOAS: Traipu, 09°52′S, 37°01′W, *da Fonseca 40* (MG); *Gardner 1257* (BM, G, GH, K, NY, P, US, W).—BAHIA: Salvador, Lago Abaeté, *Belém & Mendes 307* (NY, UB, US); Serra do Sincorá, 9 km SW of Mucugê, ca. 13°02′S, 41°25′W, *Harley 16098* (K, MICH, MO, NY, P, U, US); Monte Santo, 10°27′S, 39°20′W, *Harley 16411* (K, MICH, MO, NY, P, U, US); 5 km S of Santa Cruz Cabrália, ca. 16°19′S, 39°01′W, *Harley 17124* (K, MICH, MO, NY, P, U, US); Serra Geral de Caitité, ca. 5 km S from Caitité along the Brejinhos das Ametistas rd, 14°07′S, 42°29′W, *Harley 21110* (CEPEC, MICH); Mpio. Rio de Contas, Pico das Almas, 13°32′S, 41°54′W, *Harley 27091* (MICH); Serra do Tombador, ca. 5 km S of town of Morro do Chapéu, *Irwin et al. 32560* (COL, F, K, MICH, MO, NY, UB, US); Mpio. Mucuri, Km 6 da Rod. Mucuri/Nova Visçosa, *Mattos Silva et al. 749* (MICH).—CEARÁ: E end of Avenida Heraclito Graça, Fortaleza, *Drouet 2560* (F, GH, MICH, MO, NY, R, S, US); *Gardner 1487* (G, NY, P, W); Mpio. Crato, Chapada do Araripe, Parque Nacional de Araripe, ca. 15 km SW of Crato, ca. 07°30′S, 39°35′W, *Plowman 12703* (MICH, NY).—ESPÍRITO SANTO: entre Vitoria & Guarapari, *Duarte 4174* (RB); Rod. BR-101, Res. da Sooretama, Mpio. Linhares, *Hatschbach 46960* (MICH); Reserva de Linhares, *Sucre 8386* (MICH, SP).—GOIÁS: Rod. BR-020, 10 km S de Posse, *Hatschbach 39058* (MICH).—MARANHÃO: Mpio. Lorêto, "Ilha de Balsas" region, 15 km S of Lorêto, ca. 70°12′S, 45°7–8′W, *Eiten & Eiten 4202* (G, K, NY, SP, US); rd between Barreirinhas and Urbano Santos, 10 km W from Sobradinho, *Prance & Henriques R. 29916* (MICH).—PARÁ: Tucuruí, margem direita do Rio Tocantins, *Lisboa et al. 1296* (MG, NY), *M. G. Silva & Rosario 5220* (MG, MICH, NY).—PARAÍBA: Areia, Chã do Jardim, Campo de Aviação, *Perrazo Barbosa 124* (RB).—PERNAMBUCO: Rest. Rio Doce, Olinda, *Leal & da Silva 43* (RB, SP); Pesqueira, Serra de Ororubá, *Mattos 9783* (SP); praia da Conceição, *Tavares 846* (US).—PIAUÍ: Sete Cidades, Parque Nacional, Boqueirão, *Barroso 34* (COL, RB); Oeiras, *Jobert 1069* (P, R).—RIO DE JANEIRO: entre Barra de Tijuca y Recreio dos Bandeirantes, *Krapovickas et al. 23208* (MICH, P); Restinga da Barra da Tijuca, *J. Kuhlmann RB26347* (MICH, RB, SP); Restinga de Jacarepaguá, *Pereira et al. 4466* (COL, F, MICH, RB, SP).—RIO GRANDE DO NORTE: Parque Estadial das Dunas, *Araujo 7630* (GUA); S. José de Mipibú, *Emygdio 1714* (R); Lagoa Boqueirão, *Castellanos 23041* (R).—SERGIPE: Mpio. Estância, cerca de 19.4 km da BR-101 em direção à Praia Abais, *Amorim et al. 1533* (MICH).

Stigmaphyllon paralias, an easily recognized, highly atypical species, is a small shrub of sandy areas of eastern Brazil from Maranhão to Rio de Janeiro; recent collections from Tucuruí, Pará, and Posse, Goiás indicate that it may also be expected farther inland in sandy habitats. The usually elliptical to lanceolate leaves have short petioles (up to 1.5 cm

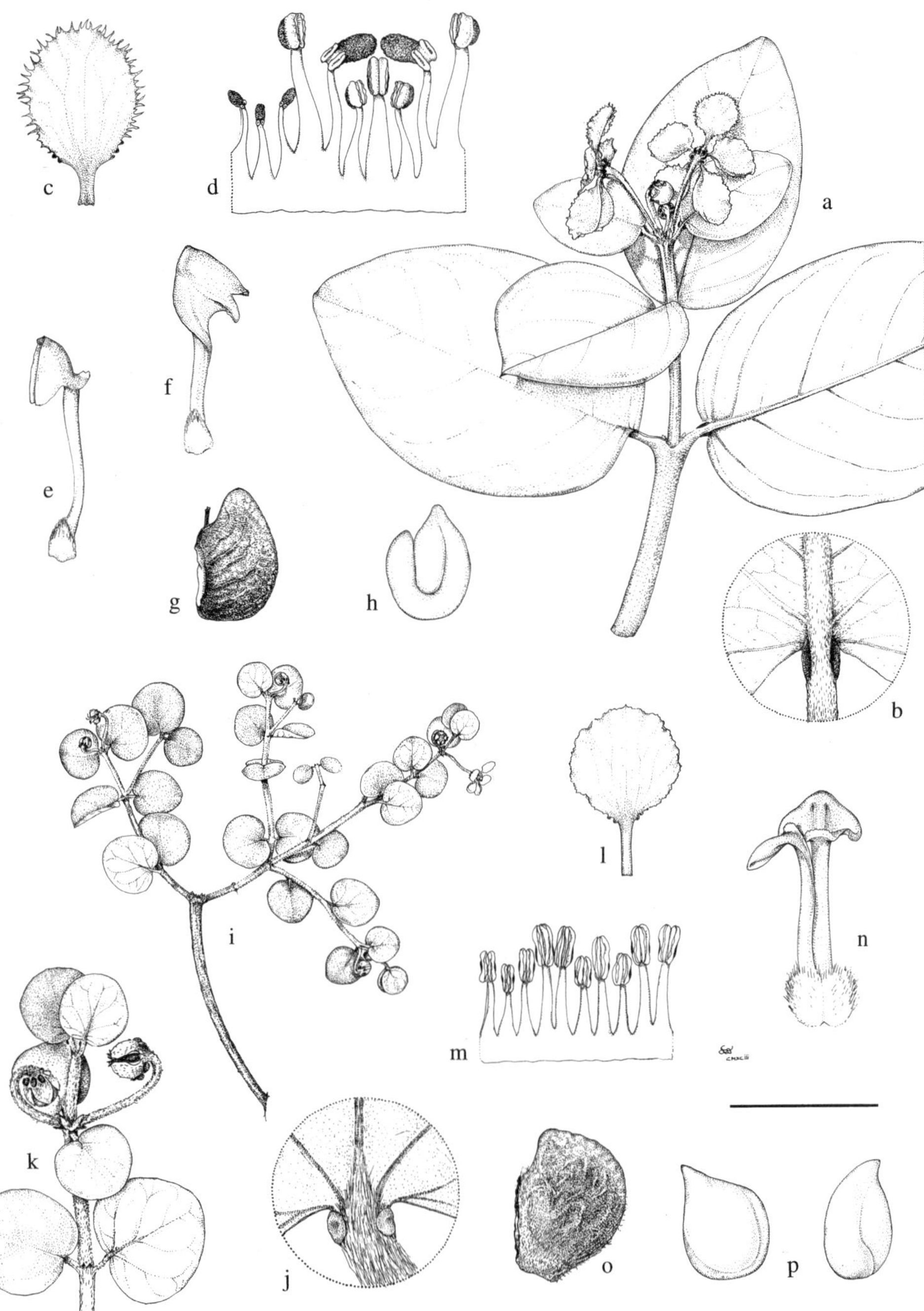

FIG. 58. *Stigmaphyllon paralias* and *S. harleyi*. a–h, *S. paralias*. a. Flowering branch. b. Base of leaf, abaxial view. c. Posterior petal (the "flag"). d. Androecium; stamen second from left opposes the posterior petal. e. Anterior style. f. Posterior style. g. Samara. h. Embryo. i–p, *S. harleyi*. i. Flowering branch. j. Base of leaf, abaxial view. k. Branchlet with axillary 2-flowered inflorescence. l. Posterior petal (the "flag"). m. Androecium; stamen second from left opposes the posterior petal. n. Gynoecium, the second posterior style removed to show anterior style. o. Samara. p. Two views of an embryo. Scale: for a, i, bar = 4 cm; for b, c, k, bar = 1.3 cm; for d–f, bar = 5 mm; for g, h, l, bar = 8 mm; for j, m, o, bar = 4 mm; for n, bar = 2.7 mm. (Based on: a, b, *Harley 19324*; c–f, *Anderson 11736*; g, h, *Harley 17124*; i–p, *Harley 19026*.)

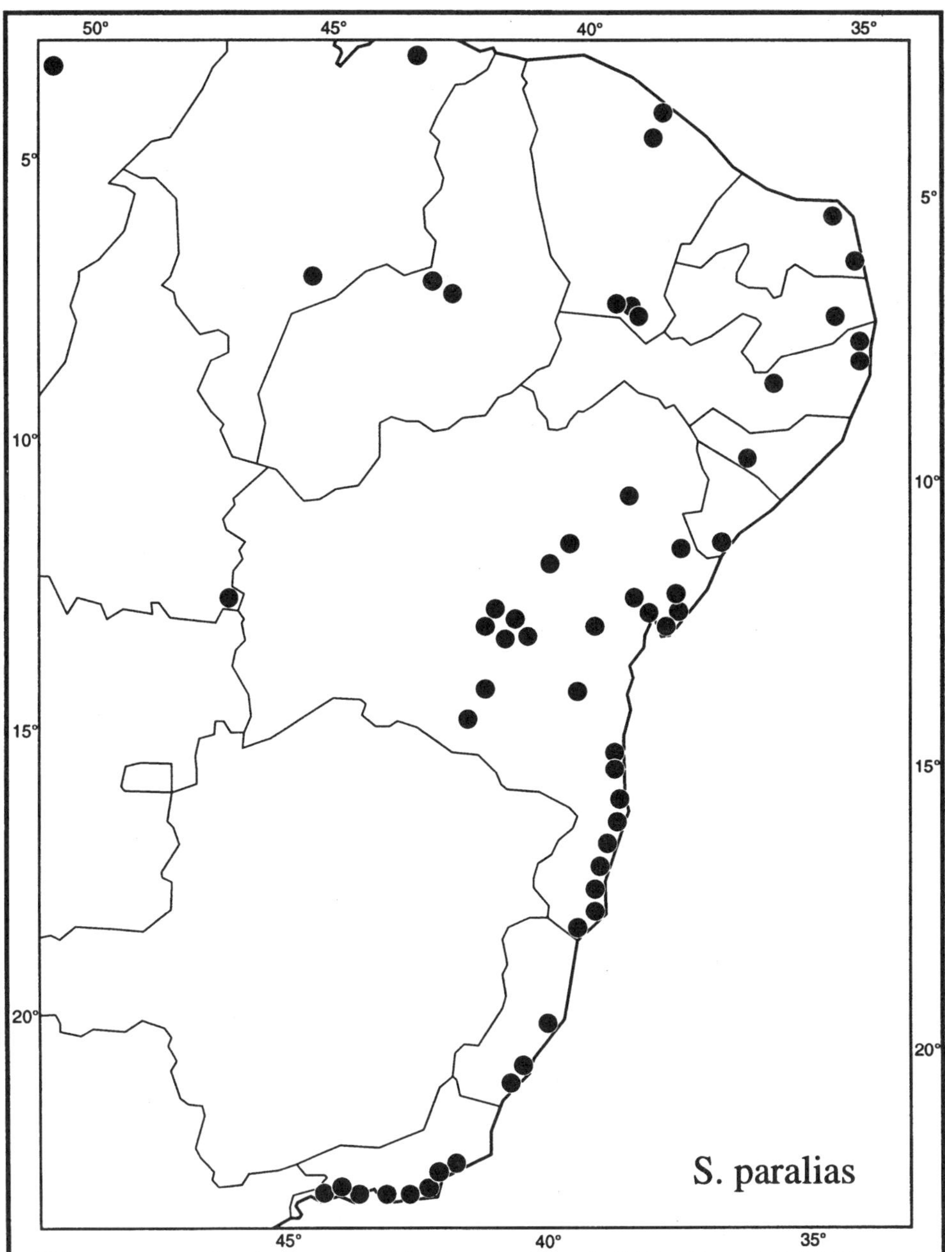

FIG. 59. Distribution of *Stigmaphyllon paralias.*

long), and each of the pair of basal glands is flush with the epidermis, rather than prominent, but has a raised margin; the stipules are often fused across the node into a bifid structure. The limb of the posterior petal bears on the margin near the base 1–3 knob-shaped glands on each side or sometimes a narrow band of glandular tissue; these glands are absent only in the collection from Goiás. The anthers of the stamens opposite the anterior-lateral sepals consist of greatly enlarged connectives that bear reduced locules and are distally recurved. The anthers of the stamens opposite the posterior-lateral sepals usually lack locules. *Stigmaphyllon paralias* is the only species in the genus in which also the stamen opposite the posterior petal has the connective enlarged and the locules reduced or absent. The "samara" is also greatly modified and consists of a nut bearing an apical crest, the rudimentary dorsal wing, and is without a carpophore; it is very similar to that of *S. harleyi* (no. 11), though larger. As in *S. bannisterioides* (no. 10), the embryo is circular to horseshoe-shaped and consists mostly of the large outer cotyledon; however, in *S. paralias*, the inner cotyledon is not rudimentary but nearly half as long as the outer.

Stigmaphyllon paralias and *S. bannisterioides* share some similarities, particularly the reduced "samara," and were segregated by Jussieu as the genus *Brachypterys;* see the discussion of *S. bannisterioides* (no. 10).

63. Stigmaphyllon lindenianum Adr. Jussieu, Arch. Mus. Hist. Nat. Paris 3: 362. 1843. *Stigmaphyllon lindenianum* var. *jussieuanum* Niedenzu, Pflanzenreich IV. 141(2): 499. 1928, nom. superfl.—TYPE: MEXICO. Tabasco: Teapa, *Linden s.n.* (holotype: P!, photos: F! MICH! US!; isotype: G! K!).

Stigmaphyllon lindenianum var. *yucatanum* Niedenzu, Ind. Lect. Lyc. Brunsberg. p. aest. 1900: 18. 1900.—TYPE: MEXICO. Yucatán, *Gaumer 408* (holotype: B, destroyed; isotypes: A! BM! BR! C! CAS! F! G! K! MICH! MO! NY! US! W!).

Stigmaphyllon tiliifolium var. *sericans* Niedenzu, Ind. Lect. Lyc. Brunsberg. p. aest. 1900: 17. 1900. *Stigmaphyllon sericans* (Niedenzu) Small, N. Amer. fl. 25(2): 144. 1910.—TYPE: HONDURAS. Santa Bárbara: Pedro Sula, 800 ft, *Thieme 5164* (lectotype, designated by C. Anderson, 1987a: US!, photo: MICH!; isolectotypes: G! GH! NY! US!).

Stigmaphyllon tiliifolium var. *sericans* f. *grandifolia* Niedenzu, Pflanzenreich IV. 141(2): 498. 1928.—TYPE: COSTA RICA. Cartago: Tuis, Turrialba, 650 m, *Tonduz 11454* (lectotype, designated by C. Anderson, 1987a: F!, photo: MICH!; isolectotypes: BR! M! MICH! US!).

Vine to 30 m. Stems and branches sericeous when young, soon becoming glabrous. Laminas (4.5–) 5–19 cm long, 4–15.5 cm wide, most commonly triangular to ovate to cordate, sometimes narrowly triangular or elliptical to suborbicular or palmately 3–7-lobed, apex mucronate to caudate, base cordate to truncate, glabrate to glabrous adaxially, sericeous to sparsely so abaxially (trabecula 0.2–0.9 mm long, straight, sessile to subsessile), margin with irregularly spaced sessile glands (0.2–0.4 mm in diameter) and/or with irregularly spaced filiform glands up to 2.5 mm long; petioles (1.1–) 1.6–8.5 (–10) cm long, densely to sparsely sericeous, not confluent across the node, with a pair of prominent but sessile glands at the apex, each gland 1.2–3.2 mm in diameter; stipules 0.4–1 mm long, 0.5–1.2 (–2) mm wide, free, triangular, eglandular. Flowers (9–) 12–35 per umbel, these borne in dichasia, compound dichasia, or small thyrses (axes to the 5th order, sericeous). Peduncles 2.7–8.5 mm long; pedicels (3.5–) 3.8–9.5 (–10.5) mm long, terete; both sericeous or densely so, peduncles usually longer but sometimes shorter than or equaling the pedicels. Bracts 0.8–15 mm long, 0.4–0.9 (–1.3) mm wide, narrowly to broadly triangular, apex acute to acuminate; bracteoles 0.7–1.4 mm long, 0.4–0.8 mm wide, oblong or sometimes broadly triangular, apex obtuse, eglandular or with a glandular area in the basal 1/3–1/2; bracts and bracteoles sericeous abaxially. Sepals 1.5–3.6 mm long, 1.5–2.5 mm wide, glands 1.3–2.2 mm long, 0.8–1.2 mm wide. All petals with the limb glabrous, yellow, margin erose, denticulate, or denticulate-fimbriate, teeth/fimbriae up to 0.2 (–0.3) mm long, limb of lateral petals obovate to orbicular; anterior-lateral petals: claw (1.3–) 1.5–2.5 mm long, limb 7–9.3 (–10) mm long, 6–8.5 (–9) mm wide, base truncate; posterior-lateral petals: claw 0.5–1.3 mm long, limb 5.5–8.5 mm long, 5–6.5 mm wide, base attenuate; posterior petal: claw 2.2–3.1 mm long, apex indented, limb 5.8–7 (–7.5) mm long, 4–5 mm wide, elliptical to obovate. Stamens unequal, those opposite the posterior-lateral petals (and the posterior styles) the stoutest, equally long or slightly longer than those opposite the lateral sepals, anthers of those opposite the lateral sepals with the connective enlarged and the locules reduced, sometimes anthers of those opposite the posterior-lateral sepals with only one locule or rarely eloculate; anthers pubescent, those with reduced locules glabrous. Stamen opposite anterior sepal: filament 1.6–2.7 mm long, anther 0.7–0.9 (–1) mm long; stamens opposite anterior-lateral petals: filaments 1.3–1.9 mm long, anthers 0.5–0.8 mm long; stamens opposite anterior-lateral sepals: filaments (1.9–) 2.1–2.6 mm long, connectives 0.5–0.8 mm long, locules 0.3–0.4 (–0.5) mm long; stamens opposite posterior-lateral petals: filaments (2–) 2.2–3 (–3.6) mm long, anthers 0.8–1.2 mm long; stamens opposite posterior-lateral sepals: filaments 1.8–2.7 mm long, connectives 0.4–0.8 mm long, locules (0.1–) 0.2–0.5 (–0.7) mm long; stamen opposite posterior petal always slightly shorter than the adjacent two: filament (1.4–) 1.6–2.5 mm long, anther 0.5–0.8 mm long. Anterior style (1.8–) 2–3 mm long, shorter than the posterior two, terete, glabrous, erect; apex 0.8–1.5 mm long, each foliole 0.5–1 mm long, 0.7–1.1 mm wide, triangular to parabolic or sometimes subsquare to narrowly trapezoidal to subrectangular, or rarely the folioles reduced to a narrow triangular or rectangular lip ca. 0.1–0.3 mm wide. Posterior styles 2.3–3.5 mm long, terete, glabrous, lyrate; folioles 0.9–1.3 (–1.7) mm long, 1–1.5 (–1.7) mm wide, subsquare to subrectangular. Dorsal wing of samara 2–4 cm long, 0.6–1.4 cm wide, upper margin usually with an obtuse or sometimes acute tooth; nut smooth or with 2–3 irregularly shaped winglets on each side, these up to 5.8 mm long and up to 4.2 mm wide, or nut bearing spurs and/or crests; nut 2.5–4 (–4.5) mm high, 2.5–4 mm in diameter, without air chambers, areole 1.8–3.3 mm long, 1.8–3 mm wide, convex, carpophore up to 1.5 mm long. Embryo 3.6–5.5 mm long, ca. 2 times as long as wide, ovoid, outer cotyledon 3.7–7.5 mm long, 1.8–3.5 mm wide, straight or the distal 1/5–2/5 folded over the inner cotyledon, inner cotyledon 2.6–4 mm long, 1–2.5 mm wide, straight or sometimes folded at the distal 1/10–1/3. Chromosome number unknown. Fig. 60a–g.

Phenology. Collected in flower and fruit throughout the year.

Distribution (Figs. 61, 62). Southern Mexico to Panama and adjacent Colombia (Chocó); in tropical deciduous forest, secondary evergreen forest, mangrove swamps, and at roadsides; sea level to 1200 m.

REPRESENTATIVE SPECIMENS. **Mexico**. CAMPECHE: 8 km al S de la carretera Escarcega–Palenque, sobre el camino a Felipe Angeles, *Cabrera 2440* (MICH).—CHIAPAS: Mpio. Ocozocoautla de Espinosa, 45 km N of Ocozocoautla, *Breedlove 20724* (CAS, ENCB, MICH, MO); Mpio. Solosuchiapa, below Ixhuatán along rd to Pichucalco, *Breedlove 34871* (CAS, ENCB, MEXU, MICH); Mpio. Las Margaritas, confluence of Río Ixcán with Río Lacantum (Río Jataté), *Breedlove 34226* (CAS, MICH, NY).—OAXACA: Mpio. Sta. María Chimalapa, Dtto. Juchitán, 4 km al NO del Ejido La Esmeralda, *Delgado S. et al. 952* (CAS, CHAPA, ENCB, F, MEXU, NY).—QUINTANA ROO: NW del entronque Chetumal–F. Carillo Puerto, *Téllez 2013* (MEXU, MO); 21 km al SE de Chunhuhub, *Téllez 2180* (MEXU, MO).—TABASCO: Mpio. Huimanguillo, Km 35 de la desviación de Huimanguillo hacia Fco. Rueda, *Cowan 2679* (CAS, CHAPA, ENCB, TEX); Mercedes, Balancán, *Matuda 3022* (A, F, MEXU, MICH, NY).—VERACRUZ: La Palma, Catemaco, *Martínez Calderón 2201* (A, CAS, MEXU);

FIG. 60. *Stigmaphyllon lindenianum, S. panamense*, and *S. tonduzii*. a–g, *S. lindenianum*. a. Flowering branch. b. Lobed leaf with detail of abaxial surface of lamina. c. Posterior petal (the "flag"). d. Portion of androecium, posterior stamen on extreme left, anterior stamen on extreme right. e. Gynoecium, anterior style to the right. f. Two samaras. g. Two embryos. h–l, *S. panamense*. h. Flowering branch, with detail of abaxial surface of lamina. i. Posterior petal (the "flag"). j. Gynoecium, anterior style to the right. k. Androecium; stamen second from left opposes the posterior petal. l. Samara. m–o, *S. tonduzii*. m. Gynoecium, anterior style to the right. n. Samara. o. Embryo. Scale: for a, b, f, h, l, n, bar = 4 cm, for detail of b, bar = 2 mm; for c, i, bar = 8 mm; for d, e, g, j, k, m, o, bar = 4 mm. (Based on: a, e, *MacDougall s.n.*; b, c, d, *Reznicek M179*; f, *Cowan 2679* (above), *Burger 10415* (below); g, *Cowan 3189* (above), *Cowan 2024* (below); h–k, *Johnston 1301*; l, *Duchassaing s.n.*; m–o, *Meerow et al. 1003*.)

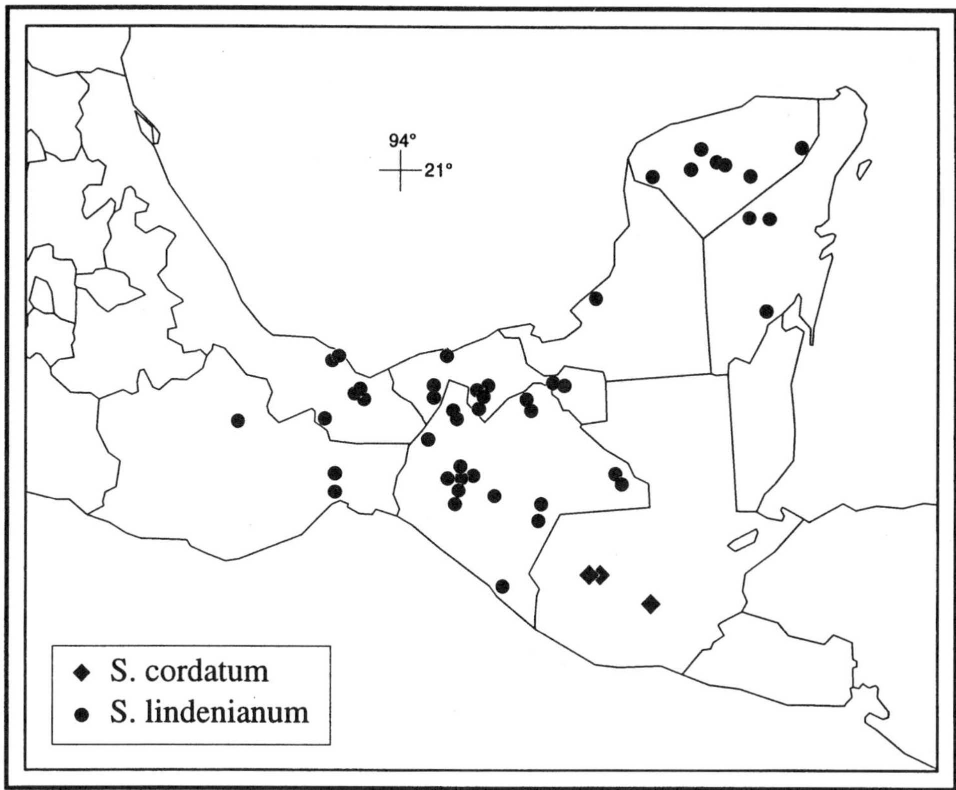

FIG. 61. Distribution of *Stigmaphyllon cordatum* and of *S. lindenianum* in Mexico.

Hidalgotitlán, *Vázquez 390* (F, MEXU, MO).—YUCATÁN: SE Kancabconot, *Gaumer 23900* (C, F, G, GH, MA, MO, NY, US); Mpio. Tunkás, Quintana Roo, *Vara & Arias 320* (CAS, CHAPA, ENCB, F, MO). **Guatemala**. ALTA VERAPAZ: along Río Sebol, downstream from Carizal, *Steyermark 45796* (A, F, LL); Cobán, *von Türckheim II-2359* (F, US).—IZABAL: vic. of Quiriguá, *Standley 24055* (GH, NY, US).—PETÉN: Poptún, carretera para San Luis, *Tun Ortíz 2137* (F, MICH, US).—ZACAPA: 7 km W of El Lobo on Hwy CA9, *Harmon & Fuentes 1843* (MO). **Belize**. BELIZE: Northern Hwy 3 mi N from Belize, *Wiley 489* (MO, US).—CAYO: Vaca, *Gentle 2481* (K).—ORANGE WALK: Roaring Creek, *Dwyer & Liesner 12259* (MICH, MO, NY).—STANN CREEK: Middlesex, *Schipp 468* (A, F, G, GH, MICH, MO, NY, S).—TOLEDO: Edwards Rd beyond Columbia, *Gentle 7312* (LL, MICH, UTD). **Honduras**. ATLÁNTIDA: along Tela River, between Peñas Gordas and Tela, *Molina R. & Molina 25678* (F, MO, NY, US).—COLÓN: 1 1/2 mi E of Trujillo, rd to Castilla, *Saunders 242* (MO).—COMAYAGUA: El Banco, *Rodríguez 2322* (F).—COPÁN: vic. of Copán ruins, *Molina R. & Molina 24605* (F, MO, NY, US).—CORTÉS: a orillas del Río Lindo, *Molina R. 11828* (F, NY, US, UTD).—ISLAS DE LA BAHÍA: Roatán Island, *Molina R. 20698* (F, NY, US).—SANTA BÁRBARA: al N de Santa Bárbara, región La Cuesta, *Molina R. 3770* (F).—YORO: near Progreso, *Standley 54968* (A, F, US). **Nicaragua**. RÍO SAN JUAN: between San Juan del Norte and Delta de San Juan, *Bunting & Licht 837* (DUKE, F, NY, US); Caño El Roble, *Moreno 23367* (MICH); Archipiélago de Solentiname, N Isla La Venada, *Sandino 3603* (MICH).—ZELAYA: Monkey Point, *Moreno & Sandino 11995* (MICH), *12031* (MO); Caño Montechristo, *Moreno 15191* (MO); Montechristo, N de Barra Punta Gorda, *Sandino 2230* (MO). **Costa Rica**. ALAJUELA: ca. 3 km NNE of Bijagua along new rd to Upala, *Burger & Baker 9877* (F, NY).—CARTAGO: Las Vueltas, Tucurrique, *Tonduz 12803* (P, US).—HEREDIA: Finca La Selva, OTS field station, *Hammel 8398, 8461, 9017, 10016, 10661, 11640* (DUKE).—LIMÓN: near Río Catarata, 09°37′N, 82°49′W, *Burger et al. 11415* (CAS, F).—PUNTARENAS: ca. 5 km W of Rincón de Osa, Osa Peninsula, *Burger & Gentry 8879* (AAU, DUKE, F, MO, U); Parque Nacional Corcovado, Sirena, 08°27–30′N,

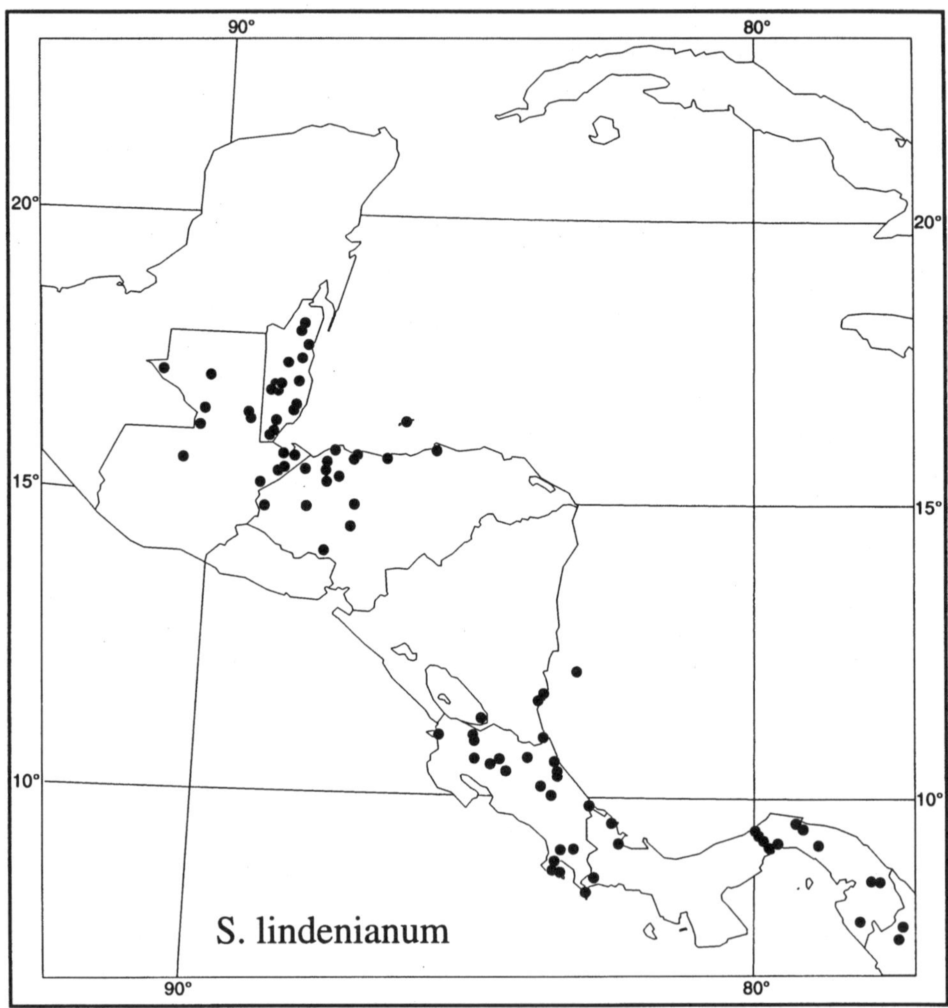

FIG. 62. Distribution of *Stigmaphyllon lindenianum* in Central America and adjacent Colombia.

83°33–38′W, *Kernan 44* (MO). **Panama.** BOCAS DEL TORO: Water Valley, vic. of Chiriquí Lagoon, *von Wedel 1525, 2754* (GH, MO, US).—CANAL ZONE: Barro Colorado Island, *Croat 7021* (F, MO, NY), *Wetmore & Abbe 220* (F, GH, MO); 4 km NW of Gamboa, *Nee 7582* (CAS, MO, TEX, U, US); NW shore of Gatún Lake, ca. 4 mi S of Río Chares, *Lewis et al. 1817* (MEXU, MO, US).—CHIRIQUÍ: Burica Peninsula, 2 km S of Puerto Armuelles, *Busey 475* (F, MO, NY); Paso Canoas to Carras Gordas, 10.5 km N of Paso Lanoas, *Busey 615* (F, MO, NY).—COLÓN: rd from Colón to Portobelo, 24.6 km E of transisthmian hwy, *Folsom 3730* (MO, MICH); at Quebrada Santa Marta on coast rd 42 km SW of Pina, *Nee 11717* (F, MO, US); Santa Rita Ridge, ca. 15 mi from hwy at Sabanita, *Wilbur & Luteyn 11845* (F, NY, US).—DARIÉN: Río Chucunaque, between Río Membrillo and Río Subcuti, *Duke 8609* (MO); Río Ucungati, *Bristan 1172* (MO).—PANAMÁ: Río Majé, above first waterfall, *Croat 34435* (MO); along headwaters of Río Corso (off Río Pacora), *Duke 11937* (MO).—SAN BLAS: El Llano–Cartí rd, 09°19′N, 78°55′W, *de Nevers & Pérez 3990* (MICH); Dubaganalla, *Duke 10199* (MO).—VERAGUAS: 5 mi W of Santa Fe on rd past Escuela Agrícola Alto Piedra, *Croat 22983* (mixed with *S. panamense*; C, CAS, F; MEXU, MO, US). **Colombia.** CHOCÓ: Río Truando, between La Nueva and La Esperanza, *Duke 9889* (ECON, MO, US); hoya del Río Atrato, Mpio. Bojayá, caño de Bojayacito, cerca de Bellavista, *Forero et al. 9226* (MO).

Stigmaphyllon lindenianum is a small-flowered species, readily distinguished by its abaxially sericeous laminas and pubescent anthers. It is common in the lowlands of southern Mexico and Central America; the only two collections from Colombia are from northern Chocó. This species is often confused with the partly sympatric *S. retusum* (no. 64; Mexico to Nicaragua) and *S. dichotomum* (no. 65; Darién, Panama, to northern Colombia and Venezuela). The leaves of *S. retusum* and *S. dichotomum* are pubescent with T-shaped hairs to tomentose abaxially; the anthers of *S. dichotomum* are glabrous. In all three, the laminas are usually entire but occasionally are 3–5-lobed.

Confusion about the identity of the three species is partly caused by their superficial similarity but also by Niedenzu's (1900, 1928) account of them. His circumscriptions of *S. lindenianum* and of *S. dichotomum* (as "*S. tiliifolium*") both include elements of *S. lindenianum, S. retusum,* and *S. dichotomum* as recognized here. His varieties *S. lindenianum* var. *jussieuanum* (based on Jussieu's *S. lindenianum*), *S. lindenianum* var. *yucatanum*, and *S. tiliifolium* var. *sericans* belong in *S. lindenianum*; his varieties *S. lindenianum* var. *lupulus* (=*S. lindenianum* var. *watsonianum* in 1928) and *S. lindenianum* var. *nicaraguense* are here included in *S. retusum*. Morton (1936), in his account of Malpighiaceae of the Yucatan Peninsula, noted that Niedenzu's separation in his key, entire vs. lobed leaves, does not serve to distinguish these species, but that the difference in abaxial laminar pubescence is a useful character. Morton assigned plants with appressed hairs to *S. lindenianum* and those with T-shaped hairs to "*S. humboldtianum*" (=*S. dichotomum*), in which he included Mexican and Central American plants here assigned to *S. retusum*.

In *S. lindenianum*, all styles are normally foliolate, but rarely the folioles of the anterior styles are very small or reduced to a narrow lip; the degree of reduction may vary even within the same umbel. Such variants occur throughout the range, e.g., *Dorantes 2979* (Veracruz, Mexico), *Laughlin 2905* (Chiapas, Mexico), *Cabrera 11347* (Yucatán, Mexico), *Tún O. 2137* (Petén, Guatemala), *Burger & Gentry 8879* (Puntarenas, Costa Rica), *de Nevers & Pérez 3990* (San Blas, Panama). Niedenzu recognized this variant as var. *yucatanum*, based on *Gaumer 408*. The holotype was destroyed at B, but of the many duplicates of this collection, only one of the three at MO also has flowers in which the anterior styles lack normal folioles. Such an occasional reduction of the folioles is also known in other species and does not merit taxonomic recognition. Although such irregularities may indicate events of hybridization, these specimens do not show any traits of intermediacy with any sympatric species. Pollen of these collections, including those duplicates of *Gaumer 408* with foliolate anterior styles, is mostly 95–96% normal; the exceptions are *Tún O. 2137* (100% of the grains small, thick-walled, misshapen), *Burger & Gentry 8879* (80% normal), and *Gaumer 408*, the efoliolate specimen at MO (38% normal). Also noteworthy are *Smith 267* and *Gómez L. 9381*, both from Costa Rica, in which the locules are all aborted and lack pollen. Perhaps they indicate the occurrence of apomixis; *Gómez L. 9381* has immature fruits. These collections also do not show any other morphological irregularities.

64. Stigmaphyllon retusum Grisebach in Oersted, Vidensk. Meddel. Dansk Naturhist. Foren. Kjøbenhavn 1853(1–2): 45. 1854.—TYPE: NICARAGUA. "Prope Granada," *Oersted s.n.* (lectotype, designated by C. Anderson, 1987a: GOET!, photo: MICH!; isolectotype: C!).

Stigmaphyllon lupulus S. Watson, Proc. Amer. Acad. Arts 21: 461. 1886. *Stigmaphyllon lindenianum* var. *lupulus* (S. Watson) Niedenzu, Ind. Lect. Lyc. Brunsberg. p. aest. 1900: 19. 1900. *Stigmaphyllon lindenianum* subsp. *lupulus* (S. Watson) Niedenzu, Pflanzenreich IV. 141(2): 499. 1928. *Stigmaphyllon lindenianum* subsp. *lupulus* var. *watsonianum* Niedenzu, Pflanzenreich IV. 141(2): 499. 1928, nom. superfl.—TYPE: GUATEMALA. Izabal: Chocón, 21 Mar 1885, *Watson 35* (holotype: GH!, photo: MICH!).

Stigmaphyllon lindenianum var. *nicaraguense* Niedenzu, Ind. Lect. Lyc. Brunsberg. p. aest. 1900: 19. 1900. *Stigmaphyllon lindenianum* subsp. *lupulus* var. *nicaraguense (*Niedenzu) Niedenzu, Pflanzenreich IV. 141(2): 499. 1928.—TYPE: NICARAGUA. *Wright s.n.* (holotype: B, destroyed; isotype: GH! GOET! US!, photo of GOET isotype: MICH!).

Vine to 20 m. Stems and branches pubescent with scalelike T-shaped hairs to sericeous when young, soon becoming glabrate to glabrous. Laminas 7–18 cm long, 5–15 cm wide, triangular to cordate to ovate to elliptical or sometimes palmately 3(–5)-lobed or rarely suborbicular, apex mucronate or acuminate-mucronate, base cordate or sometimes truncate, glabrate to glabrous adaxially, with T-shaped hairs to sometimes tomentose abaxially (trabecula 0.3–1.5 mm long, straight to wavy or sometimes curled, stalk 0.1–0.5 mm long), margin with irregularly spaced sessile glands (0.2–0.4 mm in diameter) and sometimes also with irregularly spaced filiform glands up to 2.5 (–5.5) mm long; petioles 1.6–9.5 cm long, with scalelike T-shaped hairs or sericeous, not confluent across the node, with a pair of prominent but sessile glands at the apex, each gland 1.2–2.7 mm in diameter; stipules (0.3–) 0.5–1 mm long, 0.5–1.1 mm wide, free, triangular, eglandular. Flowers 15–35 (–40) per umbel or pseudoraceme, these borne in dichasia or compound dichasia or small thyrses (axes to the 4th order, with scalelike T-shaped hairs), rarely solitary. Peduncles 2.5–8.5 (–10) mm long; pedicels 4–10 mm long, terete; both with scalelike T-shaped hairs to sericeous but the pedicels often less densely so than the peduncles, peduncles 0.5–1.1 times as long as the pedicels. Bracts 0.8–2.1 mm long, 0.5–1.2 mm wide, triangular to narrowly so, apex acute to acuminate; bracteoles 0.5–1.5 (–1.8) mm long, 0.4–1.1 mm wide, triangular to parabolic to oblong to subsquare, apex obtuse or rarely acute, eglandular or rarely each bracteole with a pair of inconspicuous glands (each gland ca. 0.2 mm in diameter); bracts and bracteoles sericeous or densely so abaxially. Sepals 1.5–3 mm long, (1.4–) 1.7–3 mm wide, glands 1.3–2.2 mm long, 0.7–1.3 mm wide. All petals with the limb orbicular to broadly obovate or limb of the posterior petal broadly elliptical, glabrous, yellow, limb of lateral petals with the margin erose or denticulate or denticulate-fimbriate or with fimbriae up to 0.3 mm long; anterior-lateral petals: claw (1.3–) 1.5–2.1 mm long, limb 7–13.5 mm long, 7–12 mm wide; posterior-lateral petals: claw 0.5–1.7 mm long, limb ca. 5.5–11 mm long, ca. 5–10.5 mm wide; posterior petal: claw 2.5–3.5 mm long, apex indented, limb 5.5–9.5 mm long, 4.5–7 mm wide, margin denticulate or denticulate-fimbriate or with fimbriae up to 0.3 (–0.4) mm long. Stamens unequal, those opposite the posterior-lateral petals (and the posterior styles) the largest, those opposite the anterior-lateral sepals sometimes equally long, anthers of those opposite the lateral sepals with the connective enlarged and the locules reduced (sometimes only slightly reduced); anthers all loculate, pubescent. Stamen opposite anterior sepal: filament (1.4–) 1.9–2.3 mm long, anther 0.8–1.2 (–1.4) mm long; stamens opposite anterior-lateral petals: filaments 1.1–2 mm long, anthers 0.5–1 mm long; stamens opposite anterior-lateral sepals: filaments (1.8–) 2–3 mm long, connectives 0.6–0.8 mm long, locules 0.2–0.7 mm long; stamens opposite posterior-lateral petals: filaments (1.8–) 2–3 mm long, anthers 0.9–1.2 mm long; stamens opposite posterior-lateral sepals: filaments 1.6–2.7 mm long, connectives 0.6–0.8 (–1) mm long, locules 0.3–0.5 (–0.7) mm long; stamen opposite posterior petal always shorter than the adjacent two: filament 1.4–2.3 mm long, anther 0.4–0.6 mm long. Anterior style 1.8–3 mm long, shorter than the posterior two or sometimes almost as long, terete, glabrous or sometimes with scattered hairs in the basal 1/4–1/2, erect or slightly recurved; apex 0.8–1.5 mm long, each foliole 0.5–1.6 mm long and wide, parabolic to rectangular to square, folioles rarely unequal, rarely one or both folioles reduced and the apex merely expanded, ca. 0.6 mm wide. Posterior styles 2.4–3.8 (–4) mm long, terete, glabrous or sometimes with scattered hairs in the basal 1/4–1/2, lyrate; folioles (0.7–) 1–2.5 mm long and wide, square to subrectangular. Dorsal wing of samara 2.5–4.5 (–4.8) cm long, 1–1.5 (–1.8) cm wide, upper margin with an obtuse or subacute tooth; nut with a pair of lateral winglets, these up to 11 mm long, up to ca. 4 mm wide, linear, oblong, triangular, parabolic, semicircular, or rectangular, and/or bearing spurs and/or crests or only prominently ribbed; nut 3.8–7.5 mm high, 3–7.3 mm in diameter, without air chambers, areole 2.5–4.5 mm long, 2–4.5 mm wide, convex or slightly so, carpophore up to ca. 4 mm long. Embryo 5.2–8.1 mm long, ca. two times as long as wide, ovoid, outer cotyledon 5.3–11.2 mm long, 2.3–4.5 mm wide, the distal 1/3 folded over the inner cotyledon, inner cotyledon 4.4–5.8 mm long, 2–3.2 mm wide, folded at the distal 1/3 or straight. Chromosome number: n = 10 (W. R. Anderson 1993; based on *Fryxell & Anderson 3485*). Fig. 63a–h.

Phenology. Collected in flower and in fruit throughout the year.

Distribution (Fig. 64, 65). Southeastern Mexico to northern Costa Rica (only three collections seen); in rain, evergreen, gallery, and scrub forests, in acahuales and matorrales, along rivers, in thickets, and at roadsides; sea level to 1300 m.

REPRESENTATIVE SPECIMENS. **Mexico.** CHIAPAS: Mpio. Palenque, 3–5 km N of Palenque along rd to Villahermosa, *Breedlove 26648* (DS, MEXU, MO); Mpio. La Independencia, valley of Santa Elena along rd to Ixcán, *Breedlove 41958* (DS, MEXU, MICH); Mpio. La Libertad, 10 km towards Chancala on rd to Bonampak, *Breedlove 57845* (CAS).—HIDALGO: Mpio. Huehuetla, 7 km al W de Acautla, *Hernández M. 7283* (MEXU).—OAXACA: Cerro Blanco, Teotitlán, *Conzatti 3437* (MEXU, US); Tuxtepec, Chiltepec, *Martínez Calderón 45* (CAS, CHAPA, ENCB, K, NY, TEX).—PUEBLA: 5 km adelante de Ceiba Grande, orillas del Río Cazones, *Riba 422B* (ENCB); adelante de Agua Fría, *Sarukhan et al. 3250* (MEXU).—QUERÉTARO: Mpio. Jalpan, al O de Tanchanaquito, punto El Sabinito, *López Ch. 582*; Mpio. Landa, cañón del Río Santa María, cerca de Tanchanaquito, *Zamudio & Carranza 7211* (MICH).—SAN LUIS POTOSÍ: Tamazunchale, *Fisher 3784* (GH, MO, NY, US); Mpio. Valles, 1 km N of La Estribera, *Fryxell & Anderson 3522* (MICH); near Tamasopo, *Pringle 4102* (BR, F, G, GH, GOET, K, LL, M, MEXU, MICH, MO, MSC, NY, P, US, W).—VERACRUZ: Mpio. Cosamaloapan, Otatitlán, *Martínez Calderón 1060* (BM, CAS, ENCB, F, GH, MEXU, MO, NY); Atoyac, *Matuda 1482* (MEXU, MICH, MO, NY, US); Mpio. Martínez de la Torre, San Carlos, *Ventura A. 1296* (DS, ENCB, F, MICH, MO, SD). **Belize.** CAYO: Vaca, *Gentle 2490A* (A, MEXU, MICH, NY, UTD).—TOLEDO: near Jacinto Hills, *Gentle 5525* (LL, MICH, UTD). **Guatemala.** ALTA VERAPAZ: Chahal, on Sebol rd, *Contreras 7759* (LL, MICH, UTD); SW of Lanquín, *Steyermark* 44074 (F, GH).—CHIQUIMULA: 2 km from Esquipulas, *Molina R. & Molina 25197* (F, NY, US).—IZABAL: Cienaga, on Petén–Guatemala rd, *Contreras 10825* (LL, MEXU, MO, S, US, UTD); 12 km N of Río Dulce on rd to Modesto Méndez, *Harmon 2487* (F, MICH, MO).—JUTIAPA: near El Molina (dept. Santa Rosa), *Standley 78475* (F).—PETÉN: Km 158 on Cadenas rd, *Contreras 6559* (LL, MICH, UTD).—RETALHULEU: vic. of Retalhuleu, *Standley 88789* (F).—SANTA ROSA: about Guazacapán, *Standley 78593* (F). **Honduras.** COMAYAGUA: vic. of Siguatepeque, *Standley & Chacón 6906* (F).—GRACIAS A DIOS: Río Plátano, *Gentry 7522* (F, MO).—MORAZÁN: drainage of Río Yeguare, 14°N, 87°W, faldas del Cerro Majicarán, *Molina R. 1756* (F, GH, MO).—OLANCHO: camino a San Francisco La Paz, matorral del Río Telica, *Molina R. 13362* (F, NY). **El Salvador.** AHUACHAPÁN: vic. of Ahuachapán, *Standley 20347* (GH, NY, US).—SAN SALVADOR: vic. of San Salvador, *Standley 19643* (GH, NY, US).—SAN VICENTE: vic. of San Vicente, *Standley 21272* (GH, NY, US). **Nicaragua.** BOACO: Camouapa, *Atwood 3513* (F, GH, NY).—CARAZO: Río Grande, ca. 4 km al N del balneario de Casares, *Grijalva & Vanegas 3416* (MICH).—CHINANDEGA: Chinandega, *C. F. Baker 2026* (A, F, G, GH, MO, MSC, NY, P, US, W).—CHONTALES: Hacienda San Martín, near confluence of Río El Jordán and Río La Pradera, 12°17′N, 85°15′W, *Stevens 22858* (MICH).—ESTELÍ: 3–7 km NW of Pueblo Nuevo, *Williams & Molina R. 42391* (F, MICH, US).—GRANADA: Volcán Mombacho, 1.3 km antes

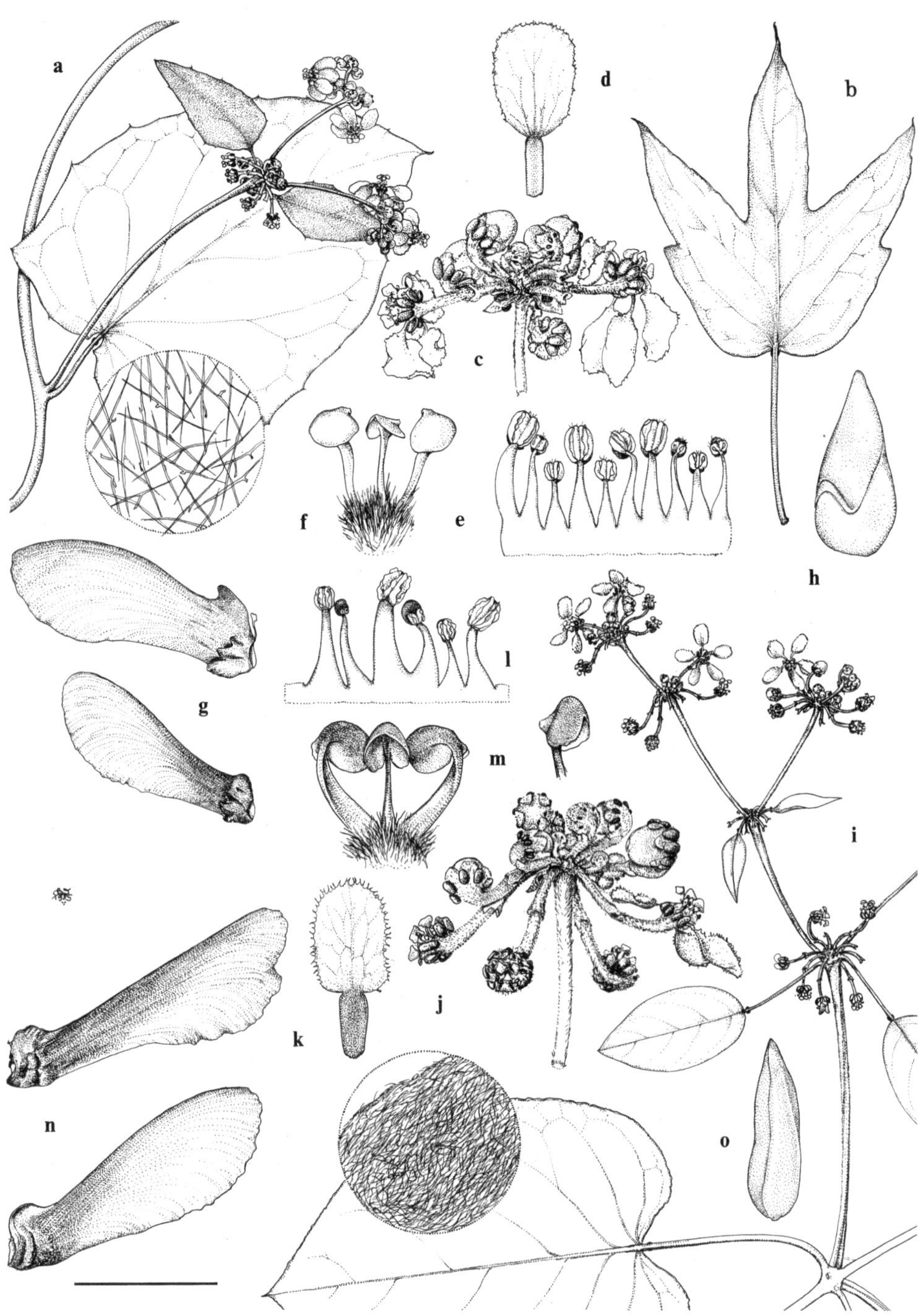

FIG. 63. *Stigmaphyllon retusum* and *S. dichotomum*. a–h, *S. retusum*. a. Flowering branch and detail of abaxial surface of lamina. b. Lobed leaf. c. Umbel. d. Posterior petal (the "flag"). e. Androecium; stamen second from right opposes the posterior petal. f. Gynoecium, anterior style in the center. g. Two samaras. h. Embryo. i–o, *S. dichotomum*. i. Flowering branch and detail of abaxial surface of lamina. j. Umbel. k. Posterior petal (the "flag"). l. Portion of androecium, posterior stamen on extreme left, anterior stamen on extreme right. m. Gynoecium (anterior style in the center) and lateral view of anterior style. n. Samaras. o. Embryo. Scale: for a, b, i, bar = 4 cm, for detail of b, bar = 1.3 mm; for c, j, bar = 1.3 cm; for d, k, bar = 8 mm; for e, f, h, l, m, bar = 4 mm; for g, n, bar = 2 cm; for o, bar = 5.7 mm. (Based on: a, c–f, *Fryxell & Anderson 3485*; g, *Sandino 2571* (above, Nicaragua), *Fryxell & Anderson 3522* (below, Mexico); h, *Pipoly 4542*; i, j, *Smith 1525*; k–m, *de Bruijn 1556*; n, *Fryxell et al. 4400* (above), *Romero C. 2045* (below); o, *Romero C. 2045*.)

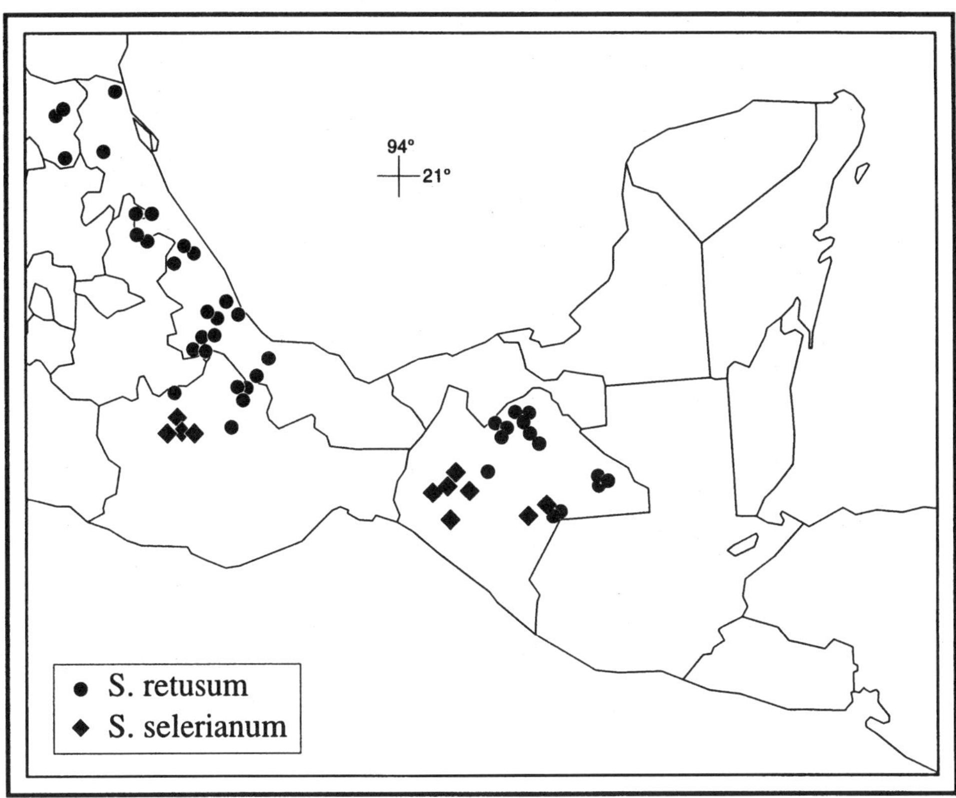

FIG. 64. Distribution of *Stigmaphyllon retusum* in Mexico and of *S. selerianum.*

de Hacienda Cutirre, "El Cacao," 11°51′N, 85°57′W, *Moreno 6326* (MICH).—JINOTEGA: al NE de Wiwilí, camino entre Carmen y Wamblán, ca. 1 km al N del Carmen, 13°43′N, 85°46′W, *Araquistain & Moreno 1505* (MICH).—LEÓN: Volcán Momotombo, alrededores del Proyecto Geotermico, *Araquistain & Moreno 1083* (MICH).—MADRIZ: a 10.5 km al S de Somoto, carretera Panamericana, en el valle de Yalaguina, 13°30′N, 86°30′W, *Moreno 5992* (MICH).—MANAGUA: ca. 5 km NNW of Hwy 12 along rd on ridge of Sierra de Mateare, ca. 12°7′N, 86°23′W, *Stevens 6198* (MICH).—MATAGALPA: along Río Las Cañas, 10–15 km NE of Matagalpa, *Williams et al. 24020* (F, NY).—MASAYA: a orillas de la Laguna Masaya, 11°58′N, 86°08′W, *Moreno 6137* (MICH).—NUEVA SEGOVIA: El Jícaro "Casas Viejas," 13°45′N, 86 °06′W, *Moreno 5716* (MICH).—RIVAS: Isla Ometepe, Volcán Concepción, poblado La Esperanza, 11°31′N, 85 °37′W, *Robleto 1613* (MICH).—SAN JUAN DEL NORTE: "El Carmen," 2 km al N de San Miguelito, 11°25′N, 84°53′W, *Moreno 23468* (MO).—ZELAYA: along new rd to Mina Nueva America, leading ca. W from 14.3 km N of El Empalme on main rd to Rosita, *Stevens 12705* (MICH). **Costa Rica.** ALAJUELA: Upala, Dos Ríos, 2.5 km al NE de la Finca Palma, 10°55′N, 85°20′W, *Herrera 1080* (MICH).—GUANACASTE: vicinity of Cañas, Finca Taboga, *Daubenmire 521* (F).—PUNTARENAS: carretera Interamericana at junction with Río Seco, *Khan et al. 1161* (BM, MICH).—PUNTARENAS: Mata de Limón, *R. Hernández 83078* (CR).

Stigmaphyllon retusum is commonly found from southern Mexico to Nicaragua and rarely in northern Costa Rica. It is distinguished by its laminas, which bear T-shaped hairs abaxially, and the pubescent anthers. The relatively small flowers are aggregated in umbels usually disposed in large compound inflorescences. Because the name *S. retusum* had been neglected, representatives of this species were traditionally assigned to the partly

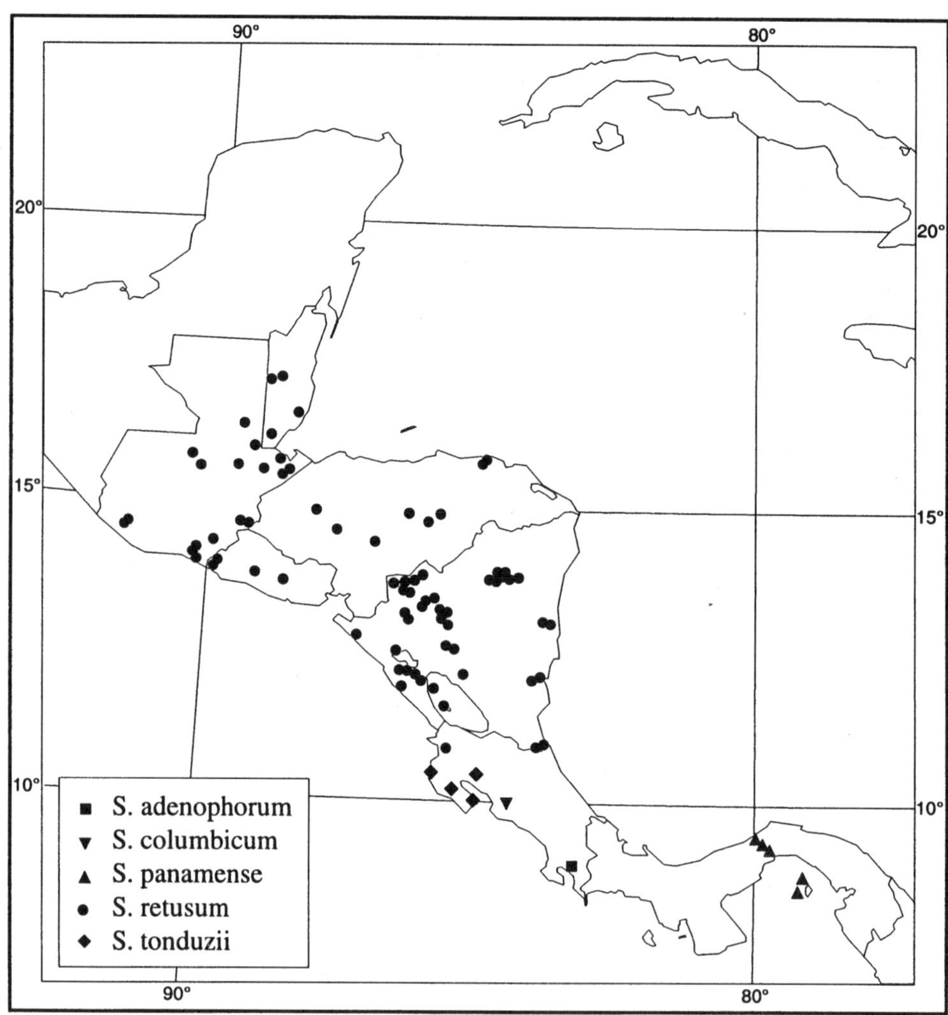

FIG. 65. Distribution of *Stigmaphyllon adenophorum, S. columbicum* in Costa Rica, *S. panamense, S. retusum* in Central America, and *S. tonduzii.*

sympatric *S. lindenianum* (no. 63) and to *S. dichotomum* (no. 65; as "*S. humboldtianum*" or "*S. tiliifolium*"), a species of western Venezuela, Colombia, and adjacent Panama (Morton 1936; C. Anderson 1987a; see also the discussion of *S. lindenianum*). *Stigmaphyllon lindenianum* is most easily separated from *S. retusum* by the sericeous abaxial vesture of its laminas (the trabecula sessile); *S. dichotomum* has glabrous anthers.

Stigmaphyllon retusum exhibits some regional variation, particularly in size of the flowers, length of the samara's dorsal wing, and length of the stalk of the T-shaped hairs of the herbage. Plants from Mexico, excluding Chiapas, and those from southern Guatemala, Nicaragua, and Costa Rica usually have lateral petals with the limbs 6.5–9.5 mm in diameter. The stalk of the T-shaped hairs is (0.1–) 0.2–0.3 (–0.4) mm long. The samaras of the Mexican plants usually have the dorsal wing 2.5–3.1 cm long, and the nut ornamented with 2–3 lateral winglets; infrequently, the nut bears only crests and/or spurs or is only prominently veined. In plants from southern Guatemala and Nicaragua, the dorsal wing is 3.4–4.5 cm long, and the nut usually has 3–5 lateral winglets per side; none of the Costa Rican collections seen have samaras. Specimens from Chiapas, Belize, and northern Guatemala generally appear more robust than plants from other parts of the range. The limbs of the lateral petals are 9.5–13.5 mm in diameter, and the T-shaped hairs have stalks up to 0.5 mm long. The samaras are similar to those from plants of southern Guatemala and Nicaragua in that the dorsal wing is 3.5–4.1 (–4.8) cm long, but differ in that the nut usually lacks lateral winglets and is only prominently ribbed or at most bears small crests/spurs. Exceptions to these patterns occur throughout the range, and taxonomic recognition of the variants is not warranted.

65. Stigmaphyllon dichotomum (L.) Grisebach, Linnaea 13: 207. 1839. *Banisteria dichotoma* L., Sp. pl. 1: 427. 1753. *Banisteria convolvulifolia* Cavanilles, Diss. 9: 428, t. 256. 1790, nom. superfl., non *Stigmaphyllon convolvulifolium* Adr. Jussieu, 1840.—TYPE: specimen in the Clifford herbarium (holotype: BM!, microfiche: MICH!, photos: MICH! P!).

Banisteria tiliaefolia H. B. K., Nov. gen. sp. 5: 162. 1822 ["1821"], non *Banisteria tiliaefolia* Ventenat, 1808. *Banisteria humboldtiana* DC., Prodr. 1: 588. 1824. *Stigmaphyllon humboldtianum* (DC.) Adr. Jussieu in St.-Hilaire, Fl. Bras. merid. 3: 56. 1833 ["1832"]. *Stigmaphyllon tiliifolium* Niedenzu, Ind. Lect. Lyc. Brunsberg. p. aest. 1900: 16. 1900, nom. superfl.—TYPE: COLOMBIA. Bolívar: ". . . inter Carthagenam et Cerro de la Popa," *Humboldt & Bonpland s.n.* (holotype: P-HBK!, photos: F! MICH! US!).

Banisteria variifolia DC., Prodr. 1: 588. 1824. *Stigmaphyllon tiliifolium* var. *berteroanum* Niedenzu, Ind. Lect. Lyc. Brunsberg. p. aest. 1900: 17. 1900.—TYPE: COLOMBIA. Magdalena: "Ad Sanctam-Martham," *Bertero s.n.* (holotype: G-DC!, photos: F! GH! MICH! NY! US!; isotype: P!, photo: MICH!).

Banisteria varia Sprengel, Syst. veg. 2: 386. 1825.—TYPE: COLOMBIA. "Ad fl. Magalen [Magdalena]," *Bertero s.n.* (holotype: B, destroyed).

Vine to 8 m. Stems and branches with scalelike T-shaped hairs when young, soon becoming glabrous. Laminas 6.2–23 cm long, 4.5–24 cm wide, usually cordate to ovate, sometimes elliptical to suborbicular, or sometimes palmately 3–5-lobed, apex mucronate, base cordate or sometimes truncate or briefly attenuate, glabrate to glabrous adaxially, densely pubescent with T-shaped hairs to tomentose abaxially (trabecula 0.6–1.3 mm long, wavy to crisped and curled, stalk 0.1–0.3 mm long), margin with irregularly spaced sessile glands (0.2–0.4 mm in diameter) and also with filiform glands up to 6 mm long; petioles 1.7–10 cm long, with scalelike T-shaped hairs or sparsely so in older leaves, with a pair of prominent but sessile glands at the apex, each gland 1–2.8 mm in diameter; stipules 0.5–1.3 mm long, 0.5–1.5 mm wide, free, triangular, glandular. Flowers 15–40 (–50) per umbel, these borne in dichasia, compound dichasia, or thyrses (axes to the 6th order, beset with scalelike T-shaped hairs). Peduncles 3.5–9 mm long; pedicels 4–11 mm long, terete; both densely sericeous (hairs subsessile), peduncles 0.7–2.1 times as long as the pedicels. Bracts 0.5–1.3 mm long, 0.5–1 (–1.2) mm wide, triangular, apex acute or acuminate; bracteoles 0.5–1.5 mm long, 0.5–1 mm wide, oblong to triangular, apex obtuse, eglandular or sometimes with one or two inconspicuous glands (each gland ca. 0.2 mm in diameter); bracts and bracteoles sericeous abaxially. Sepals 1.8–2.2 (–2.5) mm long,

(1.8–) 2–2.3 mm wide, glands 1.7–2.2 mm long, 0.9–1.2 mm wide. All petals with the limb glabrous, yellow, limb of lateral petals orbicular, margin denticulate or denticulate-fimbriate, fimbriae up to 0.3 (–0.4) mm long; anterior-lateral petals: claw (1.5–) 1.7–2.2 mm long, limb 8–8.6 mm long, 7.5–8 mm wide; posterior-lateral petals: claw 0.8–1 mm long, limb ca. 7.5 mm long, ca. 6.5–7 mm wide; posterior petal: claw 2.8–3.2 mm long, apex indented, limb 6–6.5 mm long, 4–4.5 mm wide, elliptical to oblong, margin with fimbriae up to 0.5 mm long. Stamens unequal, those opposite the posterior-lateral petals (and the posterior styles) the largest, anthers of those opposite the lateral sepals with the connective enlarged and the locules reduced or sometimes with only one reduced locule; anthers all loculate, glabrous. Stamen opposite anterior sepal: filament 1.5–2 mm long, anther 0.7–0.8 mm long; stamens opposite anterior-lateral petals: filaments 1.2–1.7 mm long, anthers 0.5–0.7 mm long; stamens opposite anterior-lateral sepals: filaments (2–) 2.2–2.6 mm long, connectives 0.5–0.7 mm long, locules 0.2–0.4 (–0.5) mm long; stamens opposite posterior-lateral petals: filaments 2.5–2.8 mm long, anthers 0.9–1 mm long; stamens opposite posterior-lateral sepals: filaments 1.8–2.4 mm long, connectives 0.6–0.7 (–0.9) mm long, locules (0.1–) 0.2–0.4 mm long; stamen opposite posterior petal shorter than or subequal to the adjacent two: filament 1.5–2.3 mm long, anther 0.5–0.6 mm long. Anterior style 2–2.7 mm long, shorter than the posterior two, terete, glabrous, erect; apex 1.2–1.5 mm long, each foliole 1.3–1.6 mm long, 1.4–2 mm wide, square to sometimes subrectangular. Posterior styles 2.7–3 mm long, terete, glabrous, lyrate; folioles 1.3–1.6 mm long, 1.4–2 mm wide, square to sometimes subrectangular. Dorsal wing of samara 3.7–4.5 cm long, 1.2–1.5 cm wide, upper margin with an obtuse or subacute tooth; nut with a pair of irregularly rectangular lateral winglets, these 4.7–5.5 mm long, 0.5–1.5 mm wide, or bearing spurs and/or crests or with only one or two lateral ridges; nut 4.5–6 mm high, 3.8–4.5 mm in diameter, without air chambers, areole 2.8–3.5 mm long, 2.2–3.3 mm wide, concave, carpophore up to 4.5 mm long. Embryo 6–7.5 mm long, ca. 3 times as long as wide, laterally flattened, outer cotyledon 5.7–6.4 mm long, 3–3.5 mm wide, inner cotyledon 4.9–6.3 mm long, 2.7–3.3 mm wide, both straight. Chromosome number unknown. Fig. 63i–o.

Phenology. Collected in flower from September through March, in fruit from November through April.

Distribution (Fig. 66). Southern Panama (Darién), northern Colombia, and northwestern Venezuela; in dry situations; sea level to 1275 m.

REPRESENTATIVE SPECIMENS. **Panama.** DARIÉN: trail between Pinogana and Yavisa, *Allen 267* (A, F, GH, MO); between Río Jesús and Sabado, *Hammel 1348* (MO); Río Jaqué valley, 07°27′N, 78°05′W, *Knapp & Mallet 3203* (MICH); Dtto. Chepijana, Tucute, *Terry & Terry 1376* (F, GH, MO); Marranganti and vicinity, *Williams 987* (NY). **Colombia.** ANTIOQUIA: Mpio. San Carlos, rd to La Calera, 2.2 km S of Puerto Nare–San Carlos rd, just E of Narices, 06°10′N, 74°49′W, *Brant et al. 1727* (K, MICH); Mpio. San Carlos, Cañón del Río Claro, sector norte, *Cogollo 932* (MICH); valley of the Río Anorí, vic. Planta Providencia, 26 km S and 23 km W (air) of Zaragoza, 07°13′N, 75°03′W, *Denslow 2537* (WIS), *2539* (MO, WIS).—ATLÁNTICO: Usiacurí, arroyo del Higuerón, *Dugand & García Barriga 2306* (COL, US); entre Baranoa y Polonuevo, *Dugand & Jaramillo 2819* (COL, US); alrededores de Tubará, *Dugand & Jaramillo 4065* (COL, US).—BOLÍVAR: Mpio. Santa Catalina, Loma Las Puas, vía Arroyo Grande a Las Canoas, *Cuadros V. 3287* (MO); San Martín de Loba, lands of Loba, *Curran 53* (GH, US); Mpio. Cartagena, Caserio Las Canoas, 10°08′N, 75°24′W, *Marulanda 801* (MO).—CESAR: entre Agua Chica y Ocaña (Norte de Santander), *Albert de Escobar et al. 3124* (MICH).—CHOCÓ: Mpio. Ríosucio, Zona Urabá, Cerros del Cuchillo, Sector "Morro Aparte," *Cárdenas 1347* (MO).—CUNDINAMARCA: Mpio. Caparrapí, estación de ferrocarril Dindal, *García Barriga 7657* (COL, US); Quebrada Carmargo, N of Apulo, *Killip et al. 33211* (COL, US).—MAGDALENA: S of Santa Marta, *Killip & Smith 21095* (A, GH, NY, US); Mpio. Ciénaga, carretera de Ciénaga a Fundación, *Romero Castañeda 8228* (COL, MO); entre Tucurinca

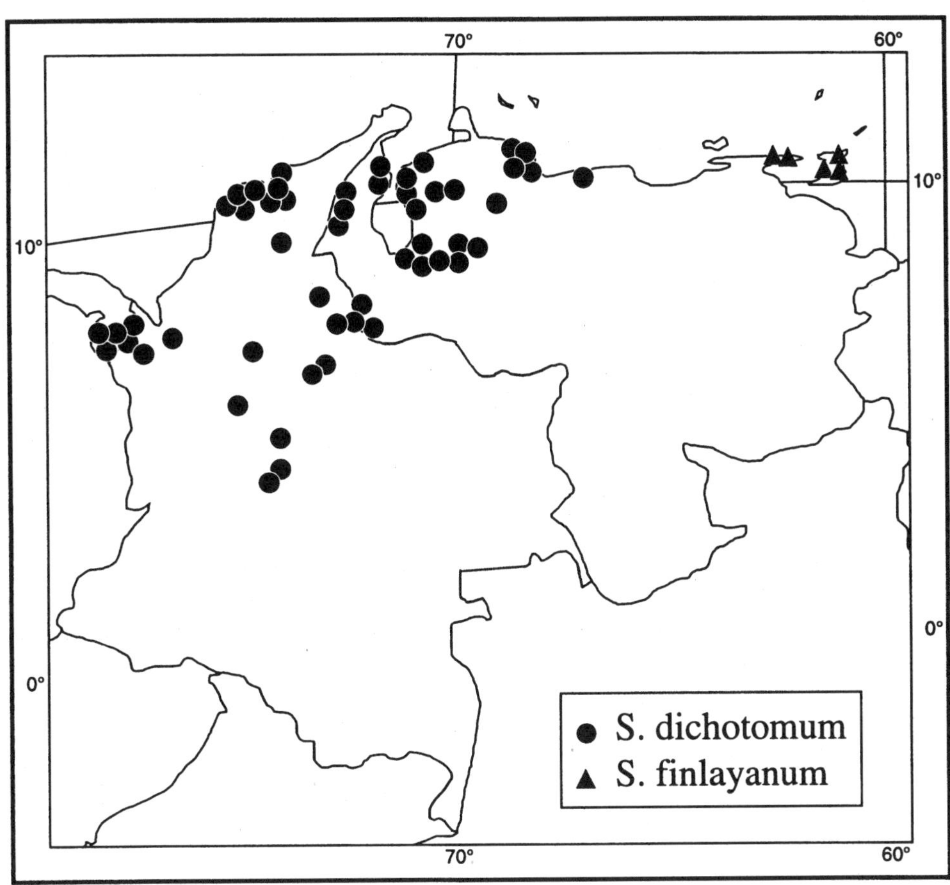

FIG. 66. Distribution of *Stigmaphyllon dichotomum* and *S. finlayanum*.

y Fundación, *Romero Castañeda 9188* (COL, NY); Santa Marta, *H. H. Smith 1525* (A, BM, BR, C, CM, COL, F, G, GH, LL, MA, MO, MT, NY, P, S, TEX, U, US, W, WIS).—NORTE DE SANTANDER: between Chinácota and La Esmeralda, *Killip & Smith 20918* (NY).—SANTANDER: orillas del Río Chicamocha, *Araque M. & Barkley 18S265* (MO, US); vicinity of Barrancabermeja, between Sogamoso and Carare rivers, *Haught 2089* (A, COL, S, US); between Nariño and El Tambor, *Killip & Smith 14956* (GH, NY, US); upper Río Lebrija Valley, NW of Bucaramanga, *Killip & Smith 16325* (GH, NY, US).—TOLIMA: Honda, *Pennell 3591* (GH, NY, US); Flandes, *Schneider 222* (COL); valle del Alto Magdalena, Guamo, Quebrada Serrezuela, *Uribe U. 4309* (COL, F, NY). **Venezuela.** APURE: Reserva Forestal San Camilo, along Río Uribante between Río Nulita and Jordan, *Steyermark 101761* (MY, VEN).—BARINAS: Dtto. Pedraza, trail from Pozo Negro (ca. 08°32′N, 70°37′W) to Mesa de Canagua (ca. 08°34′N, 70°40′W), *Dorr 7777* (MICH); Reserva Forestal de Caparo, Unidad Uno, *Jiménez Saa 1310* (NY, US); Dtto. Barinas, 10 km de Barinas hacia Corozo, (08°30′N, 70°40′W), *Rutkis 376* (MICH, VEN).—CARABOBO: between Morón and El Palito, *Alston 6090* (F, NY, P, S, U, US); Guaremales, cerca de Urama, *Pittier 13065* (G, M, MO, NY, US, VEN).—FALCÓN: Dtto. Silva, NE de La Soledad, entre La Soledad y Sanare, 10°52′N, 68°21′W, *Steyermark 100996* (MO, VEN); Dtto. Mauroa, Santo Domingo, en la vía desde hato Uverito hasta Cerro Socopo, *Flora Falcón 502* (MICH); Dtto. Silva, Reserva Forestal Río Toyuro, 9 km SSW de Riecito, *Wingfield & Smith 7961* (MICH).—LARA: cerca de Santa Rosa, *Pittier 13090* (NY, US, VEN); entre Carora y Trentino, *Saer 718* (F, VEN); El Altas, La Miel, *R. F. Smith V1270* (VEN).—MÉRIDA: Dtto. Sucre, La Gloria, carretera Canchinchi–El Rincón, *Marcano B. & Bautista 1347* (MICH); Dtto. Campo Elías, camino Estánquez–Páramo de Las Coloradas, *Quintero & Ricardi 1659* (MER).—PORTUGUESA: Dtto. Turén,

Mpio. Sta. Rosalía, La Caripucha, *Aristeguieta 1524* (VEN); Dtto. Guanare, carretera Guanare–Buscucuy, 15 km NW de Guanare, *Aymard & Cuello 3355* (MICH); en las orillas del Río Guanare, *Pittier 12054* (G, LE, M, NY, US, VEN).—TÁCHIRA: 4 km S of San Cristóbal along San Cristóbal–Barinas rd, *de Bruijn 1356* (MICH, S, US, WAG); at El Vado, along Río Lobatera, in Parcelamiento to Guarumito, 5.5 km W of La Fria (by air, 18 km by road), 08°12′N, 72°18′W, *Steyermark 120332* (MO).—TRUJILLO: subida del puente de Motatan a Carvajal, cerca de Valera, *Pittier 10759* (G, GH, NY, US, VEN).—YARACUY: Dtto. Urachiche, 20 km al N de Urachiche, 10°08′N, 69°10′W, *Aymard et al. 1616* (NY); Hacienda Iboa cerca de San Pablo, *Pittier 12606* (F, G, M, MO, NY, US, VEN); Finca Los Apamates, asentiamiento La Llanada, entre Urama y San Felipe, *T. Romero 457* (MY).—ZULIA: Dtto. Miranda, vía El Consejo–Quirós–El Pensado, ca. 5 km W de El Pensado, *Bunting 8657* (MICH); Dtto. Bolívar, entre Las Tres Marías (área 8 km E de El Pensado, 10°25′N, 70°55′W) y el Río Chiquito, *Bunting 8993* (MICH); Dtto. Lagunillas, cuenca del Embalse Burro Negro (Pueblo Viejo), a lo largo del Río Grande, ca. 13 km N del Embalse, 10°25′N, 70°49′W, *Bunting 11254* (MICH, NY); Dtto. Mana, cuenca del Río Guasare, alrededores del Destacamento Guasare No. 1 (La Yolanda), 10°52′N, 71°29′W, *Bunting 12883* (MICH); between Represa Socoi and Campo Carichuano, NW of Maracaibo, *Fryxell et al. 4400* (MICH); Perijá, *Ginés 2096* (US).

Stigmaphyllon dichotomum, a common species of southern Panama, northern Colombia, and northwestern Venezuela, is superficially similar to and has been confused with *S. lindenianum* (no. 63; Mexico, Central America, and Chocó, Colombia) and *S. retusum* (no. 64; Mexico and Central America), but is distinguished from these species by its leaf pubescence, flowers, and embryo; see *S. lindenianum* for a detailed discussion. In *S. dichotomum*, the laminas are pubescent with T-shaped hairs to tomentose abaxially (sericeous in *S. lindenianum*). The margin of the lateral petals varies from denticulate or denticulate-fimbriate, but the posterior petal is fringed with fimbriae up to 0.5 mm long. The anthers are glabrous (pubescent in *S. lindenianum* and *S. retusum*). The embryos differ in that they are laterally flattened (ca. 3 times as long as wide) and not ovoid, as in all other species except *S. ciliatum*.

Stigmaphyllon dichotomum may possibly also be confused with the sympatric *S. singulare* (no. 26) and *S. columbicum* (no. 24), whose leaves also bear T-shaped hairs abaxially. *Stigmaphyllon singulare* is readily recognized by its deciduous sepals and pubescent petals, and *S. columbicum* by its styles, the anterior style clawed but the posterior styles foliolate.

Because the identity of *Banisteria dichotoma* had been in doubt, the later names *S. tiliifolium* and *S. humboldtianum* traditionally have been applied to this species. Examination of the type of *B. dichotoma* proved it a species of *Stigmaphyllon*, and *S. dichotomum* is the correct name for it (C. Anderson 1993a).

66. Stigmaphyllon panamense C. Anderson, Contr. Univ. Michigan Herb. 16: 35. 1987.—TYPE: PANAMA. San José Island, Las Perlas Archipelago, ca. 55 mi SSE of Balboa, rd to Third Beach, *Johnston 1301* (holotype: GH!, photo: MICH!; isotype: MO!).

Vine. Stems and branches sericeous when young, soon becoming glabrous. Laminas 8.2–13 cm long, 5.5–10.5 cm wide, ovate or elliptical, apex acuminate or acuminate-mucronate, base slightly cordate to truncate, glabrous adaxially, with a mixture of sessile to subsessile to stalked T-shaped hairs abaxially (trabecula 0.4–1.8 mm long, nearly straight to wavy to crisped, stalk 0.03–0.3 mm long), the hairs sloughed off in patches and older laminas often glabrate or glabrous abaxially, margin eglandular or with scattered sessile glands; petioles 2.4–5 cm long, sericeous or sparsely so, not confluent across the node, with a pair of prominent but sessile glands at the apex, each gland 1–1.8 mm in diameter;

stipules 0.8–1.3 mm long and wide, free, broadly triangular to ovate, eglandular. Flowers 13–25 per umbel or pseudoraceme, these borne in dichasia or small thyrses (axes to the 3rd order, sericeous). Peduncles 2.5–7 mm long; pedicels 5.5–11 mm long, terete; both sericeous, peduncles (0.3–) 0.5 (–0.8) times as long as pedicels. Bracts 1–1.7 mm long, 0.7–1.2 mm wide, triangular, apex acute or obtuse; bracteoles 0.7–1.1 mm long, 0.7–1 mm wide, ovate or broadly so, apex obtuse, eglandular or sometimes with one or two inconspicuous glands (each gland ca. 0.2 mm in diameter); bracts and bracteoles sericeous abaxially. Sepals 1.8–3 mm long, 2.1–2.5 mm wide, glands 1.8–2.7 mm long, 1–1.5 mm wide. All petals with the limb glabrous, yellow, limb of lateral petals orbicular, margin erose or erose-denticulate; anterior-lateral petals: claw 1.8–2 mm long, limb 12–15 mm long and wide; posterior-lateral petals: claw (0.7–) 1–1.5 mm long, limb 10–11.5 mm long and wide; posterior petal: claw 3–3.5 mm long, apex indented, limb (8.5–) 9–11 mm long, 8–10 mm wide, elliptical to broadly obovate, margin fimbriate or denticulate-fimbriate, fimbriae (0.1–) 0.2–0.4 (–0.7) mm long. Stamens unequal, those opposite the posterior-lateral petals (and the posterior styles) the largest, anthers of those opposite the lateral sepals with the connective enlarged and the locules reduced, sometimes anthers of those opposite the posterior-lateral sepals with only one locule or rarely eloculate; anthers glabrous. Stamen opposite anterior sepal: filament 2.1–2.3 mm long, anther 1–1.3 mm long; stamens opposite anterior-lateral petals: filaments 1.4–1.9 (–2.3) mm long, anthers 0.7–0.8 (–1) mm long; stamens opposite anterior-lateral sepals: filaments 2.7–3.3 mm long, connectives 0.9–1.1 mm long, locules 0.2–0.5 (–0.8) mm long; stamens opposite posterior-lateral petals: filaments 3.1–3.6 (–4) mm long, anthers 1.1–1.2 mm long; stamens opposite posterior-lateral sepals: filaments 2.5–3 (–3.4) mm long, connectives 0.8–1 mm long, locules 0.2–0.6 (–0.8) mm long or rarely absent; stamen opposite posterior petal always shorter than the adjacent two: filament 2.1–2.8 mm long, anther ca. 0.8 mm long. Anterior style (2.7–) 3–3.6 mm long, shorter than the posterior two, terete, glabrous, erect or slightly recurved; apex 1.7–2.1 mm long, each foliole 1.4–2.1 mm long and wide, subsquare. Posterior styles (3.5–) 3.9–4.5 mm long, terete, glabrous, lyrate; folioles 2–2.8 mm long and wide, square to subrectangular. Dorsal wing of samara 3.5–3.7 cm long, 1.2–1.4 cm wide, upper margin with an obtuse tooth; nut with 1–2 rectangular to parabolic lateral winglets per side, these 1.7–2.8 mm long, 1–1.8 mm wide; nut 5–7.5 mm high, 3.5–5.5 mm in diameter, without air chambers, areole 4.5–5 mm long, 3.7–4 mm wide, deeply concave, carpophore up to 2.5 mm long. Embryo ca. 8 mm long, ca. 2 times as long as wide, ovoid, outer cotyledon ca. 9.5 mm long, 3.8 mm wide, the distal 1/4 folded over the inner cotyledon, inner cotyledon ca. 5 mm long, ca. 2.5 mm wide, folded at the distal 1/5. Chromosome number unknown. Fig. 60h–l.

Phenology. Collected in flower from December through April, in fruit in January and March.

Distribution (Fig. 65). Central Panama and islands in the Gulf of Panama; in dry forest, thickets, and at forest edge; sea level to 100 m.

ADDITIONAL SPECIMENS EXAMINED. **Panama.** CANAL ZONE: along C25E from C25 (Camino Madden) SE for 3 km toward Cerro León, *Daniel 5501* (MICH); Farfan Beach, *Dwyer 4002* (MO); near Madden Dam, *Lewis et al. 5299* (MO); hospital grounds at Ancón, *Mason 7* (US); Balboa, *Standley 25614* (US); Farfan beach, *Tyson & Blum 2614* (MO).—COLÓN: N side of Madden Dam, *Knapp 2729* (MICH).—DARIÉN: Isla Saboga, *Duke 10341, 10365* (MO), *Miller 1945, 1947* (US); Isla Casaya, *Duke 10382* (MO).—PANAMÁ: Isla Chitré, *Knapp 3221* (MICH); Isla Chapera, *Knapp 3305* (MO); Sabanas, *Bro. Paul 269, 270* (US); Isla Taboga, *Standley 27075* (US); between Las Sabanas and Matías Hernández, *Standley 31923* (US); between Matías Hernández and Juan Díaz, *Standley 32067* (US).—WITHOUT LOCALITY: *Duchassaing s.n.* (GOET).

Stigmaphyllon panamense has conspicuously large flowers borne on pedicels longer than the peduncles. The anthers are glabrous; those of stamens opposite the posterior-lateral sepals sometimes have only 0–1 locules. The pubescence of the abaxial surface of the leaves is composed of a mixture of appressed hairs and T-shaped hairs. This pubescence is sloughed off in patches, and old leaves may be glabrate abaxially. *Stigmaphyllon panamense* was reported by Johnston (1949) and Cuatrecasas and Croat (1981) as *S. lindenianum* (no. 63), a species readily separated by its sericeous laminas and pubescent anthers. A collection from Panama (Colón: Buena Vista, *Atencio 7*, MO) may represent a hybrid between *S. lindenianum* and *S. panamense*. The pollen consists of 100% thick-walled and often misshapen grains; the hairs of the lower leaf surface are not sessile or subsessile, as in *S. lindenianum,* but T-shaped.

67. Stigmaphyllon tonduzii C. Anderson, Contr. Univ. Michigan Herb. 16: 45. 1987.—TYPE: COSTA RICA. Guanacaste: Playa Tamarindo, 19 Feb 1985, *Frankie s.n.* (holotype: MICH!).

Vine to 3 m. Stems and branches sericeous when young, soon becoming sparsely so. Laminas 9.5–12 cm long, 9–12 cm wide, elliptical or sometimes palmately 3–5-lobed, apex acuminate or acuminate-mucronate, base slightly cordate to truncate, sparsely sericeous to glabrate adaxially, with T-shaped hairs to tomentose abaxially (trabecula 0.6–1.3 [–1.8] mm long, crisped and curled, stalk 0.1–0.3 mm long), margin eglandular or with irregularly spaced sessile glands; petioles 2–5 cm long, sericeous, not confluent across the node, with a pair of prominent but sessile glands at the apex, each gland 1–1.7 mm in diameter; stipules 0.5–1.1 mm long, 0.6–1.5 mm wide, free, triangular, eglandular. Flowers ca. 12–20 per pseudoraceme or umbel, these borne in dichasia or compound dichasia or small thyrses (axes to the 3rd order, sericeous). Peduncles 3.5–8.5 mm long; pedicels 4–7 mm long, terete; both densely sericeous, peduncles 0.6–1.75 times as long as pedicels. Bracts 0.8–1.5 mm long, 0.6–1 mm wide, triangular, apex acute; bracteoles 0.8–1.2 mm long, 0.6–0.8 mm wide, oblong or ovate, apex obtuse, eglandular or each bracteole with a pair of inconspicuous glands (each gland ca. 0.3 mm in diameter); bracts and bracteoles sericeous abaxially. Sepals 2.2–3 mm long, ca. 2–2.8 mm wide, glands 1.7–2 mm long, 0.9–1.3 mm wide. All petals with the limb glabrous, yellow, limb of lateral petals orbicular to suborbicular, margin erose; anterior-lateral petals: claw (1.6–) 1.8–2 mm long, limb 10–11 mm long and wide; posterior-lateral petals: claw 1–1.5 mm long, limb 7.5–9 mm long and wide; posterior petal: claw 2.5–3 mm long, apex indented, limb 6.5–8 mm long, 6–7 mm wide, ovate to broadly elliptical to orbicular, margin erose-denticulate or erose-fimbriate, fimbriae up to 0.3 (–0.5) mm long. Stamens unequal, those opposite the posterior-lateral petals (and the posterior styles) the largest, sometimes those opposite the anterior-lateral sepals equally long, anthers of those opposite the lateral sepals with the connective enlarged and the locules reduced, commonly anthers of those opposite the posterior-lateral sepals eloculate or with one reduced locule; anthers glabrous. Stamen opposite anterior sepal: filament (1.9–) 2.1–2.6 mm long, anther 0.9–1.1 mm long; stamens opposite anterior-lateral petals: filaments 1.5–2 mm long, anthers 0.7–0.9 mm long; stamens opposite anterior-lateral sepals: filaments 2.1–2.9 mm long, connectives 0.8–1 mm long, locules 0.3–0.6 (–0.7) mm long; stamens opposite posterior-lateral petals: filaments 2.8–3.3 mm long, anthers 1–1.2 mm long; stamens opposite posterior-lateral sepals: filaments 2.3–2.6 mm long, connectives 0.7–0.8 mm long, locule 0.3–0.6 mm long or absent; stamen opposite posterior petal always shorter than the adjacent two: filament 2–2.2 mm long, anther 0.6–0.7 mm long.

Anterior style 2.6–3.2 mm long, shorter than the posterior two, terete, glabrous, erect or slightly recurved; apex 1.3–1.6 mm long, each foliole 1.2–1.5 mm long, 1.4–1.6 mm wide, subsquare. Posterior styles 3.3–3.6 mm long, terete, glabrous, lyrate; folioles ca. 1.7 (–2.4) mm long, 1.8–2.3 mm wide, square to subrectangular. Dorsal wing of samara 2.3–3 cm long, 0.7–1 cm wide, upper margin with an obtuse tooth; nut smooth or bearing spurs and/or crests; nut 4.4–5.5 mm high, 3.5–3.9 mm in diameter, without air chambers, areole 3.2–3.6 mm long, 3–3.5 mm wide, convex, carpophore up to 2.5 mm long. Embryo ovoid, ca. 2 times as long as wide, outer cotyledon ca. 7.8 mm long, ca. 3.4 mm wide, the distal 1/3 folded over the inner cotyledon, inner cotyledon ca. 5 mm long, ca. 2.5 mm wide, folded at the distal 1/4. Chromosome number unknown. Fig. 60m–o.

Phenology. Collected in flower and in fruit in January, February, and April.

Distribution (Fig. 65). Costa Rica (Nicoya peninsula of Guanacaste, and northern Puntarenas); dry open woods, scrub, and thickets; sea level to ca. 100 m.

ADDITIONAL SPECIMENS EXAMINED **Costa Rica.** GUANACASTE: Playa Tamarindo, 18 Feb 1985, *Frankie s.n.* (MICH); Playa Tamarindo, 10°18′N, 85°51′W, *Haber 8973* (MO); Península de Nicoya, Playa Organos, 09°49′N, 84°54′W, *Hammel 18822* (MO); Nicoya, *Tonduz 13479, 13824* (US); San Isidro, Nicoya, *Tonduz 14008* (US).—PUNTARENAS: Hwy 1 W of San José, Km 135, *Meerow et al. 1003* (SEL).

Stigmaphyllon tonduzii is restricted to northern Costa Rica. Its leaves bear T-shaped hairs abaxially or are tomentose, and its anthers are glabrous. It is similar to *S. retusum* (no. 64), a more northern species rarely found in Costa Rica, which is readily separated by its pubescent anthers. *Stigmaphyllon tonduzii* might also be confused with *S. lindenianum* (no. 63), mostly restricted to the Atlantic lowlands but in Costa Rica also occurring on the Osa Peninsula. *Stigmaphyllon lindenianum* also has pubescent anthers; its leaves are appressed-sericeous abaxially.

68. Stigmaphyllon finlayanum Adr. Jussieu, Arch. Mus. Hist. Nat. Paris 3: 375. 1843. *Stigmaphyllon tiliifolium* var. *finlayanum* (Adr. Jussieu) Niedenzu, Ind. Lect. Lyc. Brunsberg. p. aest. 1900: 17. 1900.—TYPE: TRINIDAD [erroneously labeled "St. Thomas," see Urban, Symb ant. 7: 72. 1911], *Finlay 1* (holotype: P-JU!, photo: MICH!; isotypes: P-2 sheets!).

Vine to ca. 5 m. Stems and branches pubescent with scalelike T-shaped hairs when young, becoming glabrous. Laminas 7–14 cm long, 6–11 cm wide, ovate to cordate, apex mucronate, base cordate to deeply so, glabrous adaxially, with subsessile to T-shaped hairs (trabecula 0.5–1 mm long, straight, stalk 0.05–0.1 mm long) abaxially in younger laminas but at maturity glabrate to glabrous, the hairs most persistent on the veins, margin with irregularly spaced sessile glands (0.2–0.4 mm in diameter) and/or filiform glands up to 6.5 mm long (broken off in older leaves); petioles 1.7–8.5 cm long, sericeous to glabrate, not confluent across the node, with a pair of prominent but sessile glands at the apex, each gland 1.4–2.1 mm in diameter; stipules 0.4–0.9 mm long, 0.4–1 mm wide, free, triangular, eglandular. Flowers ca. 15–30 (–35) per pseudoraceme, these borne in thyrses (axes to the 5th order, sericeous). Peduncles 2.5–7 mm long; pedicels 6–9 mm long, terete; both pubescent with scalelike T-shaped hairs, peduncles 0.4–0.9 times as long as the pedicels. Bracts 0.7–1.5 mm long, 0.6–0.9 mm wide, triangular, apex acute; bracteoles 0.6–1 mm long, 0.5–0.7 mm wide, triangular, apex obtuse or acute, eglandular; bracts and bracteoles sericeous abaxially. Sepals 1.7–2 mm long, ca. 2 mm wide, glands 1.7–2.2 mm long, ca.

1 mm wide. All petals with the limb orbicular, glabrous, yellow, margin of the lateral petals erose, of the posterior petal denticulate-fimbriate, teeth/fimbriae up to 0.2 mm long; anterior-lateral petals: claw ca. 2 mm long, limb ca. 9 mm long and wide, base truncate; posterior-lateral petals: claw ca. 1 mm long, limb ca. 8 mm long and wide, base attenuate; posterior petal: claw ca. 3 mm long, sometimes sparsely pubescent at the strongly indented apex, limb ca. 7 mm long and wide. Stamens unequal, those opposite the posterior-lateral petals (and the posterior styles) with the longest filaments, anthers of those opposite the lateral sepals with the connective enlarged and the locules reduced, anthers of those opposite the posterior-lateral sepals with only one locule; anthers glabrous. Stamen opposite anterior sepal: filament 1.6–2 mm long, anther 0.9–1 mm long; stamens opposite anterior-lateral petals: filaments 1.5–1.6 mm long, anthers ca. 0.6 mm long; stamens opposite anterior-lateral sepals: filaments 2.3–2.4 (–2.8) mm long, connectives 0.7–0.8 mm long, locules 0.3–0.4 mm long; stamens opposite posterior-lateral petals: filaments 2.7–3 mm long, anthers 0.9–1.1 mm long; stamens opposite posterior-lateral sepals: filaments 2.7–2.8 mm long, connectives 0.7–0.8 mm long, each bearing only one locule 0.4 mm long; stamen opposite posterior petal always shorter than the adjacent two: filament 2.3–2.4 mm long, anther ca. 0.6 mm long. Anterior style ca. 2.5 mm long, shorter than the posterior two, terete, glabrous or with scattered hairs in the proximal 1/2, erect; apex 0.9–1 mm long, each foliole ca. 1.2 mm long, 0.8–0.9 mm wide, subrectangular. Posterior styles 3.1–3.2 mm long, terete, with scattered hairs in the proximal 1/2, lyrate; foliole 1.2–1.4 mm long and wide, subsquare. Dorsal wing of samara 4–4.5 cm long, 1.3–1.5 cm wide, upper margin with a blunt tooth; nut bearing 0–2 linear to rectangular lateral winglets per side, these up to 2–2.5 (–5) mm long, 2–3 mm wide, or only with spurs; nut ca. 7 mm high, 4.2–5.5 mm in diameter, without air chambers, areole 3–3.2 mm long, 3.2–3.5 mm wide, concave, carpophore up to 4 mm long. Embryo 6.5–6.8 mm long, ca. 2 times as long as wide, ovoid, outer cotyledon 6.3–6.8 mm long, ca. 4 mm wide, bent at the distal 2/3 but not folded over the inner cotyledon, inner cotyledon 4.5–5.2 mm long, 3–3.2 mm wide, bent or folded at the distal 2/3 or straight. Chromosome number unknown.

Phenology. Collected in flower from December to July, in fruit in January, February, and May.

Distribution (Fig. 66). Endemic to Trinidad and the adjacent Paria Peninsula of Venezuela, erroneously reported from St. Thomas; sea level to 900 m.

ADDITIONAL SPECIMENS EXAMINED. **St. Vincent.** *H. H. Smith & G. W. Smith 418 p.p.* (K). **Trinidad.** Los Irois Bay, *Adams 14311* (NY); Chacachacare, *R. E. D. Baker TRIN14694* (TRIN); Careange, *Britton & Broadway 2624* (NY, US); San Fernando, *Broadway 2166* (F, M); St. Anne, *Chase 90163* (MICH); *Fendler 251* (BM, K, NY, P); Morne Coco Rd, Diego Martin, *Kalloo 332* (TRIN); Fort George, *Kuntze 906* (NY); Manzanilla–Mayaro Rd, *Ramcharan 255* (TRIN). **Venezuela.** SUCRE: Península de Paria, Dtto. Valdez, Mpio. Cristóbal Colón [Macuro], *Broadway 113* (GH, NY, US), *A. Fernández 3748* (F), *3846* (F), *3849* (F); bajando hacia al Rincón, *Benítez de Rojas 523* (MY); caserio "Yoco," alrededores de Güiria, *Trujillo 9230* (MY).

Stigmaphyllon finlayanum is the only species in Trinidad and the Paria Peninsula with ovate to cordate leaves that bear subsessile to short-stalked T-shaped hairs abaxially but are usually glabrate to glabrous at maturity. The relatively small lateral petals are erose, and the anthers of the stamens opposite the posterior-lateral sepals bear only one locule. The collection from St. Vincent, *H. H. Smith & G. W. Smith 418* (K), is a mixture of the South American *S. sinuatum* and one leafy branchlet of *S. finlayanum*. Neither has been recollected on St. Vincent or elsewhere in the Antilles, and it seems likely that the specimens represent plants cultivated at the Botanical Garden in St. Vincent or perhaps escaped from it.

The names *S. convolvulifolium* and *S. adenodon* have been applied to this species, partly because the type collection was erroneously labeled to be from St. Thomas, and the name *S. finlayanum* was thought to apply to a species of the Lesser Antilles. The leaves of *S. convolvulifolium* (no. 34) are very sparsely but evenly sericeous abaxially, but the hairs are so tiny that the laminas appear glabrous to the naked eye; the anthers of stamens opposite the posterior-lateral sepals are unmodified. In *S. adenodon* var. *adenodon* (no. 28a), the laminas are always pubescent with T-shaped hairs abaxially, even at maturity, and bear stipitate nail-like glands along the margin; all anthers have two locules. This variety is also readily distinguished by its unusual samara; the nut is greatly enlarged (12–19 mm in diameter) owing to the air chambers surrounding the locule and bears a much reduced erect dorsal wing. The samara of *S. finlayanum* is like that of most species, i.e., the small nut (4.2–5.5 mm in diameter) bears a distally flared dorsal wing.

69. Stigmaphyllon cardiophyllum Adr. Jussieu, Ann. Sci. Nat. Bot., sér. 2, 13: 289. 1840.—TYPE: [BRAZIL.] "Brasilia borealis," collector unknown (holotype: P!, photos: F! MICH! US!, fragment: P-JU!).

Vine to 15 m. Stems and branches with scalelike T-shaped hairs when young, soon becoming glabrate to glabrous. Laminas 7.3–17 cm long, 3.7–13.5 cm wide, ovate to elliptical (the smaller often narrowly elliptical) to suborbicular, sometimes shallowly to deeply 2–3-lobed, apex acuminate to acuminate-caudate, base truncate to cordate or sometimes attenuate, especially in smaller laminas, very sparsely sericeous to glabrous adaxially and abaxially (trabecula 0.2–0.9 mm long, straight, sessile to subsessile), margin with irregularly spaced sessile glands (0.2–0.6 mm in diameter) and the bases of broken-off filiform glands; petioles 2.7–8 cm long, with scalelike T-shaped hairs, not confluent across the node, with a pair of prominent but sessile glands at the apex, each gland 1.1–3 mm in diameter; stipules 0.5–1 mm long, 0.5–1.4 mm wide, free, triangular, eglandular. Flowers ca. 15–25 (–30) per umbel, these borne in dichasia or compound dichasia or small thyrses (axes to the 5th order, with scalelike T-shaped hairs). Peduncles (1.5–) 2.2–5 mm long; pedicels 4–10 mm long, terete; both sericeous (trabecula subsessile), peduncles 0.3–1 times as long as the pedicels. Bracts 0.7–2 mm long, 0.5–0.8 mm wide, triangular or narrowly so, apex acute; bracteoles 0.4–1 mm long, 0.4–0.8 mm wide, triangular, apex obtuse, usually eglandular or sometimes each bracteole with a pair of inconspicuous glands (each gland ca. 0.1 mm in diameter); bracts and bracteoles sericeous abaxially. Sepals 1.2–2 mm long and wide, glands 1–1.8 mm long, 0.7–1 mm wide. All petals with the limb glabrous, yellow, limb of lateral petals orbicular, margin erose; anterior-lateral petals: claw 1–1.8 (–2.1) mm long, limb (5–) 6–6.5 mm long and wide; posterior-lateral petals: claw 0.5–1 (–1.3) mm long, limb 4–5 mm long and wide; posterior petal: claw 2–2.5 mm long, apex indented, limb (3.2–) 3.5–4 mm long, (2.3–) 3–4 mm wide, broadly obovate to broadly elliptical to suborbicular, margin irregularly denticulate to denticulate-erose to sometimes erose. Stamens unequal, those opposite the posterior-lateral petals (and the posterior styles) the largest but their filaments subequal to those of stamens opposite the anterior-lateral sepals, anthers of those opposite the lateral sepals with the connective enlarged and the locules reduced, those opposite the posterior-lateral sepals usually with only one tiny locule (very rarely with two); anthers all loculate, pubescent. Stamen opposite anterior sepal: filament 1.5–2.1 mm long, anther 0.7–1 mm long; stamens opposite anterior-lateral petals: filaments 1.2–1.7 mm long, anthers 0.5–0.6 (–0.8) mm long; stamens opposite anterior-lateral sepals: filaments 2.1–2.3 (–3) mm long, connectives 0.6–0.8 mm long, locules 0.3–0.5 mm long; stamens opposite posterior-lateral petals: filaments 2.2–2.5 (–3) mm long, anthers 0.8–1.1 mm long; stamens opposite posterior-lateral sepals: filaments 1.7–2.2 mm long, connectives 0.5–0.8 mm long, locule 0.1–0.4 mm long; stamen opposite posterior petal always shorter than the adjacent two: filament 1.5–2 mm long, anther 0.4–0.6 mm long. Anterior style 2–2.3 mm long, shorter than the posterior two, terete, glabrous, erect; apex 0.9–1.1 mm long; each foliole 0.5–0.8 mm long, 0.4–0.8 mm wide, triangular to square. Posterior styles 2.2–2.6 mm long, terete, glabrous, lyrate; foliole 0.9–1.2 mm long, 0.8–1.4 mm wide, subsquare to subtrapezoidal. Dorsal wing of samara 2.7–4 cm long, 1.1–1.8 cm wide, upper margin with a blunt tooth; lateral winglets absent, nut only prominently ribbed; nut 4.1–5.5 mm high, 2.5–3.5 mm in diameter, without air chambers, areole 2.5–3.3 mm long, 2.1–2.5 mm wide, convex, carpophore up to 3 mm long. Embryo 4.5–6.4 mm long, ca. 2 times as long as wide, ovoid, outer cotyledon 4.2–5.9 mm long, 2.5–3.1 mm wide, straight, inner cotyledon 3.9–5.5 mm long, 1.8–2.8 mm wide, straight. Chromosome number unknown. Fig. 67.

Phenology. Collected in flower from May through February, in fruit from June through April.

Distribution (Fig. 68). Amazonian lowlands of Ecuador, Peru, Brazil, and Bolivia; along riverbanks in lowland and flood plain forest, at forest edge, and in secondary growth and capoeiras; sea level to 1600 m.

REPRESENTATIVE SPECIMENS. **Ecuador.** NAPO: Yasuni Forest Reserve, along Tiputini River, 1–3 km E of Pontificia Universidad Católica del Ecuador Science Station, 00°40.853′S, 76°23.697′W, *Acevedo-Rodríguez & Cedeño 7336* (MICH); Estación Biológica Jatún Sacha, 8 km al este de Misahuallí, 01°04′S, 77°36′W, *Cerón M. & Iguago 5592* (MICH); Río Napo, Puerto Napo, *Harling 3518* (S); Misahuallí, *Steiner 275* (MICH).—PASTAZA: Puerto Sarayacu, *Lugo S. 3899* (GB, MICH). **Peru.** AMAZONAS: Río Cenepa, vicinity of Humapimi, ca. 5 km E of Chávez Valdivia, ca. 04°30′S, 78°30″W, *Ancuash 1134* (F, MICH, MO); Valle de Santiago, Quebrada Caterpiza, 03°50′S, 77°40′W, *Tunqui 708* (MICH, MO); Aramango, *Woytkowski 5629* (G, GH, MO, US).—CUZCO: Prov. Paucartambo, Atalaya, near junction of Río Carbón and Río Alto Madre de Dios, *Foster 3041* (MICH).—HUÁNUCO: Tingo María, *Asplund 12103* (S); Prov. Huánuco, Maranjillo, cerca a Tingo María, *Ferreyra 2196* (MICH); 69 km NE of Tingo María on rd to Tocache, Huallaga Valley, ca. 09°S, 76°W, *Gentry 37633* (MICH).—JUNÍN: Sani Beni, *Woytkowski 5954* (GH, MO, US); Mazamari, *Woytkowski 5979* (MO, US); San Ramón, *Woytkowski 7412* (GH, MO).—LORETO: Alto Amazonas, Dtto. Pastaza, Río Pastaza, *Ayala 2295* (AMAZ); Coronel Portillo, Tournavista, márgen izquierda del Río Pachitea, *Encarnación 26052* (MO, NY, US); Prov. Coronel Portillo, Bosque Nacional de von Humboldt, Km 86, Pucallpa–Tingo María rd, 08°40′S, 75°00′W, *Gentry 29481* (AMAZ, MICH); Mishuyacu, near Iquitos, *Klug 160* (F, NY, US); Stromgebiet des Marañón von Iquitos aufwärts bis zur Santiago-Mündung am Pongo de Manseriche, 77°30′W, *Tessman 3966* (G, S); Prov. Maynas, Iquitos, Quistacocha, 03°48′S, 73°25′W, *Vásquez & Jaramillo 12072* (MICH).—MADRE DE DIOS: Prov. Tambopata, Lago Tres Chimbadas, ca. 65 river km SSW of Puerto Maldonado, ca. 10–15 air km NW effluence Río La Torre (Río D'Orbigny)/Río Tambopata, 12°49′S, 69°17′W, *Barbour 5744* (MICH); Prov. Manu, Parque Nacional del Manu, Río Manu, Cocha Cashu Station, 11°50′S, 71°25′W, *Foster 9704* (F, MICH); small tributary of Río Madre de Dios, 1 km below Puerto Maldonado, *Gentry 19654* (F, MICH).—SAN MARTÍN: Prov. Mariscal Cáceres, entre Pólvora y Chiote, valle Huallaga, *Ferreyra 4464* (MICH, US); Prov. San Martín, cerca de Shapaja, *Ferreyra 18273* (US); Tarapoto, *Ule 6438* (G, K, MG, NY); Pona to Saposoa, *Woytkowski 5442* (F, MO, US). **Brazil.** ACRE: 9 km from Rio Branco on Rio Branco–Porto Acre rd, *Lowrie 646* (INPA, MICH); Mpio. Serra Madureira, Rio Caeté, afluente do Rio Iaco, *Ramos et al. 643* (INPA); Rio Juruá–Juruá Mirim, *Ule 5593* (G, MG).—AMAZÔNAS: Bôca do Acre, Purús, *Goeldi 3969* (MG); Tonantins (Solimões), *Jobert 764* (P); Mpio. São Paulo de Olivença, near Palmares, *Krukoff 8291* (A, BR, F, G, LE, MICH, MO, NY, P, S, U, US).—MARANHÃO: between Viana and Banderante, ca. 03°00′S, 45°10′W, *Daly et al. 648* (NY); Alzilândia, Rio Pindaré, 03°45′S, 46°05′W, *Jangoux & Bahia 321* (MICH).—PARÁ: Alenquer, Colônia Lauro Sodré, Km 15, *Fróes 29378* (IAN); Ilha de Mosqueiro, near Pará, *Killip & Smith 30394* (NY, US); Rd BR-22, Capanema to Maranhão, Km 96, *Prance & Pennington 1824* (IAN, MICH, NY). **Bolivia.** BENI: Bopi River valley, *Rusby 385* (K,

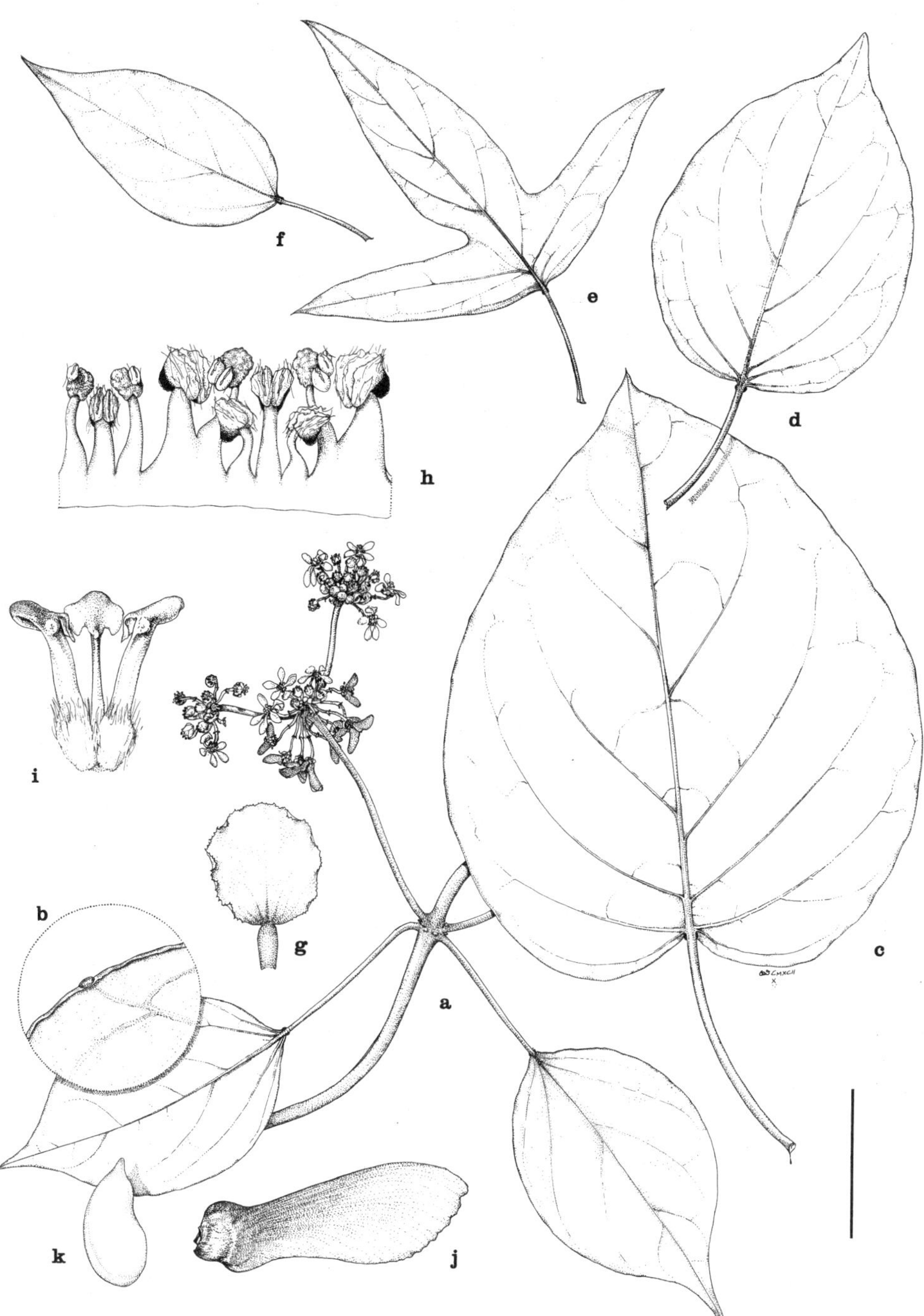

FIG. 67. *Stigmaphyllon cardiophyllum*. a. Flowering and fruiting branch. b. Detail of margin and abaxial surface of lamina. c–f. Leaves illustrating variation of laminar shape. g. Posterior petal (the "flag"). h. Androecium; stamen second from left opposes the posterior petal. i. Gynoecium, anterior style in the center. j. Samara. k. Embryo. Scale: for a, c–f, bar = 4 cm; for b, bar = 4 mm; for g, k, bar = 8 mm; for h, bar = 2 mm; for i, bar = 2.7 mm; for j, bar = 2 cm. (Based on: a, b, f–i, *Encarnación 26052*; c, *Ancuash 1134*; d, *Woytkowski 5954*; e, *Nee 36821*; j, k, *Foster 6506*.)

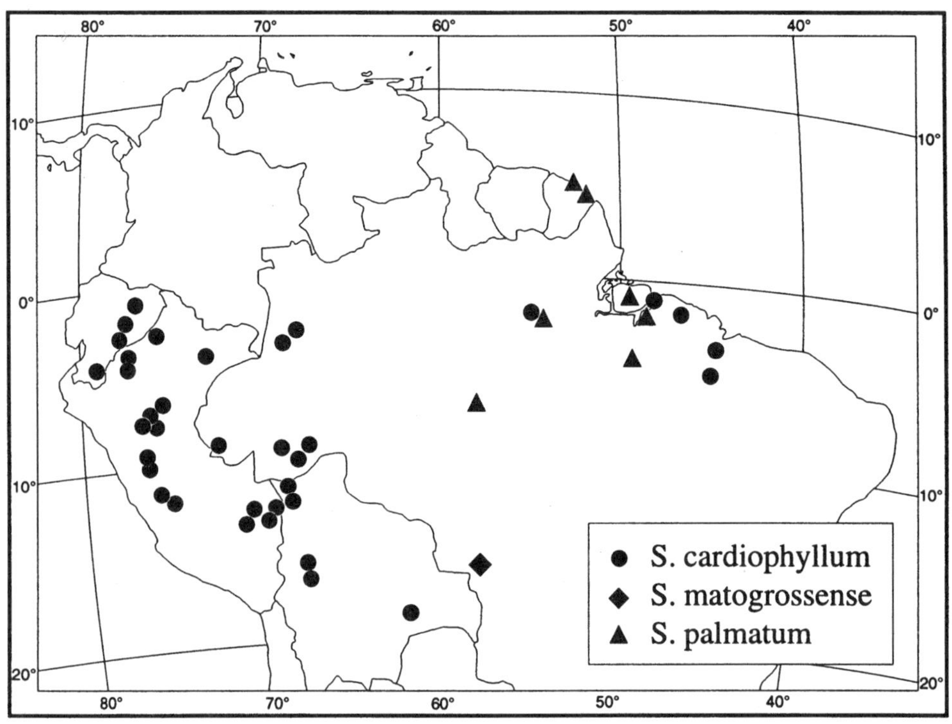

FIG. 68. Distribution of *Stigmaphyllon cardiophyllum, S. matogrossense*, and *S. palmatum.*

MICH, NY, US).—LA PAZ: Prov. Sud Yungas, Alto Beni, carretera entre Puente Sapecho y Santa Ana, *Seidel & Schulte 2301* (MICH); Prov. Nor Yungas, Alto Beni, camino del puente hacia San Antonio, *Seidel & Schulte 2320* (MICH).—PANDO: Prov. Manuripi, antes de Independencia, *Moraes 264* (MICH); Prov. Manuripi, along Río Madre de Dios, 80 km (by air) downstream from and W of Chibe, 11°54′S, 68°02′W, *Nee 31525* (MICH); Río Acre, im Walde bei Cobija, *Ule 9484* (G, K, MG).—SANTA CRUZ: Prov. Andrés Ibanez, 12 km E of center of Santa Cruz on rd to Cotoca, 17°46′S, 63°04′W, *Nee 36821* (MA, MICH).

Stigmaphyllon cardiophyllum, a common lowland species, is readily recognized by its very sparsely sericeous to glabrous leaves and its small flowers. The anthers are pubescent, and those opposite the posterior-lateral sepals bear only one tiny locule (very rarely two). The laminas are mostly ovate to elliptical (the smaller narrowly so) and only sometimes cordate, as in the type. The nut of the samara lacks lateral winglets. In the eastern part of its range *S. cardiophyllum* may be confused with *S. convolvulifolium* (no. 34), whose leaves are also very sparsely sericeous to glabrous abaxially, and in the western part with the less abundantly pubescent representatives of *S. sinuatum* (no. 33). In both of these species, the anthers are glabrous and those of stamens opposite the posterior-lateral sepals are unmodified, i.e., the two locules are not reduced; the posterior styles are usually proximally pubescent; and the nut of the samara is laterally ornamented with winglets, crests, and/or spurs.

Several collections from Ecuador and one from Peru may represent hybrids between *S. cardiophyllum* and an unknown species, possibly *S. sinuatum*. They differ from typical specimens of *S. cardiophyllum* in that the leaves are persistently sparsely sericeous abaxially (the trabecula 0.3–0.5 mm long, sessile). The anthers are sparsely pubescent or glabrous, and their locules sometimes remain closed. The anthers of the posterior-lateral stamens may have two unequal locules instead of only one. Pollen is mostly composed of misshapen, heavy-walled grains. When placed in cotton blue in lactophenol, only 0–20% of the pollen is stained; the grains that do stain are of unequal size. The posterior styles are sparsely pubescent in the proximal 1/3. Although the nut of most samaras is laterally prominently ribbed, as in typical examples of *S. cardiophyllum*, it may also have a small lateral winglet or spur (e.g., *Jaramillo 87*). Unfortunately, none of the samaras are mature, and it is unknown whether normal embryos are formed. The sympatric *S. sinuatum* is suggested as the second parent, because the putative hybrids have similar abaxial leaf pubescence. The following collections are considered putative hybrids:

Ecuador. MORONA-SANTIAGO: alrededores del puente sobre el Río Bombioza, carretera Gualaquiza–Zamora cerca la Paroquia de Bombioza, *M. A. Baker 6479* (MICH).—NAPO: Estación Biológica Jatun Sacha, 8 km al este de Misahuallí, 01°04′S, 77°36′W, *Cerón M. 2038* (MICH); Payamino, Reserva Floristica "El Chuncho," 5 km al NW de Coca, 00°30′S, 77°01′W, *Cerón M. & Neill 2367* (MICH); vía Puerto Napo–Misahuallí, *Jaramillo 87* (AAU, NY, QCA); Río Napo between Coca (Puerto Francisco de Orellana) and Armenia Vieja, *Harling & Andersson 11977* (GB, MICH); Santa Rosa at Río Napo, *Lugo 168, 1981, 2001, 2027* (GB, MICH); Misahuallí at Río Napo, *Lugo 2273* (GB, MICH).—PASTAZA: Tena, *Asplund 9373* (S).—ZAMORA-CHINCHIPE: near Méndez, *Camp E-853* (NY, US). **Peru.** AMAZONAS: Río Cenepa, vicinity of Huampami, ca. 5 km E of Chávez Valdívia, ca. 04°30′S, 78°30′W, *Ancuash 1262* (F, MICH, MO).

70. Stigmaphyllon argenteum C. Anderson, Novon 2: 302. 1992.—TYPE: PERU. Huánuco: Prov. Pachitea, Dtto. Honoria, Bosque Nacional de Iparia, a lo largo del Río Pachitea cerca del campamento Miel de Abejas, 1 km arriba del pueblo Tournavista o unos 20 km arriba de la confluencia con el Río Ucayali, 300–400 m, 30 May 1967, *Schunke V. 2018* (holotype: NY!, photo: MICH!; isotypes: COL! F! G! US!).

Vine to 14 m. Stems and branches sericeous when young, soon becoming glabrous. Laminas 2.5–15.3 cm long, 5.7–14 cm wide, triangular, ovate, elliptical to suborbicular, or sometimes palmately 3–5-lobed, apex acuminate, base truncate to cordate, sparsely sericeous to usually glabrous adaxially, sericeous abaxially (trabecula 0.2–0.5 mm long, straight, sessile), margin shallowly crenate to subentire and with irregularly spaced sessile glands (0.5–0.6 mm in diameter) in the sinuses and with filiform glands (up to 1.5 mm long); petioles 2–10+ cm long, sericeous, not confluent across the node, with a pair of prominent but sessile glands at the apex, each gland 1.5–3.5 mm in diameter; stipules 0.7–1.2 mm long and wide, free, triangular, eglandular. Flowers ca. 15–30 per umbel, these borne in dichasia or compound dichasia (axes to the 5th order, sericeous; rarely with one or two subsidiary axes). Peduncles 3–7.5 mm long; pedicels 4–8.5 mm long, terete; both densely sericeous, peduncles 0.6–1.2 times as long as the pedicels. Bracts 0.9–1.3 mm long, 0.6–1 mm wide, narrowly triangular, apex acute; bracteoles 0.7–1.2 mm long, 0.6–1 mm wide, triangular, apex obtuse, eglandular; bracts and bracteoles sericeous abaxially. Sepals 1.8–2.3 mm long, 1.5–2 mm wide, glands 1.6–2.3 mm long, 0.6–1.2 mm wide. All petals with the limb glabrous, yellow, limb of lateral petals orbicular or broadly obovate, margin erose; anterior-lateral petals: claw 1.8–2.2 mm long, limb ca. 7 mm long and wide; posterior-lateral petals: claw 0.5–1 mm long, limb 6–6.7 mm long, 4.5–6 mm wide; posterior petal: claw 2.5–2.8 mm long, apex strongly indented, limb 5–5.6 mm long, 3.5–4.8 mm wide, elliptical or broadly obovate, margin erose to fimbriate-denticulate, teeth/fimbriae up to 0.5 mm long. Stamens unequal, those opposite the posterior-lateral petals (and the posterior styles) the largest, anthers of those opposite the anterior-lateral sepals with one or two locules, those of stamens opposite the posterior-lateral sepals with only one locule; anthers glabrous. Stamen opposite anterior sepal: filament ca. 1.7 mm long, anther ca. 1 mm long; stamens opposite anterior-lateral petals: filaments ca. 1.5 mm long, anthers 0.6–0.7 mm long; stamens opposite anterior-lateral sepals: filaments 1.8–2.5 mm long, connectives 0.6–0.8 mm long, locules 0.3–0.4 mm long; stamens opposite posterior-lateral petals: filaments 2.2–2.7 mm long, anthers ca. 0.8 mm long; stamens opposite posterior-lateral sepals: filaments 2–2.6 mm long, connectives ca. 0.7 mm long, locule 0.3–0.4 mm long; stamen opposite posterior petal always shorter than the adjacent two: filament 1.5–2 mm long, anther ca. 0.6 mm long. Anterior style ca. 2.2 mm long, shorter than the posterior two, terete, glabrous, erect; apex ca. 1.5 mm long, each foliole ca. 1.4 mm long, ca. 1.2 mm wide, subsquare. Posterior styles 2.6–3 mm long, terete, glabrous, lyrate; foliole ca. 1.4–2 mm long and wide, subsquare. Dorsal wing of samara ca. 4.5 cm long, 1.4–1.7 cm wide, upper margin with a blunt tooth, lateral winglets absent, the nut only prominently ribbed; nut 4–5.5 mm high, 3.5–4.5 mm in diameter, without air chambers, areole 3–3.5 mm long, 2.5–2.8 mm wide, concave, carpophore up to 1.8 mm long. Embryo 5.8–7.3 mm long, ca. 2 times as long as wide, ovoid, outer cotyledon 6.1–8.3 mm long, 2.6–3.9 mm wide, the distal 1/6 folded over the inner cotyledon, inner cotyledon 4–6.6 mm long, 2–3.6 mm wide, straight or the tip folded back on itself. Chromosome number unknown. Fig. 69.

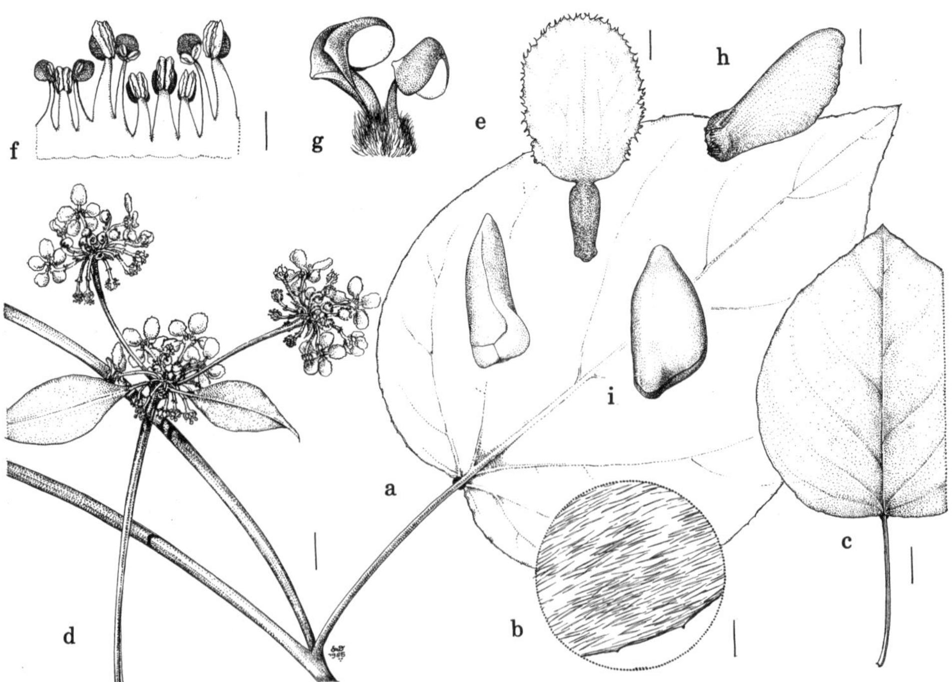

FIG. 69. *Stigmaphyllon argenteum*. a. Portion of branch with large leaf. b. Detail of margin and abaxial surface of lamina. c. Small leaf. d. Flowering branch. e. Posterior petal (the "flag"). f. Androecium; stamen second from left opposes the posterior petal. g. Gynoecium, anterior style to the right. h. Samara. i. Two views of an embryo. Scale: for a, c, d, h, bar = 1 cm; for b, bar = 0.5 mm; for e–g, bar = 1 mm. (Based on: a, b, d–g, *Schunke V. 2028*; c, h, i, *Croat 19640*.)

Phenology. Collected in flower from April through July, in fruit in May and from July through September.

Distribution (Fig. 70). Lowlands of Peru; in forests and thickets and at roadsides; 135–670 m.

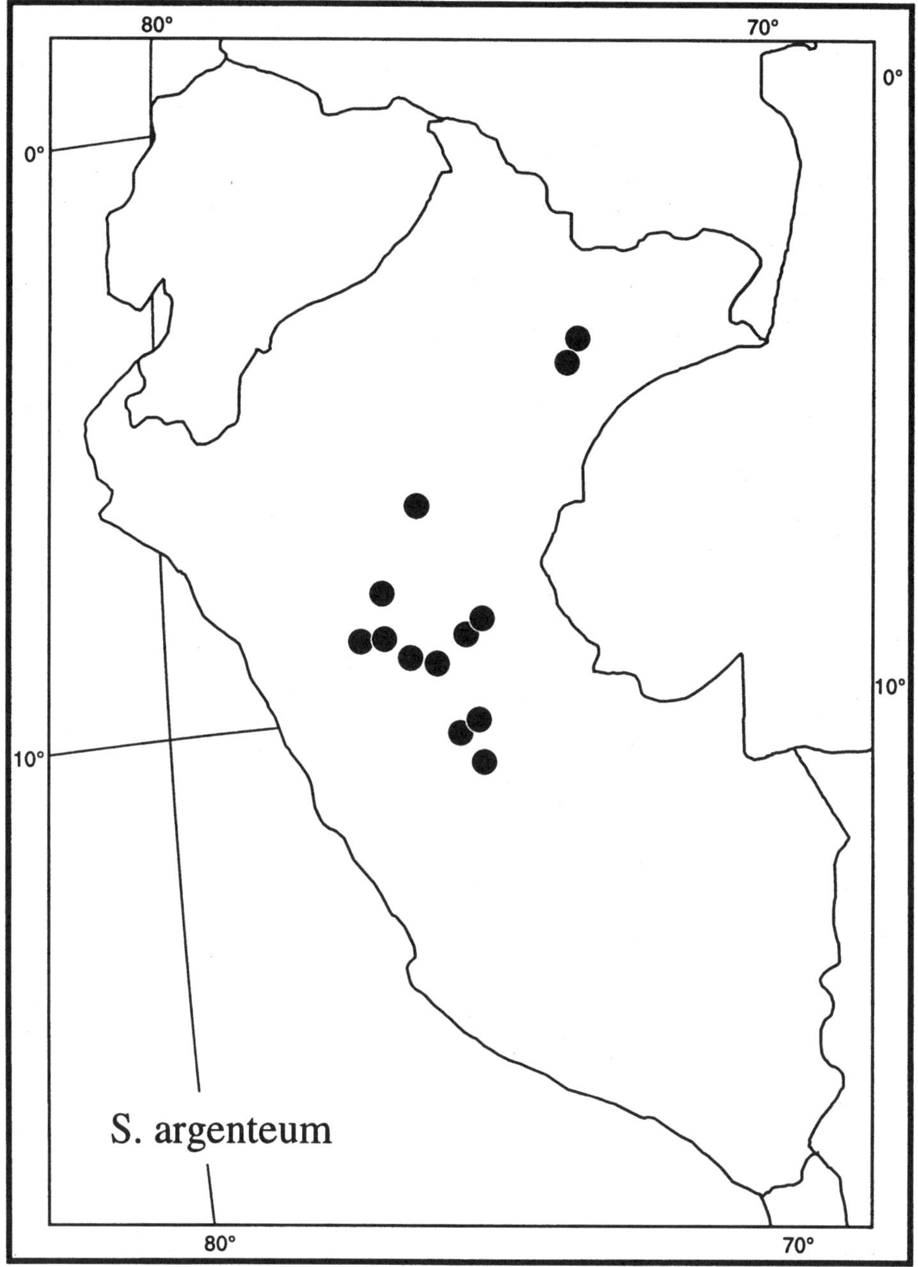

FIG. 70. Distribution of *Stigmaphyllon argenteum*.

REPRESENTATIVE SPECIMENS. **Peru.** HUÁNUCO: Prov. Pachitea, region of Pucallpa, ca. 26 km S to 24 km SSE of Puerto Inca, N of Río Yuyapichis, 09°37′–09°34′S, 74°56′–74°53′W, *Wallnöfer 11-31588* (MICH); vicinity of Tingo María, 3–5 km from Huánuco–Tingo María rd on Monzón rd, *Mathias & Taylor 3647* (F, LA).—JUNÍN: Puerto Bermúdez, *Killip & Smith 26630* (F, MA, NY, US); Prov. Satipo, E bank of Río Ene at mouth of Río Quipachiari, *Madison 10427-70* (F).—LORETO: Quebrada Shanuce above Yurimaguas, *Croat 17999* (MICH); Isla de Ushpa-cano near mouth of Río Itaya, *Croat 19640* (MICH); Ucayali, Bosque Nacional Alexander von Humboldt, between Km 90–130 of Pucallpa–Tingo María rd, 08°48′S, 75°20′W, *Gentry 41413* (MO); wooded banks on lower Río Huallaga, *Killip & Smith 29004* (F, GH, NY, US); Prov. Maynas, vicinity of Iquitos, Río Momón, quebrada Momoncillo, *McKenna et al. DMK-91* (AMAZ, F, MICH, MO).—PASCO: Oxapampa, ca. 5 km up Río Iscozacín from village of Iscozacín, 10°12′S, 75°13′W, *Knapp & Staver 7802A* (MICH); Palcazu Valley, Río San José in the Río Chuchurras drainage, 10°09′S, 75°20′W, *D. Smith 4002* (MICH).—SAN MARTÍN: between Tocache Nuevo and Juanjui, 18.7 km S of Río Pulcache, 07°55′S, 76°40′W, *Croat 58052* (MICH); vicinity of Aguaytía, Boquerón de Padre Abad, *Mathias & Taylor 3591* (F, LA, US), *6092* (F, LA); Prov. Mariscal Cáceres, Dtto. Tocache Nuevo, quebrada de Santiago, al E de Puerto Pizana, *Schunke V. 6530* (GH, MO); Prov. Mariscal Cáceres, Dtto. Tocache Nuevo, quebrada de Cachiyaca, afluente de la quebrada de Huaquista, al E de Puerto Pizana, *Schunke V. 8528* (F, MICH, MO).

Stigmaphyllon argenteum is named for the silvery pubescence on the abaxial leaf surfaces. It is similar to *S. cardiophyllum* (no. 69) in its small flowers, in which the anthers of stamens opposite the posterior-lateral sepals bear only one locule, and its samaras, which lack lateral winglets. *Stigmaphyllon cardiophyllum* is easily separated by its glabrate to glabrous leaves and abundantly pubescent anthers. In the widespread and sympatric *S. sinuatum* (no. 33), the flowers are borne in pseudoracemes, the anthers of stamens opposing the posterior-lateral petals are unmodified, i.e., the two locules are not reduced, the styles are commonly pubescent, and the nut of the samara usually bears lateral winglets.

71. Stigmaphyllon paraense C. Anderson, Contr. Univ. Michigan Herb. 17: 13. 1990.—TYPE: BRAZIL. Pará: Rio Paranapebas, control point at entrance to Serra Norte, ca. 39 km E of AMZA camp N-5, 06°04′S, 49°55′W, ca. 150 m, *Sperling 6322* (holotype: INPA!; isotypes: MICH! NY!).

Vine to 6 m. Stems and branches sericeous when young, soon becoming glabrous. Laminas 9.2–13 cm long, 5.5–9.3 cm wide, narrowly triangular to narrowly ovate to ovate, rarely 2–3-lobed, apex acuminate-mucronate to sometimes almost caudate, base truncate to cordate, sparsely sericeous to glabrous adaxially, sericeous to densely so abaxially (trabecula 0.3–0.5 mm long, wavy, sessile to subsessile), margin eglandular or subentire and with irregularly spaced sessile glands (0.3–0.8 mm in diameter) in the shallow sinuses; petioles 1.8–8 cm long, sericeous, not confluent across the node, with a pair of prominent but sessile glands at the apex, each gland 1.5–2.2 mm in diameter; stipules 0.5–1.3 mm long and wide, free, triangular, eglandular. Flowers ca. 10–20 per umbel or condensed pseudoraceme, these borne in dichasia or compound dichasia (axes to the 4th order, sericeous). Peduncles 3.5–13.3 mm long; pedicels 3.5–7 mm long, terete; both densely sericeous, peduncles 0.8–2.3 times as long as the pedicels. Bracts 1.3–3 mm long, 1–1.8 mm wide, triangular, apex acute; bracteoles 1–2.3 mm long, 1–1.6 mm wide, triangular, apex acute or obtuse, each bracteole usually with a pair of inconspicuous glands (each gland 0.2–0.5 mm in diameter), sometimes one or both glands absent; bracts and bracteoles sericeous abaxially but glabrate along the margins. Sepals 2.1–3 mm long, 2.1–3.5 mm wide, glands 2.1–2.7 mm long, 1.2–1.5 mm wide. All petals with the limb orbicular or that of the posterior petal sometimes broadly obovate, often sparsely sericeous abaxially (especially in bud) to glabrous, yellow, margin of lateral petals with fimbriae up to 0.8 mm long; anterior-lateral petals: claw 2–3.2 mm long, limb 12–15 mm long and wide; posterior-lateral petals: claw 1.5–2 mm long, limb 11–13.5 mm long and wide; posterior petal: claw 3–4 mm long, apex indented, limb 9–11 mm long and wide, margin fimbriate or fimbriate-lacerate, the fimbriae/teeth up to 0.8 (–1) mm long. Stamens unequal, those opposite the posterior-lateral petals (and the posterior styles) the largest, anthers of those opposite the lateral sepals with the connective enlarged and the locules reduced, anthers of those opposite the posterior-lateral sepals often with only one reduced closed locule or eloculate; anthers glabrous. Stamen opposite anterior sepal: filament 2.2–3 mm long, anther 1.3–1.7 mm long; stamens opposite anterior-lateral petals: filaments 2–2.5 mm long, anthers 1–1.2 mm long; stamens opposite anterior-lateral sepals: filaments 2.8–3.2 mm long, connectives 0.8–1 mm long, locules 0.3–0.7 mm long; stamens opposite posterior-lateral petals: filaments 3–3.7 mm long, anthers 1.4–1.7 mm long; stamens opposite posterior-lateral sepals: filaments 3–3.3 mm long, connectives 0.8–0.9 mm long, locules 0.3–0.6 mm long or absent; stamen opposite posterior petal always shorter than the adjacent two: filament 2.5–3 mm long, anther 0.7–1 mm long. Anterior style 3.3–4.5 mm long, shorter than or sometimes subequal to the posterior two, terete, glabrous or with a few scattered hairs in the proximal 1/4, erect; apex 1.5–2.4 mm long, each foliole 1.6–2.3 mm long, 1.5–2.4 mm wide, subsquare. Posterior styles 3.7–5 mm long, terete, with a row of hairs adaxially in the proximal 1/3–3/4, lyrate; foliole 2.3–3.2 mm long, 2.2–2.8 mm wide, subsquare to sometimes subrectangular. Dorsal wing of samara 3.7–4.5 (–5.5) cm long, 1.4–2 (–2.4) cm wide, upper margin with a blunt tooth; nut bearing a pair of rectangular to semi-circular to lunate, erose to grossly dentate lateral winglets, these 8–13 mm long, 1.5–4.5 mm wide, sometimes also with 1–2 spurs ca. 0.1 mm long, lateral winglets buttressed by 0–2 horizontal ridges (in *Malme 1736* sometimes each side of nut with a row of 3–4 winglets); nut 9–12 mm high, 5.5–7 mm in diameter, the thick-walled locule flanked by incompletely partitioned narrow air chambers, areole 3.5–4.5 mm long and wide, concave, carpophore up to 4.5 mm long. Embryo 7.5–9.4 mm long, ca. 2–3 times as long as wide, narrowly ovoid, outer cotyledon 7.1–8.6 mm long, 2.6–4 mm wide, straight, inner cotyledon 6.7–8 mm long, 2.2–3.6 mm wide, straight. Chromosome number unknown. Fig. 71.

Phenology. Collected in flower from April through September, in fruit from June through September.

Distribution (Fig. 33). Brazil (Goiás, Maranhão, Mato Grosso, western Piauí, and Pará); in woods along rivers, in wet localities in savanna and campo, várzea; sea level to 200 m.

ADDITIONAL SPECIMENS EXAMINED. **Brazil.** GOIÁS: Luziania, 15 km S da cidade no Rio Vermelho, *Heringer 18048* (MBM).—MARANHÃO: Km 176 of BR-135, 2 km N of São Mateus, 04°00′S, 44°30′W, *Daly et al. D319* (MICH, NY); Caxias, *Ducke 697* (MG); BR-10, perto de Imperatriz, *Fernandes & Matos EAC 4066* (MICH); Imperatriz, próximo do Campo de Aviação, *Pinheiro 5* (MICH); Ilha dos Botes, Rio Tocantins, perto de Carolina, *Murça Pires & Black 2076* (IAN, US).—MATO GROSSO: Barbados prope Cuiabá, *Malme 1756* (R, S).—PARÁ: Serra do Cachimbo, Mpio. Itaituba, estrada Santarém–Cuiabá, BR-163, Km 794, 09°22′S, 54°54′W, *Amaral et al. 961* (MICH, NY); Mpio. Itaituba, estrada Santarém–Cuiabá, BR-163, Km 1011, margem direita do Rio Jamaxim, 07°40′S, 55°15′W, *Amaral et al. 1230* (MICH, NY); Bôa Vista, 1931, *Carr s.n.* (F); Rio Tapajós, Fordlândia, *M. Silva 1641* (MG).—PIAUÍ: Tamboril, *das Chagas & F. Silva F12* (K, MO).

Stigmaphyllon paraense is distinguished by the sericeous vesture of the vegetative parts, and by its flowers and samaras. The laminas are sparsely sericeous to glabrous adax-

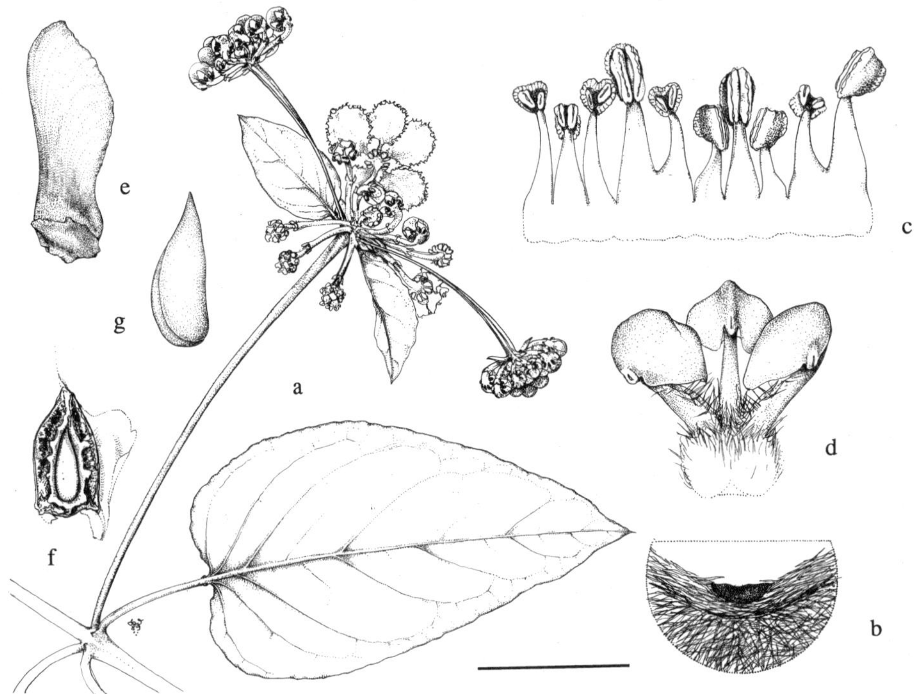

FIG. 71. *Stigmaphyllon paraense*. a. Flowering branch. b. Detail of margin and abaxial surface of lamina. c. Androecium; stamen second from left opposes the posterior petal (the "flag"). d. Gynoecium; posterior styles are bent slightly outward to show anterior style (in center). e. Samara. f. Longitudinal section through nut of samara; note the air chambers surrounding the locule. g. Embryo. Scale: for a, bar = 4 cm; for b, bar = 1.3 mm; for c, d, bar = 4 mm; for e, bar = 2.7 cm; for f, bar = 1.3 cm; for g, bar = 8 mm. (Based on: a–d, *Sperling 6322*; e–g, *M. Silva 1641*.)

ially and usually densely sericeous abaxially. The relatively large petals bear fimbriae up to 0.8 (–1) mm long. The anthers of stamens opposite the posterior-lateral sepals have an enlarged connective and reduced locules, though often one or even both locules are absent. The anterior style is glabrous or at most sparsely pubescent, but the posterior two are bearded adaxially; all bear large folioles. The samara is distinctive in that the nut contains air chambers that surround the locule, and the embryo is relatively large (7.5–9.4 mm long). The presence of air chambers probably aids in dispersal by water, and *S. paraense*, like other species with such samaras, occurs in wet areas and along margins of rivers.

The collection *Malme 1736* from central Mato Grosso is placed here with some hesitation. The four sheets at S and the duplicate at R are all in fruit. Vegetatively they match the other collections of *S. paraense*, but the samaras differ somewhat. Although none are quite mature, some of them have a dorsal wing slightly larger than those of other collections; the nut, instead of bearing a pair of lateral winglets, has a row of 3–4 lateral winglets on each side. Such variation may only reflect the diversity of samara ornamentation in this species, though flowering material may indicate that *Malme 1736* represents an undescribed species. The disjunction from the rest of the range is bridged by the recent collection in Goiás (*Heringer 18048*). Central Brazil is botanically only poorly known, and future collections may reveal that *S. paraense* occurs in wet localities throughout the region.

Stigmaphyllon paraense might be confused with the widespread, polymorphic *S. sinuatum* (no. 33), in which the laminas are also sericeous abaxially; however, the trabecula is straight rather than wavy. *Stigmaphyllon sinuatum* also differs in its erose petals, the unmodified anthers of stamens opposing the posterior-lateral sepals, and its samara, which lacks air chambers (the nut only 2.8–4.4 mm in diameter).

72. Stigmaphyllon palmatum (Cavanilles) Adr. Jussieu, Ann Sci. Nat. Bot., sér. 2, 13: 288. 1840. *Banisteria palmata* Cavanilles, Diss. 9: 430. t. 257. 1790.—TYPE: "Habitat in Sancto Domingo," *Desportes s.n.* (holotype: P-JU!, photos: A! MICH! US!).

Banisteria sagittata Cavanilles, Diss. 9: 430. t. 257. 1790. *Stigmaphyllon sagittatum* (Cavanilles) Adr. Jussieu, Ann. Sci. Nat. Bot., sér. 2, 13: 288. 1840.—TYPE: "Habitat cum praecedenti [Sancto Domingo]," *Desportes s.n.* (holotype: P-JU!, photos: F! MICH! US!).

Stigmaphyllon angustilobum var. *burchellii* Niedenzu, Ind. Lect. Lyc. Brunsberg. p. aest. 1900: 26. 1900.—TYPE: BRAZIL. Pará: prope Pará, 1830, *Burchell 10025* (holotype: GOET!, photo: MICH!; isotypes: GH! K! LE! P!).

Vine to 6 m. Stems and branches sericeous when young, becoming glabrate. Laminas 7–18 cm long, 5–14 cm wide, lanceolate or ovate to cordate to suborbicular but often palmately 3–5 (–7)-lobed, apex mucronate to sometimes caudate, base deeply cordate to auriculate, in larger leaves the basal lobes sometimes overlapping, glabrous adaxially, tomentose abaxially (trabecula 0.4–1.6 mm long, straight to wavy or crisped and curled, stalk 0.1–0.4 mm long), margin entire to grossly dentate, the teeth and lobes usually terminating in a filiform gland up to 0.2 mm long, and with irregularly spaced sessile glands (0.2–0.5 mm in diameter) or rarely eglandular; petioles 2.2–9.5 cm long, sericeous, not confluent across the node, with a pair of prominent but sessile glands at the apex, each gland 1.5–2.5 mm in diameter; stipules 0.6–1.6 mm long, 0.5–1.5 mm wide, free, triangular, eglandular. Flowers ca. 10–20 (–25) per umbel or (sometimes) pseudoraceme, these solitary or borne in dichasia or compound dichasia (axes to the 3rd order, sericeous). Peduncles 6–13 mm long; pedicels 3.5–10.5 mm long, terete; both densely sericeous, peduncles 0.7–1.7 times as long as the pedicels. Bracts 1.2–2.5 mm long, 1.1–1.8 mm wide, triangular or narrowly so, apex acute; bracteoles 1.4–2.1 mm long, 1.2–1.4 mm wide, oblong to ovate, apex obtuse, eglandular or each bracteole with a pair of inconspicuous glands (each gland 0.2–0.5 mm in diameter); bracts and bracteoles sericeous abaxially. Sepals 3–3.5 mm long, 3–3.7 mm wide, glands 2–2.7 (–3.5) mm long, 1–1.5 (–1.8) mm wide. All petals with the limb orbicular, glabrous, yellow, margin with fimbriae up to 1.5 (–1.8) mm long; anterior-lateral petals: claw 2.5–3 mm long, limb ca. 15 mm long and wide; posterior-lateral petals: claw 1.5–2 (–2.5) mm long, limb ca. 13–14 mm long and wide; posterior petal: claw 3.5–4 mm long, apex indented or only slightly so, limb ca. 11–12 mm long and wide, sometimes at the base with 1–2 stout gland-tipped fimbriae per side (ca. 0.4 mm long, ca. 0.2 mm wide). Stamens unequal, those opposite the posterior-lateral petals (and the posterior styles) the largest, anthers of those opposite the anterior-lateral sepals often with the connective slightly enlarged and the locules slightly reduced, anthers of those opposite the posterior-lateral sepals eloculate or rarely with 1–2 reduced locules, such locules closed or rarely one or both open; anthers glabrous. Stamen opposite anterior sepal: filament ca. 2.5 mm long, anther 1.2–1.4 mm long; stamens opposite anterior-lateral petals: filaments 2–2.2 mm long, anthers 0.9–1.1 mm long; stamens opposite anterior-lateral sepals: filaments 3–3.5 mm long, connectives 0.6–0.8 mm long, locules 0.4–0.8 mm long; stamens opposite posterior-lateral petals: filaments (3.6–) 4–4.2 mm long, anthers 1.3–1.6 mm long; stamens opposite posterior-lateral sepals: filaments 2.8–3.1 mm long, connectives 0.5–0.6 (–0.9) mm long, locules usually absent, if present 0.3–0.6 mm long, closed; stamen opposite posterior petal always shorter than the adjacent two: filament 2.5–2.7 mm long, anther 0.8–1 mm long. Anterior style 4–5 mm long, shorter than the posterior two, terete, erect, glabrous or with a few scattered hairs in the proximal 1/4; apex 2–2.3 mm long, each foliole 2–2.4 mm long, 1.8–2.1 mm wide, subrectangular to subsquare. Posterior styles 4.8–5.5 mm long, terete, lyrate, with scattered hairs in the proximal 1/2; foliole 2.4–3 mm long, 2.5–2.8 mm wide, suborbicular to subsquare. Dorsal wing of samara ca. 3.3 cm long, ca. 1 cm wide, upper margin with a blunt tooth; nut bearing a pair of semicircular and erose to grossly dentate lateral winglets, these up to 1.3 mm long, up to 0.5 mm wide; nut ca. 11 mm high, ca. 6.5 mm in diameter, without air chambers, areole ca. 0.5 mm long and wide, concave, carpophore up to 5 mm long. Only one almost mature seed seen; embryo 9.5 mm long, 2 times as long as wide, ovoid, outer cotyledon 15 mm long, the distal 2/5 folded over the inner cotyledon, inner cotyledon ca. 5 mm long, straight. Chromosome number unknown.

Phenology. Collected in flower throughout the year (but no records from February, April, and September); in fruit in February, August, September, and December.

Distribution (Fig. 68). Coastal French Guiana: in sandy soils at roadsides, in savannas, and at beaches; and Brazil (Pará): in sandy soils at riverbanks, sandbanks, and in scrub forest; sea level to 90 m.

REPRESENTATIVE SPECIMENS. **French Guiana.** Environs de Cayenne, *Billiet & Jadin 998* (CAY, NY); vicinity of Cayenne, *Broadway 481* (GH, NY, US); Rorota, *Feuillet 541* (CAY, MICH); Pointe Diamante, 04°53′N, 52°17′W, *Hahn 3581* (MICH); route entre Tonate et Kourou, Km 9, *Prévost 703* (CAY, MICH); Mana, *unknown collector ex Herb. Maire* (P). **Brazil.** PARÁ: Belém, 4–6 km NW of Instituto Agronomico do Norte, near São Joaquim, *Barbosa da Silva 151* (NY, US); Santarém, Rio Curuá Una, junto a cachoeira do Palhão, *Cavalcante 1651* (MG); Tucuruí, Breú Branco, margem izquierda do Rio Tocantins, *Lisboa et al. 1203* (MG, NY); Tucuruí, margem direita do Rio Tocantins, *Lisboa et al. 1503* (MG, NY); Rio São Manoel abaixo do Igarapé Preto, *Murça Pires 3771* (IAN, US).

Stigmaphyllon palmatum is a coarse, large-flowered vine named for the palmately lobed laminas; however, leaf shape is polymorphic in this species. It varies from lanceolate or ovate to cordate to suborbicular to palmately 3–5 (–7)-lobed, and several forms may be found on the same specimen. In unlobed laminas, the margins are often grossly and/or shallowly dentate. The limb of the lateral petals is 13–15 mm in diameter and fringed with fimbriae up to 1.5 (–1.8) mm long. The stamens opposing the posterior-lateral sepals are essentially sterile; the enlarged connectives either lack locules or bear 1–2 greatly reduced ones, which remain closed.

For a comments about Jussieu's (1840, 1843) misapplication of the name *S. sinuatum* to *S. palmatum*, see the discussion of *S. sinuatum* (no. 33).

73. Stigmaphyllon strigosum Adr. Jussieu, Ann. Sci. Nat. Bot., sér. 2, 13: 289. 1840.— TYPE: PERU. San Martín: Tocache, Jul 1830, *Poeppig 1941* (holotype: P!, photo: MICH!; isotypes: F! G! NY! W-3 sheets!).

Vine to 20 m. Stems and branches bearing scalelike T-shaped hairs, becoming glabrate. Laminas 7–11 cm long, 5–8.5 cm wide, ovate, apex mucronate, base cordate to truncate, glabrous adaxially, with T-shaped hairs abaxially (trabecula 0.2–0.7 mm long, wavy or crisped, stalk 0.1–0.2 mm long), margin with irregularly spaced glands (0.2–0.4 mm in diameter); petioles 2–5.5 cm long, bearing scalelike T-shaped hairs, not confluent across the node, with a pair of prominent but sessile glands at the apex, each gland elliptical, 2–2.8 mm long; stipules 0.5–1.2 mm long, 0.7–1.3 mm wide, free, triangular, eglandular. Flowers ca. 12–20 per umbel, these borne in compound dichasia or small thyrses (axes to the 3rd order, bearing scalelike T-shaped hairs). Peduncles 5.5–11 mm long; pedicels 6–9 mm long, terete; both bearing scalelike T-shaped hairs, peduncles 0.8–1.6 times as long as the pedicels. Bracts 1–2 mm long, 0.7–1.1 mm wide, triangular, apex acute; bracteoles (1–) 1.4–2.3 mm long, 0.9–1.3 mm wide, parabolic or broadly triangular, apex obtuse, each bracteole with a pair of inconspicuous glands (each gland 0.3–0.5 mm in diameter); bracts and bracteoles sericeous (hairs subsessile) abaxially. Sepals 2.5–3 mm long, 2–2.8 mm wide, glands 2–2.9 mm long, 1–1.2 mm wide. All petals with the limb orbicular, glabrous, yellow and suffused with red (?), margin with fimbriae up to 1 mm long; anterior-lateral petals: claw 2–2.5 mm long, limb ca. 13–14 mm long and wide; posterior-lateral petals: claw 1–1.5 mm long, limb ca. 11–12 mm long and wide; posterior petal: claw ca. 3 mm long, the distal 1/5–1/3 laterally-adaxially bearded, apex indented, limb ca. 9 mm long and wide, often at the base with a pair of stalked glands (0.3–0.4 mm long). Stamens unequal, those opposite the posterior-lateral petals (and the posterior styles) with the longest filaments, anthers of those opposite the lateral sepals with the connective enlarged and the locules reduced; anthers all loculate, glabrous. Stamen opposite anterior sepal: filament 2.5–2.6 mm long, anther 0.9–1 mm long; stamens opposite anterior-lateral petals: filaments 2–2.3 mm long, anthers 0.6–0.8 mm long; stamens opposite anterior-lateral sepals: filaments 3–3.1 mm long, connectives 0.5–0.6 mm long, locules 0.3–0.5 mm long; stamens opposite posterior-lateral petals: filaments 3.4–3.8 mm long, anthers ca. 1 mm long; stamens opposite posterior-lateral sepals: filaments ca. 3 mm long, connectives 0.5–0.6 mm long, locules 0.2–0.3 mm long; stamen opposite posterior petal always shorter than the adjacent two: filament 2.5–2.7 mm long, anther 0.5–0.6 mm long. Anterior style 3.2–3.3 mm long, shorter than the posterior two, terete, glabrous or with scattered hairs in the proximal 1/2, erect; apex 1–1.5 mm long, each foliole 1.4–1.7 mm long, 0.9–1.6 mm wide, parabolic to subsquare. Posterior styles 3.7–4 mm long, terete, glabrous or with scattered hairs in the proximal 1/4–1/3, lyrate; foliole 1.6–2 mm long, 1.3–1.8 mm wide, subsquare. Dorsal wing of samara ca. 5.5 cm long, ca. 2 cm wide, upper margin with an acute or blunt tooth; nut bearing 3–4 semicircular to rectangular lateral winglets per side, these up to 8.5 mm long, up to 5 mm wide; nut 5.5–6 mm high, 4–4.5 mm in diameter, without air chambers, areole ca. 3 mm long and wide, concave, carpophore up to 3.5 mm long. Embryo ca. 7 mm long, ca. 1 1/2 times as long as wide, broadly ovoid, outer cotyledon ca. 9 mm long, ca. 4.5 mm wide, the distal 1/3 folded over the inner cotyledon, inner cotyledon ca. 6.2 mm long, ca. 3.3 mm wide, folded at the distal 1/3. Chromosome number unknown.

Phenology. Collected in flower July through September, in fruit in July and September.

Distribution (Fig. 72). Peru (San Martín and adjacent Loreto, Huánuco, Junín, Cuzco, Madre de Dios), and adjacent Brazil (Acre) and Bolivia (Beni, Pando); in forests and secondary growth; 180–800 m.

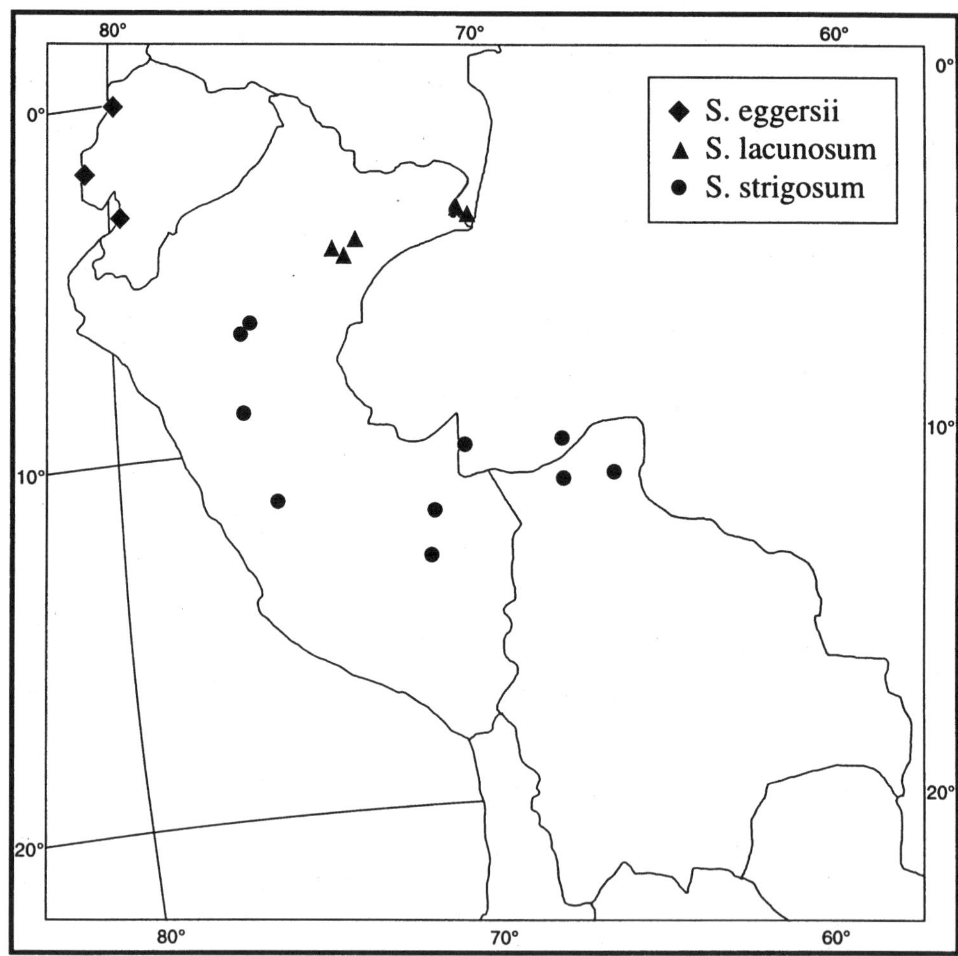

FIG. 72. Distribution of *Stigmaphyllon eggersii, S. lacunosum,* and *S. strigosum.*

ADDITIONAL SPECIMENS EXAMINED. **Peru.** CUZCO: Prov. Paucartambo, alrededores de Pilcopata, *Vargas 14670* (US).—HUÁNUCO: Prov. Pachita, region of Pucallpa, ca. 26 km S to ca. 24 km SSE of Puerto Inca, *Gottsberger & Döring G14-26888* (WU).—JUNÍN: Río Negro, *Woytkowski 5792* (GH, MO, US).—LORETO: zwischen El Sacramento und Pucallpa, *Ellenberg 2588* (U); ad Yurimaguas, *Poeppig s.n.* (W).—MADRE DE DIOS: Parque Nacional Manu; far up Quebrada Fierro - first significant tributary of Río Manu upriver from Tayakome to West, 11°22′S, 71°58′W, *Shepard 2156* (F).—SAN MARTÍN: Prov. Marscal Cáceres, Dtto. Tocache Nuevo, Bambamarca, a 20 km de Tocache, *Schunke V. 7199* (GH, MO); Prov. Marscal Cáceres, Dtto. Tocache Nuevo, camino antiguo a Limón, cerca a La Granja San Isabel, *Schunke V. 10315* (MICH). **Brazil.** ACRE: Km 4 on Rio Branco–Porto Velho rd, *Lowrie et al. 292* (INPA, MICH); Rio Acre, Seringal S. Francisco, *Ule 9485* (MG). **Bolivia.** BENI: Prov. Vaca Diez, Riberalta, 11°00′S, 66°08′W, *Gentry 77559* (MICH); Prov. Vaca Diez, Riberalta, 11°01′S, 66°04′W, *Moraes 181* (MICH); Prov. Vaca Diez, E side of Riberalta, near Experimental Station, 11°00′S, 66°05′W, *Solomon 6256* (MICH).—PANDO: Manuripi, 35 km al N de Puerto America, *Jardim & Cuellar 998* (MICH).

Stigmaphyllon strigosum is a distinctive species readily recognized by its large flowers with long-fimbriate petals. It is the only species in the genus in which the claw of the posterior petal is bearded. On all herbarium labels the petals are said to be yellow or greenish yellow; however, the dried limbs all show pink to purple areas, as is found in petals which are suffused with red. The samaras have a large dorsal wing, and the nut bears 3–4 large, overlapping, semicircular to rectangular lateral winglets per side. An unusual aspect of the development of the samara, which Jussieu (1843) also noted, is the early maturation of these lateral winglets (up to 8.5 mm long and 3 mm wide), while the dorsal wing is still quite small. *Stigmaphyllon strigosum* might be confused with *S. florosum* (no. 42), which is easily separated by its efoliolate styles; in *S. strigosum* all styles bear large folioles.

Two anomalous collections from Peru may represent hybrids, *Schunke V. 10572* (fruits; MICH; Huánuco: Prov. Leoncio Prado, Dtto. Padre Luyando) and *Woytkowski 5902* (flowers and fruits; G, GH, MO, US; Junín: Satipo). They differ in that the leaves are glabrate to glabrous abaxially; the hairs present, similar to those typical of *S. strigosum*, are concentrated along the midrib and major veins. The flowers differ in two aspects: the fimbriae of the petals are shorter, only up to 0.5 mm long instead of up to 1 mm long, and the unmodified anthers are sparsely pubescent instead of glabrous. Pollen from *Woytkowski 5902*, stained with cotton blue in lactophenol, shows ca. 50% misshapen, thick-walled, non-staining grains. The duplicate at GH is a mixture; it consists of three large glabrate leaves and an inflorescence typical of *S. strigosum*. Pollen from buds of this inflorescence is ca. 97% normal. The samaras differ in the lateral ornamentation of the nut. The winglets are not as large, only up to 3 mm wide, and often do not overlap, or they may be reduced to spurs. The putative hybrids perhaps resulted from crosses between *S. strigosum* and the sympatric *S. cardiophyllum* (no. 69). In that species, the leaves are glabrous to very sparsely sericeous, the petals erose to denticulate, the anthers abundantly pubescent, and the nut is laterally only ribbed instead of winged.

74. Stigmaphyllon stylopogon C. Anderson, Contr. Univ. Michigan Herb. 17: 17. 1990.—TYPE: BRAZIL. Rondônia: Mpio. Pimenta Bueno, rodovia Cuiabá–Pôrto Velho, BR-364, Km 188, 11°12′S, 61°62′W, 19 Jun 1984, *Cid et al. 4648* (holotype: INPA!, photo: MICH!; isotypes: MG! MICH! NY!).

Vine. Stems and branches sericeous and with scalelike T-shaped hairs when young, soon becoming glabrate. Laminas 7.5–13.5 cm long, 5–10 cm wide, ovate, apex mucronate or emarginate-mucronate, base deeply cordate to auriculate, sparsely tomentulose to glabrate adaxially, very densely appressed-tomentulose abaxially (trabecula 0.2–0.7 mm long, crisped and curled, stalk 0.03–0.1 mm long) and the epidermis hidden, margin subentire and with irregularly spaced sessile glands (0.4–0.6 mm in diameter) in the sinuses; petioles 2.2–6.5 cm long, densely sericeous to glabrate in older leaves, not confluent across the node, with a pair of prominent but sessile glands at the apex, each gland 1.5–2.5 mm in diameter; stipules 1–1.7 mm long, 1.2–2 mm wide, free, triangular, eglandular. Flowers ca. 10–16 per umbel, these borne in dichasia or compound dichasia or small thyrses (axes to the 4th order, with scalelike T-shaped hairs). Peduncles 5.5–17 mm long; pedicels 3–12 mm long, terete; both densely sericeous, peduncles 0.6–2.2 times as long as the pedicels. Bracts 1.6–2.2 mm long, 1.2–1.7 mm wide, triangular, apex acute; bracteoles 1.5–2.3 mm long, 1–2 mm wide, triangular, apex obtuse, eglandular or each bracteole with a narrow band of glandular tissue along the margin; bracts and bracteoles densely sericeous abaxially. Sepals 2.7–3.5 mm long, 2.5–3 mm wide, glands 2.6–3 mm long, 1.3–1.5 mm wide. All petals with the limb orbicular, glabrous, yellow, limb of lateral petals with the margin fimbriate, fimbriae up to 0.5 mm long; anterior-lateral petals:

claw 3–3.5 mm long, limb 12–13 mm long and wide; posterior-lateral petals: claw 2.2–2.5 mm long, limb 11–12 mm long and wide; posterior petal: claw 3.5–3.7 mm long, apex indented or only slightly so, limb ca. 10 mm long and wide, margin with fimbriae up to 0.8 mm long and commonly at the base each side with 2–3 stout gland-tipped fimbriae (0.6–1 mm long, ca. 0.2 mm wide). Stamens unequal, those opposite the posterior-lateral petals (and the posterior styles) the largest, anthers of those opposite the lateral sepals with the connective enlarged and the locules reduced, anthers of those opposite the posterior-lateral sepals with only one locule; anthers all loculate, glabrous. Stamen opposite anterior sepal: filament 2.5–3 mm long, anther 1.5–2 mm long; stamens opposite anterior-lateral petals: filaments 1.8–2.3 mm long, anthers 1.3–1.6 mm long; stamens opposite anterior-lateral sepals: filaments 3–3.3 mm long, connectives 0.8–0.9 mm long, locules 0.5–0.8 mm long; stamens opposite posterior-lateral petals: filaments 3.5–3.8 mm long, anthers 1.5–2 mm long; stamens opposite posterior-lateral sepals: filaments 3–3.6 mm long, connectives 0.6–0.8 mm long, locule 0.4–0.5 mm long; stamen opposite posterior petal always shorter than the adjacent two: filament 2.5–2.7 mm long, anther ca. 1 mm long. Anterior style 3.7–4.7 mm long, shorter than the posterior two, terete, adaxially with a row of hairs in the proximal 1/2–3/4, erect; apex 2.2–2.5 mm long, each foliole 2–2.3 mm long, 2.2 mm wide, subsquare. Posterior styles 4.5–5 mm long, terete, adaxially with a row of hairs from the base nearly to the stigma, lyrate; foliole 3–3.7 mm long, 3–3.2 mm wide, subsquare. Samara not seen. Chromosome number unknown. Fig. 73.

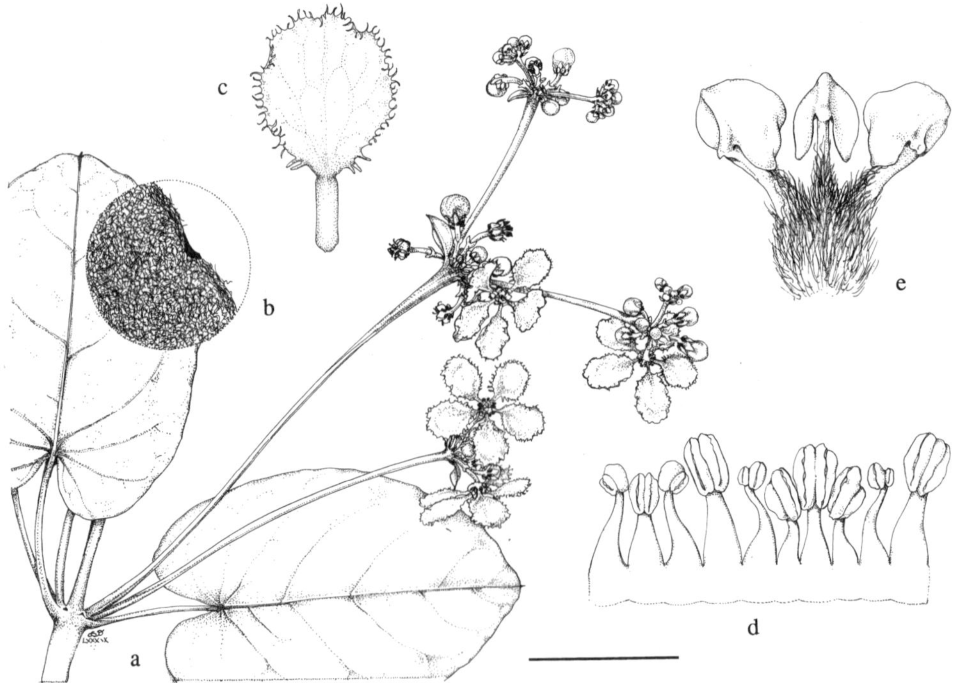

FIG. 73. *Stigmaphyllon stylopogon*. a. Flowering branch. b. Detail of margin and abaxial surface of lamina. c. Posterior petal (the "flag"). d. Androecium; stamen second from left opposes the posterior petal. e. Gynoecium; posterior styles are bent slightly outward to show anterior style (in center). Scale: for a, bar = 4 cm; for b, bar = 2 mm; for c, bar = 8 mm; for d, e, bar = 4 mm. (Based on *Cid et al. 4648*.)

Phenology. Collected in flower in May and June.

Distribution (Fig. 33). Brazil (Mato Grosso, Rondônia); at forest margin, along riverbanks, and in "cerrado degradado."

ADDITIONAL SPECIMENS EXAMINED. **Brazil.** MATO GROSSO: Mpio. Cáceres, Gleba Facão, rodovia para o Rio Jacobina, *Hatschbach 62296* (MICH); Mpio. Alto Paraguai, Rod. BR-246, Km 47, *Hatschbach 62781* (MICH); Rio Juruena, Aripuanã, Fontanilha, RADAM SC-21-YD, descampado da beira do rio, *M. G. Silva 3192* (MG, NY).

Stigmaphyllon stylopogon is named for the bearded styles, each with an adaxial row of hairs. The anthers are glabrous, and those opposing the posterior-lateral sepals bear only one reduced locule. The abaxial laminar pubescence is so dense that the epidermis and the marginal glands are hidden. This species is similar to *S. matogrossense* (no. 75), which also has the flag petal bearing gland-tipped fimbriae at the base of the limb and abaxially tomentose laminas. *Stigmaphyllon matogrossense* differs in that the abaxial laminar pubescence is less dense (the epidermis not hidden) and not appressed, and that the laminar margin is eglandular. Also, the anthers of stamens opposing the lateral sepals are eloculate (sterile), and the anterior style is glabrous.

75. Stigmaphyllon matogrossense C. Anderson, Contr. Univ. Michigan Herb. 17: 11. 1990.—TYPE: BRAZIL. Mato Grosso: Mpio. Cáceres, 9 km ENE de Porto Esperidião, BR-174, 150 m, 18 May 1985, *Krapovickas et al. 40114* (holotype: MICH!; isotype: CTES!).

Vine. Stems and branches sericeous when young, soon becoming glabrous. Laminas 5.5–9.6 cm long, 3–5.8 cm wide, ovate or narrowly so, apex mucronate or emarginate-mucronate, base cordate or shallowly cordate, glabrous adaxially, tomentose abaxially (trabecula 0.5–0.7 mm long, crisped and curled, stalk 0.2–0.3 mm long), margin eglandular; petioles 1.3–4.5 cm long, sericeous, not confluent across the node, with a pair of prominent but sessile glands at the apex, each gland 1.4–2 mm in diameter; stipules 1–1.4 mm long, 0.8–1.3 mm wide, free, triangular, eglandular. Flowers ca. 15 per umbel, these solitary or borne in dichasia or compound dichasia (axes to the 3rd order, sericeous). Peduncles 3.5–4.5 mm long; pedicels 5–8.5 mm long, terete; both sericeous, peduncles 0.5–0.8 times as long as the pedicels. Bracts 1.5–2.2 mm long, 1.2–1.6 mm wide, triangular, apex acute; bracteoles 1.6–2 mm long, 1.2–1.5 mm wide, triangular, apex acute, eglandular or each bracteole with a pair of inconspicuous glands (each gland 0.1–0.2 mm in diameter); bracts and bracteoles sericeous abaxially. Sepals 2.8–3 mm long and wide, glands ca. 2.2 mm long, 1–1.1 mm wide. All petals with the limb orbicular, glabrous, yellow, margin with gland-tipped fimbriae up to 0.5 mm long; anterior-lateral petals: claw 2–2.3 mm long, limb ca. 13 mm long and wide; posterior-lateral petals: claw ca. 1.5 mm long, limb ca. 10 mm long and wide; posterior petal: claw 3–3.2 mm long, apex indented, limb 9–9.5 mm long and wide, at the base with 1–3 stout glandular fimbriae per side (0.6–0.7 mm long, ca. 0.2 mm wide). Stamens unequal, those opposite the posterior-lateral petals (and the posterior styles) the largest, anthers of those opposite the lateral sepals eloculate; anthers glabrous. Stamen opposite anterior sepal: filament ca. 3 mm long, anther ca. 1.7 mm long; stamens opposite anterior-lateral petals: filaments ca. 2.2 mm long, anthers ca. 1.4 mm long; stamens opposite anterior-lateral sepals: filaments ca. 3 mm long, connectives ca. 0.7 mm long, locules absent; stamens opposite posterior-lateral petals: filaments ca. 4

mm long, anthers ca. 1.5 mm long; stamens opposite posterior-lateral sepals: filaments ca. 3 mm long, connectives ca. 0.8 mm long, locules absent; stamen opposite posterior petal always shorter than the adjacent two: filament ca. 2.7 mm long, anther ca. 0.8 mm long. Anterior style ca. 3.5 mm long, shorter than the posterior two, terete, glabrous, erect or slightly recurved; apex ca. 2.4 mm long, each foliole ca. 1.7 mm long, ca. 2.2 mm wide, subrectangular. Posterior styles ca. 4.3 mm long, terete, adaxially with a row of hairs in the proximal 2/3–3/4, lyrate; foliole ca. 2.2 mm long, ca. 2.7 mm wide, suborbicular. Samara not seen. Chromosome number unknown. Fig. 74.

Stigmaphyllon matogrossense is known only from the type, collected in western Brazil (Fig. 68). It is distinguished by its flowers and leaves. The petals are all glandular-fimbriate, but the posterior petal also bears 1–3 stout glandular fimbriae per side at the base of the limb. The anthers of stamens opposing the lateral sepals lack locules, and the posterior styles are bearded. The laminas are tomentose abaxially (but the epidermis visible) and lack marginal glands. *Stigmaphyllon stylopogon* (no. 74), also from Mato Grosso, has similarly ornamented petals, but is readily separated by its laminas, which are so densely appressed-tomentulose abaxially that the epidermis is hidden. The anthers of stamens opposing the lateral sepals bear at least one locule, and all the styles are bearded.

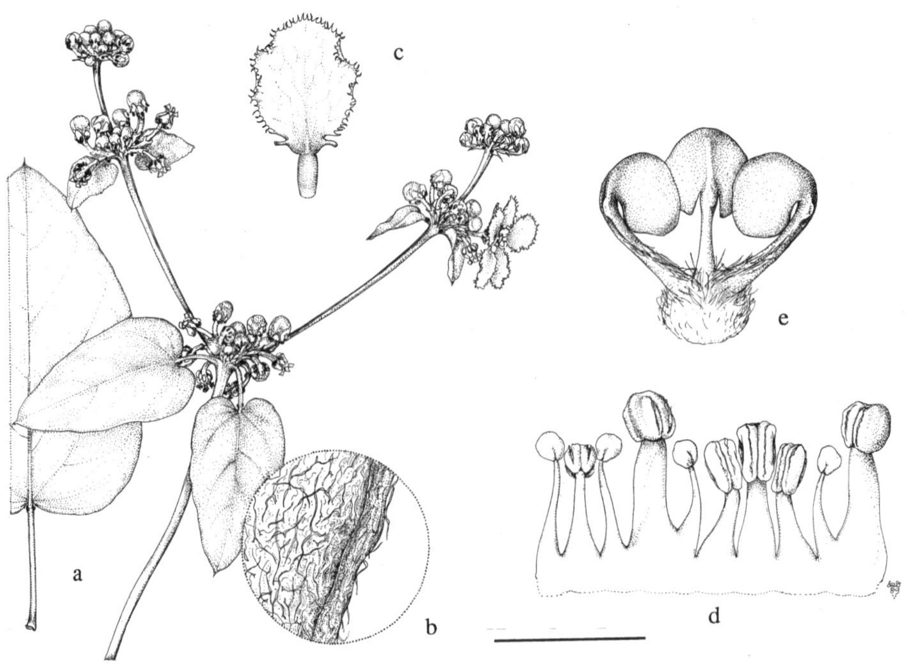

FIG. 74. *Stigmaphyllon matogrossense*. a. Flowering branch and portion of large leaf. b. Detail of margin and abaxial surface of lamina. c. Posterior petal (the "flag"). d. Androecium; stamen second from left opposes the posterior petal. e. Gynoecium; posterior styles are bent slightly outward to show anterior style (in center). Scale: for a, bar = 4 cm; for b, bar = 1.3 cm; for c, bar = 1 cm; for d, e, bar = 4 mm. (Based on *Krapovickas et al. 40114.*)

76. Stigmaphyllon angustilobum Adr. Jussieu in St.-Hilaire, Fl. bras. mer. 3: 53. 1833. ["1832"].—TYPE: BRAZIL. São Paulo: prés Bananal, *St.-Hilaire D786* (holotype: P!, photo: MICH!; isotype: P!).

Stigmaphyllon multilobum Miquel, Linnaea 19: 433. 1846.—TYPE: BRAZIL. *Claussen 2026* (holotype: U!, photo: MICH!; isotypes: G! MO! P! W!).

Vine. Stems and branches with scalelike T-shaped hairs. Laminas in outline 9.3–15.5 cm long, 8–16 cm wide, pinnately 5–7-lobed, terminal lobe usually the largest (6.5–9.5 cm long, 1.6–3.8 cm wide), lateral lobes 3.5–10 cm long, 1.1–2 cm wide, the basalmost lobes usually smaller than the other lateral lobes, apex of all lobes acuminate-mucronate, base of lamina truncate to slightly cordate, with T-shaped hairs on the major veins but otherwise glabrous adaxially, with T-shaped hairs abaxially (trabecula 0.9–1.7 mm long, straight or slightly wavy, stalk 0.2–0.4 mm long), the abaxial pubescence sometimes dense, margin eglandular or sometimes with irregularly spaced prominent glands (0.1–0.3 mm in diameter) borne adjacent to the margin abaxially; petioles 1–4.5 cm long, densely pubescent with T-shaped hairs, not confluent across the node, with a pair of prominent but sessile glands at the apex or up to 12 mm below the base of the lamina, each gland 1–2.3 mm in diameter; stipules 0.5–1.2 mm long, 0.4–1.1 mm wide, free, triangular, eglandular. Flowers ca. 15–30 (–35) per pseudoraceme, these solitary or borne in dichasia or compound dichasia or large thyrses (axes to the 3rd order, densely pubescent with scalelike T-shaped hairs). Peduncles 7–16.5 mm long; pedicels (6.5–) 7–11 mm long, terete; both densely sericeous, peduncles 1–2 times as long as the pedicels. Bracts 1.2–3.5 mm long, 0.6–1.4 (–2) mm wide, triangular, apex acute; bracteoles 1.2–2.2 mm long, 0.8–1.2 mm wide, triangular to sometimes oblong, apex obtuse, eglandular or each bracteole with a pair of inconspicuous glands (each gland 0.2–0.4 mm in diameter); bracts and bracteoles densely sericeous abaxially. Sepals 2.5–3.5 mm long, 2.5–3 mm wide, glands 2.3–3 mm long, ca. 1 mm wide. All petals with the limb orbicular or limb of the posterior petal broadly obovate, glabrous, yellow, margin with fimbriae up to 1 mm long; anterior-lateral petals: claw 2–2.5 mm long, limb 14–17 mm long and wide; posterior-lateral petals: claw 1–1.5 mm long, limb 13–16.5 mm long and wide; posterior petal: claw 3–4 mm long, apex indented or not, limb 12–14.5 mm long, 11–14 mm wide. Stamens unequal, those opposite the posterior-lateral petals (and the posterior styles) the largest but those opposite the lateral sepals usually with equally or subequally long filaments, anthers of those opposite the lateral sepals with the connective enlarged and the locules reduced, frequently with only one locule or eloculate; anthers pubescent. Stamen opposite anterior sepal: filament 2.1–3.1 mm long, anther 1.1–1.3 mm long; stamens opposite anterior-lateral petals: filaments 1.8–2.5 mm long, anthers 0.8–0.9 mm long; stamens opposite anterior-lateral sepals: filaments 3–4 mm long, connectives 0.8–1 mm long, locules 0.3–0.6 mm long or absent; stamens opposite posterior-lateral petals: filaments 3.1–4 mm long, anthers 1.1–1.6 mm long; stamens opposite posterior-lateral sepals: filaments 2.7–3 mm long, connectives 0.8–0.9 mm long, locules 0.2–0.4 mm long or absent; stamen opposite posterior petal always shorter than the adjacent two: filament 2.5–2.7 mm long, anther 0.6–0.8 mm long. Anterior style 3–3.8 mm long, shorter than the posterior two, terete, glabrous, erect or slightly recurved; apex 1.8–2 mm long, each foliole 1.3–1.5 mm long, 1.3–1.7 mm wide, subsquare to subrectangular. Posterior styles 3.8–4.5 mm long, terete, pubescent in the proximal 1/2–2/3, lyrate; folioles 2.5–3.3 mm long, 2.5–3 mm wide, subsquare. Samara not seen. Chromosome number unknown.

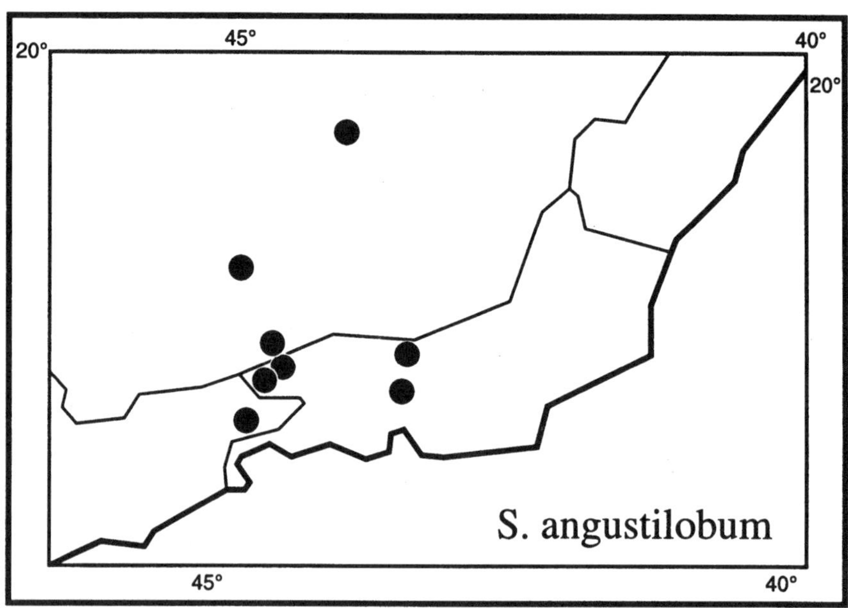

FIG. 75. Distribution of *Stigmaphyllon angustilobum.*

Phenology. Collected in flower from April through July and from October through December.

Distribution (Fig. 75). Brazil (southeastern Minas Gerais, Rio de Janeiro, and adjacent São Paulo); in woods and disturbed areas, such as fields and roadsides; 300–1600 m.

REPRESENTATIVE SPECIMENS. **Brazil.** MINAS GERAIS: Serraria, *Motta R20847* (R); Serra de Carrancas, *Rabelo R147058* (R).—RIO DE JANEIRO: Serra do Itatiaia, *Brade 10489* (R); Parque Nacional de Itatiaia, Km 1, estrada Maromba, *Ducke 1185* (RB, SP); près de la station de Boa Vista à la base de la Serra do Picú, *Glaziou 10360* (C, K, LE, P); Resende, Faz. de Barra, *J. Kuhlmann RB26353* (RB); Vale das Videiras, Petrópolis, *Occhioni 6466* (MICH); Parque Nacional de Itatiaia, Maromba, *Plowman 2865* (GH, US), *Sucre 5165* (K, RB).—SÃO PAULO: S. José do Barreiro, *Hoehne & Gehrt SP17660* (GH, NY, SP); Serra da Bocina, Faz. Bonito, *Lutz R147049* (R).

Stigmaphyllon angustilobum is readily recognized by its pinnately lobed leaves and large flowers, aggregated into pseudoracemes. The limb of the lateral petals is 14–17 mm in diameter and fringed with fimbriae up to 1 mm long. *Stigmaphyllon carautae* (no. 21), *S. urenifolium* (no. 7), and *S. vitifolium* (no. 51) are the only sympatric species with lobed leaves. *Stigmaphyllon carautae* and *S. urenifolium* are readily separated by their efoliolate styles and *S. vitifolium* by its glabrate to glabrous leaves.

77. Stigmaphyllon alternifolium Adr. Jussieu in St.-Hilaire, Fl. Bras. mer. 3: 54. 1833. ["1832"].—TYPE: BRAZIL. Rio de Janeiro: "prope Sebastianopolis," June, *St.-Hilaire A1, no. 17* (holotype: P!, photos: F! MICH! US!).

Stigmaphyllon alulatum Niedenzu, Ind. Lect. Lyc. Brunsberg., p. aest. 1900: 16. 1900.—TYPE: BRAZIL. Rio de Janeiro: Teresópolis, *de Moura 636* (holotype: B, destroyed, photos: A! F! MICH! NY! US!).

Vine. Stems and branches commonly with scalelike T-shaped hairs. Laminas 5.3–12 cm long, 2.8–5.5 cm wide, narrowly to broadly lanceolate to sometimes ovate, apex acuminate or mucronate or rarely caudate, base truncate or sometimes cordate, glabrate to glabrous adaxially, with T-shaped hairs to tomentose abaxially (trabecula 0.2–1.1 mm long, wavy to crisped and curled, stalk 0.1–0.2 mm long), margin eglandular or sometimes with irregularly spaced sessile glands (ca. 0.2 mm in diameter) borne adjacent to the margin abaxially; petioles 1–4.8 cm long, with scalelike T-shaped hairs, not confluent across the node, with a pair of prominent but sessile glands at the apex, each gland 0.6–1.2 mm in diameter; stipules 0.2–0.8 mm long, 0.2–1 mm wide, free, triangular, eglandular. Flowers ca. 10–25 per umbel or (sometimes) pseudoraceme, these solitary or borne in simple or compound dichasia or small thyrses (axes to the 3rd order, with scalelike T-shaped hairs). Peduncles 2.5–8.5 mm long; pedicels (3–) 5–9 mm long, terete; both densely sericeous, peduncles 0.3–1.5 times as long as the pedicels. Bracts 1.2–2.5 mm long, 0.6–1.1 mm wide, triangular or narrowly so, apex acute; bracteoles 1–1.7 mm long, 0.6–1.1 mm wide, triangular to parabolic, apex obtuse, eglandular; bracts and bracteoles densely sericeous abaxially. Sepals 1.8–2.6 mm long, 1.6–2.4 mm wide, glands 1.3–2 mm long, 0.8–1 mm wide. All petals with the limb glabrous, yellow, orbicular, margin digitate-fimbriate, fimbriae up to 0.5 (–0.6) mm long; anterior-lateral petals: claw 1.2–2 mm long, limb 8.5–11 mm long and wide; posterior-lateral petals: claw 1–1.5 mm long, limb 8–10 mm long and wide; posterior petal: claw 2.5–2.8 mm long, apex indented, limb 6.5–7.5 mm long and wide. Stamens unequal, those opposite the posterior-lateral petals (and the posterior styles) the largest (or sometimes stamen opposite the anterior style equally large) but those opposite the anterior-lateral and posterior-lateral sepals with the filaments subequally long, anthers of those opposite the anterior-lateral sepals with the connective enlarged and the locules reduced, those opposite the posterior-lateral sepals eloculate; all loculate anthers pubescent. Stamen opposite anterior sepal: filament 1.6–2.4 mm long, anther 0.9–1 mm long; stamens opposite anterior-lateral petals: filaments 1.5–2 mm long, anthers 0.7–0.8 mm long; stamens opposite anterior-lateral sepals: filaments 2–2.9 mm long, connectives 0.7–0.9 mm long, locules 0.2–0.3 mm long; stamens opposite posterior-lateral petals: filaments 2–3.2 mm long, anthers 1–1.1 mm long; stamens opposite posterior-lateral sepals: filaments 1.8–2.4 mm long, connectives 0.6–0.8 mm long, locules absent; stamen opposite posterior petal always at least slightly shorter than the adjacent two: filament 1.6–2.2 mm long, anther 0.5–0.7 mm long. Anterior style 2.2–3.3 mm long, shorter than the posterior two, terete, glabrous, erect; apex 1.2–1.7 mm long, each foliole 1–1.4 mm long, 0.6–1.2 mm wide, subrectangular or subsquare. Posterior styles 2.5–3.7 mm long, terete proximally, flattened distally, pubescent in the proximal 1/2, lyrate; foliole 1.2–2 mm long, 1.4–2 mm wide, square or sometimes subtrapezoidal. Dorsal wing of samara 3.9–4.7 cm long, 1.2–2.1 cm wide, upper margin with a blunt tooth; nut bearing on each side a row of ca. 5–7 rectangular to semi-circular to lunate lateral winglets, these 1–6.5 mm long, 1–1.7 mm wide; nut 4–7 mm high, 5–8 mm in diameter, areole 3.5–4.5 mm long, 3–3.5 mm wide, concave, carpophore up to 2 mm long. Embryo ca. 7 mm long, ca. 1 1/2 times as long as wide, ovoid, outer cotyledon ca. 5.5 mm long, ca. 4.3 mm wide, the distal 1/3 folded over the inner cotyledon, inner cotyledon ca. 4.9 mm long, ca. 3.5 mm wide, folded at the middle. Chromosome number unknown.

Phenology. Collected in flower in March through June, and in August, October, and December, in fruit in April, July, and September.

Distribution (Fig. 76). Brazil (Rio de Janeiro, and adjacent Espírito Santo and Minas Gerais); at roadsides and forest edges; 660–2000 m.

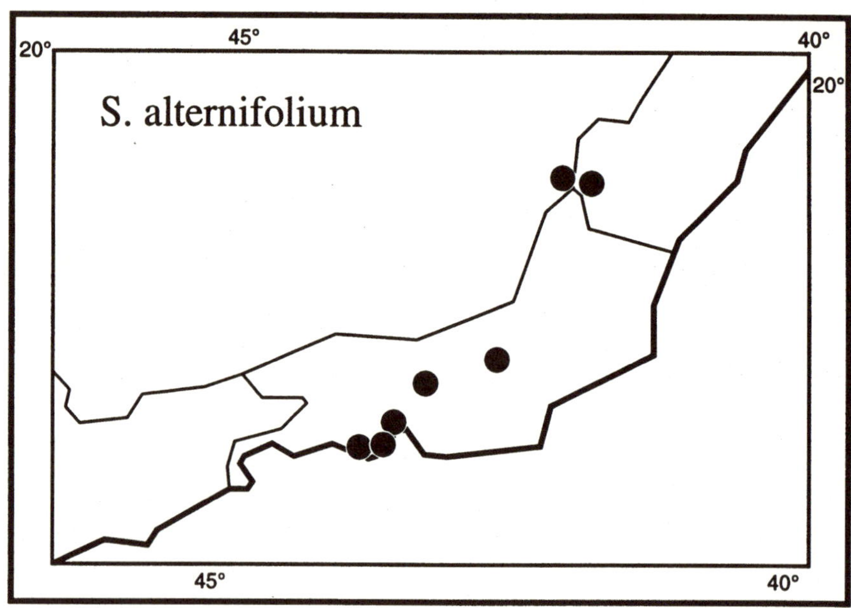

FIG. 76. Distribution of *Stigmaphyllon alternifolium.*

REPRESENTATIVE SPECIMENS. **Brazil.** RIO DE JANEIRO: Pico do Papagaio, *Almeida de Jesus 1892* (RB); Teresópolis, Serra Carvalho, *Brade 9841* (R); Serra da Carioca, *Brade R20868* (R); Nova Friburgo, 25 Oct 1953, *Capell s.n.* (MA-265752, RB-85161); Pedra da Gávea, entre a Mesa e a Cabeça, *Carauta 2372* (GUA); Corcovado, junto do Cristo, *Duarte 1130* (RB, SP); Petrópolis, Carangola, *Goés & Constantino 1059* (RB); Serra dos Orgâos, Retiro, *Lützelburg 12511* (M); Serra do Mendanha, *Sucre 8879* (MICH, SP).—ESPÍRITO SANTO: Mpio. Iuna, Pouso Alto, *Hatschbach 61121* (MICH); Serra do Caparão, *Occhioni 7181* (MICH).—MINAS GERAIS: Rancho das Tres Barras, Serra do Caparão, *N. Santos & Campos R52102* (R).

Stigmaphyllon alternifolium is a relatively small-leaved vine with a distinctive samara in which each side of the nut bears 5–7 lateral winglets. The flowers are also small, and have fimbriate petals and pubescent anthers. The stamens opposing the posterior-lateral sepals are sterile, i.e., the enlarged connectives lack locules. The epithet reflects the commonly alternate arrangement of the leaves in the distal axes, particularly near the inflorescences. Such a shift from opposite to alternate phyllotaxy is exhibited by many species of *Stigmaphyllon* as well as other vines.

78. Stigmaphyllon alternans Triana & Planchon, Ann. Sci. Nat. Bot., sér. 4, 18: 316. 1862.—TYPE: COLOMBIA. Meta: Villavicencio, 450 m, Jan 1856, *Triana s.n.* (holotype: P!, photo: MICH!; isotypes: BM! BR! COL! G! K! P! US!; photo of BM isotype: MICH!, photos of G isotype: F! GH! MICH! MO!).

Vine to 15 m. Stems and branches with scalelike T-shaped hairs when young, soon becoming glabrate. Leaves alternate or rarely subopposite. Laminas 6.5–21.5 cm long, 5.5–18 cm wide, broadly ovate to elliptical to broadly so to orbicular, apex mucronate, base cordate to auriculate, glabrous adaxially, tomentose or sparsely so abaxially (trabecula 0.5–1.6 mm long, crisped and curled, stalk 0.2–0.5 mm long); margin with irregularly spaced raised to short-stipitate glands (0.3–0.5 mm in diameter, 0.1–0.2 mm long) borne adjacent to the margin abaxially; petioles 2.6–10.5 cm long, densely pubescent with scalelike T-shaped hairs, not confluent across the node, with a pair of prominent but sessile glands at apex, each gland 1.8–3.5 mm in diameter; stipules 0.8–1 mm long, 0.5–1.1 mm wide, free, triangular or narrowly so, eglandular. Flowers ca. 10–15 per umbel, these borne in dichasia and compound dichasia or thyrses (axes to the 5th order, sometimes alternate, with scalelike T-shaped hairs). Peduncles 3.5–6.5 mm long; pedicels 4.5–8.5 mm long, terete; both densely pubescent with scalelike T-shaped hairs, peduncles 0.6–1.2 times as long as the pedicels. Bracts 1.1–2.7 mm long, 0.9–2 mm wide, triangular, apex acute; bracteoles 1–1.8 mm long, 0.8–1.5 mm wide, triangular, apex obtuse, eglandular or sometimes each bracteole with a pair of inconspicuous glands (each gland ca. 0.3 mm in diameter); bracts and bracteoles sericeous (hairs subsessile) abaxially. Sepals 2.5–3 mm long and wide, glands 2–3 mm long, 1.2–1.5 mm wide. All petals with the limb orbicular, glabrous, yellow, margin erose or denticulate-erose or only the apex erose and the sides denticulate-fimbriate, teeth/fimbriae up to 0.5 mm long; anterior-lateral petals: claw (1.5–) 2–3 mm long, limb 11–12.5 mm long and wide; posterior-lateral petals: claw 1.2–1.8 mm long, limb ca. 10 mm long and wide; posterior petal: claw 4–4.3 mm long, apex indented, limb 7.5–9 mm long and wide. Stamens unequal, those opposite the posterior-lateral petals (and the posterior styles) the largest, anthers of those opposite the lateral sepals with the connective enlarged and the locules reduced; anthers all loculate, pubescent. Stamen opposite anterior sepal: filament 2.4–2.8 mm long, anther 1.2–1.4 mm long; stamens opposite anterior-lateral petals: filaments 1.8–2 mm long, anthers 0.8–1.1 mm long; stamens opposite anterior-lateral sepals: filaments 2.5–3.2 mm long, connectives 0.6–0.8 mm long, locules 0.3–0.4 mm long; stamens opposite posterior-lateral petals: filaments 3.3–3.7 mm long, anthers 1.2–1.3 mm long; stamens opposite posterior-lateral sepals: filaments 2.3–2.5 mm long, connectives 0.5–0.6 mm long, locules 0.3–0.4 mm long; stamen opposite posterior petal always shorter than the adjacent two: filament ca. 2 mm long, anther ca. 0.8 mm long. Anterior style 3.3–3.8 mm long, shorter than the posterior two, terete, glabrous, erect, apex 1.3–1.7 mm long, each foliole ca. 2 mm long, 1.8–2.2 mm wide, subsquare, fused for 50–90% of their length and cucullate. Posterior styles 4–4.2 mm long, terete, glabrous or with scattered hairs in the proximal 1/2, lyrate; foliole 2.2–2.9 mm long, 2.1–2.3 mm wide, subsquare. Dorsal wing of samara 3.8–4.6 cm long, 1.7–1.9 cm wide, upper margin with a blunt tooth; nut bearing a pair of rectangular lateral winglets, these 8–9.5 mm long, 2–2.5 mm wide, or only with spurs and/or crests and/or lateral ridges; nut ca. 8.5 mm high, ca. 5 mm in diameter, without air chambers, areole ca. 4.5 mm long, ca. 5 mm wide, strongly concave, carpophore up to 4.5 mm long. Mature embryo not seen. Chromosome number unknown.

Phenology. Collected in flower in January, February, June, July, and September; in fruit in February and July.

Distribution (Fig. 5). Colombia (Caquetá, Meta), eastern Ecuador (Napo), and northern Peru (Loreto, San Martín); in lowland rainforest and at forest margin; 200–450 m.

ADDITIONAL SPECIMENS EXAMINED. **Colombia.** CAQUETÁ: Mpio. Vicente del Caguán, vereda Puerto Losada, vía de Mina Blanca–Puerto Losada (16 km), *Callejas & Marulanda 5892* (MO).—META: Villavicencio, Llano de San Martín, *Karsten s.n.* (LE, W). **Ecuador.** NAPO: Yasuní Forest Reserve, along Tiputini River, E of PUCE Scientific Station, 00°40.853′S, 76°23.697′W, *Acevedo-Rodríguez & Cedeño 7466* (MICH); Yasuni Forest Reserve, along rd between Km 70 and 100, E of PUCE Scientific Station, 00°50.014′S, 76°20.518′W–00°54.730′S, 76°13.304′W, *Acevedo-Rodríguez & Cedeño 7627* (MICH); Aguarico, Reserva Etnica Huarorani, carretera y oleoducto de Maxus, Km 111–112, 01°00′S, 76°11′W, *Duk & Ahue 1569* (MICH); road Coca (Puerto Francisco de Orellana)–Curaray, 20–30 km S of Coca, *Harling & Andersson 11942* (GB, MICH); Estación Experimental INIAP–Napo Payamino, 5 km al N Coca, Reserva Floristica "El Chuncho," 00°25′S, 77°00′W, *Palacios & Neill 1225* (MO); Aguarico Cantón, Reserva Faunistica Cuyabeno, 00°29′S, 73°32′W, *Palacios et al. 7729* (MICH). **Peru.** LORETO: Prov. Loreto, Marsella, Río Tigre, 02°30′S, 75°50′W, *Lewis 12691* (MO).—SAN MARTÍN: Alto Río Huallaga, Tarapoto, *Ll. Williams 5606* (F), *6499* (F, US), *6725* (F, US).

Stigmaphyllon alternans is named for the alternate disposition of its leaves in a family characterized by opposite leaves. In most specimens seen, the leaves are alternate; the exceptions are *Ll. Williams 5606, Ll. Williams 6499* (F; one inflorescence shoot with two pairs of subopposite leaves), and *Ll. Williams 6725* (F; one inflorescence shoot with three pairs of subopposite leaves), all collected in December, 1929, and labeled with the same locality. It is not uncommon in other species to see an alternate arrangement at nodes near or in the inflorescence; however, consistently alternate leaves on vegetative axes apparently occur only in *S. alternans*. Unfortunately, none of the collections include leaves from near the base of the vine. This species may be confused with the widely distributed (but not sympatric) *S. dichotomum* (no. 65) of southern Panama, Venezuela, and Colombia. *Stigmaphyllon alternans* is readily separated by its marginal leaf glands, which are borne on the blade adjacent to the margin, and its pubescent anthers. In *S. dichotomum*, the laminas bear sessile glands and also filiform glands up to 6 mm long on the margin, and the anthers are glabrous. The position of the marginal leaf glands in *S. alternans* is noteworthy, because glands on the lower surface adjacent to the margin are otherwise known only in species of eastern Brazil.

79. Stigmaphyllon tomentosum Adr. Jussieu in St.-Hilaire, Fl. bras. mer. 3: 53. 1833 ["1832"], non *Stigmaphyllon tomentosum* (Desfontaines ex DC.) Niedenzu, 1899.—TYPE: BRAZIL. Minas Gerais: prope Itajurú praedium non procul a vico S. Miguel de Mato Dentro, Feb, *St.-Hilaire B1 643* (holotype: P!, photo: MICH!; isotypes: P!, photos: MICH!).

Stigmaphyllon affine var. *paulinum* Niedenzu, Ind. Lect. Lyc. Brunsberg. p. aest. 1900: 14. 1900.—TYPE: BRAZIL. São Paulo: Rio Pardo, *Commiss. geogr. e geol. da Prov. de S. Paulo 110* (holotype: B, destroyed, photos: A! F! MICH! NY! US!).

Stigmaphyllon psilocardium Niedenzu, Ind. Lect. Lyc. Brunsberg. p. aest. 1900: 15. 1900. *Stigmaphyllon eriocardium* Niedenzu, Verz. Vorles. Ak. Braunsberg W.-S. 1912–1913: 29. 1912, nom. superfl.—TYPE: BRAZIL. *Glaziou 10362* (holotype: B, destroyed, photos: A! F! MICH! NY! US!; isotypes: C! LE! P!, photo of P isotype: MICH!).

Vine to 8 m. Stems and branches with scalelike T-shaped hairs when young, soon becoming glabrous. Laminas 6.2–31 cm long, 5–19 cm wide, lanceolate to ovate to triangular to cordate to elliptical or sometimes palmately 2–5-lobed, apex mucronate or emarginate-mucronate, base truncate to cordate or deeply so, glabrate to glabrous adaxially, with T-shaped hairs to sometimes tomentose abaxially (trabecula 0.5–1.5 mm long, straight to wavy to crisped and curled, stalk 0.2–0.4 mm long), margin with irregularly spaced sessile to slightly raised glands (0.2–0.3 mm in diameter, up to 0.1 mm long) borne adjacent to the margin abaxially; petioles 1.5–9.5 cm long, densely pubescent with scalelike T-shaped hairs, not confluent across the node, with a pair of prominent but sessile glands at the apex, each gland 1–2.8 mm in diameter; stipules 0.5–1.8 mm long, 0.5–1.2 mm wide, free, triangular, eglandular. Flowers ca. 10–40 per umbel or pseudoraceme, these solitary or borne in dichasia or compound dichasia or small thyrses (axes to the 3rd order, with scalelike T-shaped hairs). Peduncles 3–25.5 mm long; pedicels 3–14 mm long, terete; both densely sericeous, peduncles 0.6–2.5 times as long as the pedicels. Bracts (1.2–) 1.5–2.5 mm long, 0.6–1.8 mm wide, triangular or narrowly so, apex acute; bracteoles 1.2–2.5 mm long, 0.6–1.2 (–2) mm wide, oblong or triangular or narrowly so, apex obtuse, eglandular or each bracteole with a pair of inconspicuous glands (each gland 0.3–0.4 mm in diameter); bracts and bracteoles sericeous or densely so abaxially. Sepals (1.5–) 1.8–2.7 mm long, 2–3 mm wide, glands 1.8–3 mm long, 0.8–1.5 mm wide. All petals with the limb orbicular, glabrous, yellow or sometimes suffused with red, margin fimbriate or sometimes denticulate–fimbriate, fimbriae of the lateral petals up to 0.5 mm long, sometimes gland-tipped; anterior-lateral petals: claw (1.5–) 2–2.5 (–3.2) mm long, limb (9–) 11–15 mm long and wide; posterior-lateral petals: claw (0.8–) 1–1.5 (–2.5) mm long, limb 8–15 mm long and wide; posterior petal: claw 2.5–3.8 mm long, apex indented or slightly so, limb (7–) 9–11 mm long and wide, fimbriae up to 0.7 mm long, sometimes gland-tipped, often near the base with 1–3 stout gland-tipped fimbriae per side (0.3–0.5 mm long, ca. 0.2 mm wide). Stamens unequal, those opposite the posterior-lateral petals (and the posterior styles) the largest, sometimes those opposite the lateral sepals with the filaments subequally long, anthers of those opposite the lateral sepals with the connective enlarged and the locules reduced or commonly with only one locule or eloculate; anthers glabrous or sometimes pubescent. Stamen opposite anterior sepal: filament 2–2.5 mm long, anther 0.8–1.3 mm long; stamens opposite anterior-lateral petals: filaments 1.5–2 mm long, anthers 0.7–1 mm long; stamens opposite anterior-lateral sepals: filaments (2.2–) 2.5–3.2 mm long, connectives 0.7–0.9 mm long, locules 0.2–0.5 mm long or one or both locules absent; stamens opposite posterior-lateral petals: filaments (2.5–) 2.8–3.5 mm long, anthers 1–1.3 (–1.6) mm long; stamens opposite posterior-lateral sepals: filaments 2–3.2 mm long, connectives 0.7–1 mm long, locules 0.1–0.5 mm long or commonly one or both locules absent; stamen opposite posterior petal always shorter than the adjacent two: filament (1.8–) 2.3–2.7 mm long, anther 0.6–0.9 mm long. Anterior style 2.5–3.8 (–4.1) mm long, shorter than the posterior two or sometimes only slightly so, terete, glabrous, erect or slightly incurved; apex 1.3–2.2 mm long, each foliole 1.1–1.8 mm long, (0.8–) 1–2 mm wide, subsquare to subrectangular to parabolic. Posterior styles 2.9–4.5 mm long, terete, glabrous or with scattered hairs in the proximal 1/4–1/2, lyrate; foliole 1.6–2.8 mm and wide, subsquare to subtrapezoidal. Dorsal wing of samara 2.8–5.2 cm long, 1.3–3.6 (–4.1) cm wide, 1.7–2.7 times as long as wide, upper and lower margin with a blunt tooth; nut bearing 1–3 triangular to rectangular lateral winglets per side, these 1–12 (–20) mm long, 2–3.5 (–5) mm wide, sometimes also with one or two spurs or tubercles; nut 6.5–10 mm high, 5–6.3 (–8) mm in diameter, without air chambers, areole 3.5–6 mm long, 3.5–5.5 mm wide, concave, carpophore up to 3.5 mm long. Embryo ovoid, ca. 2 times as long as wide, outer cotyledon 11–11.5 mm long, 3.2–4.8 mm wide, the distal 2/5–3/5 folded over the inner cotyledon, inner cotyledon 5.2–7.5 mm long, 2.5–4.3 mm wide, straight or folded at the distal 1/10–1/3. Chromosome number unknown.

Phenology. Collected in flower throughout the year, in fruit from December through February and in July and August.

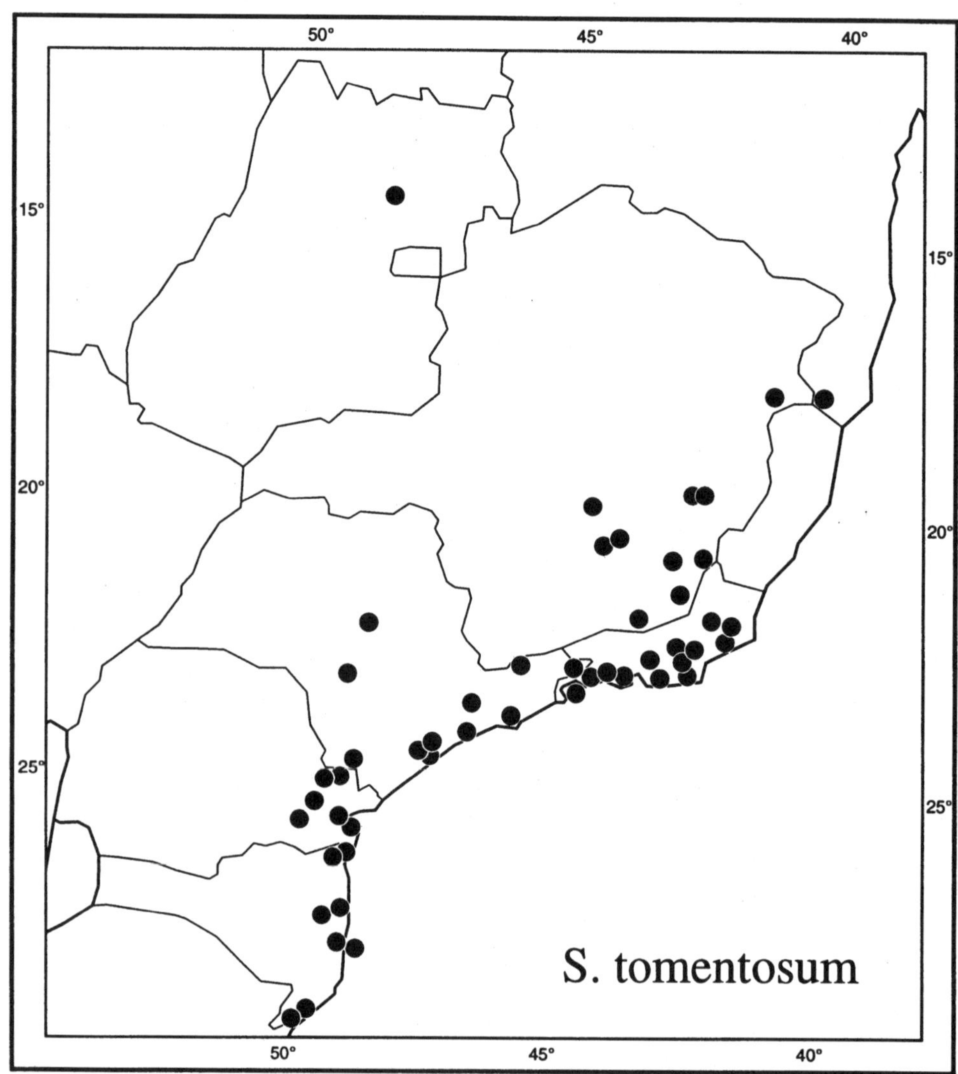

FIG. 77. Distribution of *Stigmaphyllon tomentosum.*

Distribution (Fig. 77). Southeastern Brazil (Rio de Janeiro, Minas Gerais and adjacent Bahia, central Goiás, São Paulo, Paraná, Santa Catarina); in cerrado, capoeiras, and forest, also in secondary growth, at forest edge and roadsides; sea level to 670 m.

REPRESENTATIVE SPECIMENS. **Brazil.** BAHIA: Argôlo, *Hatschbach 47805* (MICH).—GOÍAS: Alto Paraíso, *Heringer 18162* (K, MG, MICH, MO, US).—MINAS GERAIS: Mpio. Tombos, Fazenda do Cachoeira, *Barreto 1594* (F, R, UB); Belo Horizonte, Estação Experimental, *Barreto 8682* (R); Serra do Ouro Branco, *Glaziou 12481* (BR, C, F, G, K, LE, NY, P, R); Rod. BR-11, 10 km N de Mucuri, *Hatschbach 44290* (MICH); Viçosa, Agricultural College, *Mexia 4338* (A, BM, CAS, F, G, GB, GH, K, MICH, MO, NY, S, US, WIS); Juiz de Fora, *Roth 1593* (RB).—PARANÁ: Mpio. Paranaguá, Saquarema, *Cordeiro 38* (MBM, MICH); Jacarehy, *Dusén 17212* (G, GH, MO, S); Mpio. Antonina, Morro Grande, *Hatschbach 18108* (MICH, P); Mpio. Rio Branco do Sul, Serra do Voturuva, *Hatschbach 35703* (MICH); Mpio. Morrotes, Rio Cruzeiro, *Hatschbach 39873* (MICH); BR-

11, 10 km N de Mucuri, *Hatschbach 44290* (MICH); Adrianópolis, BR-373, *Krapovickas et al. 23268* (CTES); Mpio. Campo Largo, Açungui, *Kummrow 2731* (MICH); Mpio. Cerro Azul, Barra do Rio Ponta Grossa, *J. M. Silva 118* (MICH).—RIO DE JANEIRO: 25 km by rd SW of Sta. Maria Madalena, *Anderson 11715* (MBM, MICH); perto de Maricá, Lagoa do Padre, *Araujo 688* (MICH, RB); pr. Araruama, *Fromm et al. 1318* (R); *Glaziou 10361* (C, K, NY, P); Nova Friburgo, *Leite 4269* (F); Serra dos Orgãos, *Occhioni 6675* (MBM, MICH); Teófilo Otoni, *Trinta 738* (NY, R).—SANTA CATARINA: Alto Rio d'Una, Imarui, *Bresolin 926* (MICH); Rio Vermelho, Ilha de S. Catarina, *Bresolin 937* (MICH); Turvo–Arar, a margem do Rio A. Faca, *Reitz C87* (RB); Campo do Massiambú, Palhoça, *Reitz & Klein 1300* (S, US); Três Barras, Garuva, S. Francisco do Sul, *Reitz & Klein 6233* (MICH, P).—SÃO PAULO: Iporanga, *Dial FUEL2067* (MICH); Moji das Cruces, *Gehrt SP5485* (NY, SP); Piassaguéra, Santos, *Gehrt SP8237* (NY, SP); Alto da Serra Velha, *Gehrt SP12036* (GH, NY, SP); Mpio. Paraibuna, 10 km NW de Paraibuna, camino a Caraguatatuba, *Krapovickas et al. 14377* (CTES, P, WIS); entre Piedad e Juquiá, *M. Kuhlmann SP45744* (SP); Registro, *Moura SP123438* (SP).

Stigmaphyllon tomentosum is a widespread and variable species. The laminas are abundantly pubescent with T-shaped hairs abaxially; often the trabecula is crisped or curled and the vesture is then tomentose. The marginal glands are sessile to slightly raised and borne adjacent to the margin, buried in the vesture. The petals are fimbriate. Some regional tendencies may be discerned. Plants from Paraná and Santa Catarina in general have the flowers [(10–) 15–40] arranged in pseudoracemes, and the limbs of lateral petals are commonly 14–15 mm in diameter. The posterior petal (flag) often has stout gland-tipped fimbriae (0.3–0.5 mm long, ca. 0.2 mm in diameter) at the base of the limb. The enlarged connectives of stamens opposing the anterior-lateral sepals bear one or two reduced locules and those opposing the posterior-lateral sepals are usually eloculate or bear only one reduced locule; all anthers are usually glabrous. Plants from Rio de Janeiro and eastern Minas Gerais most commonly have 10–30 flowers disposed in umbels, and the limbs of lateral petals vary from 9–12 mm in diameter. The posterior petal is fimbriate, but stout gland-tipped fimbriae at the base of the limb are infrequent. The enlarged connectives of stamens opposing the anterior-lateral sepals usually bear two reduced locules, rarely only one. In specimens from Rio de Janeiro, the enlarged connectives of stamens opposing the posterior-lateral sepals are usually eloculate or sometimes bear one, rarely two, reduced locules; in specimens from Minas Gerais, they mostly bear one reduced locule though sometimes two or lack them entirely. The unmodified anthers may be glabrous or pubescent. Because all variants occur throughout the range, none are given taxonomic recognition here.

80. Stigmaphyllon saxicola C. Anderson, Contr. Univ. Mich. Herb. 17: 16. 1990.—TYPE: BRAZIL. Espirito Santo: Mpio. Ibatiba, Corrego S. João, sobre bloco de rochas, 14 Jun 1985, *Hatschbach 49397* (holotype: MBM!; isotype: MICH!).

Vine. Stems and branches bearing scalelike T-shaped hairs. Laminas 9–17 cm long, 8–11 cm wide, elliptical to ovate to suborbicular, apex emarginate-mucronate, base truncate to cordate, tomentulose to glabrate adaxially, densely tomentose abaxially (trabecula 0.6–1.3 mm long, crisped and curled, stalk 0.2–0.6 mm long), margin with irregularly spaced prominent to raised glands (up to 0.1 mm high, 0.2–0.3 mm in diameter) borne adjacent to the margin abaxially; petioles 2.5–4.5 cm long, densely pubescent with scalelike T-shaped hairs, not confluent across the node, with a pair of flush glands at the apex, each gland 1.5–2.3 mm in diameter; stipules 0.6–1.2 mm long, 0.5–1.1 mm wide, free, narrowly triangular, glabrous abaxially, eglandular. Flowers ca. 15–25 per umbel, these borne in dichasia or compound dichasia arranged in thyrses (axes to the 3rd order, bearing T-shaped hairs), the distal secondary axes sometimes alternate and divaricate and the inflorescence then appearing racemiform. Peduncles 6–15 mm long; pedicels 4.5–7.5 mm long, terete; both densely pubescent with T-shaped hairs, peduncles 0.5–3 times as long as the pedicels. Bracts 1–1.6 (–3) mm long, 0.4–1 mm wide, triangular or narrowly so, apex acute; bracteoles 1.1–2 mm long, 0.6–1 mm wide, triangular, apex obtuse or sometimes acute, eglandular; bracts and bracteoles sericeous abaxially. Sepals 2–2.6 mm long, (1.8–) 2.2–2.5 mm wide, glands (1.5–) 2–2.5 mm long, ca. 1 mm wide. All petals with the limb orbicular, glabrous, yellow, margin erose-denticulate to denticulate-fimbriate, teeth/fimbriae up to 0.3 mm long; anterior-lateral petals: claw (1.2–) 1.5–2.2 mm long, limb 8–9 mm long and wide; posterior-lateral petals: claw 0.7–1.5 mm long, limb 7–8.5 mm long and wide; posterior petal: claw 2–3 mm long, apex indented or slightly so, limb 6–7.5 mm long and wide, sometimes margin near the base with 1–3 stout gland-tipped fimbriae per side (0.3–0.6 mm long, ca. 0.2 mm wide). Stamens unequal, those opposite the posterior-lateral petals (and the posterior styles) with the longest filaments, anthers of those opposite the anterior-lateral sepals with the connective enlarged and the locules reduced, anthers of those opposite the posterior-lateral sepals with one or sometimes two reduced locules; anthers glabrous or sometimes sparsely pubescent. Stamen opposite anterior sepal: filament 1.9–2.2 mm long, anther 1.1–1.3 mm long; stamens opposite anterior-lateral petals: filaments 1.5–1.8 mm long, anthers 0.7–0.9 mm long; stamens opposite anterior-lateral sepals: filaments 2.3–2.5 mm long, connectives (0.6–) 0.9 mm long, locules 0.3–0.6 mm long; stamens opposite posterior-lateral petals: filaments 2.5–2.8 mm long, anthers ca. 1.2 mm long; stamens opposite posterior-lateral sepals: filaments 2–2.5 mm long, connectives 0.6–0.9 mm long, locules 0.1–0.5 mm long; stamen opposite posterior petal shorter than or subequal to the adjacent two: filament 1.7–2.3 mm long, anther 0.6–0.8 mm long. Anterior style 2.7–3 mm long, shorter than or sometimes subequal to the posterior two, terete, glabrous or with a few scattered hairs in the proximal 1/2, erect; apex 1.2–1.5 mm long, each foliole 1.1–1.3 mm long and wide, subsquare to suborbicular. Posterior styles (2.8–) 3.2–3.5 mm long, terete, pubescent in the proximal 1/2–3/4, lyrate; foliole 1.4–2 mm long and wide, subsquare to suborbicular. Dorsal wing of samara 5.5–6.2 cm long, 1.3–1.8 cm wide, 3.5–4.7 times as long as wide, upper margin with a slight blunt tooth; nut bearing one grossly dentate lateral winglet per side, these up to 1.2 mm long, and/or spurs and crests, these 1.5–6 mm long, 1.5–3.3 mm wide; nut 9.5–10 mm high, 5.5–6 mm in diameter, areole ca. 5.5 mm long and wide, concave, carpophore up to 2.5 mm long. Mature seed not seen. Chromosome number unknown. Fig. 78.

Phenology. Collected in flower in May through July, and in fruit in June.

Distribution (Fig. 9). Brazil (Bahia, Espírito Santo, eastern Minas Gerais); in campo, on rocks, along small streams; 950 m (one report).

ADDITIONAL SPECIMENS EXAMINED. **Brazil.** BAHIA: Jucarí, plantação de cacau, *Belém & Pinheiro 2349* (CEPEC, IAN, MICH, NY).—ESPÍRITO SANTO: Mpio. Ibatiba, Corrego S. João, *Hatschbach 49399* (MBM, MICH); Serra do Caparão, *Occhioni 7193* (MICH).—MINAS GERAIS: perto de Realeza, *Castellanos 24979* (GUA); Mpio. Congonhas do Norte, Alves, *Hatschbach 52962* (MICH).

Stigmaphyllon saxicola has relatively small flowers borne in umbels arranged in thyrses or sometimes in eccentrically branched inflorescences. The limbs of the petals are less than 1 cm in diameter. The samara is among the largest in the genus; the dorsal wing is ca. 6 cm long, and the nut is ca. 1 cm high. The abaxial surfaces of the laminas are tomentose and bear glands adjacent to the margin.

The collection from Minas Gerais, *Hatschbach 52962*, is placed here with some hesitation. The specimen consists of a length of mature stem with a young inflorescence and

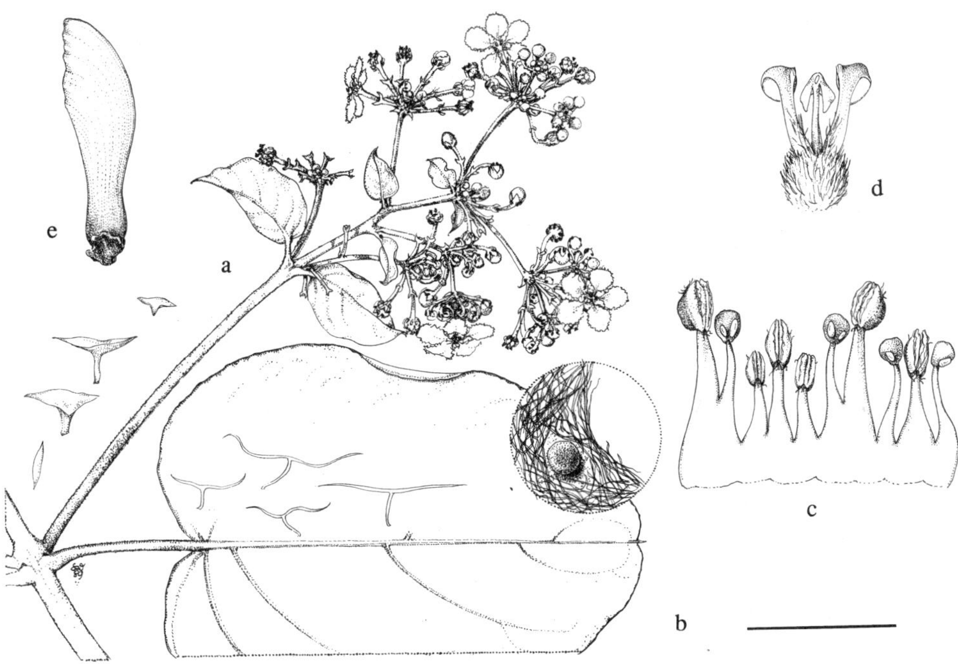

FIG. 78. *Stigmaphyllon saxicola*. a. Flowering branch and several stem hairs. b. Detail of margin and abaxial surface of lamina, and several laminar hairs. c. Androecium; stamen second from right opposes the posterior petal (the "flag"). d. Gynoecium, anterior style in the center. e. Samara. Scale: for a, e, bar = 4 cm; for b, bar = 0.8 mm; for c, bar = 2.7 mm; for d, bar = 4 mm; for individual hairs, bar = 1 mm. (Based on: a–d, *Hatschbach 49397*; e, *Hatschbach 49399*.)

four large old leaves that apparently have lost much of their vesture. It may represent yet another new species from an area noted for the high incidence of endemism.

81. Stigmaphyllon bonariense (Hooker & Arnott) C. Anderson, Brittonia 48: 543. 1997 ["1996"]. *Banisteria bonariensis* Hooker & Arnott, Bot. Misc. 3: 157. 1832.—TYPE: URUGUAY. *Tweedie s.n.* (lectotype, designated by C. Anderson, 1997: K!, photo: MICH!).

Stigmaphyllon littorale Adr. Jussieu in St.-Hilaire, Fl. bras. mer. 3: 55. 1833 ["1832"].—TYPE: URUGUAY. Colonia: "Ad littora fluminis Río de la Plata, prope urbem Colonia del Sacramento, in parte australi prov. Cisplatinae," *St.-Hilaire C2 2368* (holotype: P!, photos: F! MICH! US!; isotype: P!, photo: MICH!).

Stigmaphyllon heterophyllum Hooker, Bot. Mag. 69: t. 4014. 1843.—TYPE: based on cultivated material raised from seed sent by Tweedie from Buenos Aires, Argentina (holotype: K!, photo: MICH!).

Stigmaphyllon littorale var. *trilobum* Niedenzu, Ind. Lect. Lyc. Brunsberg. p. aest. 1900: 28. 1900.—TYPE: BRAZIL. São Paulo: Rio Sorocaba, 21 Oct 1887, *Löfgren 238* (holotype: B, destroyed; isotypes: C! P!, photo of C isotype: MICH!).

Vine to 10 m. Stems and branches sericeous and also with scalelike T-shaped hairs when young, becoming glabrate to glabrous. Laminas 6.2–16 cm long, 5–16.5 cm wide, triangular to cordate to broadly ovate to orbicular or sometimes 2–3-lobed, apex acuminate or obtuse- to emarginate-mucronate, base cordate to truncate or sometimes nearly attenuate, sparsely pubescent with T-shaped hairs to sparsely tomentulose to glabrous adaxially but always with some hairs on the major veins, with T-shaped hairs to sometimes tomentose abaxially (trabecula 0.5–1.9 mm long, straight to wavy to sometimes crisped and curled, stalk 0.2–0.4 mm long), margin eglandular or sometimes with irregularly spaced sessile glands (0.1–0.3 mm in diameter) borne adjacent to the margin abaxially; petioles 1.4–6.3 cm long, sericeous or sometimes also with scalelike T-shaped hairs, not confluent across the node, with a pair of prominent but sessile glands at the apex, each gland 1.2–2.5 mm in diameter; stipules 0.5–2 mm long, 0.4–1.5 mm wide, free, triangular or narrowly so, eglandular. Flowers ca. 8–25 per umbel, these solitary or borne in dichasia or rarely in compound dichasia (axes to the 3rd order, densely pubescent with scalelike T-shaped hairs to sericeous). Peduncles 3–17.5 mm long; pedicels 6.5–12.5 mm long, terete; both densely pubescent with scalelike T-shaped hairs to sericeous, peduncles 0.3–2.3 times as long as the pedicels. Bracts 1–2.5 mm long, 0.6–1.4 mm wide, triangular or narrowly so, apex acute; bracteoles 1–2 mm long, 0.7–1.2 mm wide, broadly triangular, apex obtuse, eglandular or each bracteole with a pair of inconspicuous glands (each gland 0.2–0.6 mm in diameter); bracts and bracteoles sericeous abaxially. Sepals 2–3.1 mm long, 2.1–3 mm wide, glands 1.9–2.7 mm long, 1–1.4 mm wide. All petals with the limb glabrous, yellow, limb of the lateral petals orbicular, margin irregularly denticulate or denticulate-fimbriate or fimbriate, teeth/fimbriae up to 0.3 (–0.5) mm long; anterior-lateral petals: claw 2.6–3.3 mm long, limb 10.5–12 mm long and wide; posterior-lateral petals: claw 1–2.2 mm long, limb 9–11.5 mm long and wide; posterior petal: claw 3.1–3.5 mm long, apex indented, limb 8–10 mm long, 7–9.5 mm wide, obovate to orbicular, margin with fimbriae up to 0.6 (–1) mm long, rarely denticulate-fimbriate. Stamens unequal, those opposite the posterior-lateral petals (and the posterior styles) the largest, sometimes those opposite the lateral sepals with the filaments nearly as long, anthers of those opposite the lateral sepals with the connective enlarged and the locules reduced, or sometimes anthers of stamens opposite the anterior-lateral sepals with the locules as long as the connective; anthers all loculate, glabrous or sometimes pubescent. Stamen opposite anterior sepal: filament 2–2.7 mm long, anther 1.1–1.5 mm long; stamens opposite anterior-lateral petals: filaments 1.6–22.5 mm long, anthers 0.8–1.2 mm long; stamens opposite anterior-lateral sepals: filaments 2.3–3.3 mm long, connectives (0.5–) 0.7–1 mm long, locules 0.3–1 mm long; stamens opposite posterior-lateral petals: filaments 2.4–3.3 mm long, anthers 1.2–1.5 (–1.8) mm long; stamens opposite posterior-lateral sepals: filaments 2–3.1 mm long, connectives 0.7–0.9 mm long, locules 0.3–0.6 mm long; stamen opposite posterior petal always at least slightly shorter than the adjacent two: filament 1.8–2.6 mm long, anther 0.7–0.9 mm long. Anterior style 3.1–3.9 mm long, shorter than the posterior two, terete, glabrous or with scattered hairs in the proximal 1/3, erect or slightly incurved; apex 1.2–3 mm long, each foliole 1.2–1.7 mm long, (0.6–) 0.8–1.8 mm wide, commonly subsquare to subrectangular or sometimes triangular. Posterior styles 3.7–4.2 mm long, terete, glabrous or sometimes with scattered hairs in the proximal 1/3, lyrate; foliole 1.8–2.3 mm long, (1.5–) 1.7–2.5 mm wide, subsquare to subrectangular and them sometimes slightly curved, rarely parabolic. Dorsal wing of samara encircling the nut, 1.6–2.9 cm high from base of nut, 1.7–2.7 cm wide; nut bearing 1–4 triangular to semi-circular to lunate lateral winglets per side, these up to 5.5 mm long, up to 3.5 mm wide, and/or 1–5 tubercles or a lunate ridge or sometimes the nut only prominently ribbed; nut 8.1–12 mm high, 5.5–6.2 mm in diameter, without air chambers, areole 3.3–6.5 mm long, 2.5–4.6 mm wide, concave, carpophore up to 5.5 mm long. Embryo 7.2–9.5 mm long, ca. 2 times as long as wide, ovoid, outer cotyledon 9.5–12 mm long, 4.3–5.2 mm wide, the distal 1/3–1/2 folded over the inner cotyledon, inner cotyledon 4–7 mm long, 2.6–3.7 mm wide, straight or folded only at the tip or folded at the distal 1/7–1/3. Chromosome number unknown.

Phenology. Collected in flower from September through May, in fruit from November through May and in August.

Distribution (Fig. 79). Riverbanks along the Río Paraná, Río Uruguay, the southern Río Paraguay, and their tributaries, in Argentina (Buenos Aires, Chaco, Corrientes, Entre Ríos, Misiones, Santa Fe), Brazil (Paraná, Rio Grande do Sul, São Paulo), Paraguay (Alto Paraná, Pilar), and Uruguay; sea level to 250 m.

REPRESENTATIVE SPECIMENS. **Argentina.** BUENOS AIRES: Tuyuparí, Delta inferior, *Burkart 20972* (MICH); Campana, *Krapovickas 3266* (MO, W); Punta Lara, along Río de la Plata near the city of La Plata, ca.

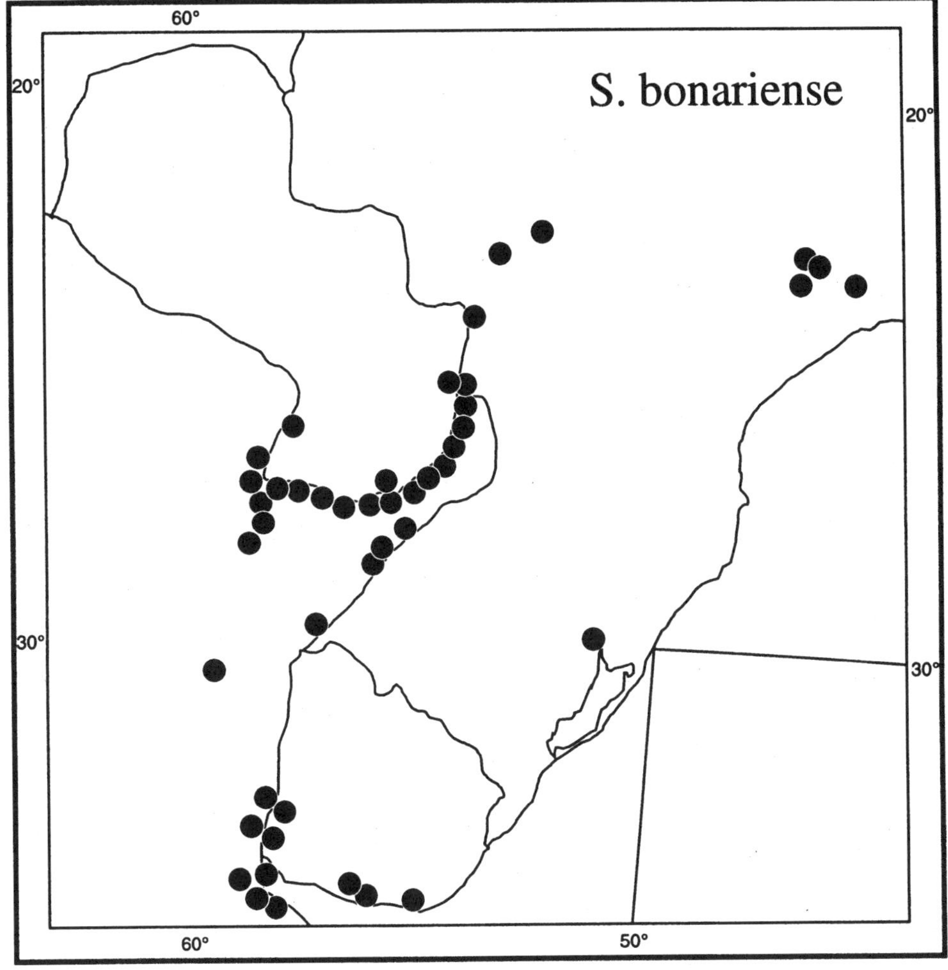

FIG. 79. Distribution of *Stigmaphyllon bonariense*.

35°S, 58°W, *Landrum 3067* (MICH).—CHACO: Depto. Resistencia, Isla Anterqueras, *Krapovickas & Cristóbal 12720* (CTES); Campo Antiguira, *Schulz 2336* (CTES).—CORRIENTES: Depto. San Cosme, Paso de la Patria, *Cristóbal et al. 1527* (CTES, F, MICH); Depto. Paso de los Libres, Laguna Mansa, *Krapovickas & Cristóbal 21660* (CTES, MICH, P); Depto. Empedrado, Estancia "Las Tres Marías," *Pedersen 4533* (BR, C, G, GH, K, MO, NY, P, S, U, US; herb. Pedersen); Depto. Santo Tomé, Río Aguapay y Ruta 14, *Quarin et al. 2692* (CTES, F, MICH); Depto. Itatí, Itatí, orillas del Río Paraná, *Schinini & Mroginski 4442* (CTES, MICH, P).—ENTRE RÍOS: Delta del Paraná, Isla La Chilena, *Burkart 8906* (MICH, US); Depto. Gualeguaychú, Punta Caballo, *Burkart et al. 30906* (MICH); islands of Cambaá Cuá in the Río Uruguay, near Concepción, *Pedersen 4749* (BR, C).—MISIONES: San Ignacio, Santo Pipó, ribera Alto Paraná, *Schulz 7221* (CTES, MO); Depto. Iguazú, cataratas, *Krapovickas & Cristóbal 13692* (CTES, P); Depto. San Ignacio, Arroyo Yabebiú, *Krapovickas et al. 15714* (C, CTES).—SANTA FE: Reconquista, Río Paraná, Isla Mascota, *Job 968* (F). **Brazil.** PARANÁ: Parque Nacional do Ihguasú, margem de Tamanduá, *Duarte 1803* (MICH, SP); Mpio. Sto. Inácio, Rio Paranapanema, *Hatschbach 51921* (MICH); Mpio. Icaraima, Pôrto Camargo, *Hatschbach 15826* (MICH, P).—RIO GRANDE DO SUL: Porto Alegre, *Malme 841* (S).—SÃO PAULO: Atibaia, *Duarte SP12035* (SP); Juquery, *Holway & Holway 1529* (US); Piraciaba, *Puttemans 2848 = SP12033* (SP); Campinas, *Viégas 6815* (SP). **Paraguay.** ALTO PARANÁ: 7 km al S de Villa Fortuna, *Casas & Molero 5785* (NY); in regione fluminis Alto Paraná, *Fiebrig 5459* (G, GH, K, US), *Fiebrig 5862* (BM, E, G, P).—ITAPUÁ: Pirapó, Escuela Técnica Forestal, *Pérez 165* (CTES).—NEEMBUCÚ: Depto. Pilar, Itá-pirú, *Meyer 15893* (CTES). **Uruguay.** COLONIA: Isla San Gabriel, *Berro 2308* (P).—MALDONADO: Piriápolis, *García Z. 1656* (P).—MONTEVIDEO: Punta Gorda, *Osten 5312* (GB, GH, US).—RÍO NEGRO: San Javier, *Herter 692* (F, G, GH, LE, M, MO, NY, S, U, WIS).

Stigmaphyllon bonariense is a common and variable species of the riverbanks of the Río Paraná, Río Uruguay, the southern Río Paraguay, and their tributaries. This showy species is also cultivated (W. R. Anderson, pers. comm.), and it is likely that the collections from Rio Grande do Sul and São Paulo represent escapes from gardens. *Stigmaphyllon bonariense* is readily identified by its unique samara, in which the short and broad, broadly triangular to nearly square, dorsal wing encircles the nut. The inflorescences are characteristically borne on long, axillary primary axes. The laminas are mostly cordate to broadly ovate to orbicular though sometimes are 2–3-lobed; this diversity is reflected in Hooker's epithet *heterophyllum* and Niedenzu's *trilobum*. In Brazil, this species might be confused with the also polymorphic *S. tomentosum*, but only in flower. The samara of *S. tomentosum* bears the flared dorsal wing typical for the genus. In the region of sympatry, *S. tomentosum* usually has the posterior petal bearing stout, gland-tipped fimbriae at the base of the limb, and at least the stamens opposite the posterior-lateral sepals are eloculate. The abaxial laminar vesture is never sparse, and the adaxial surfaces are usually glabrous. The flowers are in umbels but more commonly arranged in condensed pseudoracemes, which are usually disposed in large secondary inflorescences. The flowers of *S. bonariense* are always borne in umbels, either solitary or in a dichasium; compound dichasia are infrequent. The limb of the posterior petal never has stout glandular fimbriae, and the anthers opposing the lateral sepals always bear two, usually reduced locules.

Because earlier workers had assumed that Jussieu's name *Stigmaphyllon littorale* predates Hooker's *Banisteria bonariensis* this species traditionally has been known by Jussieu's name. I also annotated the material studied as *S. littorale* and only recently became aware that volume 3, part 8, of the *Botanical Miscellany* was published already on 1 August 1832 (C. Anderson 1997). Thus, the correct name for this species is *Stigmaphyllon bonariense*.

82. Stigmaphyllon affine Adr. Juss. in St.-Hilaire, Fl. bras. mer. 3: 57. 1833 ["1832"].—TYPE: BRAZIL. Rio de Janeiro: "circa Sebastianopolim a Cl. Gaudichaud Decemberi lecta putamus" [fide Jussieu] (holotype: P!, photos: F! MICH! US!).

Vine to 15 m. Stems and branches with scalelike T-shaped hairs when young, eventually becoming glabrate. Laminas 8–14 cm long, 4.3–9 cm wide, lanceolate to elliptical to ovate or sometimes 2–3-lobed, apex mucronate, base acute to truncate, glabrous adaxially, with T-shaped hairs abaxially (trabecula 0.5–1 mm long, straight or wavy, stalk 0.15–0.3 mm long), with irregularly spaced prominent to raised glands (0.2–0.4 mm in diameter, up to 0.2 mm long) borne adjacent to the margin abaxially; petioles 1.2–3.5 cm long, with scalelike T-shaped hairs, not confluent across the node, with a pair of prominent but sessile glands at the apex or up to 0.2 mm below the base of the lamina, each gland 1.2–1.8 mm in diameter; stipules 0.6–1.2 mm long, 0.4–0.9 mm wide, free, triangular, eglandular. Flowers ca. 15–25 (–30) per umbel, these solitary or borne in dichasia or small thyrses (axes to the 2nd order, with scalelike T-shaped hairs). Peduncles 3.5–13.5 mm long; pedicels 5–12 mm long, terete; both with short-stalked T-shaped hairs, peduncles 0.4–1.3 times as long as the pedicels. Bracts 1–2 mm long, 0.5–1 mm wide, triangular, apex acute; bracteoles 1–1.8 mm long, 0.6–1 mm wide, triangular or narrowly so or sometimes oblong, apex obtuse or sometimes acute, eglandular or each bracteole with a pair of inconspicuous glands (each gland 0.3–0.5 mm in diameter); bracts and bracteoles sericeous abaxially. Sepals 2.5–2.8 mm long, 2.2–2.6 mm wide, glands 2–2.3 mm long, 1–1.2 mm wide. All petals with the limb glabrous, yellow, margin with fimbriae up to 0.5 mm long, limb of lateral petals orbicular; anterior-lateral petals: claw 1.6–2.5 mm long, limb 9.5–12 mm long and wide; posterior-lateral petals: claw 1–1.2 mm long, limb 8.5–10 mm long and wide; posterior petal: claw 3–3.5 mm long, apex indented, limb ca. 8–9 mm long, ca. 7.5–8.5 mm wide, broadly elliptical to broadly obovate to orbicular. Stamens unequal, those opposite the posterior-lateral petals (and the posterior styles) the largest but those opposite the anterior-lateral sepals with the filaments subequally long, anthers of those opposite the lateral sepals with the connective enlarged and the locules reduced or rarely those opposite the posterior-lateral sepals eloculate; anthers glabrous or sometimes sparsely pubescent. Stamen opposite anterior sepal: filament (1.8–) 2.4–2.7 mm long, anther 1–1.4 mm long; stamens opposite anterior-lateral petals: filaments (1.3–) 1.5–2.2 mm long, anthers 0.7–1 mm long; stamens opposite anterior-lateral sepals: filaments 2.5–3.1 mm long, connectives 0.8–1 mm long, locules 0.3–0.6 mm long; stamens opposite posterior-lateral petals: filaments 2.7–3.2 mm long, anthers 1.2–1.4 mm long; stamens opposite posterior-lateral sepals: filaments 2.5–3.1 mm long, connectives 0.6–0.8 mm long, locules 0.3–0.4 mm long or rarely absent; stamen opposite posterior petal always shorter than the adjacent two: filament 2.1–2.5 mm long, anther 0.7–0.9 mm long. Anterior style (2.6–) 3–3.6 mm long, shorter than the posterior two, terete, glabrous, slightly incurved; apex 1.4–1.9 mm long, each foliole 1–1.7 mm long, 1–1.5 mm wide, usually subsquare, sometimes triangular. Posterior styles (3.1–) 3.7–4.2 mm long, terete, glabrous or sometimes with scattered hairs in the proximal 1/3, lyrate; foliole (1.8–) 2–2.5 mm long and wide, subsquare. Dorsal wing of samara triangular and encircling the nut, 3–3.7 cm high from base of nut, 2–2.7 cm wide, nut bearing several lateral crests and winglets (these sometimes interconnected) and sometimes also one large lateral winglet per side, all these up to 4 mm long, up to 3 mm wide; nut ca. 2 cm high, ca. 1.3–1.5 cm in diameter, locule surrounded by air chambers, areole 4–5.5 mm long, 4.8–5.2 mm wide, concave, carpophore up to 2 mm long. Embryo ca. 13 mm long, ca. 2 times as long as wide, ovoid, outer cotyledon ca. 16.5 mm long, 5.7 mm wide, the distal 1/3 folded over the inner cotyledon, inner cotyledon ca. 18 mm long, ca. 5.5 mm wide, folded at the distal 1/3. Chromosome number unknown. Fig. 80a–k.

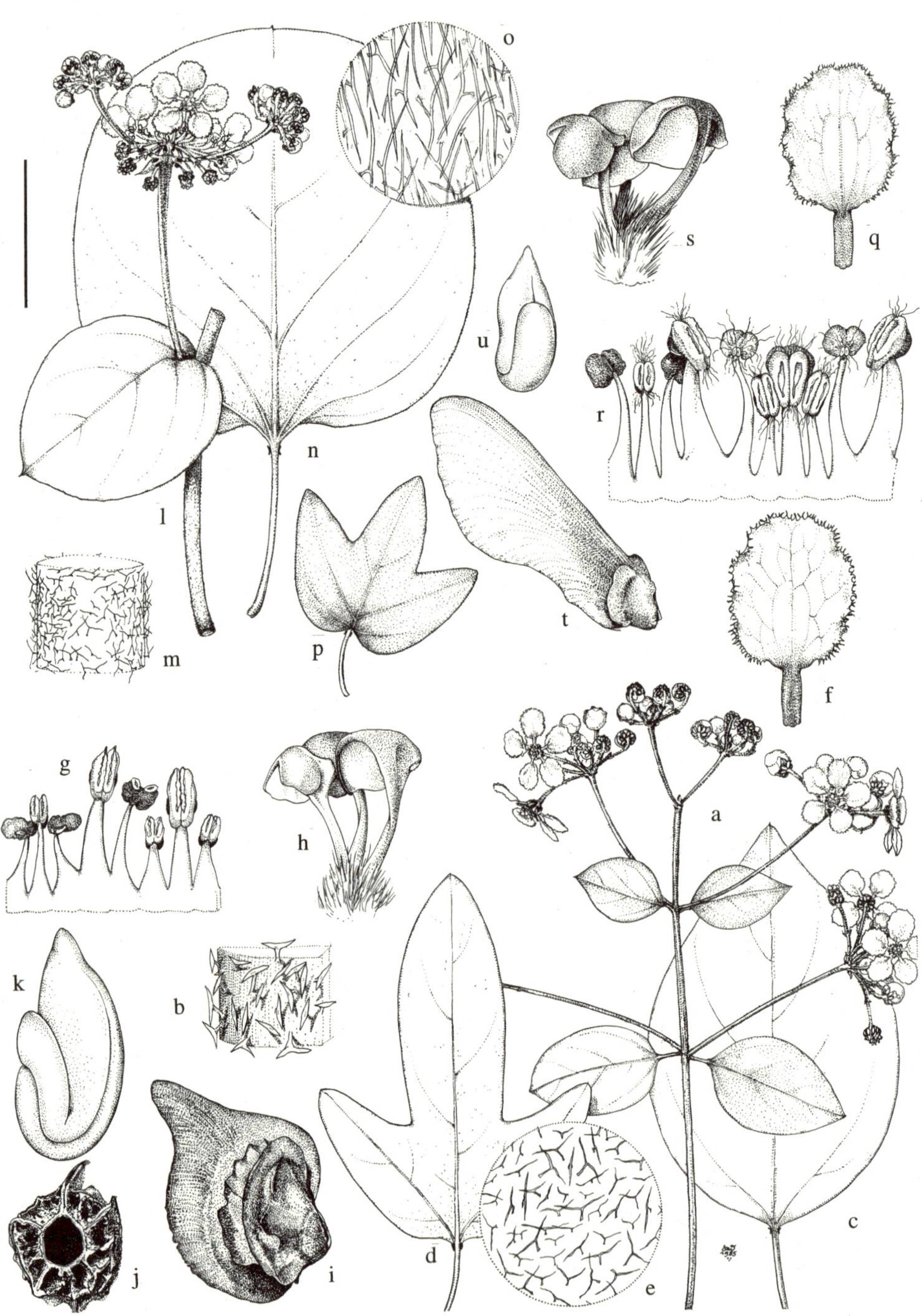

FIG. 80. *Stigmaphyllon affine* and *S. gayanum*. a–k, *S. affine*. a. Flowering branch. b. Detail of stem showing pubescence. c. Leaf with unlobed lamina. d. Leaf with lobed lamina. e. Detail of abaxial surface of lamina. f. Posterior petal (the "flag"). g. Androecium; stamen second from left opposes the posterior petal. h. Gynoecium, anterior style to the left. i. Samara. j. Longitudinal section through nut of samara; note the air chambers surrounding the locule. k. Embryo. l–u, *S. gayanum*. l. Flowering branch. m. Detail of stem showing pubescence. n. Large leaf with unlobed lamina. o. Detail of abaxial surface of lamina. p. Leaf with lobed lamina. q. Posterior petal (the "flag"). r. Androecium; stamen second from left opposes the posterior petal. s. Gynoecium, anterior style to the left. t. Samara. u. Embryo. Scale: for a, c, d, i, l, n, p, bar = 4 cm; for b, bar = 2 mm; for e, o, bar = 1.3 mm; for f, k, q, u, bar = 8 mm; for g, h, r, s, bar = 4 mm; for j, t, bar = 2 cm; for m, bar = 1 mm. (Based on: a, b, *Kardel s.n. R19564*; c, *Glaziou 5767*; d, e, *Passarelli 143*; f–h, *Occhioni 3852*; i–k, *Weddell 597*; l, m, q, s, *Hoehne 6082*; n, o, *Brade 132*; p, *Azevedo s.n. RB48467*; r, *Carauta 3531*; t, u, *Mosén 2439*).

Phenology. Collected in flower in April, May, August, October, and December, in fruit in February and August.

Distribution (Fig. 81). Brazil (Rio de Janeiro); along rivers and canals, in swales of marshes and mangrove swamps, and in wet situations in secondary growth and along roadsides; sea level to 50 m.

REPRESENTATIVE SPECIMENS. **Brazil.** RIO DE JANEIRO: Mpio. Angra do Reis, Ilha Grande, Res. Biol. Est. da Praia do Sul, *Araujo 5831* (GUA); São Christovão, *Glaziou 5767* (C, GOET, K, P, R); Morro de São João, *Hoehne SP24697* (SP); Barra do Pirahy, *Hoehne & Gehrt SP17302* (GH, SP); *Lund 561* (C); Rio Garea, *Lutz 231-85* (R); Duas Barras, *Occhioni 3852* (MICH); Florestal da estrada Rio/Petrópolis, *Passarelli 143* (R); *Riedel & Luschnath 177* (A, LE, NY, S, US); *Schüch s.n.* (F, M, NY, US, W); Guaratiba, *Sucre 6167* (RB); *Weddell 597* (F, G, P).

Stigmaphyllon affine is endemic to the region of Rio de Janeiro, where it is found in wet places, marshes, and mangrove swamps. The unique samara, adapted for water dispersal, separates it from all other species. The large nut (ca. 1.5 cm in diameter) bears a greatly modified, short, triangular dorsal wing and is elaborately ornamented laterally. The locule is surrounded by air chambers and contains an unusually large embryo (ca. 13 mm long). The laminas are most commonly lanceolate to elliptical but occasionally also 2–3-lobed (*Passarelli 143*, R; *Schüch s.n.*, NY, W). The petals are all fringed with fimbriae up to 0.5 mm long.

For an account concerning the confusion of *S. affine* with *S. rotundifolium* (no. 84; as "*S. irregulare*"), see that species.

83. Stigmaphyllon cavernulosum C. Anderson, Syst. Bot. 14: 513. 1989.—TYPE: BRAZIL. Bahia: Ilhéus, Centro de Pesquisas do Cacau, CEPEC, CEPLAC, *Belém & Magalhães 598* (holotype: UB!, photo: MICH!; isotypes: CEPEC! NY!).

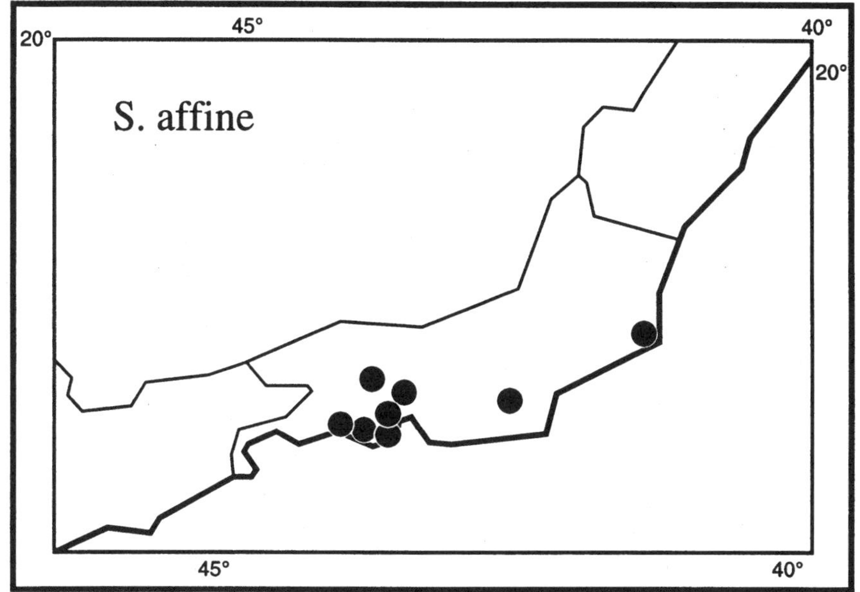

FIG. 81. Distribution of *Stigmaphyllon affine.*

Vine. Stems and branches with scalelike T-shaped hairs when young, soon becoming glabrate. Leaves often alternate, especially near and in the inflorescences. Laminas 5.8–16.5 cm long, 3.4–9.5 cm wide, narrowly to broadly elliptical, apex obtuse-mucronate or rarely acute, base truncate to very slightly cordate, glabrous adaxially, pubescent with T-shaped hairs abaxially (trabecula 0.3–1.4 mm long, straight to wavy, stalk 0.1–0.5 mm long), with irregularly spaced prominent glands (0.1–0.2 mm in diameter, up to 0.1 mm long) borne adjacent to the margin abaxially; petioles 1–6.3 cm long, with scalelike T-shaped hairs to glabrate, not confluent across the node, with a pair of prominent but sessile glands at the apex, each gland 1.2–2 mm in diameter; stipules 0.8–1.3 mm long, 0.5–1.2 mm wide, free, triangular or narrowly so, eglandular. Flowers ca. 15–25 per umbel, these borne in dichasia (axes to the 3rd order, sericeous), the dichasia borne on axes arranged alternately on a primary inflorescence axis. Peduncles 4.5–12.5 mm long; pedicels 4.5–10.5 mm long, terete; both densely sericeous, peduncles 0.6–1.6 times as long as the pedicels. Bracts 1.5–2.3 mm long, 0.5–1 mm wide, narrowly triangular, apex acute; bracteoles 1.1–1.7 mm long, 0.8–1.1 mm wide, triangular to oblong, apex obtuse, each bracteole with a pair of inconspicuous glands (each gland up to 0.4 mm in diameter) or sometimes eglandular; bracts and bracteoles sericeous abaxially. Sepals 2.3–2.5 (–3) mm long, 2–2.3 (–2.8) mm wide, glands 2.1–2.5 (–3) mm long, ca. 1.2–1.5 mm wide. All petals with the limb orbicular, glabrous, yellow, margin denticulate to fimbriate, teeth/fimbriae up to 0.3 mm long; anterior-lateral petals: claw (1.5–) 2–2.8 mm long, limb 12–13 mm long and wide; posterior-lateral petals: claw 1–2 mm long, limb ca. 11 mm long and wide; posterior petal: claw 3–4 mm long, apex indented, limb 9–10 mm long and wide. Stamens unequal, those opposite the posterior-lateral petals (and the posterior styles) the largest, anthers of those opposite the lateral sepals with the connective enlarged and the locules reduced; anthers all loculate, glabrous. Stamen opposite anterior sepal: filament 1.8–2.8 mm long, anther 1.1–1.4 mm long; stamens opposite anterior-lateral petals: filaments 1.6–1.9 (–2.4) mm long, anthers 0.9–1.1 mm long; stamens opposite anterior-lateral sepals: filaments 2.7–3.1 mm long, connectives 0.8–0.9 mm long, locules 0.2–0.6 mm long; stamens opposite posterior-lateral petals: filaments (2.5–) 3–3.5 long, anthers 1.1–1.5 mm long; stamens opposite posterior-lateral sepals: filaments 2.6–3 mm long, connectives 0.7–0.8 mm long, locules 0.2–0.5 mm long; stamen opposite posterior petal always shorter than the adjacent two: filament 2.2–2.3 (–2.8) mm long, anther 0.7–1 mm long. Anterior style 2.7–3.6 mm long, shorter than the posterior two, terete, glabrous, erect; apex (1.3–) 1.7–2 mm long, each foliole 1.1–1.6 mm long, 1–1.4 mm wide, triangular to trapezoidal to subsquare. Posterior styles (3.2–) 3.5–4.3 mm long, terete, glabrous or with a few scattered hair in the basal 1/3, lyrate; foliole 1.7–2.5 mm long, 1.5–2 mm wide, subsquare to subrectangular. Dorsal wing of samara erect, 3.7–4.6 cm high measured from base of nut, 1.7–2 cm wide; nut with one lunate lateral winglet per side, these up to 13 mm long and up to 3 mm wide, and usually with a vertical row of ca. 3–5 crests and spurs next to the winglets; nut 11.5–12.5 mm high, 5.5–6 mm in diameter, the locule surrounded by air chambers, areole ca. 4 mm long and wide, deeply concave, carpophore up to 2.5 mm long. Embryo ca. 8.2 mm long, embryo ca. 2 times as long as wide, ovoid, outer cotyledon ca. 11.5 mm long, ca. 3.8 mm wide, the distal 1/3 folded over the inner cotyledon, inner cotyledon ca. 6.7 mm long, ca. 2.5 mm wide, straight. Chromosome number unknown. Fig. 83h–m.

Phenology. Collected in flower in March, July, September, and October, and fruit in March, July, August, and September.

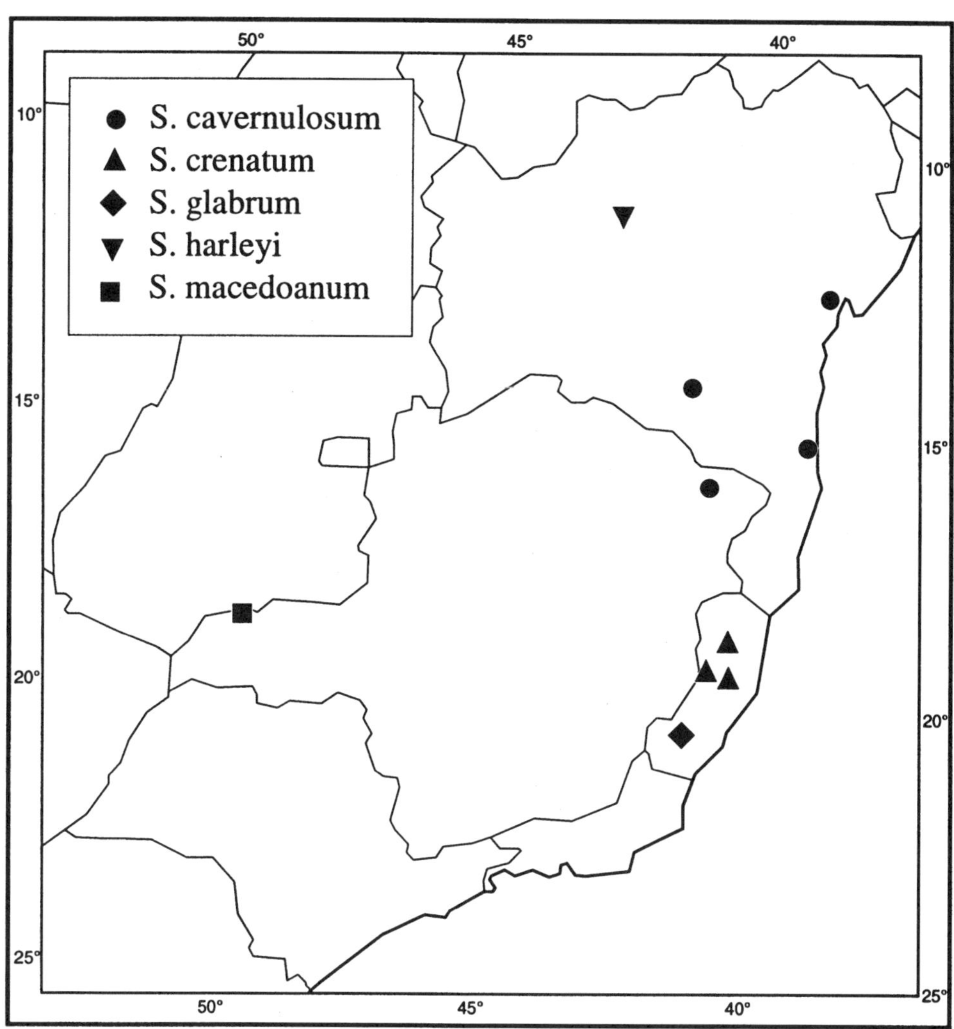

FIG. 82. Distribution of *Stigmaphyllon cavernulosum, S. crenatum, S. glabrum, S. harleyi,* and *S. macedoanum.*

Distribution (Fig. 82). Brazil (eastern Bahia and adjacent Minas Gerais); along rivers; sea level to 120 m.

ADDITIONAL SPECIMENS EXAMINED. **Brazil.** BAHIA: cerca Bahia, *Blanchet 1722* (BM); *Blanchet 2203* (F, G); Ilhéus, *Blanchet 3027* (BM, F, G, LE); Iguaçu, Cachoeira/Bahia, vale dos rios Paraguaçu e Jacuipe, ca. 12°32′S, 39°05′W, *do Cavalo 717* (CEPEC, MICH); Mpio. Ilhéus, area do CEPEC, Km 22 da rodovia Ilhéus/Itabuna (BR-415), *dos Santos 3654* (CEPEC, MICH), *Hage 2160* (MICH), *Hage & dos Santos 1416* (MICH), *Mori & dos Santos 11725a* (CEPEC, MICH); Mpio. Rio de Contas, Rio Brumado, *Hatschbach 46537* (CAS, MBM, MICH); Mpio. Ilhéus, Rio do Braço, *Velloso 717* (R); Mpio. Ilhéus, Faz. Pirataquisse, *Velloso 923* (R).—MINAS GERAIS: Mpio. Medina, BR-116, *Hatschbach 45025* (MBM, MICH).

Stigmaphyllon cavernulosum is distinguished by its large samaras. The dorsal wing is erect and tapers gradually from the base, and the locule is surrounded by air chambers and contains a relatively large embryo. These traits are also exhibited by other species probably dispersed by water; *S. cavernulosum* has been reported to grow along riverbanks. The distal leaves and the units of the inflorescence are commonly alternately arranged. Such a change from opposite phyllotaxy is not unusual in large vines, and it is very strongly expressed in this species.

Although easily recognized in fruit, in flower this species might be confused with the sympatric *S. blanchetii* (no. 89) and possibly *S. salzmannii* (no. 90), in which the leaves also bear T-shaped hairs abaxially. These two species are readily separated from *S. cavernulosum* by stamen and leaf characters. In both, the anthers are abundantly pubescent rather than glabrous, and the marginal glands are stipitate instead of sessile; in *S. salzmannii* the laminas are only very sparsely pubescent to glabrous abaxially.

84. Stigmaphyllon rotundifolium Adr. Jussieu, Ann. Sci. Nat. Bot., sér. 2, 13: 289. 1840. *Stigmaphyllon rotundifolium* f. *suborbiculare* Niedenzu, Pflanzenreich IV. 141(2): 491. 1928, nom. superfl.—TYPE: BRAZIL. Bahia, *Salzmann 98* (holotype: G!, photos: F! GH! MICH!, fragment: P-JU!, photo: MICH!).

Stigmaphyllon irregulare Adr. Jussieu, Ann. Sci. Nat. Bot., sér. 2, 13: 288. 1840.—TYPE: BRAZIL. Bahia, *Blanchet s.n.* (holotype: P-JU!, photo: MICH!; isotypes?: G-3 sheets!, photos of one possible isotype at G: F! GH! MICH!). [Note: the samaras shown in the photo, Field Museum negative no. 24243, belong to *Weddell 597*, which is *S. affine* Adr. Juss.]

Stigmaphyllon lalandianum var. *grisebachianum* Niedenzu, Ind. Lect. Lyc. Brunsberg. p. aest. 1900: 6. 1900.—TYPE: BRAZIL. "Espirito Santo: zwischen Vittoria und Bahia" (fide Niedenzu, 1928), *Sellow s.n.* (holotype: B, destroyed, photos: A! F! MICH! NY! US!; lectotype, here designated: NY!, fragments of holotype, the leaf and samara but excluding the buds).

Stigmaphyllon rotundifolium f. *ovatum* Niedenzu, Pflanzenreich IV. 141(2): 491. 1928.—TYPE: BRAZIL. Bahia, *Blanchet 130* (lectotype, designated by C. Anderson, 1989: G!, photo: MICH!).

Vine. Stems and branches sericeous when young, soon becoming glabrate. Laminas 7.8–14 cm long, 5.8–10 cm wide, ovate to elliptical to suborbicular or rarely 2–3-lobed, apex mucronate, base truncate to slightly cordate, glabrate to glabrous adaxially, sericeous or sparsely so abaxially (trabecula 0.3–0.7 mm long, straight, sessile), eglandular or sometimes with a few irregularly spaced sessile glands (0.1–0.3 mm in diameter) borne adjacent to the margin abaxially; petioles 3–6 cm long, sericeous or densely so, not confluent across the node, with a pair of prominent but sessile glands at the apex, each gland 1.6–2.1 mm in diameter; stipules 0.7–1 mm long, 0.6–0.9 mm wide, free, triangular, eglandular. Flowers ca. 10–20 per umbel or condensed pseudoraceme, these borne in dichasia or irregular compound arrangements (axes to the 2nd order, sericeous), the 2nd order axes sometimes suppressed and then the lateral umbels sessile. Peduncles 5–9.5 mm long; pedicels 4–12 mm long, terete; both densely pubescent with scalelike T-shaped hairs, peduncles 0.5–1.7 times as long as the pedicels. Bracts 1–1.6 mm long, 0.6–1.3 mm wide, triangular, apex acute; bracteoles 0.8–1.5 mm long, 0.6–1 mm wide, parabolic, apex obtuse, each bracteole with two inconspicuous glands (each gland 0.2–0.5 mm in diameter); bracts and bracteoles sericeous abaxially. Sepals ca. 2 mm long, ca. 2.2–2.5 mm wide, glands ca. 1.8–2 mm long, ca. 1 mm wide. All petals with the limb orbicular, glabrous, yellow, limb of the lateral petals with the margin erose or with fimbriae up to 0.2 mm long; anterior–lateral petals: claw 2–2.5 mm long, limb 10–11 mm long and wide; posterior-lateral petals: claw ca. 1 mm long, limb 7–8 mm long and wide; posterior petal: claw ca. 3 mm long, apex indented, limb ca. 7 mm long and wide, margin with fimbriae up to 0.3 mm long. Stamens unequal, those opposite the posterior-lateral petals (and the posterior styles) the largest, anthers of those opposite the lateral sepals with the connective enlarged and the locules reduced; anthers all loculate, glabrous. Stamen opposite anterior sepal: filament 2.1–2.3 mm long, anther 1.1–1.3 mm long; stamens opposite anterior-lateral petals: filaments 1.5–1.8 mm long, anthers 1–1.1 mm long; stamens opposite anterior-lateral sepals: filaments (2.2–) 2.7–3 mm long, connectives 0.7–0.8 mm long, locules 0.3–0.4 mm long; stamens opposite posterior-lateral petals: filaments (2.6–) 3–3.3 mm long, anthers 1–1.3 mm long; stamens opposite posterior-lateral sepals: filaments 2–2.3 mm long, connectives 0.7–0.8 mm long, locules 0.3–0.4 mm long; stamen opposite posterior petal always shorter than the adjacent two: filament 1.8–2 mm long, anther 0.5–0.7 mm long. Anterior style 3–3.3 mm long, shorter than the posterior two, terete, glabrous, erect; apex ca. 1.7 mm long, each foliole 1.4–1.7 mm long, 1–1.3 mm wide, broadly parabolic to subsquare. Posterior styles 3.6–4 mm long, terete, glabrous, lyrate; foliole ca. 2 mm long and wide, subsquare. Dorsal wing of samara 3.7–3.9 cm long, 1.2–1.3 cm wide, upper margin with a blunt tooth; nut bearing one rectangular to lunate lateral winglet on each side, these 4.5–8.5 mm long, ca. 3 mm wide, extending below the nut, and sometimes also with 1–2 spurs; nut ca. 6.5 mm high, ca. 3.5 mm in diameter, the locule flanked by two narrow air chambers, areole ca. 2.5–3 mm long and wide, concave; carpophore ca. 1.5 mm long. Mature seed not seen. Chromosome number unknown. Fig. 89h–k.

Phenology. Collected in flower in June (*de Carvalho 6195*; other collections without date).

Distribution (Fig. 52). Brazil (Bahia, Espírito Santo?); wet forest near rivers; probably sea level to ca. 100 m.

ADDITIONAL SPECIMENS EXAMINED. **Brazil.** BAHIA: Mpio. Una, Reserva Biológica Mico-Ieão (IBAMA), entrada no Km 46 da Rod. BA-001 Ilhéus–Una, 8–10 km na estrada que margeia o Rio Maruim, 15°09′S, 39°05′W, *de Carvalho 6195* (MICH); without locality, *Blanchet 381* (BM, G), *698* (G), *Bondar 1710 p.p.* (F), *Glocker s.n.* (BM), *Salzmann s.n.* (CGE, G, K, LE, P, R, W).

Stigmaphyllon rotundifolium is a little-known species of Bahia and perhaps Espírito Santo, Brazil. It is named for its broadly elliptical leaves, which are sericeous abaxially and only infrequently bear marginal sessile glands. These laminar characters serve to distinguish it from *S. salzmannii* (no. 90) and the widespread *S. blanchetii* (no. 89). In both of these species, the laminas bear stalked marginal glands and T-shaped hairs abaxially (or are glabrous in *S. salzmannii*). *Stigmaphyllon macropodum* (no. 85) also has large roundish laminas, but they are appressed-tomentose to densely sericeous abaxially.

As in other species with usually unlobed leaves, individuals of *S. rotundifolium* with lobed laminas are occasionally encountered. Jussieu described a Blanchet collection of *S. rotundifolium* with 2–3-lobed laminas as "*S. irregulare*." Niedenzu mistakenly assigned several fruiting collections of *S. affine* to "*S. irregulare*," and this misinterpretation of "*S. irregulare*" has been perpetuated by the photo of the "type" at G (Field Museum neg. 24243). The photo shows a probable isotype of *S. irregulare,* supplemented with a fruiting branch of *S. affine* (*Weddell 597*) (C. Anderson 1989).

Niedenzu based *S. lalandianum* var. *grisebachianum* on an unnumbered Sellow collection, apparently a mixed gathering that is no longer extant. *Stigmaphyllon lalandianum* (no. 14) is quite different from *S. rotundifolium*, notably in that the stamens are subequal (none of the anthers are modified) and the styles lack large folioles. The photos of the type show a fruiting specimen of *S. rotundifolium* and a closed mounted packet. There is at NY a sheet from the Niedenzu herbarium with fragments of this type. These fragments consist of a leaf and samara of *S. rotundifolium* as well as of three buds of *S. lalandianum* (two partially dissected). Niedenzu's variety is here lectotypified on the basis of the leaf and samara, and the name *S. lalandianum* var. *grisebachianum* is therefore included in the synonymy of *S. rotundifolium*.

85. Stigmaphyllon macropodum Adr. Jussieu, Ann. Sci. Nat. Bot., sér. 2, 13: 289. 1840. *Stigmaphyllon tomentosum* β *pubescens* Grisebach in Martius, Fl. bras. 12(1): 39. 1858. *Stigmaphyllon megacarpon* var. *macropodum* (Adr. Jussieu) Niedenzu, Ind. Lect. Lyc. Brunsberg. p. aest. 1900: 28. 1900. *Stigmaphyllon fulgens* var. *macropodum* (Adr. Jussieu) Niedenzu, Pflanzenreich IV. 141(2): 493. 1928.—TYPE: BRAZIL. Bahia, *Blanchet 715* (holotype: F!, photo: MICH!, fragment: P-JU!, photo: MICH!; isotype: G!, photos: F! GH! MICH! NY! US!).

Stigmaphyllon bahiense C. Anderson, Syst. Bot. 14: 509. 1989.—TYPE: BRAZIL. Bahia: Bom Gosto e Olivença [near Ilhéus], 15 Mar 1943, *Fróes 20021* (holotype: IAN!; isotypes: F! K! MICH! NY!).

Vine to 20 m. Stems and branches densely (golden-) sericeous when young, less densely so in older parts. Laminas 8.8–27 cm long, 5.8–22.5 cm wide, ovate to broadly ovate to elliptical to very broadly elliptical to suborbicular, apex obtuse-mucronate to acuminate-mucronate, base attenuate to truncate to cordate, glabrous adaxially or sometimes with appressed hairs on the major veins near the base, moderately to densely appressed-tomentose or rarely sericeous abaxially (trabecula 0.2–0.9 mm long, commonly crisped and curled or sometimes wavy or rarely straight, sessile or subsessile with a stalk up to 0.1 mm long), with irregularly spaced sessile glands (0.3–0.5 mm in diameter) borne on the margin or adjacent to the margin abaxially or sometimes in both places; petioles 1.8–9 cm long, densely (golden-) sericeous, not confluent across the node, with a pair of prominent but sessile glands at the apex, each gland 1.3–3.2 mm in diameter; stipules (0.3–) 0.5–0.8 mm long, (0.3–) 0.5–1 mm wide, free, triangular, eglandular. Flowers ca. (10–) 15–25 per umbel or sometimes arranged in a pseudoraceme, these borne in dichasia, compound dichasia, or small thyrses (axes to the 5th order, densely sericeous). Peduncles 4–15 mm long; pedicels 4–10.5 mm long, terete; both densely (golden-) sericeous, peduncles 0.8–2 times as long as the pedicels. Bracts 1.2–2.5 mm long, 1–1.8 mm wide, triangular or narrowly so, apex acute; bracteoles 1.2–2.2 mm long, 0.8–1.5 mm wide, oblong to ovate, apex obtuse, each bracteole with a pair of inconspicuous glands (each gland 0.2–0.3 mm in diameter); bracts and bracteoles sericeous abaxially. Sepals 2.3–2.6 mm long, 2–2.4 mm wide, glands (2–) 2.3–3 mm long, 1–1.4 mm wide. All petals with the limb orbicular, glabrous, yellow, limb of lateral petals with the margin fimbriate or denticulate, fimbriae/teeth up to 0.4 mm long; anterior-lateral petals: claw (2–) 2.5–3 mm long, limb 12–14 mm long and wide; posterior-lateral petals: claw 1–1.5 (–1.8) mm long, limb ca. 11 mm long and wide; posterior petal: claw ca. 3 mm long, apex indented, limb 8–10 mm long and wide, margin fimbriate and rarely with 1–2 stalked glands (0.3–0.4 mm long) per side at the base. Stamens unequal, those opposite the posterior-lateral petals (and the posterior styles) the largest, anthers of those opposite the anterior-lateral sepals with the connective enlarged and the locules reduced, anthers of those opposite the posterior-lateral sepals usually eloculate or sometimes with one reduced locule or rarely with two reduced locules; loculate anthers glabrous or rarely very sparsely pubescent. Stamen opposite anterior sepal: filament (1.8–) 2–2.5 mm long, anther 1.2–1.4 mm long; stamens opposite anterior-lateral petals: filaments 1.4–2.2 mm long, anthers 0.9–1.2 mm long; stamens opposite anterior-lateral sepals: filaments 1.8–2.6 mm long, connectives 0.7–0.9 mm long, locules 0.2–0.6 mm long; stamens opposite posterior-lateral petals: filaments 2.7–3.8 mm long, anthers 1.2–1.5 mm long; stamens opposite posterior-lateral sepals: filaments 2–3 mm long, connectives 0.6–0.9 mm long, locules 0.2–0.4 mm long or absent; stamen opposite posterior petal always shorter than the adjacent two: filament 1.8–2.5 mm long, anther 0.8–1 mm long. Anterior style 2.7–3.6 mm long, shorter than the posterior two, terete, glabrous, erect; apex 1.4–2.3 mm long, each foliole 1.2–1.6 mm long, 1.2–2 mm wide, subsquare to suborbicular to subrectangular. Posterior styles 3.7–4.3 mm long, terete, with scattered hairs in the proximal 1/3, lyrate; foliole 2.2–2.5 mm long and wide, subsquare or suborbicular. Dorsal wing of samara 4.3–6.7 cm long, 1.5–2 cm wide, upper margin with a blunt tooth; nut bearing a lunate to rectangular lateral winglet per side, these up to 10 mm long, up to 3 mm wide, sometimes the winglet interrupted, sometimes also with one or two additional winglets or spurs (up to 1 mm long and wide); nut 8.5–9 mm high, 4.5–6.5 mm in diameter, wall containing spongy tissue, without air chambers, areole 4–5 mm long and wide, concave, carpophore up to 4.5 mm long. Embryo ca. 9.2 mm long, ca. 2 times as long as wide, ovoid, outer cotyledon ca. 13.5 mm long, ca. 4.7 mm wide, the distal 1/3 folded over the inner cotyledon, inner cotyledon ca. 5.5 mm long, ca. 3.4 mm wide, straight. Chromosome number unknown. Fig. 83a–g.

Phenology. Collected in flower from January through June, November, and December, in fruit in January, March through May, November, and December.

Distribution (Fig. 84). Brazil (Bahia); coastal wet forest and restinga forest; sea level to 100 m.

REPRESENTATIVE SPECIMENS. **Brazil.** BAHIA: Mpio. Ilhéus, ramal no 7 km da estrada Ilhéus/Olivença, *Carvalho & Plowman 1618* (F, MICH); ca. 5 km SW of Itacaré, on side rd from main Itacaré–Ubaitaba rd, S of mouth of Rio de Contas, ca. 14°20′S, 39°03′W, *Harley 17486* (K, MICH, NY, P, U); Itacaré, near mouth of Rio de Contas, ca. 14°18′S, 38°59′W, *Harley 17573* (K, MICH); ca. 5 km N from turning to Maraú along the Campinho rd, 14°04′S, 38°58′W, *Harley 22167* (MICH); estr. Itabuna–Una, *Heringer et al. 3278* (MICH, US); Mpio. Una, estrada Una–Canavieiras, Km 25, *Martinelli et al. 8892* (CEPEC); Mpio. Uruçuca, nova estrada que liga Uruçuca a Serra Grande, a 45 km de Uruçuca, *Mori & Kallunki 9898* (CEPEC, MICH); Mpio. Itacaré, a 1–3 km ao S de Itacaré, *Mori & dos Santos 10145* (CEPEC, MICH); Rio Branco, plantação de cacau, *Pinheiro 59* (CEPEC, UB); Mpio. Ilhéus, 10 km S of Ilhéus (Pontal) on rd to Olivença, then 3 km W, Fazenda Manquinho, 14°55′S, 39°05′W, *Thomas et al. 10193* (NY).

Stigmaphyllon macropodum is a large-leaved vine of the coastal wet forest of Bahia, Brazil. It is distinctive in its generally golden, densely sericeous pubescence, which covers the axes, petioles, abaxial surfaces of the laminas, peduncles, pedicels, and sepals (the color often fades in age).

Jussieu based the name *S. macropodum* on a Blanchet collection in bud in which the abaxial surfaces of the laminas are densely sericeous (all trabeculas straight). The anthers of stamens opposite the posterior-lateral sepals consist of enlarged connectives bearing two reduced locules; the unmodified anthers are pubescent. In contrast, the laminas of collections assigned to "*S. bahiense*" are mostly densely to moderately appressed-tomentose

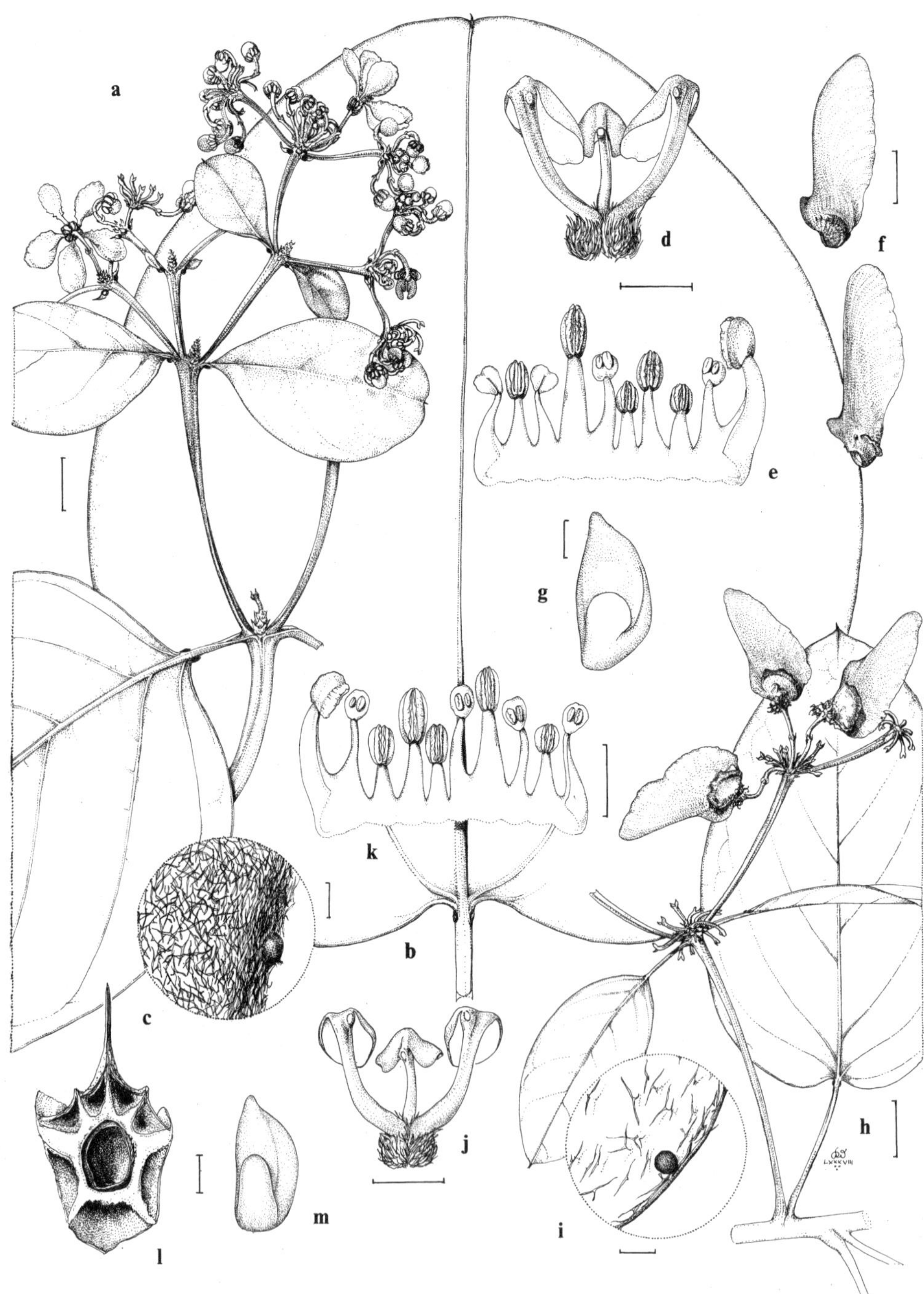

FIG. 83. *Stigmaphyllon macropodum* and *S. cavernulosum*. a–g, *S. macropodum*. a. Flowering branch. b. Large leaf. c. Detail of margin and abaxial surface of lamina. d. Gynoecium, anterior style in the center. e. Androecium; stamen second from left opposes the posterior petal (the "flag"). f. Two samaras. g. Embryo. h–m, *S. cavernulosum*. h. Fruiting branch. i. Detail of margin and abaxial surface of lamina. j. Gynoecium, anterior style in the center. k. Androecium; stamen second from right opposes the posterior petal (the "flag"). l. Longitudinal section through nut of samara; note the air chambers surrounding the locule. m. Embryo. Scale: for a, b, f, h, bar = 1.5 cm; for c, i, bar = 0.5 mm; for d, e, g, j–m, bar = 2 mm. (Based on: a–c, *Froés 20021*; d, e, *Pinheiro 59*; f, *Heringer et al. 3278* (above), *Mori & Kallunki 9898* (below); g, *Heringer et al. 3278*; h–m, *Belém & Magalhães 598*.)

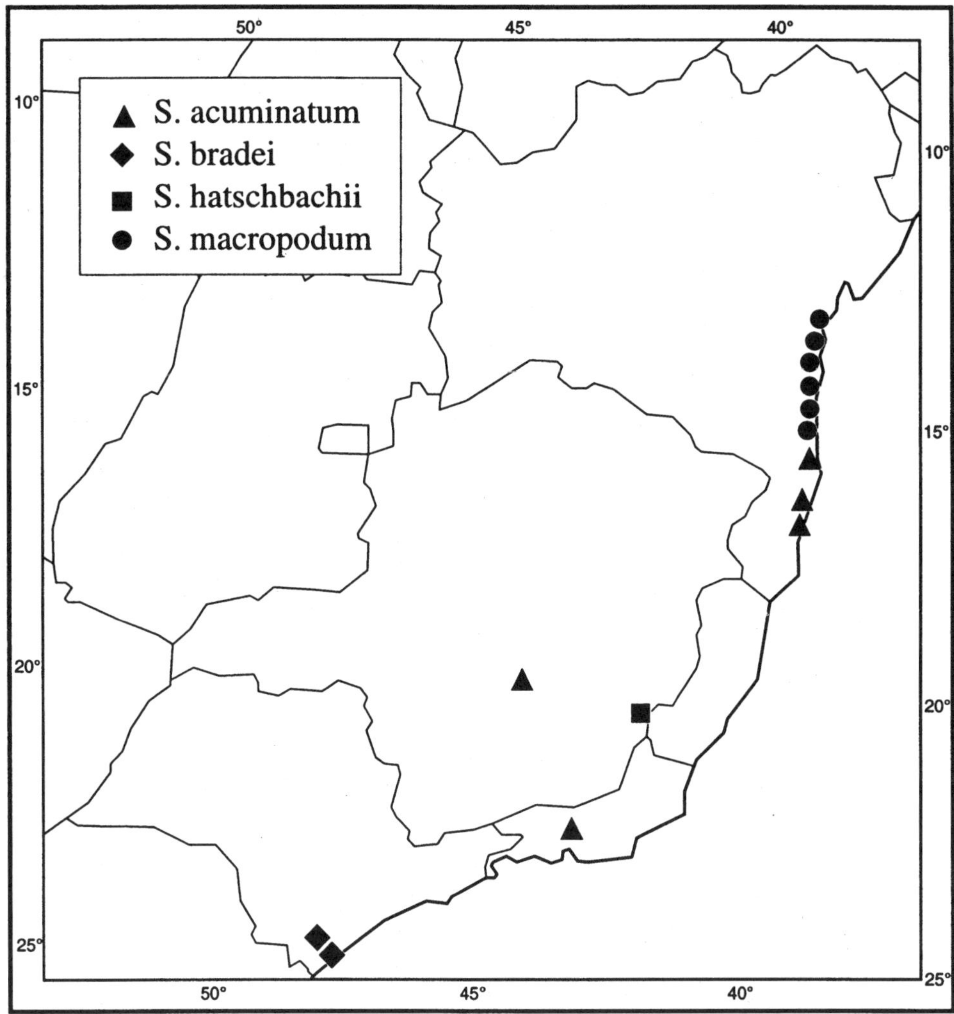

FIG. 84. Distribution of *Stigmaphyllon acuminatum, S. bradei, S. hatschbachii,* and *S. macropodum.*

abaxially. The pubescence may consist of hairs with the trabecula curled and crisped or mixed with some with the trabecula only wavy. The anthers of stamens opposite the posterior-lateral sepals consist of enlarged connectives usually lacking locules, and the unmodified anthers are glabrous or rarely sparsely pubescent. More recent collections, e.g., *Mattos Silva & dos Santos 1907* in which the laminas are moderately sericeous abaxially, bridge the distinctions between these two forms. It now seems that the Blanchet specimens exhibit an extreme condition in the variation of leaf vesture; the tomentose vesture, in which the trabecula is curled or crisped, is the common condition. The nature of the anthers also appears to be more variable than previously realized. The enlarged connectives of stamens opposing the posterior-lateral sepals are mostly eloculate but sometimes bear one or rarely even two tiny locules. The unmodified anthers are usually glabrous though sometimes bear a few hairs, occasionally only one hair at the apex and/or one at the base.

Blanchet specimens of *S. macropodum* are deposited at BM, F, G, P-JU (one leaf and one inflorescence branch), and SP (ex BM). They appear to constitute the same gathering and thus may all be part of the type collection, although they are variously numbered: at BM "1090" and "2206," at F "715" and "1090," at G "715.1090" and "2206," at P-JU "715," at SP "1090." In the *Monographie* of 1843, Jussieu notes that *S. macropodum* is based on *Blanchet 715*, which he saw in the Delessert herbarium; the fragment at P-JU is also labeled *Blanchet 715*. Neither Blanchet specimen now at G is the one seen by Jussieu; both are labeled to be part of the Moricand herbarium and lack Jussieu's annotations. On the other hand, the sheet at F, labeled "Duplicatum ex Conservatorio Botanico Genevensi," is numbered "715"and also annotated as "*Stigmaphyllon macropodum* Adr. Juss." in Jussieu's characteristic hand. I believe this specimen was the one originally in the Delessert herbarium and thus is the holotype of *S. macropodum*.

86. Stigmaphyllon arenicola C. Anderson, Contr. Univ. Michigan Herb. 19: 416. 1993.—TYPE: BRAZIL. São Paulo: Mpio. Iguape, ca. 1 km WSW of city of Iguape, 24°43′S, 47°34′W, *Eiten & Clayton 6202* (holotype: UB!, photo: MICH!; isotypes: K! NY! SP! US!).

Stigmaphyllon martianum var. *variabile* Niedenzu, Pflanzenreich IV. 141(2): 491. 1928.—TYPE: BRAZIL. São Paulo: Praia Grande near Santos, *Mosén 3350* (holotype: S!, photo: MICH!; isotypes: C! P!). [The duplicates at C and P are additionally labeled *Glaziou 13604*; the duplicate at K and the fragments at NY (ex B) labeled *Glaziou 13604* are *S. urenifolium*.]

Vine to 4 m. Stems and branches with scalelike T-shaped hairs and/or sericeous when young, soon becoming glabrous. Laminas 5.8–14.5 cm long, 2.7–9.5 cm wide, lanceolate to triangular to ovate to elliptical but often 2–3-lobed, apex acute-mucronate to obtuse-mucronate or emarginate-mucronate, base attenuate or truncate to cordate, glabrous or glabrate adaxially, and glabrous or glabrate or sometimes sparsely to rarely moderately sericeous abaxially (trabecula 0.3–0.8 mm long, straight or wavy, mostly sessile to subsessile but sometimes with a stalk up to 0.1 mm long) but with abundant hairs along the margin (the hairs sometimes sloughed off in older laminas), margin eglandular or sometimes with irregularly spaced sessile glands (0.2–0.3 mm in diameter) borne adjacent to the margin abaxially, the lateral lobes sometimes terminating in a filiform gland up to 0.1 mm long; petioles 1–11 cm long, with scalelike T-shaped hairs or sericeous or densely so, not confluent across the node, with a pair of prominent but sessile glands at the apex, each gland 1.1–2.5 mm in diameter; stipules 0.5–1.7 mm long, 0.7–1.2 mm wide, free, triangular or narrowly so, eglandular. Flowers ca. 10–30 per umbel, these solitary or borne in dichasia or compound dichasia (axes to the 3rd order, with scalelike T-shaped hairs or sericeous). Peduncles 3.5–13.5 mm long; pedicels 4–10.5 mm long, terete; both densely sericeous, peduncles 0.6–1.6 times as long as the pedicels. Bracts 1–2 mm long, 0.5–1.4 mm wide, triangular or narrowly so, apex acute; bracteoles 1–1.8 mm long, 0.8–1.1 mm wide, triangular, apex obtuse, eglandular or each bracteole with a pair of inconspicuous glands (each gland 0.3–0.5 mm in diameter); bracts and bracteoles sericeous abaxially. Sepals 2.1–2.5 mm long, 2–2.7 mm wide, glands 1.7–2.7 (–3.2) mm long, 1–1.5 mm wide. All petals with the limb orbicular, glabrous, yellow, margin fimbriate or denticulate-fimbriate, teeth/fimbriae up to 0.3 (–0.5) mm long; anterior-lateral petals: claw 2–3 mm long, limb 12–14 mm long and wide; posterior-lateral petals: claw 1–1.8 mm long, limb 10–12.5 mm long and wide; posterior petal: claw 2.5–3.7 mm long, apex usually indented, limb 9–10 mm long and wide, sometimes near the base with 1–2 stout gland-tipped fimbriae per side (ca. 0.5 mm long, ca. 0.2 mm wide). Stamens unequal, those the posterior-lateral petals (and the posterior styles) the largest, anthers of those opposite the lateral sepals eloculate or sometimes the enlarged connective with one or rarely two reduced locules; anthers glabrous. Stamen opposite anterior sepal: filament 1.8–2.6 mm long, anther 1–1.3 mm long; stamens opposite anterior-lateral petals: filaments 1.2–2.2 mm long, anthers 0.8–1 mm long; stamens opposite anterior-lateral sepals: filaments 2.2–3 mm long, connectives 0.8–1 mm long, locules usually absent, sometimes one or two present, 0.1–0.6 mm long; stamens opposite posterior-lateral petals: filaments 2.7–3.3 mm long, anthers 1.2–1.5 mm long; stamens opposite posterior-lateral sepals: filaments 2.2–2.7 mm long, connectives 0.7–1 mm long, locules usually absent, sometimes one or rarely two present, 0.2–0.5 mm long; stamen opposite posterior petal always shorter than the adjacent two: filament 1.9–2.5 mm long, anther 0.7–0.9 mm long. Anterior style 3–4 mm long, shorter than the posterior two or sometimes subequal, terete, glabrous, erect or slightly recurved; apex 1.3–1.9 mm long, each foliole 1.2–1.6 mm long, 1.1–1.8 mm wide, subsquare or suborbicular. Posterior styles 3.6–4.5 mm long, terete, glabrous or with scattered hairs in the proximal 1/3–1/2, lyrate; foliole 2–3 mm long, (1.7–) 2–2.6 mm wide, subsquare to sometimes subrectangular. Dorsal wing of samara 2.8–3 cm long, 0.9–1 cm wide, upper margin with a blunt tooth; nut smooth or bearing a lunate to rectangular lateral winglet per side, these up to 10 mm long, up to 4 mm wide, sometimes also with one or two additional winglets or spurs (up to 2 mm long and wide) or only with a few spurs; nut 6.5–7.5 mm high, ca. 4.5 mm in diameter, without air chambers, areole 2.7–4.5 mm long, 3–3.5 mm wide, concave, carpophore up to 3.5 mm long. Embryo 6.5–8.5 mm long, 1 1/2–2 times as long as wide, ovoid, outer cotyledon 8–15 mm long, 4.1–5 mm wide, the distal 2/3 folded over the inner cotyledon, inner cotyledon 4–5.5 mm long, 2.9–3.5 mm wide, straight. Chromosome number unknown. Fig. 85.

Phenology. Collected in flower throughout the year, in fruit in February and in May through July.

Distribution (Fig. 86). Southeastern Brazil (Rio de Janeiro to Paraná); along beaches, on dunes, in restingas, one report from vicinity of coastal swamp forest; sea level to 340 m.

REPRESENTATIVE SPECIMENS. **Brazil.** PARANÁ: Mpio. Guaraqueçaba, Tagaçaba, *Hatschbach 42721* (MICH); Ilha do Mel, Prainha, *Kummrow 1334* (MBM); Ilha do Mel, Mpio. Panaguá, *Ribas 9* (MICH).—RIO DE JANEIRO: Lagôa de Marapendi, estrada do Autódromo, *Almeida de Jesus 1564* (RB); restinga ca. 11 km W of Barra da Tijuca, *Anderson 11197* (MBM, MICH); restinga de Jacarepaguá, *Angeli 475* (MICH); restinga de Jacarepaguá, Pedro de Itauna, *Araujo 657 & Peixoto 490* (MICH, RB); Mpio. Angra dos Reis, Ilha Grande, Reserva Biológica Estadual da Praia do Sul, 23°10′S, 44°17′W, *Araujo 6133, 7305* (GUA); Recreio do Bandeirantes, *Lutz R23973* (R); restinga de Sernambetiba, *Markgraf 3783* (RB); Recreio dos Bandeirantes, Jacarepaguá, *Palacios et al. 4061* (R); 1–3 km S of Lidice, *Smith & McWilliams 15354* (MICH, R); pr. Bonfim, *Trinta 882 & Fromm 1958* (R).—SÃO PAULO: Mpio. Itanhaém, Km 118 da Rodovia SP-55, *Araujo 6552* (GUA); Iguape, Morro das Pedras, *Brade 7898* (R); Santos, *Curran 13* (US); between Ubatuba and Caraguatuba, *Davis et al. D59894* (E, MBM); Mpio. Mongaguá, at Mongaguá, *Eiten & Eiten 2542* (SP); Mpio. Caraguatatuba, SW of Caraguatatuba, *Eiten & Eiten 2800* (SP); Mongaguá, praia de Suarão, *Kirizawa 67* (SP); Mpio. Ilha do Cardoso, *Leitão Filho 10792* (SP); Praia Grande, Praia Piassabusú, *Löfgren 4160* (SP); Ubatuba, *C. Smith 4844* (SP); rd between São Vicente and Itaipu, 24°00′S, 46°24′W, *L. B. Smith 2000* (GH); Ilha Bela, Serra do Castelhanos, *Sucre 6982* (MICH, SP).

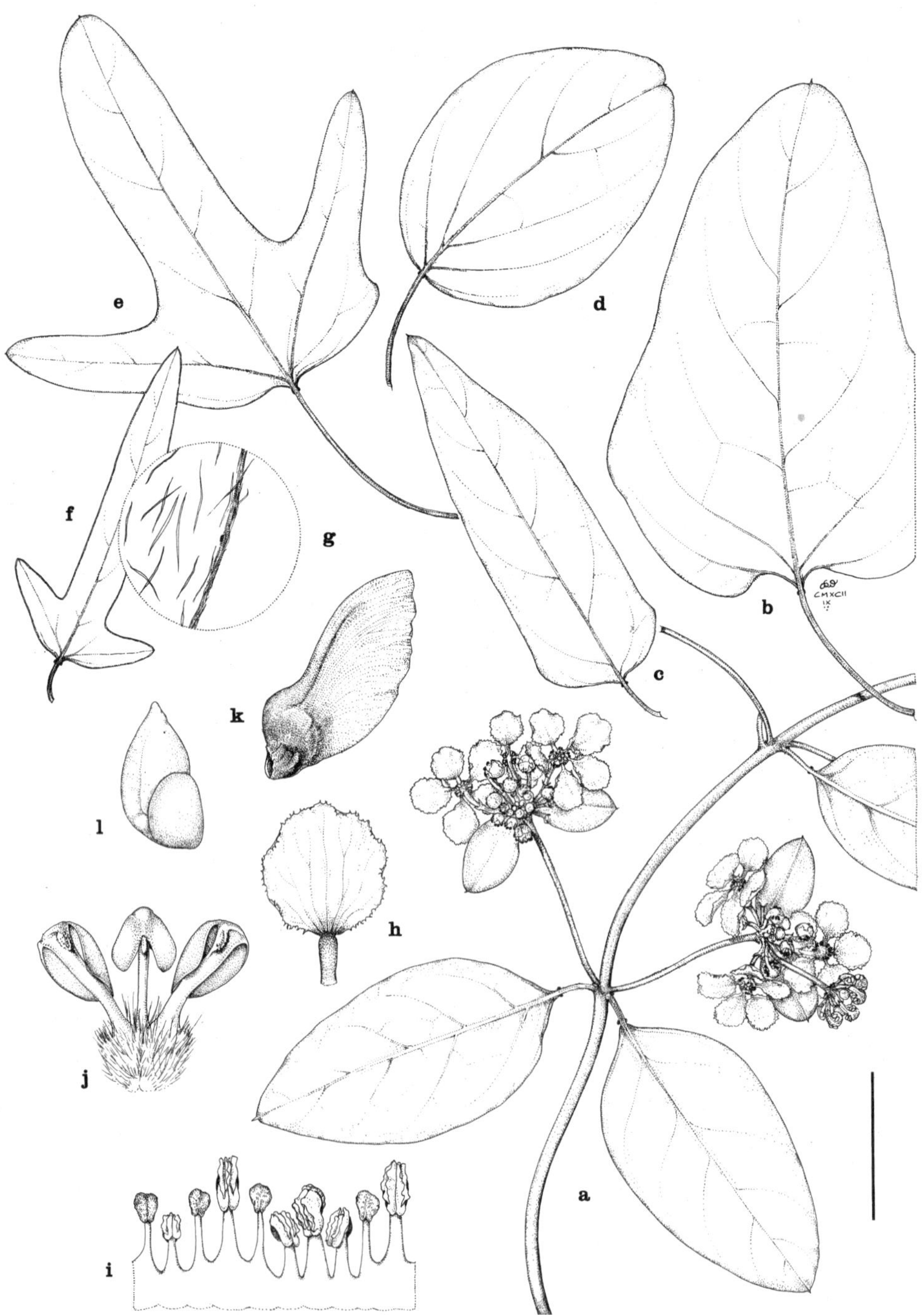

FIG. 85. *Stigmaphyllon arenicola*. a. Flowering branch. b–f. Leaves illustrating variation of laminar shape. g. Detail of margin and abaxial surface of lamina. h. Posterior petal (the "flag"). i. Androecium; stamen second from left opposes the posterior petal. j. Gynoecium; posterior styles are bent slightly outward to show anterior style (in center). k. Samara. l. Embryo. Scale: for a–f, bar = 4 cm; for g, bar = 2 mm; for h, bar = 1 cm; for i, j, bar = 4 mm; for k, bar = 2 cm; for l, bar = 8 mm. (Based on: a, *Eiten & Clayton 6202*; b, *Markgraf 3783*; c, f, g, *Eiten & Eiten 2800*; d, k, l, *Almeida de Jesus 1564*; e, *Anderson 11197*; h, i, *Angeli 475*; j, *Smith & McWilliams 15354*).

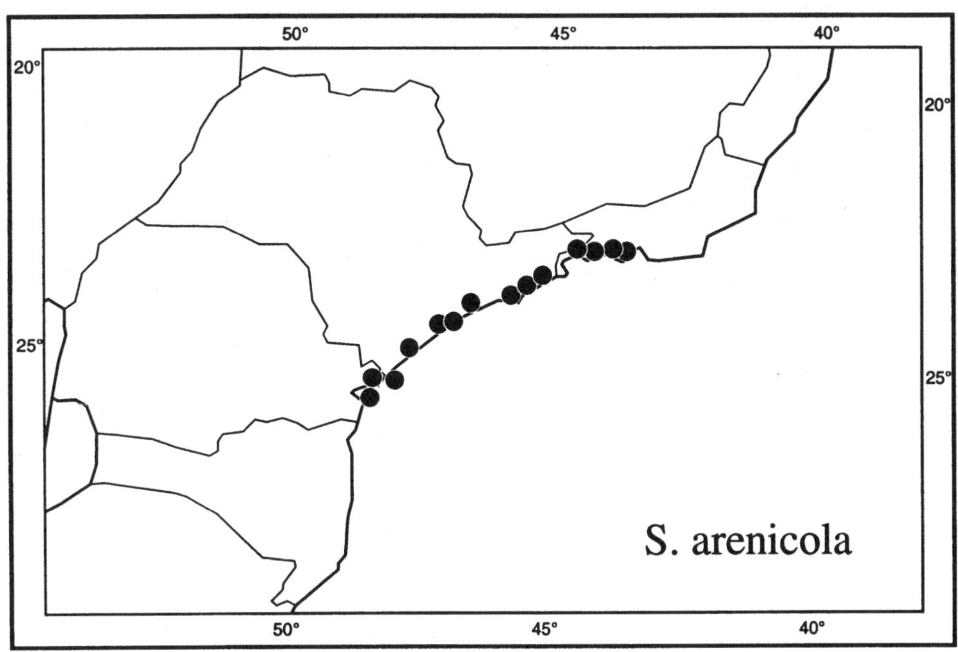

FIG. 86. Distribution of *Stigmaphyllon arenicola.*

Stigmaphyllon arenicola, a species of sandy areas of coastal southeastern Brazil, is notable for its polymorphic leaves. The laminas vary from lanceolate to triangular to ovate to elliptical but may also be 2–3-lobed. They are most commonly very sparsely sericeous to glabrate abaxially but sometimes are glabrous or sericeous. Mostly, the mature laminas are glabrous to glabrate, whereas the younger laminas and the reduced ones near and in the inflorescences are at least sparsely sericeous (the hairs sessile to subsessile); however, plants with all of the laminas abaxially sericeous or glabrous also occur. The anthers are glabrous. Those of stamens opposite the lateral sepals usually lack locules, but sometimes the enlarged connectives bear one or rarely two reduced locules. *Stigmaphyllon arenicola* might be confused with the partly sympatric *S. lalandianum* (no. 14), in which the laminas are sericeous abaxially. *Stigmaphyllon lalandianum* is readily separated by its small flowers with erose petals (the lateral limbs only up to 10 mm in diameter); all stamens bear locules, and the styles are efoliolate or bear only tiny folioles less than 1 mm long.

87. Stigmaphyllon gayanum Adr. Jussieu in St.-Hilaire, Fl. bras. mer. 3: 56. 1833 ["1832"].—TYPE: BRAZIL. Rio de Janeiro: "circa Sebastianopolim," *Gay s.n.* (holotype: P, photos: F! MICH! US!; isotype: P!).

Stigmaphyllon gayanum var. *prostratum* Niedenzu, Ind. Lect. Lyc. Brunsberg. p. aest. 1900: 10. 1900.—TYPE: BRAZIL. Rio de Janeiro, *Martius s.n.* (holotype: M!, photos: F! GH! MICH! NY! US!).

Stigmaphyllon hirsutum Niedenzu, Ind. Lect. Lyc. Brunsberg. p. aest. 1900: 25. 1900.—TYPE: BRAZIL. Rio de Janeiro: Santa Cruz, 1832, *Lhotzky s.n.* (lectotype, here designated: G!, photo: MICH!).

Stigmaphyllon gayanum f. *parvifolium* Niedenzu, Pflanzenreich IV. 141(2): 492. 1928.—TYPE: BRAZIL. Rio de Janeiro: Corcovado, *Mosén 2439* (lectotype, here designated: S!, photo: MICH!).

Vine. Stems and branches with scalelike T-shaped hairs when young, sparsely so in older parts. Laminas 4.7–14 cm long, 6–13 cm wide, broadly elliptical to cordate to suborbicular or sometimes trifid, apex mucronate or emarginate-mucronate, base cordate to truncate or sometimes attenuate, velutinous (the trabecula V-shaped) and sometimes also with some T-shaped hairs adaxially, with T-shaped hairs abaxially (trabecula 0.9–1.6 mm long, straight, stalk 0.1–0.3 mm long), the abaxial pubescence sometimes sparsely but evenly distributed in older leaves, with irregularly spaced raised to stipitate glands (0.1–0.5 mm high, 0.2–0.3 mm in diameter) borne adjacent to the margin abaxially; petioles 1.1–6.2 cm long, with scalelike T-shaped hairs, not confluent across the node, with a pair of prominent but sessile glands at the apex or more commonly up to 7 mm below the base of the lamina, each gland 1.4–2.4 mm in diameter; stipules 0.6–1.3 mm long, 0.4–1 mm wide, free, triangular or narrowly so, eglandular. Flowers ca. 15–35 per umbel, these borne in dichasia or small thyrses (axes to the 3rd order, densely pubescent with scalelike T-shaped hairs). Peduncles 5–11 mm long; pedicels 3.5–9 mm long, terete; both densely pubescent with T-shaped hairs, peduncles 0.9–2 times as long as the pedicels. Bracts 1.5–2.5 mm long, 0.8–1.5 mm wide, triangular or narrowly so, apex acute; bracteoles 1.1–1.8 mm long, 0.8–1.3 mm wide, triangular, apex acute or obtuse, with two inconspicuous glands (each gland 0.3–0.5 mm in diameter); bracts and bracteoles sericeous abaxially. Sepals 2.5–3 mm long, 2.2–2.8 mm wide, glands 1.8–2.5 mm long, 1–1.2 mm wide. All petals with the limb glabrous, yellow, margin with fimbriae up to 0.5 mm long, limbs of lateral petals orbicular; anterior-lateral petals: claw (1.2–) 1.8–2.4 mm long, limb 11–14 mm long and wide; posterior-lateral petals: claw 0.5–1.5 mm long, limb 9–13 mm long and wide; posterior petal: claw (2.6–) 3–4 mm long, apex indented, limb 8–11 mm long, 6.5–9.5 mm wide, broadly elliptical. Stamens unequal, those opposite the posterior-lateral petals (and the posterior styles) the largest, those opposite the anterior-lateral sepals with the filaments subequally long, anthers of those opposite the lateral sepals with the connective enlarged, those opposite the anterior-lateral sepals with the locules reduced, those opposite the posterior-lateral sepals eloculate; loculate anthers pubescent. Stamen opposite anterior sepal: filament 2–2.5 mm long, anther 0.9–1.4 mm long; stamens opposite anterior-lateral petals: filaments 1.5–2 mm long, anthers 0.7–1.2 mm long; stamens opposite anterior-lateral sepals: filaments 2.7–3.5 mm long, connectives (0.6–) 0.8–1 mm long, locules 0.1–0.3 (–0.4) mm long; stamens opposite posterior-lateral petals: filaments 3–3.5 mm long, anthers 1–1.5 mm long; stamens opposite posterior-lateral sepals: filaments 2.5–3.1 mm long, connectives 0.6–1 mm long, locules absent; stamen opposite posterior petal always shorter than the adjacent two: filament 1.7–2.5 mm long, anther 0.6–0.8 mm long. Anterior style 2.7–3.3 mm long, shorter than the posterior two, terete, glabrous, erect or slightly incurved; apex 1.5–2 mm long, each foliole 1.3–2 mm long, 1–1.8 mm wide, subsquare to rectangular. Posterior styles 3.5–4.1 mm long, terete, the proximal 1/3–1/2 pubescent, glabrous distally, lyrate; foliole 2.1–2.7 mm long, 2.3–3 mm wide, subsquare. Dorsal wing of samara 4–4.3 cm long, 1.2–1.4 cm wide, upper margin with a blunt tooth; nut bearing a lunate lateral winglet per side and often also a small spur at the base; winglets 8–9.5 mm long, 1.5–3 mm wide; nut 7.5–9 mm high, 4.5–6 mm in diameter, without air chambers, areole 4–4.5 mm long, 3.5–4.8 mm wide, concave, carpophore up to 3.5 mm long. Embryo ca. 7.6 mm long, ca. 2 times as long as wide, ovoid,

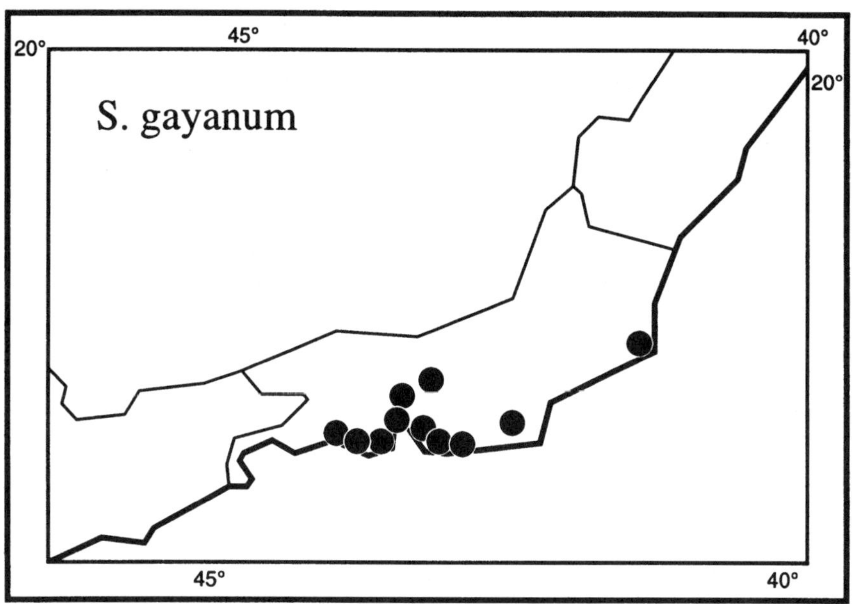

FIG 87. Distribution of *Stigmaphyllon gayanum.*

outer cotyledon ca. 11 mm long, ca. 3.5 mm wide, the distal 2/5 folded over the inner cotyledon, inner cotyledon ca. 6 mm long, ca. 2.5 mm wide, straight. Chromosome number unknown. Fig. 80l–u.

Phenology. Collected in flower from May through December, in fruit in June, September, and December.

Distribution (Fig. 87). Brazil (Rio de Janeiro); in forest, capoeiras, and along roadsides; 150–900 m.

REPRESENTATIVE SPECIMENS. **Brazil.** RIO DE JANEIRO: Teresópolis, *Brade 9675* (R); Petrópolis, *Brade 10533* (R); Alto da Boa Vista, *Brade 11186* (CAS, GH, R); Jacarapaguá, Maciço da Pedra Branca, Pau-da-Fome, *Casari 337* (GUA); Dois Irmãos, *Duarte 164* (COL, MICH, SP); Corcovado, *Flaster et al. 1146* (R); Petrópolis, Carangola, *Goés & Constantino 203* (RB); Serra da Estrella, estrada de Rodagem, Rio–Petrópolis, *J. Kuhlmann 125* (RB); Silva Jardim, Poço d'Anta, *Martinelli et al. 2851* (GUA); Serra dos Orgãos, *Occhioni 6276* (MICH); Parque Nacional da Tijuca, *Schettino RB179623* (MICH); Mata do Rumo, *Sucre 9167* (RB, SP); entre Rio Bonito e Casimiro de Abreu, *Trinta 917 & Fromm 1993* (R); Cachoẹiras de Macacu, *Vianna 1633* (GUA).

Stigmaphyllon gayanum is endemic to the state of Rio de Janeiro. The stem and foliage are abundantly pubescent. Distinctively, the laminas are velutinous adaxially, which imparts a scabrous texture, and bear T-shaped hairs abaxially (the straight trabeculas usually aligned roughly parallel). The marginal leaf glands are usually stipitate, up to 0.5 mm high. The petiole glands may be borne at the apex of the petiole or, more commonly, up to 7 mm below the base of the lamina. The stamens opposite the posterior-lateral sepals bear greatly enlarged connectives that lack locules. The petals are all fimbriate. *Stigmaphyllon gayanum* is most commonly confused with the sympatric *S. affine* (no. 82), which differs most strikingly in its greatly modified samaras. These have an enlarged nut containing air chambers and bearing a reduced triangular dorsal wing. The samaras of *S.*

gayanum are typical for the genus; the nut lacks air chambers and bears a slightly flared, elongate dorsal wing. The laminas of *S. affine* are glabrous adaxially and bear sessile marginal glands. The connectives of stamens opposing the lateral sepals usually all bear reduced locules; only rarely do those of stamens opposing the posterior-lateral sepals lack locules. For a comparison with the somewhat similar *S. hatschbachii* (no. 88), see that species.

88. Stigmaphyllon hatschbachii C. Anderson, Contr. Univ. Michigan Herb. 17: 9. 1990.—TYPE: BRAZIL. Minas Gerais: Realeza, orla da mata, 15 Oct 1983, *Hatschbach 46863* (holotype: MBM!; isotype: MICH!).

Vine. Stems and branches pubescent with scalelike T-shaped hairs. Laminas ca. 10 cm long, 6–7 cm wide, narrowly ovate to elliptical, apex mucronate, base attenuate to truncate, glabrous to glabrate adaxially but with T-shaped hairs on the midvein, very densely pubescent with golden T-shaped hairs abaxially (trabecula 1.1–2.3 mm long, straight, stalk 0.2–0.4 mm long), margin with irregularly spaced stalked glands (ca. 0.4–0.5 mm high, ca. 0.3 mm in diameter) borne adjacent to the margin abaxially; petioles 2.5–2.7 cm long, densely pubescent with T-shaped hairs, not confluent across the node, with a pair of prominent but sessile glands at the apex, each gland 1.8–2 mm in diameter; stipules 0.5–1 mm long and wide, free, triangular, eglandular. Flowers ca. 20–30 per umbel, these borne in small thyrses (axes to the 3rd order, pubescent with subsessile T-shaped hairs). Peduncles 7–9 mm long; pedicels 7–10 mm long, terete; both densely pubescent with golden T-shaped hairs, peduncles 0.7–1 times as long as the pedicels. Bracts 2.8–3.2 mm long, 1–1.3 mm wide, narrowly triangular, apex acute; bracteoles 1.8–2.2 mm long, 1–1.2 mm wide, narrowly triangular, apex acute or obtuse, eglandular; bracts and bracteoles densely pubescent with golden T-shaped hairs abaxially. Sepals ca. 3 mm long, ca. 2.5 mm wide, glands ca. 1.8 mm long, ca. 1 mm wide. All petals with the limb orbicular, glabrous, yellow, margin with fimbriae up to 0.5 (–0.6) mm long; anterior-lateral petals: claw ca. 1.5 mm long, limb ca. 13 mm long and wide; posterior-lateral petals: claw ca. 1.3 mm long, limb ca. 11 mm long and wide; posterior petal: claw 3.5 mm long, apex not indented, limb ca. 10 mm long and wide. Stamens unequal, those opposite the posterior-lateral petals (and the posterior styles) the largest, anthers of those opposite the anterior-lateral sepals with the connective enlarged and bearing only one or rarely two reduced locules, anthers of those opposite the posterior-lateral sepals eloculate; loculate anthers pubescent. Stamen opposite anterior sepal: filament ca. 3 mm long, anther ca. 1.1 mm long; stamens opposite anterior-lateral petals: filaments ca. 2.5 mm long, anthers ca. 1 mm long; stamens opposite anterior-lateral sepals: filaments ca. 3 mm long, connectives ca. 0.8 mm long, locule ca. 0.4 mm long; stamens opposite posterior-lateral petals: filaments ca. 3.5 mm long, anthers ca. 1.4 mm long; stamens opposite posterior-lateral sepals: filaments ca. 3 mm long, connectives ca. 0.8 mm long, locules absent; stamen opposite posterior petal always shorter than the adjacent two: filament ca. 2.5 mm long, anther ca. 0.7 mm long. Anterior style ca. 3.5 mm long, shorter than the posterior two, terete, glabrous, erect; apex ca. 1.9 mm long, each foliole ca. 1.8 mm long, ca. 1.4 mm wide, parabolic. Posterior styles ca. 4 mm long, terete, glabrous, lyrate; foliole ca. 2.7 mm long and wide, suborbicular. Samara unknown. Chromosome number unknown. Fig. 88.

Stigmaphyllon hatschbachii, known only from the type collected in Minas Gerais, Brazil (Fig. 84), is distinguished by the copious golden pubescence that covers most of the

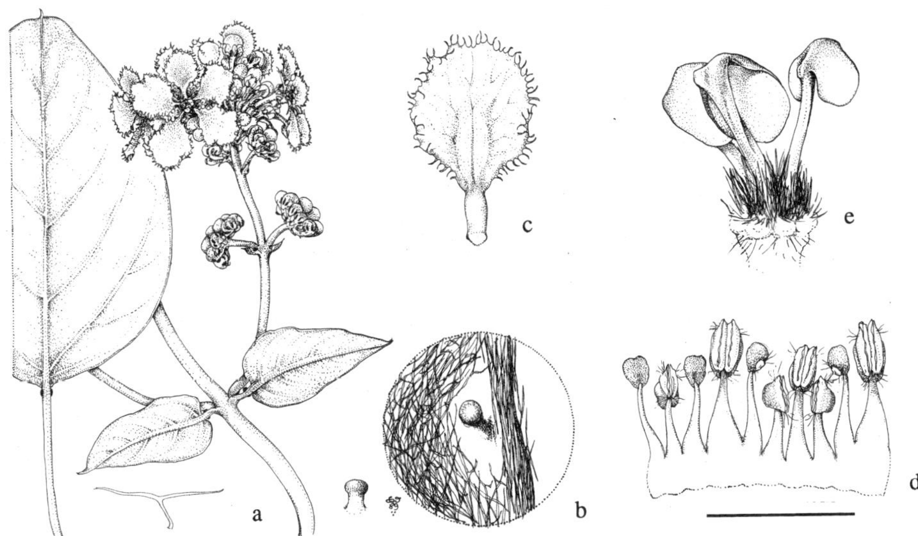

FIG. 88. *Stigmaphyllon hatschbachii*. a. Flowering branch and portion of large leaf. b. Detail of marginal region of abaxial surface of lamina with some of the pubescence removed to show position of gland, lateral view of marginal gland, and a laminar hair. c. Posterior petal (the "flag"). d. Androecium; stamen second from left opposes the posterior petal. e. Gynoecium, anterior style to the right. Scale: for a, bar = 4 cm; for b and individual hair and gland, bar = 2 mm; for c, bar = 1 cm; for d, e, bar = 4 mm. (Based on *Hatschbach 46863*.)

vegetative parts. The laminas are so densely pubescent abaxially that the epidermis is obscured and the stalked marginal glands are hidden. The petals are fringed with fimbriae up to 0.5 (–0.6) mm long. Anthers of the stamens opposite the anterior-lateral sepals have an enlarged connective bearing only one (rarely two) reduced locules, and those of the stamens opposite the posterior-lateral sepals lack locules. All loculate anthers are pubescent.

In some aspects, this species is similar to *S. gayanum* (no. 87) of Rio de Janeiro, which has similar flowers and also abundantly pubescent vegetative parts. *Stigmaphyllon gayanum* differs in that the laminas are adaxially velutinous (sometimes mixed with T-shaped hairs); in *S. hatschbachii*, the laminas are glabrous adaxially except for some T-shaped hairs concentrated on the midvein. The lower surfaces of the laminas of *S. gayanum* are also moderately to abundantly pubescent, but the epidermis and the marginal glands are always visible. The bracteoles of *S. gayanum* bear a pair of small glands, whereas those of *S. hatschbachii* are eglandular. The androecium of *S. gayanum* differs from that of *S. hatschbachii* in that the enlarged connectives of the anthers of the stamens opposite the anterior-lateral sepals always bear two tiny locules [0.1–0.3 (–0.4) mm long] rather than only one locule.

89. Stigmaphyllon blanchetii C. Anderson, Syst. Bot. 14: 511. 1989.—TYPE: BRAZIL. Bahia: Fonte dos Protomartires do Brasil, Porto Seguro, 0–10 m, ca. 16°26′S, 39°05′W, *Harley 17244* (holotype: RB!, photo: MICH!; isotypes: MICH! MO! NY! P! U!).

Vine to 8 m. Stems and branches with scalelike T-shaped hairs when young, becoming glabrate to glabrous. Laminas 5.2–14.5 cm long, 3.2–11 cm wide, narrowly elliptical to elliptical to narrowly lanceolate to lanceolate to narrowly ovate to ovate to sometimes suborbicular or sometimes palmately 3–5-lobed, apex mucronate or emarginate-mucronate or sometimes acuminate to rarely caudate, base usually truncate to cordate but sometimes attenuate or auriculate, glabrous or sometimes glabrate adaxially, densely pubescent with T-shaped hairs abaxially (trabecula 0.3–1.5 mm long, straight or crisped, stalk 0.1–0.2 mm long), with irregularly spaced stalked glands (0.1–0.2 mm in diameter, 0.1–0.4 mm long) borne adjacent to the margin abaxially or rarely the margin eglandular; petioles 1–4.7 cm long, pubescent with scalelike T-shaped hairs, not confluent across the node, with a pair of prominent but sessile glands at the apex, each gland 1–2.5 mm in diameter; stipules 0.5–1.7 mm long, 0.5–1.1 mm wide, free, triangular or narrowly so, eglandular. Flowers ca. (10–) 15–25 per umbel, these borne in dichasia or compound dichasia (axes densely pubescent with scalelike T-shaped hairs). Peduncles (2.5–) 4.5–12 (–15) mm long; pedicels (2.5–) 4–11 mm long, terete; both densely pubescent with scalelike T-shaped hairs, peduncles 0.5–2 times as long as the pedicels. Bracts 0.8–2.1 (–3) mm long, 0.7–1.3 (–1.6) mm wide, triangular or narrowly so, apex acute; bracteoles 0.8–1.4 (–1.9) mm long, 0.8–1.2 mm wide, triangular to ovate, apex obtuse, eglandular or each bracteole with one or two inconspicuous glands (each gland 0.2–0.4 mm in diameter); bracts and bracteoles sericeous abaxially. Sepals 2–2.5 mm long, 1.8–2.6 mm wide, glands (1.2–) 1.5–2.3 mm long, 0.8–1.1 mm wide. All petals with the limb orbicular or limb of the posterior petal sometimes broadly obovate, glabrous, yellow, margin fimbriate, fimbriate-denticulate, denticulate, erose-denticulate, or sometimes erose, teeth/fimbriae up to 0.4 mm long; anterior-lateral petals: claw (1.5–) 2–2.5 (–3.5) mm long, limb ca. 11–15 mm long and wide; posterior-lateral petals: claw 1–1.5 (–2) mm long, limb ca. 10–14 mm long and wide; posterior petal: claw (2.2–) 2.5–3 mm long, apex indented, limb 10–13 mm long, 8.5–13 mm wide. Stamens unequal, those opposite the posterior-lateral petals (and the posterior styles) the largest, anthers of those opposite the anterior-lateral sepals with the connective enlarged and the locules reduced or commonly with only one reduced locule or rarely eloculate, anthers of those opposite the posterior-lateral sepals usually eloculate or sometimes with one reduced locule; loculate anthers pubescent. Stamen opposite anterior sepal: filament 1.8–2.5 mm long, anther 1–1.2 mm long; stamens opposite anterior-lateral petals: filaments 1.3–2 mm long, anthers 0.7–1 mm long; stamens opposite anterior-lateral sepals: filaments 2–3.5 mm long, connectives 0.7–0.9 mm long, locules 0.1–0.6 mm long or rarely absent; stamens opposite posterior-lateral petals: filaments 2.5–4.3 mm long, anthers 1.1–1.4 mm long; stamens opposite posterior-lateral sepals: filaments 2–3 mm long, connectives 0.6–0.9 mm long, locule 0.1–0.2 mm long or absent; stamen opposite posterior petal always shorter than the adjacent two: filament 1.7–2.9 mm long, anther 0.6–0.8 mm long. Anterior style 2.8–3.6 mm long, shorter than the posterior two, terete, glabrous or with some scattered hairs in the proximal 1/2, apex 1.4–2.2 mm long, each foliole 1.2–2.3 mm long, 1.3–2.2 mm wide, subsquare to suborbicular. Posterior styles 3.5–4.8 mm long, terete, glabrous or with scattered hairs in the proximal 1/2, lyrate; foliole 2–3.4 mm long, 1.8–2.7 mm wide, subsquare to suborbicular to subrectangular. Dorsal wing of samara 4–5.5 cm long, 1–1.7 cm wide, upper margin with a blunt tooth, nut bearing a lunate or rectangular lateral winglet per side, these up to 10 mm long and up to 2.5 mm wide or also with an additional spur or winglet; nut 7.5–9 mm high, 4–5.5 mm in diameter, without air chambers, areole 3.5–5 mm long, 3–4 mm wide, deeply concave, carpophore up to 3.5 mm long. Embryo ca. 8.6 mm long, ca. 2

times as long as wide, ovoid, outer cotyledon ca. 13 mm long, ca. 4.6 mm wide, the distal 2/3 folded over the inner cotyledon, inner cotyledon ca. 6 mm long, ca. 4 mm wide, straight. Chromosome number unknown. Fig. 89a–g.

Phenology. Collected in flower throughout the year, in fruit in August, September, November, and December.

Distribution (Fig. 90). Eastern Brazil (Paraíba south to Espírito Santo and eastern Minas Gerais); in wet forest and moist forest ("tabuleiro"), at forest edge, in riverside marsh, at roadsides, near cultivated and otherwise disturbed area, reported from sandy soils; sea level to 400 m.

REPRESENTATIVE SPECIMENS. **Brazil.** ALAGOAS: Maceió, *Campêlo et al. 1569* (MICH, RB); 15 km E de Boca de Mata, 09°40′S, 36°06′W, *Kirkbride 4614* (UB); Maceió, Faz. Santa Luzia, próximo a Riacho Doce, *de Lyra et al. 17* (MO).—BAHIA: Ilhéus, Centro de Pesquisas de Cacau, CEPEC, CEPLAC, *Belém & Magalhães 577* (CEPEC, IAN, NY, UB); Mpio. Santa Cruz de Cabrália, Estação Ecológica do Pau-brasil, cerca 16 km a W de Porto Seguro, *Brito & da Vinha 222* (CEPEC); estrada de Bom Gosto a Pontal, Ilhéus, *Fróes 20018* (F, IAN, K, MICH, NY); [Mpio. Belmonte] Barrôlandia–Estação Experimental Gregório Bondar, *Hage 15* (CEPEC, MICH); BA-360 hwy just E of Itambé, N of the Rio Pardo, 15°15′S, 04°36′W, *Harley 15017* (K, MICH, NY, P, U); 20 km from Una and 10 km from Nova Colonial, W along rd to Rio Branco, 15°15′S, 39°13′W, *Harley 18212* (CEPEC, MICH); Mpio. Nova Visosa, Dois Irmãos, *Hatschbach 47754* (MICH); Mpio. Caravelas, Corrego Taquaral, *Hatschbach 49484* (MICH); Mpio. Prado, Prado, Km 15–25 na rodovia para Itamaraju, *Hatschbach 62998* (MICH); Mpio. Jandaíra, rodovia Linha Verde, 10–20 km S de Abadia, *Hatschbach 63167* (MICH); estr. Itabuna a Una, *Heringer 3267* (MICH, MO, US); Mpio. Porto Seguro, 16°32′S, 39°08′W, *Lima 37* (CEPEC, MG); Mpio. Ilhéus, estrada entre Sururú e Vila Brasil, a 6–14 km de Sururú, a 12–20 km ao SE de Buerarema, *Mori & Benton 12872* (CEPEC, MICH); Mpio. Una, UNACAU, Fazenda Basilândia, Rodovia São José/Una, a 12 km do entroncamento coma BR-101, *dos Santos 156* (MICH); Mpio. Prado, Reserva Florestal da Brasil de Holanda Industrias S.A., the entrance at Km 18 E of Itamaraju on rd to Prado, 17°11′S, 39°20′W, *Thomas et al. 10113* (NY).—ESPÍRITO SANTO: Mpio. Jacaré, Agua Limpa, *Hatschbach 46976* (MICH); Mpio. São Mateus, Ligação Rod. BR-101 a Ponta da Ipiranga, Sede, *Hatschbach 60066* (MICH); Reserva Florestal de Linhares, CVRD, próximo Estrada 221 Talhão 203, *Lino 68* (RB, SP); entre Morro d'Anta e Santana, a 10 km de Santana [near Vitória], *Mattos 10758* (SP); Derrubada dos Paulistas corrego Dourado, *Mattos Filho & Magnanini 31* (RB); Reserva Florestal de Linhares, CVRD, próximo Estrada 142 Talhão 403, *Spada 142* (RB, SP).—MINAS GERAIS: Estacão Biologica de Caratinga, *Andrade 321* (MICH); Salto da Divisa, 15–20 km W, *Hatschbach 52208* (MICH); Serra do Cipó, *Sena 12863 (RB 61947)*.—PARAÍBA: Remígio, *Barbosa 151* (RB); Areia, Eugenho Bom-Fim, *Fevereiro* 282 (RB).—PERNAMBUCO: Recife, *Falcão et al. 776* (RB, SP); Goiânia, *Falcão et al. 1145* (SP); Sítio Agua Comprida, *Guades 45* (US); *Pickel 254, 986* (SP); *Pickel 2436* (NY-fragment); Praia do Guaibu, 35 km S from Recife, *Tsugaru et al. B-1365* (MO, NY).—SERGIPE: Mpio. Santa Luzia do Itanhi, ca. 2 km na estrada de Crasto para Santa Luzia, *de Carvalho et al. 4319* (NY), *Amorim et al. 1495* (NY).

Stigmaphyllon blanchetii is a common species of easternmost Brazil from Paraíba south to Espírito Santo and adjacent Minas Gerais. The laminas are abaxially abundantly pubescent with T-shaped hairs and bear stipitate glands adjacent to the margin abaxially. The anthers are pubescent. The connectives of anthers of stamens opposite the anterior-lateral petals are enlarged and usually bear only one reduced locule, those of stamens opposite the posterior-lateral sepals usually lack locules. *Stigmaphyllon blanchetii* is most similar to the less common *S. salzmannii* (no. 90), in which the laminas are glabrous abaxially or sometimes very sparsely pubescent with T-shaped hairs with a straight trabecula. *Stigmaphyllon cavernulosum* (no. 83) of coastal Bahia also has the laminas with T-shaped hairs abaxially, but the marginal glands are sessile, and the anthers are glabrous.

The collection *Sena 12863 (RB 61947)* recorded from the Serra do Cipó, part of the Serra do Espinhaço of central Minas Gerais, is a very poor specimen and placed here ten-

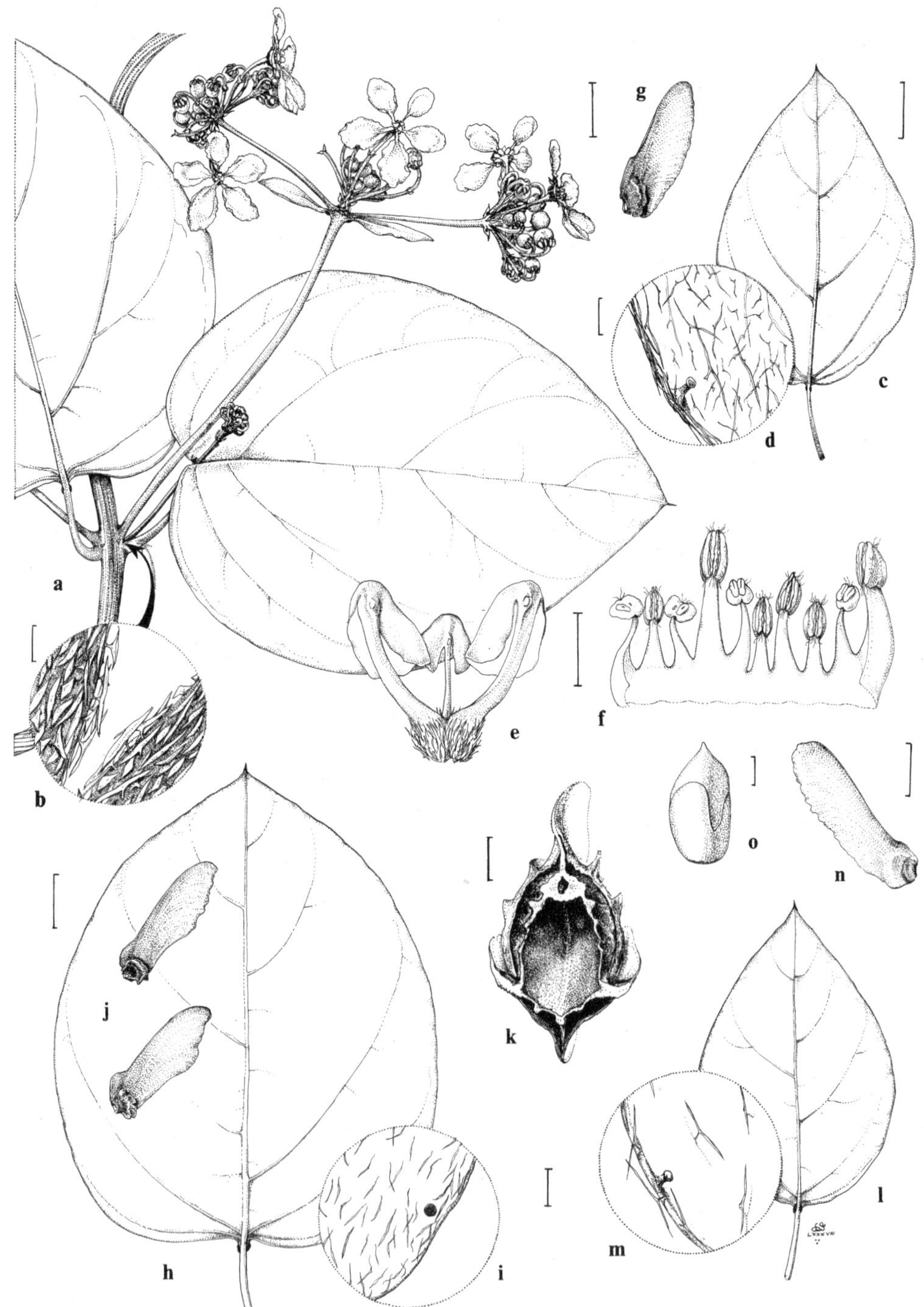

FIG. 89. *Stigmaphyllon blanchetii, S. rotundifolium,* and *S. salzmannii.* a–g, *S. blanchetii.* a. Flowering branch. b. Detail of pubescence of stem and petiole. c. Small leaf. d. Detail of margin and abaxial surface of lamina. e. Gynoecium, anterior style in the center. f. Androecium; stamen second from left opposes the posterior petal (the "flag"). g. Young samara. h–k, *S. rotundifolium.* h. Leaf. i. Detail of margin and abaxial surface of lamina. j. Two samaras. k. Longitudinal section through nut of samara; note the two air chambers flanking the locule. l–o, *S. salzmannii.* l. Leaf. m. Detail of margin and abaxial surface of lamina. n. Samara. o. Embryo. Scale: for a, c, g, h, j, l, n, bar = 1.5 cm; for b, bar = 0.2 mm; for d, i, m, bar = 0.5 mm; for e, f, k, o, bar = 2 mm. (Based on: a–d, *Harley 18212*; e, f, *Harley 17244*; g, *Campêlo et al. 1569*; h–k, *Blanchet 381*; l, m, *N. T. Silva 58340*; n, o, *Blanchet s.n.*)

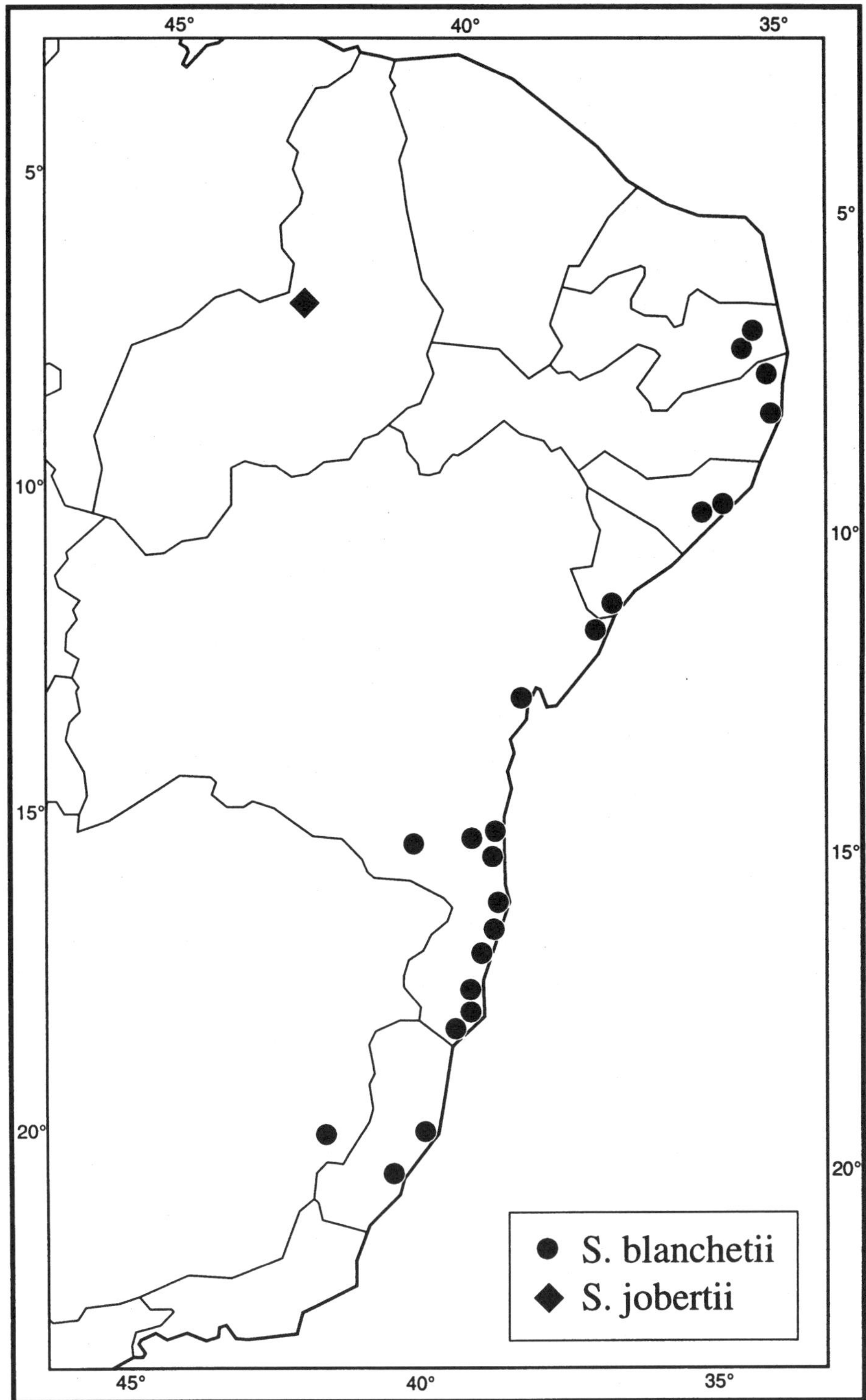

FIG. 90. Distribution of *Stigmaphyllon blanchetii* and *S. jobertii*.

tatively. Perhaps it represents an undescribed species from this region noted for its great number of endemics.

90. Stigmaphyllon salzmannii Adr. Jussieu, Ann. Sci. Nat. Bot., sér. 2, 13: 288. 1840.—TYPE: BRAZIL. Bahia, *Salzmann 95* (holotype: G!, photo: MICH!, fragment: P-JU!, photo: MICH!).

Stigmaphyllon trifidum Niedenzu, Bot. Jahrb. Syst. 14, Beibl. 30: 4. 1891. *Stigmaphyllon digitatum* Niedenzu, Repert. Spec. Nov. Regni Veg. 33: 70. 1933, nom. superfl.—TYPE: BRAZIL. Rio de Janeiro: Carmo de Cantagalo, Jun 1888, *Neves Armond* (*herb. Schrenck 307*) (holotype: B, destroyed; isotypes: R-72139! R-19565! R-19566!; Niedenzu based *Stigmaphyllon digitatum* on R-19565, photo: MICH!).

Vine. Stems and branches with scalelike T-shaped hairs and/or T-shaped hairs when young, often becoming glabrate to glabrous. Laminas 6–13 cm long, 3–7 cm wide, elliptical to lanceolate to ovate or sometimes 3-lobed, apex mucronate or acuminate or sometimes caudate, base truncate or attenuate, glabrate to glabrous adaxially, glabrous to very sparsely pubescent with T-shaped hairs abaxially (trabecula 0.6–1.8 mm long, straight, stalk 0.1–0.3 mm long), with irregularly spaced stalked glands (0.2–0.3 mm in diameter, 0.1–0.7 mm long) borne adjacent to the margin or rarely on the margin abaxially; petioles 0.8–3.5 cm long, bearing scalelike T-shaped hairs, not confluent across the node, with a pair of prominent but sessile glands at the apex or up to 4 mm below the base of the lamina, each gland 0.9–1.9 mm in diameter; stipules 0.6–1 mm long, 0.4–1 mm wide, free, triangular, eglandular. Flowers ca. 15–25 per umbel, these solitary or borne in dichasia or small thyrses but commonly on long axes in irregular compound arrangements (axes pubescent with scalelike T-shaped hairs). Peduncles (4.2–) 4.5–12 mm long; pedicels 5–8.5 mm long, terete; both densely sericeous, peduncles (0.5–) 0.8–1.6 times as long as the pedicels. Bracts 1–2.1 mm long, 0.6–1 mm wide, triangular or narrowly so, apex acute to acuminate; bracteoles 1–1.5 mm long, 0.6–1 mm wide, triangular to ovate, apex acute to obtuse, each bracteole eglandular or with two inconspicuous glands (each gland 0.2–0.3 mm in diameter); bracts and bracteoles sericeous abaxially. Sepals (2–) 2.6–2.7 mm long and wide, glands 1.5–2.3 mm long, 0.8–1.1 mm wide. All petals with the limb orbicular, glabrous, yellow, margin of lateral petals fimbriate or denticulate-fimbriate, of the posterior petal fimbriate, teeth/fimbriae up to 0.6 mm long; anterior-lateral petals: claw (1.5–) 2–2.5 mm long, limb ca. (12–) 14–16 mm long and wide; posterior-lateral petals: claw 1–1.2 mm long, limb ca. (10–) 11–14 mm long and wide; posterior petal: claw (2.6–) 3–3.2 mm long, apex strongly indented, limb 11–12 mm long and wide. Stamens unequal, those opposite the posterior-lateral petals (and the posterior styles) the largest, anthers of those opposite the anterior-lateral sepals with the connective enlarged and the locules reduced and unequal or with only one locule, those opposite the posterior-lateral sepals eloculate or sometimes with one tiny locule or rarely with two tiny locules; loculate anthers pubescent. Stamen opposite anterior sepal: filament 1.8–2.5 mm long, anther 1–1.2 mm long; stamens opposite anterior-lateral petals: filaments 1.5–1.7 mm long, anthers 0.7–0.8 mm long; stamens opposite anterior-lateral sepals: filaments 2.3–3 mm long, connectives 0.7–0.9 mm long, locules 0.1–0.5 mm long; stamens opposite posterior-lateral petals: filaments 2.7–3.5 mm long, anthers 1.1–1.2 mm long; stamens opposite posterior-lateral sepals: filaments 2–2.8 mm long, connectives 0.6–0.8 mm long, locules 0.3–0.4 mm long but usually absent; stamen opposite posterior petal always shorter than the adjacent two: filament 1.8–2.3 mm long, anther 0.6–0.7 mm long. Anterior style 3–3.6 mm long, shorter than the posterior two, terete, glabrous or with a few scattered hairs proximally, erect; apex (1.4–) 1.6–2 mm long, each foliole (1.2–) 1.5–2.1 mm long and wide, subsquare. Posterior styles 3.8–4.3 mm long, terete, glabrous or with scattered hairs in the proximal 1/3, lyrate; foliole 2–3 mm long and wide, subsquare to subrectangular. Dorsal wing of samara 3.8–4.7 (–5.5) cm long, 1.2–1.6 (–2) cm wide, upper margin with a blunt tooth; nut bearing 1–2 lunate to rectangular lateral winglets per side, these up to 8 (–12) mm long and up to 3 mm wide, or sometimes only with a lateral tooth (ca. 1.2 mm long and 1.8 mm high); nut 7.8–9 (–12) mm high, 5–5.7+ mm in diameter, without air chambers, areole 3.5–5 mm long and wide, concave, carpophore up to 3 mm long. Embryo 7.5–8 mm long, ca. 2 times as long as wide, ovoid, outer cotyledon 12–14 mm long, 4.2–4.5 mm wide, the distal 1/2–2/3 folded over the inner cotyledon, inner cotyledon 5.4–5.7 mm long, 3.2–3.5 mm wide, straight or the tip folded back. Chromosome number unknown. Fig. 89l–o.

Phenology. Collected in flower from October through July, in fruit in May, July, October, and November.

Distribution (Fig. 91). Brazil (Bahia, southeastern Minas Gerais, and adjacent Rio de Janeiro; one report each from Espírito Santo and Pernambuco; from moist woods ("tabuleiro" forest), capoeiras, campo sujo, coffee plantations, at forest edge, along roadsides, in secondary woods near varzea, and other disturbed areas; sea level to 500 m.

ADDITIONAL SPECIMENS EXAMINED. **Brazil.** BAHIA: Mpio. Itabuna, 10 km S de Pontal (Ilhéus), camino a Olivença, 14°54′S, 39°02′W, *Arbo et al. 5567* (MICH); Mpio. Ilhéus, 16 km de Itabuna, *Belém & Mendes 174* (CEPEC, NY, UB); *Bierens de Haan 112* (U); Ilhéus, *Blanchet s.n.* (G); forests of the Gongoji basin, *Curran 151* (US); Mpio. Ilhéus, Fazenda Theobroma, prox. a margem do Rio Santana, 14°52′S, 39°04′W, *Ginzbarg 776* (MICH); Mpio. Ilhéus, Olivença, 2–3 km N, *Hatschbach 57005* (MICH); Mpio. Buerarema, S. José, *Magalhães 55* (CEPEC); *Martius 1163* (G, M, NY); *Salzmann s.n.* (K, LE, P); Itabuna, rd Itabuna–Buerarema, Km 25, *N. T. Silva 58340* (MICH, NY, UB); Ilhéus, rd from Olivença to Marium, 5 km W of Olivença, 14°59′S, 39°03′W, *Thomas et al. 8997* (NY); Mpio. Prado, 21 km E of Itamaraju on rd to Prado, 17°15′S, 39°22′W, *Thomas et al. 9960* (MICH); *Wawra & Maly 570* (W).—ESPÍRITO SANTO: Mpio. Aracruz, Estação de Biologia Marinha Mello Leitão, *Araujo 250 & Peixoto 120* (MICH, SP).—MINAS GERAIS: Mpio. Leopoldina, Fazenda Nyagara–Santa Isabel, *Barreto 7623* (F); Mpio. Leopoldina, próx. Rio Pomba, Rod. BR-116, *Hatschbach 52167* (MICH); Teixeira Soares, *Sampaio 724* (R), *838* (R).—PERNAMBUCO: varzea nächst Recife, *Zerny 1426* (WU).—RIO DE JANEIRO: Serra dos Orgãos, *Occhioni 6001* (MICH); Cantagallo, *Peckolt 39* (W), *289* (W); by the railroad between Dois Irmãos and Caxaya, 2 Aug 1887, *Ridley et al. s.n.* (BM).

Stigmaphyllon salzmannii is very similar to the partially sympatric *S. blanchetii* (no. 89). Both species have stipitate glands on the laminas abaxially adjacent to the margin and pubescent anthers. In stamens opposite the anterior-lateral sepals, the connectives are enlarged and bear one or two unequal, reduced locules; in those opposite the posterior-lateral sepals, the enlarged connectives lack locules (sometimes one or very rarely two locules present in *S. salzmannii* in Minas Gerais and Rio de Janeiro). The two species are readily separated by the vesture of the abaxial surface of the laminas. In *S. blanchetii* the laminas are abundantly pubescent with T-shaped hairs with a wavy to crisped trabecula, whereas in *S. salzmannii* they are glabrous or only very sparsely beset with T-shaped hairs with a straight trabecula.

Niedenzu based the names *S. trifidum* and *S. digitatum* on specimens with 3-lobed leaves. Such forms with lobed leaves in species characterized by unlobed leaves are not unusual in *Stigmaphyllon* and do not merit taxonomic recognition. A duplicate of the type

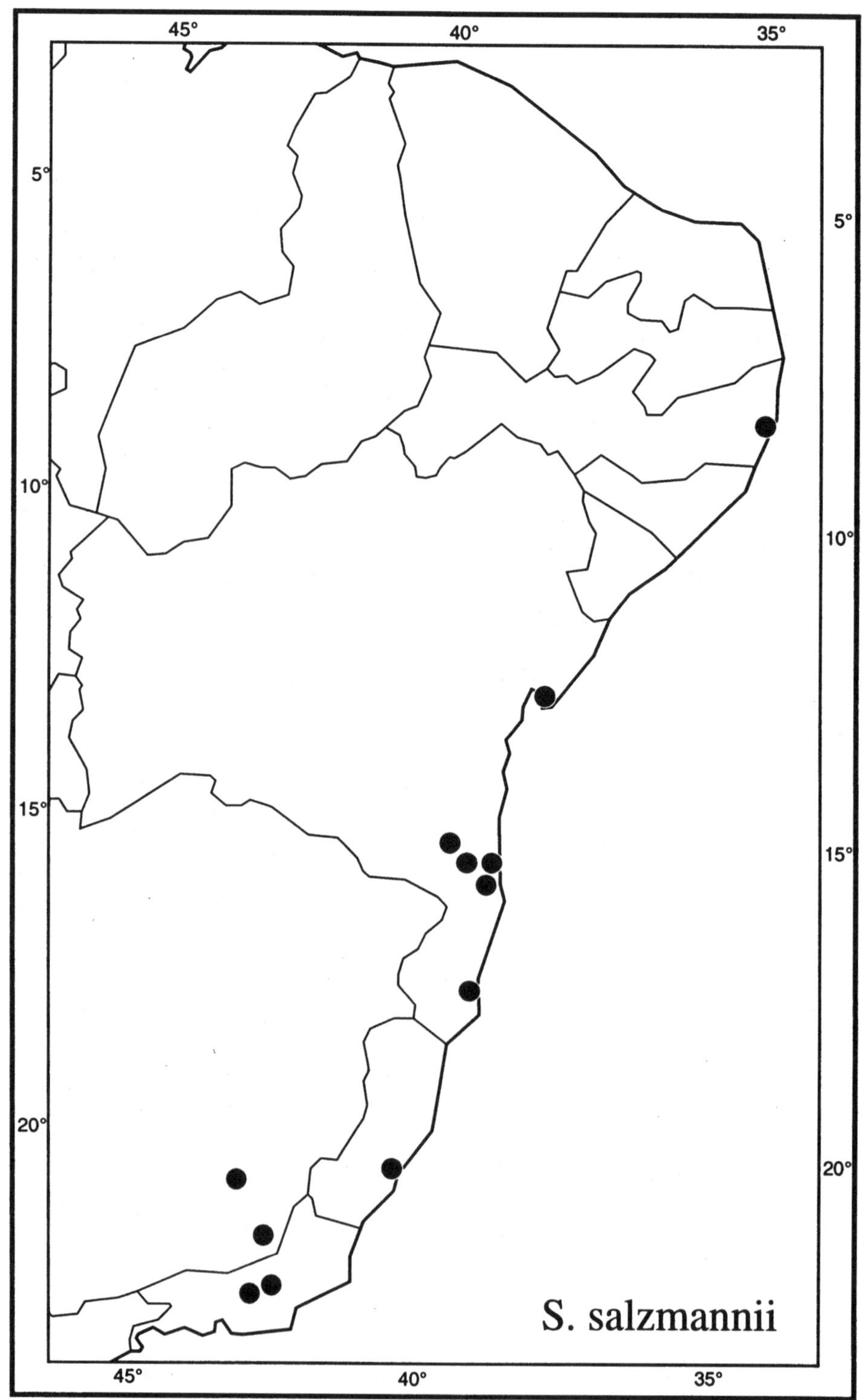

FIG. 91. Distribution of *Stigmaphyllon salzmannii.*

collection of Niedenzu's names at R, which Niedenzu did not see, consists of a large inflorescence with one unlobed leaf.

Doubtful and Excluded Names

Banisteria megacarpos Vellozo, Fl. flum. 4: 189, t. 150. 1829. *Stigmaphyllon megacarpon* (Vellozo) Grisebach, Linnaea: 209. 1839.—Type: Brazil. Rio de Janeiro (holotype: not located).—Vellozo's name may apply to one of the many species of *Stigmaphyllon* of Rio de Janeiro, but the description is too vague and the illustration too stylized to permit assigning the name with certainty.

Banisteria vitifolia Sessé & Mociño ex DC., Prodr. 1: 588. 1824, pro syn.

Stigmaphyllon sect. *Giralia* Cuatrecasas, Ciencia, México 23: 140. 1964.—Type: *Stigmaphyllon albiflorum* Cuatrecasas [=*Malpighia albiflora* (Cuatrecasas) Cuatrecasas].

Stigmaphyllon albiflorum Cuatrecasas, Ciencia, México 23: 139. 1964.—Type: Colombia. Antioquia: Hoya del Río León o Bacubá, entre Villa Arteaga y Chogorodó, 3 Oct 1961, *Cuatrecasas & Willard 26181* (holotype: US!; isotype: COL!) = *Malpighia albiflora* (Cuatrecasas) Cuatrecasas, Phytologia 40: 36. 1978.

Stigmaphyllon angustifolium Grisebach in Martius, Fl. bras. 12(1): 41. 1858, error for *Stigmaphyllon angustilobum* Adr. Jussieu.

Stigmaphyllon chiapense Lundell, Wrightia 6: 32. 1978.—Type: Mexico. Chiapas: Carelas, near Motozintla, Apr 1945, *Matuda 5510* (holotype: UTD!; isotypes: F! MICH!) = *Callaeum chiapense* (Lundell) D. M. Johnson, Syst. Bot. 11: 346. 1986.

Stigmaphyllon lineare var. *morroensis* Kitanov, God. Sofisk. Univ. 66: 4. 1974.—Type: Cuba. Oriente: Lomas cerca del Morro, Santiago de Cuba, 24 Aug 1952, *Lopes Figueiras s.n.* (SV-624).—I did not see the type of this name, which probably applies to a narrow-leaved form of *S. diversifolium*. Only leaf characters are mentioned in the very brief description; the name cannot be assigned with certainty without characterization of the androecium and styles.

Stigmaphyllon lupinus S. Watson; typographical error in Gray Herbarium Card Index for *Stigmaphyllon lupulus* S. Watson.

Stigmaphyllon mathiasiae W. R. Anderson, Bol. Mus. Bot. Munic. Curitiba 43: 3. 1980.—Type: Peru. Loreto: Pucallpa–Aguaytía Rd, Km 34 just W of Tournavista, 31 Jul 1962, *Mathias & Taylor 6078* (holotype: MICH!; isotypes: COL, F, LA, MBM, MO, NY! RB, U, USM) = *Banisteriopsis mathiasiae* (W. R. Anderson) W. R. Anderson, Contr. Univ. Michigan Herb. 20: 15. 1995.

Stigmaphyllon malpighioides Turczaninow, Bull. Imp. Soc. Naturalistes Moscou 36: 582. 1863.—Type: Mexico. Veracruz: Orizaba, Jul 1855, *Botteri 1073* (lectotype, designated by Johnson, 1986: G!) = *Callaeum malpighioides* (Turczaninow) D. M. Johnson, Syst. Bot. 11: 344. 1986.

Stigmaphyllon nigrescens (Adr. Jussieu) Kuntze, Rev. gen. pl. 1: 89. 1891. = *Banisteriopsis lucida* (Richard) Small, N. Amer. fl. 25(2): 133. 1910; fide Gates (1982).

Stigmaphyllon olivaceum Cuatrecasas, Ciencia, México 23: 141. 1964.—TYPE: COLOMBIA. Vaupés: Río Apaporis, Raudal de Jirijirimo, below mouth of Río Kananarí, ca. 900 ft, 00°05′N, 70°40′W, *Schultes & Cabrera 14591* (holotype: US!) = *Heteropterys olivacea* (Cuatrecasas) W. R. Anderson, Mem. New York Bot. Gard. 32: 178. 1981.

Stigmaphyllon sagraeanum var. *angustiifolium* Kitanov, God. Sofisk. Univ. 66: 4. 1974.—TYPE: CUBA. Mantanzas: Cuabal del Espinal, Canasi, *Acuña & León s.n.* (SV-N22847).—I did not see the type of this name, which probably applies to a narrow-leaved form of *S. sagraeanum*. Only leaf characters are mentioned in the very brief description; the name cannot be assigned with certainty without characterization of the androecium and styles.

ACKNOWLEDGMENTS

I am grateful to William R. Anderson for his generous advice and assistance and his valuable comments throughout this study. I thank Thomas F. Daniel for his careful editing, and Shirley A. Graham and an anonymous reviewer for their thoughtful suggestions. I am indebted to many colleagues for their help, especially with bibliographic and nomenclatural matters, and thank them all for their many kindnesses: C. D. Adams, Rupert C. Barneby, John H. Beaman, Alan J. Bornstein, Richard K. Brummitt, Hervé M. Burdet, Ray Desmond, Charles E. Jarvis, Guiliana Forneris, Gordon W. Frankie, Bronwen Gates, Shirley A. Graham, Alicia Lourteig, Roger Lundin, Antonio M. Regueiro, James C. Solomon, John L. Strother, and Edward G. Voss. Karin Douthit drew the handsome illustrations. Figure 31 is reprinted with permission from *Memoirs of the New York Botanical Garden* 32: 169, fig. 5, copyright 1981, The New York Botanical Garden. I thank the curators of the following herbaria for making their material of *Stigmaphyllon* available to me and for their assistance and courtesies during visits: A, AAU, AMAZ, BM, BR, C, CAS, CAY, CEPEC, CHAPA, CM, COL, CTES, DAV, DS, DUKE, ENCB, F, G, GB, GH, GOET, GUA, IAN, INPA, JBSD, K, LE, LL, M, MA, MBM, MER, MEXU, MG, MICH, MSC, MO, MT, MY, NY, P, QCA, R, RB, S, SEL, SP, TEX, U, UB, UPR, US, UTD, VEN, W, WAG, WIS. I also examined some specimens of *Stigmaphyllon* on loan to W. R. Anderson from CEN, E, ECON, SD, SI, TRIN, UC, LA, and from the private herbarium of T. M. Pedersen. The early phase of this study was supported by a grant from the National Science Foundation to the University of Michigan (DEB-8215276).

LITERATURE CITED

Adams, C. D. 1972. *Flowering plants of Jamaica*. Mona: University of the West Indies.

Anderson, C. 1982. A monograph of the genus *Peixotoa* (Malpighiaceae). Contr. Univ. Michigan Herb. 15: 1–92.

———. 1986. Novelties in Stigmaphyllon (Malpighiaceae) Syst. Bot. 11: 120–130.

———. 1987a. *Stigmaphyllon* (Malpighiaceae) in Mexico, Central America, and the West Indies. Contr. Univ. Michigan Herb. 16: 1–48.

———. 1987b. Two new species of *Stigmaphyllon* (Malpighiaceae) from South America. Contr. Univ. Michigan Herb. 16: 49–53.

———. 1989. Salzmann's collections of *Stigmaphyllon* (Malpighiaceae) from Bahia, Brazil. Syst. Bot. 14: 506–515.

———. 1990. Seven new species of *Stigmaphyllon* (Malpighiaceae) from Brazil. Contr. Univ. Michigan Herb. 17: 7–19.

———. 1992a. *Stigmaphyllon bannisterioides,* the correct name for a well-known species of *Stigmaphyllon* (Malpighiaceae). Taxon 41: 327–328.

———. 1992b. Two new species of *Stigmaphyllon* (Malpighiaceae) from Peru. Novon 2: 302–305.

———. 1993a. The identity of *Stigmaphyllon dichotomum* (L.) Griseb. (Malpighiaceae). Brittonia 45: 34–38.

———. 1993b. The identities of the sericeous-leaved species of *Stigmaphyllon* (Malpighiaceae) in the Amazon region. Contr. Univ. Michigan Herb. 19: 393–413.

———. 1993c. Novelties in *Stigmaphyllon* (Malpighiaceae) from South America. Contr. Univ. Michigan Herb. 19: 415–429.

———. 1997 ["1996"]. A new combination in *Stigmaphyllon* (Malpighiaceae), and notes on the publication dates of Hooker's *Botanical Miscellany*. Brittonia 48: 542–543.

Anderson, W. R. 1978 ["1977"]. Byrsonimoideae, a new subfamily of the Malpighiaceae. Leandra 7: 5–18.

———. 1979a. *Mcvaughia*, a new genus of Malpighiaceae from Brazil. Taxon 28: 157–161.

———. 1979b. Floral conservatism in neotropical Malpighiaceae. Biotropica 11: 219–233.

———. 1980. *Ectopopterys*, a new genus of Malpighiaceae from Colombia and Peru. Contr. Univ. Michigan Herb. 14: 11–15.

———. 1981. Malpighiaceae. In "The botany of the Guayana Highland—Part XI" by Bassett Maguire and collaborators. Mem. New York Bot. Gard. 32: 21–305.

———. 1993. Chromosome numbers of neotropical Malpighiaceae. Contr. Univ. Michigan Herb. 19: 341–354.

Barneby, R. C. 1991. Sensitivae Censitae. Mem. New York Bot. Gard. 65: 1–835.

Cavanilles, A. J. 1790. *Nona dissertatio botanica.* Madrid: Typographia regia.

Cuatrecasas, J. 1958. Prima Flora Colombiana. 2. Malpighiaceae. Webbia 13: 343–664.

Cuatrecasas, J., and T. B. Croat. 1981 ["1980"]. *Stigmaphyllon.* In "Flora of Panama," ed. W. G. d'Arcy, Ann. Missouri Bot. Gard. 67: 924–933.

Desfontaines. R. L. 1829. *Tableau de l'école de botanique*, ed. 3. Paris: J. A. Brosson.

Fawcett, W., and A. B. Rendle. 1920. *Flora of Jamaica*, vol. 4. London: Trustees of the British Museum.

Gates, B. 1982. A monograph of *Banisteriopsis* and *Diplopterys*, Malpighiaceae. Fl. Neotropica 30: 1–237.

Grisebach, A. 1839. Malpighiacearum brasiliensium centuriam. Linnaea 13: 155–259.

———. 1858. Malpighiaceae. In *Flora brasiliensis* by K. F. P. von Martius, 12(1): 1–123.

Hensold, N. 1988. Morphology and systematics of *Paepalanthus* subgenus *Xeractis* (Eriocaulaceae). Syst. Bot. Monog. 23: 1–150.

Hooker, J. D. 1862. Malpighiaceae. In *Genera plantarum*, by G. Bentham and J. D. Hooker, 1: 247–262. London.

Johnson, D. M. 1986. Revision of the neotropical genus *Callaeum* (Malpighiaceae). Syst. Bot. 11: 335–353.

Johnston, I. M. 1949. Botany of San José Island (Gulf of Panama). Sargentia 8: 1–298.

Jussieu, Adr. de. 1833 ["1832"]. Malpighiaceae. In *Flora brasiliensis meridionalis* by A. de Saint-Hilaire, 3: 5–86, t. 161–180. Paris: A. Belin.

———. 1838 ["1837"]. *Brachypterys australis*. In *Icones selectae plantarum*, ed. J. P. B. Delessert, 3: 20, t. 34.

———. 1840. Malpighiacearum synopsis. Ann. Sci. Nat. Bot., sér. 2, 13: 247–291, 321–338.

———. 1843. Monographie des malpigiacées. Arch. Mus. Hist. Nat. Paris 3: 5–151, 255–626, t. 1–23.

Lamarck, J. B. A. P. M. de. 1785. *Encyclopédie méthodique. Botanique.* Vol. 1(2). Paris: Panckoucke.

Linnaeus, C. 1737. *Genera plantarum.* Leiden.

———. 1738 ["1737"]. *Hortus cliffortianus*. Amsterdam.

———. 1753. *Species plantarum.* Stockholm.

Liogier, A. 1963. Novitates Antillanae I. Bull. Torrey Bot. Club 90: 186–192.

Lowrie, S. R. 1982. The palynology of the Malpighiaceae and its contribution to family systematics. University of Michigan, Ann Arbor, Ph.D. thesis.

Morton, C. V. 1936. Enumeration of the Malpighiaceae of the Yucatan Peninsula. Carnegie Inst. of Washington Publ. No. 461: 127–140.

———. 1968. A typification of some subfamily, sectional, and subsectional names in the family Malpighiaceae. Taxon 17: 314–324.

Niedenzu, F. 1896. Malpighiaceae. In *Die natürlichen Pflanzenfamilien*, ed. A. Engler and K. Prantl, 3(4): 41–74. Leipzig: Wilhelm Engelmann.

———. 1899. De genere Stigmatophyllo (pars prior). Ind. Lect. Lyc. Brunsberg. p. hiem. 1899–1900: 1–13.

———. 1900. De genere Stigmatophyllo (pars posterior). Ind. Lect. Lyc. Brunsberg. p. aest. 1900: 1–32.

———. 1912. Stigmatophyllon. Malpighiaceae americanae II. Verz. Vorles. Akad. Braunsberg W.–S. 1912–1913: 23–32.

———. 1928. Malpighiaceae. In *Das Pflanzenreich,* ed. A. Engler, IV. 141: 1–870. Leipzig: Wilhelm Engelmann.

Ormond, W. T., M. T. Alves da Silva, and A. R. Corella de Castells. 1981. Contribução ao estudo citológico de Malpighiaceae. Arq. Jard. Bot. Rio de Janeiro 25: 169–173.

Pohl, R. W. 1965. Dissecting equipment and materials for the study of minute plant structures. Rhodora 67: 95–96.

Small, J. K. 1910. Malpighiaceae. N. Amer. Fl. 25(2): 117–171.

Triana, J., and J. É. Planchon. 1862. Prodromus florae novo-granatensis. Ann. Sci. Soc. Nat., Bot., sér. 4, 17: 5–190, 319–382; 18: 258–382.

Urban, I. 1893. Biographische Skizzen [1. Friedrich Sellow (1789–1831)]. Bot. Jahrb. Syst. 17: 177–198.

———. 1906. Vitae itineraque collectorum botanicorum. In *Flora brasiliensis* by K. F. P. von Martius (ed. I. Urban), 1: 1–268. Munich and Leipzig: R. Oldenbourg.

———. 1911–1913. *Symbolae antillanae*, vol. 7. Leipzig: Gebrüder Borntraeger; Paris: Paul Klincksieck; London: Williams & Norgate.

APPENDIX

The regional keys to *Stigmaphyllon* are presented in the following sequence: 1) West Indies, and Trinidad and Tobago; 2) Mexico and Central America; 3) Venezuela and the Guianas; 4) Colombia; 5) Ecuador; 6) Peru; 7) Bolivia; 8) Argentina, Paraguay, and Uruguay; 9) Brazil.

1. Key to Stigmaphyllon in the West Indies, and Trinidad and Tobago

1. Laminas greatly dissected or sinuate-lobate or at least the margin sinuate.
   2. Laminas greatly dissected, laciniate; dorsal wing of samara 1.5–1.8 cm long; Île de la Gonâve. 58. *S. laciniatum.*
   2. Laminas sinuate-lobate with 5–7 (–9) lobes (rarely ovate to suborbicular, but then the margin sinuate); dorsal wing of samara 2.8–4.5 cm long; Hispaniola. 36. *S. angulosum.*
1. Laminas unlobed, the margin even.
   3. Anterior style without folioles (very rarely so in *S. microphyllum* and then the flowers always in solitary 4-flowered umbels).
      4. Flowers (3–) 4 (–6) per umbel; anthers subequal in shape; "samara" with a dorsal crest 4–9 mm high, the nut 8–11 mm in diameter; Cuba, Jamaica, Hispaniola, Puerto Rico, Guadeloupe, Martinique, St. Lucia, Barbados, Trinidad, Tobago. 10. *S. bannisterioides.*
      4. Flowers (5–) 8–50 per umbel or pseudoraceme; anthers unequal, those of stamens opposite the lateral sepals with the connective enlarged and 0–2 reduced locules; samara with a dorsal wing 1.5–3.2 cm long, the nut 1.7–3.5 mm in diameter.
         5. Peduncles well developed, 1.5–25 mm long, 0.3–1 times as long as the pedicels; posterior styles canaliculate-complicate, terete (Fig. 36m, n; in dried material the posterior styles sometimes appearing flattened); Jamaica, Hispaniola, Puerto Rico, the Virgin Islands, Anguilla to Martinique except Dominica. 37. *S. emarginatum.*
         5. Peduncles absent to 9 mm long, less than 0.3 times as long as the pedicels; posterior styles terete or distally laterally flattened (Fig. 41h).
            6. Stamens opposing the lateral sepals with the connective enlarged and bearing 2 reduced locules; pair of glands at the apex of the petiole commonly peg-shaped, up to 2 mm long, or sometimes subsessile or 1 or both glands absent; the Bahamas, Cuba. 39. *S. sagraeanum.*
            6. Stamens opposing the lateral sepals bearing only a glandular connective without locules, or rarely the anthers of stamens opposing the anterior-lateral sepals with 1 or 2 minute locules, ca. 0.1 mm long; pair of glands at the apex of the petiole sessile (sometimes glands absent in *S. diversifolium*).
               7. Flowers 8–18 (–27) per umbel (sometimes a pseudoraceme), the umbels usually solitary but sometimes borne in dichasia or compound dichasia or rarely in a small thyrse; apex of anterior style 0.9–1.7 mm long, 0.3–1.2 mm wide, linear with a spur 0.6–1.4 mm long or triangular or rhombic; laminas extremely variable, linear to suborbicular, 0.3–7 cm wide; Cuba and the Lesser Antilles south to Martinique. 40. *S. diversifolium.*
               7. Flowers (10–) 20–25 (–40) per congested or interrupted pseudoraceme (sometimes an umbel), the pseudoracemes usually borne in large compound inflorescences, rarely solitary; apex of anterior style 0.6–0.7 (–1.2) mm long, 0.1–0.2 mm wide, linear with a spur 0.2–0.3 (–0.6) mm long; laminas elliptical to broadly so to oblong, or sometimes orbicular or lanceolate, 2.5–15.5 cm wide; Puerto Rico, Virgin Gorda, St. John. 41. *S. floribundum.*

3. Anterior style with well-developed folioles.
   8. Umbels solitary, with 3–8 flowers (if more than 4, the laminar margin fringed with filiform glands).
      9. Laminas less than 4 cm long, the margin not ciliate, eglandular; petioles up to 0.4 cm long; flowers 4 per umbel; limb of petals erose; nut of samara ovoid to subspherical; Cuba. 38. *S. microphyllum.*
      9. Laminas 4.3–9.5 cm long, the margin fringed with filiform glands; petioles 1.6–5.1 cm long; flowers 3–8 per umbel; limb of petals fimbriate; samara lenticular, the nut laterally compressed; Barbados (naturalized), Trinidad. 55. *S. ciliatum.*
   8. Umbels or pseudoracemes usually borne in compound inflorescences, rarely solitary, with 8–50 flowers (if fewer than 10, the laminar margin not fringed with filiform glands).
      10. Flowers 8–15 per umbel; lateral petals digitate-fimbriate; anterior style and its opposing stamen longer than the posterior styles and their opposing stamens; Jamaica (rare), Hispaniola, Puerto Rico, Guadeloupe, Dominica, Martinique, St. Vincent, Trinidad, Tobago. 17. *S. puberum.*
      10. Flowers 15–40 per pseudoraceme; lateral petals erose or erose-denticulate; anterior style and its opposing stamen shorter than the posterior styles and their opposing stamens.
         11. Marginal glands of lamina nail-like, i.e., stipitate and with a disklike apex; anthers pubescent; nut of samara 12–19 mm in diameter, the locule surrounded by air chambers, the dorsal wing reduced and encircling the nut; Grenada, Trinidad, Tobago. 28a. *S. adenodon* var. *adenodon.*
         11. Marginal glands of lamina sessile and/or filiform; anthers glabrous; nut of samara 3.5–5.5 mm in diameter, without air chambers, the dorsal wing elongate.
            12. Laminas abaxially sparsely and in age patchily pubescent with subsessile hairs to glabrate, the trabecula 0.5–1 mm long; anthers of stamens opposing the posterior-lateral sepals with 1 reduced locule; peduncles 0.4–0.9 times as long as pedicels; St. Vincent (rare), Trinidad. 68. *S. finlayanum.*
            12. Laminas abaxially appearing glabrous to the naked eye, but usually very sparsely and evenly sericeous, the hairs sessile, ca. 0.1 (–0.2) mm long and widely spaced; anthers of stamens opposing the posterior-lateral sepals unmodified and bearing 2 locules, i.e., like that of the stamen opposing the posterior petal; peduncles 0.7–2 times as long as pedicels; South America, but recorded from Martinique and Trinidad. 34. *S. convolvulifolium.*

2. Key to Stigmaphyllon in Mexico and Central America

1. Anterior style without folioles, the apex at most on each side with a narrow lip, or with very small folioles less than 1 mm in diameter and unlike the larger folioles of the posterior styles (if present).
   2. Posterior styles without folioles or with a narrow lateral lip; stamens subequal or differing only in size, connectives not enlarged.
      3. Flowers (3–) 4 (–6) per umbel; peduncles rudimentary to 2.5 mm long, pedicels 15–30 mm long; "samara" with a dorsal crest 0.4–0.9 cm high, the nut 8–11 mm in diameter; Atlantic lowlands from Mexico (Veracruz) to Panama, not reported from Honduras and Costa Rica but to be expected there. 10. *S. bannisterioides.*
      3. Flowers (5–) 8–50 per umbel or pseudoraceme; peduncles 3–9 mm long, pedicels 2.5–13.5 mm long; samara with a dorsal wing 3.1–5.4 cm long, the nut 3.5–6 mm in diameter.
         4. Laminas abaxially sericeous to densely so, base attenuate to truncate or slightly cordate; limb of petals 3.5–7 mm in diameter; samara with the dorsal wing 4–5.4 cm long, 1.5–2.3 cm wide; Mexico (Chiapas); Guatemala (Alta Verapaz, Huehuetenango, Quezaltenango, Suchitepéquez). 1. *S. pseudopuberum.*
         4. Laminas abaxially glabrous but young laminas abaxially sparsely sericeous, base auriculate; limb of petals 10–13 mm in diameter; samara with the dorsal wing ca. 3 cm long, ca. 1.2 cm wide; Guatemala (Guatemala, Huehuetenango). 12. *S. cordatum.*
   2. Posterior styles with folioles; stamens unequal, those opposing the anterior-lateral sepals or also those opposing the posterior-lateral sepals with the connective enlarged and bearing 0–2 reduced locules.

5. Laminas abaxially pubescent with T-shaped hairs; anthers of stamens opposing the posterior-lateral sepals unmodified, i.e., like that of the stamen opposing the posterior petal; Costa Rica (San José). 24. *S. columbicum.*
5. Laminas abaxially sericeous to glabrous, the hairs sessile and appressed; anthers of stamens opposing the posterior-lateral sepals with the connective enlarged and the locules reduced.
6. Anthers pubescent; flowers (9–) 12–35 per umbel; limb of lateral petals 5.5–10 mm long, limb of posterior petal 5.8–7.5 mm long; Mexico to Panama and adjacent Columbia. 63. *S. lindenianum.*
6. Anthers glabrous (rarely the largest anthers of *S. selerianum* with a few hairs); flowers 3–12 per umbel; limb of lateral petals 8.5–17 mm in diameter, limb of posterior petal 8–14.5 mm in diameter.
7. Laminas lanceolate to elliptical to ovate, rarely suborbicular, base truncate to cordate, margin eglandular, entire; petioles 0.6–2.8 cm long; peduncles well developed and usually longer than the pedicels; pedicels glabrous, distally expanded and widest below the calyx; Mexico to northern South America. 46. *S. ellipticum.*
7. Laminas cordate, base auriculate, margin with filiform glands to 0.8 mm long (these often broken off in the larger leaves but the bases remaining) and sometimes dentate, petioles 1.5–7 cm long; peduncles rudimentary to 3 mm long, always shorter than the pedicels; pedicels pubescent, of uniform diameter; Mexico (Oaxaca, Chiapas). 44. *S. selerianum.*
1. Anterior styles with well-developed folioles ca. 1 mm or more long and/or wide.
8. Laminas abaxially glabrous or sericeous (the hairs all sessile, appressed, and evenly distributed).
9. Laminas abaxially glabrous, the margin evenly fringed with filiform glands; flowers 3–8 per solitary umbel; samara lenticular, the reduced dorsal wing encircling the laterally compressed nut; Atlantic lowlands from Belize to Nicaragua, not reported but to be expected in Costa Rica and Panama. 55. *S. ciliatum.*
9. Laminas abaxially sericeous, the margin with scattered sessile and/or filiform glands; flowers 8–35 per umbel, these usually borne in compound inflorescences; samara with an elongate dorsal wing, the nut ovoid.
10. Anthers pubescent; petals yellow; anterior style and its opposing stamen shorter than or rarely subequal to the posterior styles and their opposing stamens; Mexico to Panama and adjacent Colombia. 63. *S. lindenianum.*
10. Anthers glabrous; petals red with a yellow border or yellow but suffused with red; anterior style and its opposing stamen longer than the posterior styles and their opposing stamens.
11. Laminas abaxially very densely white-sericeous, the epidermis hidden; petals erose to denticulate or sometimes with tapered fimbriae to 0.2 (–0.3) mm long; dorsal wing of the samara distally flared, abruptly narrowed at the nut to only 0.3–0.4 cm wide; Panama (Canal Zone, Darién, Los Santos, Panamá, San Blas). 19. *S. hypargyreum.*
11. Laminas abaxially abundantly to sparsely sericeous, the epidermis visible; petals digitate-fimbriate, the fimbriae up to 0.5 mm long; dorsal wing of samara tapered from the nut, widest at the nut; Guatemala (Izabal, Petén), Belize (Toledo), Honduras (Atlántida, Colón, Cortés), Nicaragua (Río San Juan, Zelaya), Costa Rica (Herédia, Limón, Osa Peninsula), Panama (Bocas del Toro, Canal Zone, Colón, Darién, Panamá, San Blas). 17. *S. puberum.*
8. Laminas abaxially sparsely to densely pubescent with T-shaped hairs to tomentose (in *S. panamense,* the vesture a mixture of subsessile hairs and stalked hairs, in mature leaves often sloughed off in patches).
12. Each bracteole with a pair of prominent glands, each gland 0.6–0.8 mm in diameter; stipules consisting of a prominent gland, ca. 0.8 mm in diameter, with a membranous tip; Costa Rica (Osa Peninsula). 27. *S. adenophorum.*
12. Bracteoles eglandular or each bracteole with 1 or 2 inconspicuous glands, each gland 0.2–0.3 mm in diameter; stipules herbaceous, eglandular.
13. Anthers pubescent; Mexico to northern Costa Rica (Guanacaste, Puntarenas). 64. *S. retusum.*
13. Anthers glabrous.
14. Abaxial vesture of lamina a mixture of subsessile and stalked hairs, in mature leaves often sloughed off in patches; limb of lateral petals 10–15 mm in diameter, limb of posterior petal 9–11 mm in diameter; Panama (Canal Zone, Colón, Darién, islands in the Gulf of Panama). 66. *S. panamense.*

14. Abaxial vesture of lamina composed of stalked hairs only, always evenly distributed; limb of lateral petals 7.5–11 mm in diameter, limb of posterior petal 6–8 mm in diameter.
   15. Anthers of stamens opposing the posterior-lateral sepals with 2 reduced locules; limb of anterior-lateral petals 8–8.6 mm in diameter; flowers 15–40 (–50) per umbel; embryo laterally flattened, 3 times as long as wide; Panama (Darién). 65. *S. dichotomum.*
   15. Anthers of stamens opposing the posterior-lateral sepals with 0–1 reduced locule; limb of anterior-lateral petals 10–11 mm in diameter; flowers 12–20 per umbel; embryo ovoid, 2 times as long as wide; Costa Rica (Nicoya Peninsula, northern Puntarenas). 67. *S. tonduzii.*

3. Key to Stigmaphyllon in Venezuela and the Guianas

1. All styles without folioles.
   2. Flowers (3–) 4 (–6) per umbel; petals erose; laminas abaxially sparsely sericeous, the hairs sessile and appressed; "samara" with a dorsal crest 0.4–0.9 cm high, the nut 8–11 mm in diameter; Venezuela (Delta Amacuro, Miranda, Sucre, Yaracuy) and coastal areas of the Guianas. 10. *S. bannisterioides.*
   2. Flowers 10–35 per umbel or pseudoraceme; petals digitate-fimbriate; laminas abaxially with T-shaped hairs to tomentose; samara with a dorsal wing 2.5–5.5 cm long, the nut 2.5–5.5 mm in diameter; Venezuela (Barinas, Distrito Federal, Falcón, Lara, Mérida, Portuguesa, Táchira, Trujillo). 2. *S. bogotense.*
1. All styles or only the posterior two with folioles.
   3. Laminas with the marginal glands stipitate, 0.2–0.6 mm long, 0.2–0.5 mm in diameter at the apex; anthers pubescent.
      4. Sepals deciduous; each bracteole with a pair of prominent glands, each gland 0.5–0.6 mm in diameter; petals abaxially pubescent; nut of samara 3.3–4 mm in diameter, without air chambers; Venezuela (Táchira, Zulia). 26. *S. singulare.*
      4. Sepals persistent; bracteoles eglandular or each bracteole with 1 or 2 inconspicuous glands hidden by the pubescence, each gland 0.2–0.4 mm in diameter; petals glabrous; nut of samara 12–19 mm in diameter, the locule surrounded by air chambers; Venezuela (Delta Amacuro, Paria Peninsula). 28a. *S. adenodon* var. *adenodon.*
   3. Laminas with the marginal glands sessile and/or filiform; anthers glabrous.
      5. Laminas abaxially abundantly and evenly pubescent with T-shaped hairs to tomentose (the stalk up to 0.4 mm long).
         6. Lateral petals 13–15 mm in diameter, posterior petal 11–12 mm in diameter, with fimbriae up to 1.8 mm long; base of the larger laminas deeply cordate to auriculate; stamens opposing the posterior-lateral sepals bearing only a glandular connective without locules (rarely with 1–2 tiny closed locules); nut of samara ca. 11 mm high, ca. 6.5 mm in diameter, embryo ovoid; French Guiana. 72. *S. palmatum.*
         6. Lateral petals 7.5–8.6 mm in diameter, posterior petal 6–6.5 mm long, 4–4.5 mm wide, denticulate or with fimbriae up to 0.5 mm long; base of the larger laminas cordate to truncate or sometimes attenuate; anthers of stamens opposing the posterior-lateral sepals with the connective enlarged and bearing 2 reduced locules; nut of samara 4.5–6 mm high, 3.8–4.5 mm in diameter, embryo laterally compressed; Venezuela (Apure, Barinas, Carabobo, Falcón, Lara, Mérida, Portuguesa, Táchira, Trujillo, Yaracuy, Zulia). 65. *S. dichotomum.*
      5. Laminas abaxially glabrous or sericeous to sparsely so, the hairs appressed and sessile, or the mature leaves sparsely and patchily pubescent with subsessile hairs (the stalk less than 0.1 mm long; *S. finlayanum*).
         7. Anterior style without well-developed folioles, the apex extended into a claw, or somewhat laterally expanded and thus with a lip on each side, or with very small folioles less than 1 mm wide.
            8. Flowers 3–9 (–12) per umbel; pedicels unlike the peduncles, glabrous, distally expanded and widest below the calyx; petals with fimbriae up to 1.1 mm long; Venezuela (Zulia). 46. *S. ellipticum.*

8. Flowers 15–35 (–40) per pseudoraceme; pedicels like the peduncles, pubescent and of uniform diameter; petals erose to denticulate; Venezuela and the Guianas. 33. *S. sinuatum.*
7. Anterior style with well-developed folioles.
9. Margin of lamina fringed with filiform glands; flowers 3–8 per solitary umbel; samara lenticular, the nut laterally compressed; not recorded from Venezuela and the Guianas, but should be expected there in the coastal lowlands. 55. *S. ciliatum.*
9. Margin of lamina eglandular or with sessile and/or scattered filiform glands; flowers 8–40 per pseudoraceme, arranged in compound inflorescences (rarely solitary); nut of samara spherical to ovoid.
10. Flowers 8–15 per umbel; petals with fimbriae up to 0.6 mm long; anterior style and its opposing stamen longer than the posterior styles and their opposing stamens; Venezuela (Delta Amacuro, Monagas) and the Guianas. 17. *S. puberum.*
10. Flowers 15–40 per pseudoraceme; petals erose to denticulate; anterior style and its opposing stamen shorter than the posterior styles and their opposing stamens.
11. Mature laminas abaxially sparsely and unevenly pubescent with subsessile hairs (the trabecula 0.5–1 mm long), especially on the veins, to glabrate; anthers of stamens opposing the posterior-lateral sepals with the connective enlarged and bearing 1 reduced locule; Venezuela (Paria Peninsula). 68. *S. finlayanum.*
11. Mature laminas abaxially very densely sericeous to glabrous, the hairs sessile and evenly distributed, 0.1–0.5 (–0.7) mm long; anthers of stamens opposing the posterior-lateral sepals unmodified, i.e., like that of the stamen opposing the posterior petal.
12. Laminas abaxially appearing glabrous to the naked eye, but usually very sparsely sericeous, the hairs ca. 0.1 (–0.2) mm long and widely spaced, never touching; laminas ovate to cordate or narrowly so, the base cordate; the Guianas. 34. *S. convolvulifolium.*
12. Laminas abaxially sparsely to very densely sericeous, the hairs (0.2–) 0.3–0.5 (–0.7) mm long, usually touching to overlapping; laminas triangular to ovate to cordate to elliptical to broadly so to orbicular to oblate to reniform, rarely 3–5-lobed, the base acute to truncate to cordate to deeply auriculate; Venezuela and the Guianas. 33. *S. sinuatum.*

4. Key to Stigmaphyllon in Colombia

1. Posterior styles without well-developed folioles, at most with a narrow lateral lip.
2. Flowers 3–9 (–12) per umbel; limb of lateral petals 9–17 mm in diameter.
3. Peduncles rudimentary to 2.5 mm long, pedicels 15–30 mm long; petals erose; stamens subequal or differing only in size, connectives not enlarged; "samara" with a dorsal crest 0.4–0.9 cm high, the nut 8–11 mm in diameter; Antioquia, Atlántico, Bolívar, Chocó. 10. *S. bannisterioides.*
3. Peduncles well developed, 2.5–34 mm long, pedicels 2–23 mm long; petals fimbriate (-denticulate); stamens unequal, those opposing the lateral sepals with the connective enlarged and the locules reduced; samara with a dorsal wing 2.1–3.5 cm long, the nut 4–6 mm in diameter; Antioquia, Atlántico, Chocó, Magdalena, Nariño, Valle. 46. *S. ellipticum.*
2. Flowers 13–60 per umbel or pseudoraceme (rarely as few as 10 in *S. bogotense* and 12 in *S. romeroi*); limb of lateral petals 6–10 mm in diameter.
4. Mature laminas adaxially abundantly pubescent; peduncles 0.9–2.3 times as long as the pedicels; Antioquia, Cundinamarca, Norte de Santander, Quindío, Risaralda. 22. *S. velutinum.*
4. Mature laminas adaxially glabrous or at most glabrate; peduncles 0.1–1 times as long as the pedicels.
5. Laminas abaxially densely (often white-) sericeous-tomentulose, the vesture appressed and obscuring the epidermis; Cundinamarca. 3. *S. sarmentosum.*
5. Laminas abaxially sparsely sericeous to glabrate, or pubescent with T-shaped hairs to tomentose but the epidermis always visible.
6. Glands at apex of petiole peg-shaped, 0.4–0.8 mm long; laminas 3.2–7 cm long, abaxially sparsely sericeous (the hairs sessile) to glabrous; petals erose, yellow; Bolívar. 13. *S. romeroi.*
6. Glands at apex of petiole circular to elliptical, prominent but sessile; laminas 7.2–24 cm long, abaxially sparsely to abundantly pubescent with T-shaped hairs, the hairs with a

stalk 0.1–0.4 mm long (in *S. suffruticosum* the hairs with a stalk 0.05–0.1 mm long), to glabrate; petals fimbriate, at least the posterior petal suffused with red.

7. Laminas linear-lanceolate, 1.7–3 cm wide; peduncles rudimentary to 3 mm long, 0.1–0.2 times as long as the pedicels; Antioquia. 5. *S. stenophyllum.*
7. Laminas elliptical to ovate to cordate to suborbicular, 4.3–24 cm wide, peduncles well developed, 3.5–10.5 mm long, 0.3–1 times as long as the pedicels.
    8. Laminas abaxially with T-shaped hairs to tomentose, the hairs with a stalk 0.1–0.4 mm long, the trabecula 0.9–1.3 mm long; usually the lowest two (sometimes more) flowers of an umbel or pseudoraceme separated a short distance from the rest; Antioquia Boyacá, Caldas, Cauca, Cundinamarca, Huila, Magdalena, Nariño, Norte de Santander, Putumayo, Risaralda, Santander, Tolima, Valle. 2. *S. bogotense.*
    8. Laminas of mature leaves abaxially very sparsely pubescent to glabrate, the hairs with a stalk 0.05–0.1 mm long, the trabecula 1–2.3 mm long; the lowest pair of flowers of a pseudoraceme not separated a short distance from the rest; Caquetá, Valle. 4. *S. suffruticosum.*

1. Posterior styles with well-developed folioles.
  9. Laminas abaxially very sparsely to densely pubescent with T-shaped hairs to tomentose, the hairs with a stalk and not appressed, sometimes glabrate in the largest laminas of *S. adenodon.*
    10. Sepals deciduous; petal limbs abaxially pubescent; each bracteole with a pair of conspicuous prominent glands, each gland 0.5–0.6 mm in diameter; Norte de Santander. 26. *S. singulare.*
    10. Sepals persistent; petal limbs glabrous; bracteoles eglandular or with 1 or 2 inconspicuous glands, each gland 0.1–0.3 mm in diameter.
      11. Anterior style without folioles, the apex distally extended into a spur; Antioquia, Bolívar, Boyacá, Chocó, Cundinamarca, Magdalena, Santander, Sucre, Tolima. 24. *S. columbicum.*
      11. Anterior style with folioles, these narrow or broad, but the apex never spurred.
        12. Limb of lateral petals 15–19 mm in diameter, limb of posterior petal 13–15 mm in diameter, all limbs fimbriate, the fimbriae up to 2.2 mm long; petioles joined across the node, the opposing stipules commonly fused into a bifid structure.
          13. Laminas abaxially abundantly pubescent but the epidermis visible; limb of anterior-lateral petals ca. 15 mm in diameter, with fimbriae to 2.2 mm long; Antioquia, Cauca, Cesar, Huila, Nariño, Norte de Santander, Valle. 49. *S. echitoides.*
          13. Laminas abaxially very densely woolly-tomentose, the epidermis hidden; limb of anterior-lateral petals ca. 19 mm in diameter, with fimbriae to 1 mm long; Huila. 50. *S. tergolanatum.*
        12. Limb of lateral petals 8–12 mm in diameter, erose or erose-denticulate, limb of posterior petal 6–10 mm long, 4–8 mm wide, erose or with fimbriae/teeth up to 0.5 mm long; petioles and stipules free.
          14. Laminas with the marginal glands borne abaxially adjacent to the margin, the glands raised or substipitate, sometimes hidden by the vesture; flowers 10–15 per umbel; anthers pubescent; Caquetá, Meta. 78. *S. alternans.*
          14. Laminas with the margin eglandular or with glands borne on the margin, the glands sessile, stipitate, or filiform; flowers 15–50 per umbel or pseudoraceme; anthers glabrous or pubescent (*S. adenodon*).
            15. Anthers pubescent; margin of lamina with nail-like glands, i.e., stalked with a disklike apex, up to 0.5 mm long; samara with the dorsal wing encircling the nut, 3–4.4 cm high from base of nut, the nut 12–19 mm in diameter; Amazonas, Putumayo. 28a. *S. adenodon* var. *adenodon.*
            15. Anthers glabrous; margin of lamina with sessile/filiform glands (*S. dichotomum*), or eglandular or with elongate glands up to 0.2 mm long and in diameter but the apex not disklike (*S. goudotii*); samara with an elongate dorsal wing 3.7–5.5 cm long, the nut 3.8–10 mm in diameter.
              16. Laminas abaxially densely pubescent with T-shaped hairs to tomentose, the margin with irregularly spaced sessile glands and often also with filiform glands up to 6 mm long; peduncles 3.5–9 mm long, 0.7–2.1 times as long as the pedicels; nut of samara 3.8–4.5 mm in diameter, without air chambers, the embryo laterally compressed; Antioquia, Atlántico, Bolívar, Cesar, Chocó, Cundinamarca, Magdalena, Norte de Santander, Santander, Tolima. 65. *S. dichotomum.*

16. Laminas abaxially sparsely pubescent with T-shaped hairs, the margin eglandular or with irregularly spaced prominent glands up to 0.2 mm long and in diameter; peduncles 15–21 mm long, (1.5–) 3.4–4.2 times as long as the pedicels; nut of samara ca. 10 mm in diameter, with air chambers surrounding the locule, the embryo ovoid; Antioquia, Tolima, Valle. 25. *S. goudotii.*

9. Laminas abaxially glabrous to sericeous, the hairs sessile or subsessile, appressed.

17. Laminas abaxially densely white sericeous-tomentulose, the epidermis hidden; Antioquia. 19. *S. hypargyreum.*

17. Laminas abaxially moderately to sparsely sericeous, the epidermis always visible and not obscured, or glabrate to glabrous.

18. Mature laminas abaxially glabrate to glabrous, or very sparsely sericeous and the hairs unevenly and patchily distributed.

19. Base of lamina deeply auriculate, the basal lobes usually overlapping, the margin fringed with filiform glands; umbels solitary; samara lenticular, the nut laterally compressed; Atlantic lowlands of Antioquia. 55. *S. ciliatum.*

19. Base of lamina truncate to cordate or sometimes attenuate, the margin eglandular or with scattered sessile glands; umbels and pseudoracemes solitary or in compound inflorescences; nut of samara ovoid.

20. Apex of anterior style with distal 1/2–3/4 (–4/5) elliptically to ovately to suborbicularly (rarely triangularly) expanded or sometimes with tiny folioles, each up to 1 mm long and up to 0.8 mm wide, anterior style shorter than or at most subequal to the posterior two; limb of petals with fimbriae 0.4–2.2 mm long.

21. Pedicels entirely glabrous; laminas usually narrowly to broadly elliptical (rarely lanceolate, ovate, or suborbicular); petioles 0.6–2.8 cm long, free; stipules free; peduncles (1.5–) 2.5–34 mm long; nut of samara without air chambers; Antioquia, Atlántico, Chocó, Magdalena, Nariño, Valle. 46. *S. ellipticum.*

21. Pedicels glabrous except for a row of hairs extending from the base of the pedicel to the base the posterior petal; laminas cordate to narrowly ovate to triangular; petioles 2.5–5.3 cm long, confluent across the node and forming a band bearing the stipules, these commonly fused across the node into a bifid structure; peduncles 3.2–5.5 mm long; nut of samara with air chambers surrounding the locule; Antioquia, Caldas, Risaralda, Valle. 48. *S. venulosum.*

20. Apex of anterior style with well-developed folioles, each up 1.1–1.8 mm long and wide, anterior style longer than the posterior styles; limb of petals erose or those of the posterior and posterior-lateral petals digitate-fimbriate, the fimbriae up to 0.2 mm long.

22. Laminas 3.3–7.8 cm wide, elliptical or narrowly so (sometimes ovate); limb of all petals erose; anthers of anterior-lateral stamens with the connective enlarged and the locules reduced, posterior-lateral stamens bearing very small unmodified anthers 0.3–0.5 mm long; Antioquia, Bolívar, Boyacá, Caldas, Cauca, Chocó, Santander, Valle. 20. *S. herbaceum.*

22. Laminas 7–10 cm wide, cordate; limb of posterior and posterior-lateral petals digitate-fimbriate, limb of anterior-lateral petals erose; all anthers unmodified, varying only slightly in size; Meta. 16. *S. orientale.*

18. Mature laminas abaxially sericeous, the hairs evenly distributed.

23. Pedicels entirely glabrous (*S. ellipticum*) or with a row of hairs extending from the base of the pedicel to the base the posterior petal (*S. venulosum*), distally much wider than at the base, unlike the peduncles in shape and vesture; margin of lamina eglandular; limb of petals with fimbriae up to 2.2 mm long.

24. Pedicels entirely glabrous; laminas usually narrowly to broadly elliptical (rarely lanceolate, ovate, or suborbicular); petioles 0.6–2.8 cm long, free; stipules free; peduncles (1.5–) 2.5–34 mm long; nut of samara without air chambers; Antioquia, Atlántico, Chocó, Magdalena, Nariño, Valle. 46. *S. ellipticum.*

24. Pedicels glabrous except for a row of hairs extending from the base of the pedicel to the base the posterior petal; laminas cordate to narrowly ovate to triangu-

lar; petioles 2.5–5.3 cm long, confluent across the node and forming a band bearing the stipules, these commonly fused across the node into a bifid structure; peduncles 3.2–5.5 mm long; nut of samara with air chambers surrounding the locule; Antioquia, Caldas, Risaralda, Valle. 48. *S. venulosum.*

23. Pedicels pubescent, distally not much wider than at the base, similar to the peduncles in shape and vesture; margin of lamina with irregularly spaced sessile and/or filiform or stipitate glands; limb of petals erose, denticulate, or with fimbriae up to 0.6 (–0.8) mm long.
    25. Anterior style and its opposing stamen longer than the posterior styles and their opposing stamens; flowers 8–15 per umbel; limb of petals with fimbriae up to 0.6 (–0.8) mm long; laminas usually lanceolate to elliptical, to 12.7 cm wide; Antioquia, Chocó, Putumayo. 17. *S. puberum.*
    25. Anterior style and its opposing stamen shorter than or subequal to the posterior styles and their opposing stamens; flowers (9–) 12–40 per umbel or pseudoraceme; limb of petals erose, denticulate, or with fimbriae up to 0.3 mm long; laminas usually ovate to cordate to suborbicular, to 20 cm wide.
        26. Anthers glabrous; margin of lamina grossly and shallowly crenate to subentire and with irregularly spaced sessile glands in the sinuses, sometimes also with filiform glands; apex of anterior style extended into a claw to 1.3 mm long or rarely with well-developed folioles; Amazonas, Antioquia, Caquetá, Meta, Vaupés. 33. *S. sinuatum.*
        26. Anthers pubescent; margin of the lamina with prominent to stipitate glands and/or filiform glands; anterior style with two well-developed folioles (in *S. lacunosum* rarely only laterally expanded into a narrow lip, never extended into a claw).
            27. Margin of lamina usually with nail-like glands, i.e., stipitate with a disk-like apex, filiform glands absent; anthers of stamens opposing the posterior-lateral sepals unmodified, i.e., like that of the stamen opposing the posterior petal; dorsal wing of samara a reduced triangular crest encircling the nut, the nut ca. 12.5 mm in diameter, the locule surrounded by air chambers; Amazonas. 29. *S. lacunosum.*
            27. Margin of lamina with sessile and/or filiform glands; anthers of stamens opposing the posterior-lateral sepals with the connective enlarged and the locules reduced; samara with an elongate dorsal wing, the nut 2–4 mm in diameter, without air chambers; Chocó. 63. *S. lindenianum.*

5. Key to Stigmaphyllon in Ecuador

1. Flowering and fruiting plants leafless or rarely with a few young leaves; Guayas. 32. *S. nudiflorum.*
1. Flowering and fruiting plants leaf-bearing.
    2. Laminas abaxially very densely white-sericeous or densely white sericeous-tomentulose, the epidermis hidden.
        3. Styles efoliolate, subequal in length; pedicels 8.5–27 mm long; nut of samara without lateral ornamentation; Loja, Pichincha. 3. *S. sarmentosum.*
        3. Styles with well-developed folioles, the anterior style longer than the posterior two; pedicels 2.3–7 mm long; nut of samara with 3–4 lateral winglets per side; Napo, Pastaza, Zamora-Chinchipe. 18. *S. maynense.*
    2. Laminas abaxially glabrous to pubescent, the epidermis always visible.
        4. Posterior styles without folioles.
            5. Petals marked with red; anthers glabrous, all unmodified and subequal; usually the lowest two (sometimes more) flowers of an umbel or pseudoraceme separated a short distance from the rest; pedicels 6.2–19 mm long, peduncles 0.3–0.8 (–1) times as long as the pedicels; Azuay, Bolívar, Carchi, Chimborazo, Esmeraldas, Imbabura, Morona-Santiago, Napo, Pastaza, Pichincha, Tungurahua, Zamora-Chinchipe. 2. *S. bogotense.*

5. Petals yellow; anthers pubescent, those of stamens opposing the lateral sepals with the connective enlarged and the locules reduced; the lowest pair of flowers of an umbel or pseudoraceme not separated a short distance from the rest; pedicels 3.2–8.5 mm long, peduncles 0.9–2.5 times as long as the pedicels; Napo. 42. *S. florosum.*

4. Posterior styles with well-developed folioles.

6. Laminas abaxially pubescent with T-shaped hairs to tomentose.

7. Peduncles densely pubescent, pedicels glabrate to glabrous; limb of petals fimbriate-lacerate, the fimbriae up to 1.6 mm long; margin of lamina eglandular; Guayas, Manabí. 47. *S. eggersii.*

7. Peduncles and pedicels equally densely pubescent; limb of petals erose to denticulate (-fimbriate), teeth/fimbriae up to 0.5 mm long; margin of lamina with sessile and/or filiform or stipitate glands, or with raised glands borne abaxially adjacent to the margin (*S. alternans*).

8. Anterior style without folioles, the apex distally extended into a spur; Napo. 24. *S. columbicum.*

8. Anterior style with well-developed folioles.

9. Marginal glands of lamina borne abaxially adjacent to the margin, sometimes hidden by the vesture; flowers 10–15 per umbel; anthers pubescent; leaves alternate; Napo. 78. *S. alternans.*

9. Marginal glands of lamina borne on the margin; flowers 15–40 per pseudoraceme; anthers glabrous or pubescent; leaves opposite.

10. Marginal glands of lamina nail-like, i.e., stipitate with a disklike apex; anthers pubescent; limbs of petals 8–11 (–13) mm in diameter; nut of samara 12–19 mm in diameter, the locule surrounded by air chambers; Morona-Santiago, Napo, Pastaza. 28. *S. adenodon.*

10. Marginal glands of lamina sessile and/or filiform and/or stipitate with a capitate apex; anthers glabrous; limbs of petals 12–15 mm in diameter; (completely mature samara not seen) nut probably much less than 10 mm in diameter, the locule not surrounded by air chambers; Manabí. 31. *S. ecuadorense.*

6. Laminas abaxially moderately to sparsely sericeous to glabrous.

11. Pedicels entirely glabrous, distally much wider than at the base, unlike the peduncles in shape and vesture; margin of lamina eglandular; flowers 3–9 (–12) per umbel; limb of petals with fimbriae up to 1.1 mm long; Azuay, Bolívar, Cotopaxi, El Oro, Esmeraldas, Guayas, Loja, Los Ríos, Manabí, Pichincha. 46. *S. ellipticum.*

11. Pedicels pubescent, distally not much wider than at the base, similar to the peduncles in shape and vesture; margin of lamina with irregularly spaced sessile and/or filiform glands; flowers 10–40 per umbel or pseudoraceme; limb of the petals erose to denticulate, with teeth up to 0.3 mm long.

12. Mature laminas abaxially sericeous, the hairs evenly distributed; apex of anterior style extended into a claw to 1.3 mm long or with folioles.

13. Flowers 15–35 (–40) per pseudoraceme; apex of anterior style extended into a claw to 1.3 mm long (rarely bearing folioles); anterior style and its opposing stamen subequal to or shorter than the posterior two and their opposing stamens; nut of samara usually with 1–3 lateral winglets per side and/or spurs and crests (rarely without lateral ornamentation), 2.8–4.4 mm in diameter, locules not surrounded by spongy tissue; Morona-Santiago, Napo, Pastaza. 33. *S. sinuatum.*

13. Flowers 8–15 per umbel; apex of anterior style with folioles; anterior style and its opposing stamen longer than the posterior styles and their opposing stamens; nut of samara without lateral ornamentation, 5–7 mm in diameter, locule embedded in spongy tissue; not reported from Ecuador but to be expected in the wet lowlands. 17. *S. puberum.*

12. Mature laminas abaxially glabrate to glabrous, or very sparsely sericeous and the hairs unevenly and patchily distributed; anterior style with folioles.

14. Flowers 15–30 per umbel; anterior style and its opposing stamen shorter than the posterior two and their opposing stamens; anthers pubescent, those of stamens opposing the posterior-lateral sepals bearing only 1 reduced locule; Napo, Pastaza. 69. *S. cardiophyllum.*

14. Flowers 10–15 per umbel; anterior style and its opposing stamen longer than the posterior styles and their opposing stamens; anthers glabrous, all with two locules; Esmeraldas. 20. *S. herbaceum.*

6. Key to Stigmaphyllon in Peru

1. Laminas abaxially pubescent with T-shaped hairs to tomentose, the hairs with a stalk 0.1–0.4 mm long.
  2. Anterior and posterior styles with well-developed folioles.
    3. Laminas with nail-like glands borne on the margin, the glands stipitate (0.2–0.6 mm long) with a disklike apex; limb of lateral petals 8–11 (–13) mm in diameter; nut of samara 12–19 mm in diameter, the locule surrounded by air chambers; Amazonas, Loreto. 28. *S. adenodon.*
    3. Laminas with the margin eglandular, or with prominent but sessile glands on the margin, or with raised to substipitate glands (up to 0.1 mm high) abaxially adjacent to the margin; limb of lateral petals 10–18 mm in diameter; nut of samara 4–5.5 mm in diameter, the locule not surrounded by air chambers.
      4. Lamina with the marginal glands borne abaxially adjacent to the margin, sometimes hidden by the vesture; anthers pubescent; leaves alternate; Loreto, San Martín. 78. *S. alternans.*
      4. Laminas eglandular or with the marginal glands borne on the margin; anthers glabrous (rarely sparsely pubescent in *S. cuzcanum*); leaves opposite (rarely alternate in or near the inflorescence).
        5. Petals marked or suffused with red; claw of posterior petal bearded; limb of petals with fimbriae up to 1 mm long, limb of lateral petals 11–14 mm in diameter, limb of posterior petal ca. 9 mm in diameter; anthers of stamens opposing the lateral sepals with the connective enlarged and the locules reduced; nut of samara with 3–4 lateral winglets per side; Cuzco, Huánuco, Junín, Loreto, San Martín. 73. *S. strigosum.*
        5. Petals yellow; claw of posterior petal glabrous; limb of petals with teeth/fimbriae up to 0.5 mm long, limb of lateral petals 15–18 mm in diameter, limb of posterior petal ca. 14 mm in diameter; only stamens opposing the anterior-lateral sepals with the connective enlarged and the locules reduced, those of stamens opposing the posterior-lateral sepals unmodified, i.e., like anther of stamen opposing the posterior petal; nut of samara with 1–2 lateral winglets per side and/or spurs; Cuzco. 30. *S. cuzcanum.*
  2. Anterior and posterior styles without folioles (or in *S. tarapotense* the posterior styles varying from efoliolate to bearing a narrow lip or a small to large foliole).
    6. Apex of anterior style extended into a claw 0.6–1 mm long.
      7. Flowers 15–35 per pseudoraceme; one bracteole of each pair with one lateral prominent gland, 0.5–0.8 mm in diameter; limb of petals fimbriate or fimbriate-denticulate, the teeth/fimbriae up to 0.3 mm long; Junín, Pasco. 43. *S. aberrans.*
      7. Flowers (4–) 6–8 per umbel; bracteoles eglandular; limb of petals with fimbriae up to 0.7 mm long; Amazonas, Cajamarca. 9. *S. peruvianum.*
    6. Apex of anterior style blunt or extended into a spur up to 0.2 mm long.
      8. Anthers glabrous, all unmodified and subequal in shape; styles subequal in length; usually the lowest two (sometimes more) flowers of an umbel or pseudoraceme separated a short distance from the rest; Amazonas, Ayacucho, Cajamarca, Huánuco, Junín, Pasco. 2. *S. bogotense.*
      8. Anthers pubescent, those of stamens opposing the anterior-lateral sepals only or also those opposing the posterior-lateral sepals with the connective enlarged and the locules reduced; anterior style shorter than the posterior two; lowest pair of flowers of an umbel or pseudoraceme not separated from the rest.
        9. Laminas abaxially abundantly pubescent but the vesture never so dense as to hide the epidermis, each hair individually visible; anthers of stamens opposing the lateral sepals with the connective enlarged and the locules reduced; apex of posterior styles with a lateral lip to 0.2 mm wide; nut of samara with only a lateral ridge per side or with 1–2 spurs or winglets projecting from the ridge; Huánuco, Junín, Loreto, Madre de Dios, San Martín. 42. *S. florosum.*
        9. Laminas abaxially densely silvery-pubescent, the vesture almost obscuring the epidermis, the hairs densely overlapping; only anthers of stamens opposing the anterior-lateral sepals with the connective enlarged and the locules reduced, those of stamens opposing

the posterior-lateral sepals unmodified, i.e., like anther of stamen opposing the posterior petal; posterior styles usually with folioles of variable shape and size (variable even within a flower) or rarely with only a lateral lip; nut of samara with 2–4 lateral winglets set in two rows per side; San Martín. 23. *S. tarapotense.*

1. Laminas abaxially sericeous to glabrous, the hairs appressed (or in *S. sarmentosum* some hairs also subsessile with a stalk up to 0.075 mm long).
   10. Laminas abaxially very densely (white)-sericeous, the epidermis hidden.
      11. Styles efoliolate, subequal in size; pedicels 8.5–27 mm long; nut of samara without lateral ornamentation; Piura. 3. *S. sarmentosum.*
      11. Styles with well-developed folioles, the anterior style larger than the posterior two; pedicels 2.3–7 mm long; nut of samara with 3–4 lateral winglets per side; Amazonas, Huánuco, Loreto, Madre de Dios, Pasco, San Martín. 18. *S. maynense.*
   10. Laminas abaxially sericeous to glabrous, the epidermis always visible.
      12. Laminas abaxially glabrous or glabrate (with some scattered, mostly isolated hairs).
         13. Flowers 3–9 (–12) per umbel; limb of lateral petals 10–17 mm in diameter, with fimbriae up to 1.1 mm long; anthers glabrous; pedicels distally expanded and widest below the calyx, unlike the peduncles; Amazonas, Cajamarca, La Libertad, Lambayeque, Piura, Tumbes. 46. *S. ellipticum.*
         13. Flowers 15–25 (–30) per umbel; limb of lateral petals 4–6.7 mm in diameter, erose; anthers pubescent; pedicels of uniform diameter, like the peduncles; Amazonas, Cuzco, Huánuco, Junín, Loreto, Madre de Dios, San Martín. 69. *S. cardiophyllum.*
      12. Laminas abaxially sericeous to densely so, the pubescence evenly distributed.
         14. Flowers 8–15 per umbel; anterior style and its opposing stamen longer than the posterior styles and their opposing stamens; limb of petals with fimbriae up to 0.6 mm long; Huánuco, Loreto. 17. *S. puberum.*
         14. Flowers 15–40 per umbel or pseudoraceme; anterior style and its opposing stamen shorter than or subequal to the posterior styles and their opposing stamens; limb of petals erose or denticulate-erose, the teeth up to 0.3 mm long.
            15. Lamina abaxially densely sericeous, the surface appearing white-silvery; anthers of stamens opposing the posterior-lateral sepals with the connective enlarged and bearing 1 reduced locule; lateral petals 6–7 mm in diameter; anterior style with folioles; nut of the samara without lateral ornamentation, rarely with tiny spurs; Huánuco, Junín, Loreto, Pasco, San Martín. 70. *S. argenteum.*
            15. Lamina abaxially sparsely to moderately sericeous, the surface never white-silvery; anthers of stamens opposing the posterior-lateral sepals unmodified, i.e., like that of the stamen opposing the posterior petal; lateral petals 7–15 mm in diameter; anterior style with or without folioles; nut of the samara usually with 1–3 lateral winglets per side and/or spurs and crests, rarely without lateral ornamentation.
               16. Anthers glabrous; lamina with the margin grossly and shallowly crenate to subentire and with irregularly spaced sessile glands in the sinuses and sometimes also with filiform glands; nut of samara 2.8–4.4 mm in diameter, without air chambers; Amazonas, Huánuco, Loreto, San Martín. 33. *S. sinuatum.*
               16. Anthers pubescent; lamina with the margin entire, the marginal glands raised to stipitate, up to 0.5 mm long; nut of samara ca. 12.5 mm in diameter, the locule surrounded by air chambers; Loreto. 29. *S. lacunosum.*

7. Key to Stigmaphyllon in Bolivia

1. Anterior sepal with a pair of glands like the lateral sepals; petiole with a pair of shallowly cupulate glands borne 1.5–18 mm below the base of the lamina; stamens opposing the lateral sepals bearing only a glandular connective without locules.
   2. Mature laminas abaxially very sparsely pubescent with T-shaped hairs to glabrous; La Paz, Santa Cruz. 60. *S. coloratum.*
   2. Mature laminas abaxially densely pubescent with T-shaped hairs; Santa Cruz. 59. *S. boliviense.*
1. Anterior sepal eglandular; petiole with a pair of prominent but sessile glands borne at the base of the lamina; anthers of stamens opposing the lateral sepals with the connective enlarged and bearing 1 or 2 reduced locules.

3. Laminas abaxially sericeous to glabrous.
   4. Mature laminas abaxially very sparsely and unevenly sericeous to glabrous; limb of lateral petals 4–6.5 mm in diameter; anthers pubescent, those of stamens opposing the posterior-lateral sepals with the connective enlarged and bearing 1 reduced locule; nut of samara without lateral ornamentation; Beni, La Paz, Pando, Santa Cruz. 69. *S. cardiophyllum.*
   4. Mature laminas abaxially sericeous, the hairs evenly distributed; limb of lateral petals (7–) 9–15 mm in diameter; anthers glabrous, those of stamens opposing the posterior-lateral sepals unmodified, i.e., like those of the stamen opposing the posterior petal; lateral petals (7–) 9–15 mm in diameter; anterior style with or without folioles; nut of the samara usually with 1–3 lateral winglets per side and/or spurs and crests, rarely without lateral ornamentation; Beni. 33. *S. sinuatum.*
3. Laminas abaxially pubescent with T-shaped hairs to tomentose.
   5. Styles with well-developed folioles; petals marked or suffused with red; claw of posterior petal bearded; flowers 12–20 per umbel; Beni. 73. *S. strigosum.*
   5. Styles without folioles; petals yellow; claw of posterior petal glabrous; flowers (16–) 20–50 per umbel or pseudoraceme.
      6. Margin of lamina with sessile and/or filiform glands; anthers pubescent, those of stamens opposing the lateral sepals with the connective enlarged and the locules reduced; posterior styles longer than the anterior one; Beni, La Paz, Pando, Santa Cruz. 42. *S. florosum.*
      6. Margin of lamina eglandular; anthers glabrous, all unmodified, unequal in size but equal in shape; posterior styles and anterior style about equally long; La Paz. 6. *S. yungasense.*

8. Key to Stigmaphyllon in Argentina, Paraguay, and Uruguay

1. Laminas abaxially pubescent with T-shaped hairs to tomentose; along the Río Paraná and Río Uruguay and their tributaries. 81. *S. bonariense.*
1. Laminas abaxially glabrous to sparsely sericeous, the hairs sessile or subsessile, appressed.
   2. Mature petioles and pedicels densely pubescent; margin of lamina entire, without filiform glands (or if basally cordate or hastate, sometimes each basal lobe with one filiform gland); petioles only up to 2.1 cm long; samara greatly modified, the nut with a dorsal crest and laterally covered with warty excrescences; Argentina (Chaco, Corrientes, Santa Fe) and adjacent Paraguay. 56. *S. calcaratum.*
   2. Mature petioles and pedicels glabrate to glabrous; margin of lamina nearly entire and bearing filiform glands to coarsely dentate, often each tooth ending in a filiform gland (the glands sometimes broken off in larger leaves in *S. jatrophifolium*); petioles up to 5.1 cm long; samara with a dorsal wing, the nut laterally unornamented.
      3. Laminas palmately to pedately lobed, infrequently triangular to elliptical or rarely suborbicular, the basal lobes never overlapping, the margin irregularly and commonly coarsely dentate, each tooth ending in a filiform gland; umbels usually in dichasia but sometimes solitary, flowers 8–15 per umbel; samara with an elongate dorsal wing 3–4 cm long, the nut ovoid, 6.5–7.2 mm in diameter; along the Río Paraná and Río Uruguay and their tributaries. 52. *S. jatrophifolium.*
      3. Laminas broadly ovate to cordate, the basal lobes usually overlapping; margin evenly fringed with filiform glands, nearly entire to shallowly dentate; umbels solitary, flowers 3–8; samara lenticular, the dorsal wing 2–2.5 cm long and encircling the nut, the nut laterally compressed, 3.4–4 mm in diameter; Uruguay (environs of Montevideo). 55. *S. ciliatum.*

9. Key to Stigmaphyllon in Brazil

9a. Northern and western Brazil (Piauí, Maranhão, Amapá, Pará, Amazônas, Roraima, Acre, Rondônia, Mato Grosso, and Mato Grosso do Sul)

1. Laminas abaxially very densely white-sericeous or appressed sericeous-tomentulose, the epidermis hidden.
   2. Anterior style longer than the posterior two, all glabrous; petals yellow and suffused or marked with red, the margin erose (-denticulate); nut of samara with 3–4 lateral winglets per side; Amazônas. 18. *S. maynense.*

2. Anterior styles shorter than the posterior two, each with an adaxial row of hairs; petals yellow, the margin with fimbriae to 0.8 mm long, limb of posterior petal often also with stout gland-tipped fimbriae at the base; samara unknown; Rondônia, Mato Grosso. 74. *S. stylopogon.*

1. Laminas abaxially glabrous to pubescent, the epidermis always visible.
 3. Laminas abaxially pubescent with T-shaped hairs to tomentose.
  4. Styles without folioles; Acre, Rondônia. 42. *S. florosum.*
  4. Styles with well-developed folioles.
   5. Marginal glands of lamina nail-like, i.e., stipitate with a disklike apex; petals erose (-denticulate), the teeth up to 0.2 m long; anthers pubescent; nut of samara 12–19 mm in diameter, the locule surrounded by air chambers; Amapá, Pará, Amazônas. 28a. *S. adenodon* var. *adenodon.*
   5. Marginal glands of lamina sessile or filiform, or the laminar margin eglandular; petals with fimbriae up to 1.5 (–1.8) mm long (to 0.5 mm long in *S. matogrossense*); anthers glabrous; nut of samara 4–6.5 mm in diameter, without air chambers (samara unknown in *S. matogrossense.*)
    6. Petals yellow and marked or suffused with red; claw of posterior petal bearded; laminas unlobed; anthers of stamens opposing the posterior-lateral sepals bearing 2 reduced locules; nut of samara with 3–4 lateral winglets per side; Acre. 73. *S. strigosum.*
    6. Petals yellow; claw of posterior petal glabrous; anthers of stamens opposing the posterior-lateral sepals consisting only of a glandular connective without locules; laminas lobed or unlobed; nut of samara with 1 lateral winglet per side (in *S. palmatum*; samara unknown in *S. matogrossense*).
     7. Laminas commonly palmately 3–5 (–7)-lobed or lanceolate, ovate, cordate to suborbicular, the margin with filiform and/or sessile glands; limb of posterior petal without stout gland-tipped fimbriae at the base; anthers of stamens opposing the anterior-lateral sepals with the connective enlarged and bearing 2 locules; Pará, not recorded from Amapá but to be expected there. 72. *S. palmatum.*
     7. Laminas ovate, the margin eglandular; limb of posterior petal bearing at the base 1–3 stout gland-tipped fimbriae per side at the base; anthers of stamens opposing the anterior-lateral sepals consisting only of a glandular connective without locules; Mato Grosso. 75. *S. matogrossense.*
 3. Laminas abaxially glabrous to sericeous, the hairs sessile, appressed.
  8. Styles without folioles; pedicels 15–30 mm long; anthers all unmodified and subequal; Amapá, Pará, Maranhão. 10. *S. bannisterioides.*
  8. Styles with folioles (or in *S. sinuatum* the anterior style sometimes without folioles but extended into a claw); pedicels 2.5–10.5 mm long (in *S. paralias* 8–30 mm long); anthers of stamens opposing the anterior-lateral sepals with the connective enlarged and the locules reduced or absent (unknown in *S. jobertii*).
   9. Petiole glands peg-like, 0.8–1.5 mm long, borne up to 2 mm below the base of the lamina; Piauí. 54. *S. jobertii.*
   9. Petiole glands prominent but sessile, borne at the apex of the petiole.
    10. Anterior style and its opposing stamen longer than the posterior styles and their opposing stamens; Amapá, Amazônas, Acre, Pará. 17. *S. puberum.*
    10. Anterior style and its opposing stamen shorter than or subequal to the posterior styles and their opposing stamens.
     11. Petioles of even the largest leaves only 0.4–2.1 cm long; laminas linear to narrowly elliptical to lanceolate (rarely to orbicular in *S. paralias*); flowers 3–12 (–15) per umbel; samaras with a dorsal crest rather than an elongate dorsal wing.
      12. Shrubs; peduncles rudimentary, up to 2.8 (–4.5) mm long, pedicels 8–30 mm long; stipules eglandular, commonly fused across the node into a bifid structure; anther of stamen opposing the posterior petal consisting of only a glandular connective without locules; surface of nut of samara rugose to reticulate; dry, sandy habitats in Piauí, Maranhão, Pará (Tucuruí). 62. *S. paralias.*
      12. Vines; peduncles rudimentary to 10 mm long, pedicels 1.5–10.3 mm long; stipules glandular in the proximal 1/3–1/2, the distal portion herbaceous, free; anther of stamen opposing the posterior petal with 2 locules, the connective not enlarged; nut of samara covered with numerous bulbous and warty excrescences composed of spongy tissue; in marshy areas and along rivers in Pará, Amazônas, Mato Grosso, and Mato Grosso do Sul. 56. *S. calcaratum.*

11. Petioles of the larger leaves 2.5–13 cm long; laminas triangular to ovate to cordate to orbicular to reniform; flowers 15–30 per umbel or pseudoraceme (10–20 in *S. paraense*); samaras with an elongate dorsal wing 2.7–5.5 cm long, or in *S. lacunosum* with a reduced triangular dorsal wing encircling the nut, (2.5–) 3–4.5 cm high measured from base of nut.
 13. Limb of petals fringed with fimbriae up to 0.8 (–1) mm long; laminas abaxially densely sericeous; flowers 10–20 per umbel or pseudoraceme; anthers of stamens opposing the posterior-lateral sepals with the connective enlarged and bearing 0–2 reduced locules; Brazil (Piauí, Maranhão, Pará, Mato Grosso?). 71. *S. paraense.*
 13. Limb of petals erose to erose-denticulate, the teeth up to 0.2 mm long; laminas abaxially glabrous to densely sericeous; flowers 15–35 (–40) per pseudoraceme; anthers of stamens opposing the posterior-lateral sepals unmodified, i.e., like that of the stamen opposing the posterior petal, or with the connective enlarged and bearing 1 reduced locule (*S. cardiophyllum*).
  14. Anthers pubescent; limb of anterior-lateral petals (5–) 6–9 mm in diameter, limb of posterior-lateral petals 4–8.5 mm in diameter.
   15. Mature laminas abaxially glabrate to glabrous, or sometimes very sparsely sericeous and the hairs unevenly and patchily distributed; anthers of stamens opposing the posterior-lateral sepals with the connective enlarged and bearing 1 reduced locule; nut of samara 2.5–3.5 mm in diameter, without air chambers; Maranhão, Pará, Amazônas, Acre. 69. *S. cardiophyllum.*
   15. Mature laminas evenly pubescent with appressed hairs; anthers of stamens opposing the posterior-lateral sepals unmodified, i.e., like that of the stamen opposing the posterior petal; nut of samara 13–15 mm in diameter, the locule surrounded by air chambers; collected once in Amazonian Brazil (state unknown). 29. *S. lacunosum.*
  14. Anthers glabrous; limb of anterior-lateral petals (8–) 11–15 mm in diameter, limb of posterior-lateral petals 8–13 mm in diameter.
   16. Laminas abaxially appearing glabrous to the naked eye, but usually very sparsely sericeous, the hairs ca. 0.1 (–0.2) mm long and widely spaced, never touching; laminas ovate to cordate or narrowly so, the base cordate; anterior style with folioles; Amapá, eastern Pará. 34. *S. convolvulifolium.*
   16. Laminas abaxially sparsely to very densely sericeous, the hairs (0.2–) 0.3–0.5 (–0.7) mm long, touching to overlapping; laminas triangular to ovate to cordate to elliptical to broadly so to orbicular to oblate to reniform, rarely 3–5-lobed, the base acute to truncate to cordate to deeply auriculate; anterior style with folioles or without them but extended into a long spur; Amapá, Maranhão, Pará, Amazônas, Roraima, Acre, Rondônia. 33. *S. sinuatum.*

9b. Coastal states from Ceará to Rio Grande do Sul, and Minas Gerais and Goiás

1. Leaves subsessile, petioles up to 0.3 cm long.
 2. Vines; margin of lamina deeply crenate, abaxially each sinus with a gland; flowers ca. 10 per umbel; Espírito Santo. 45. *S. crenatum.*
 2. Subshrubs; margin of lamina entire, eglandular; flowers grouped in 2's; Bahia. 11. *S. harleyi.*
1. Leaves distinctly petiolate, petioles 0.5–19 cm long.
 3. Petioles confluent across the node and forming a prominent corky ridge bearing the stipules; the larger laminas palmately 3–5 (–7)-lobed to broadly ovate or broadly elliptical, the smaller 2–3-lobed to broadly ovate; Minas Gerais. 35. *S. macedoanum.*
 3. Petioles not joined across the node, or confluent across the node to form a band or line but never a prominent corky ridge; laminas entire or lobed.
  4. Larger laminas pinnately 5–7-lobed (the smaller ones associated with the inflorescence often unlobed or 3-lobed).

5. Styles all with well-developed folioles; anthers pubescent; lamina with the margin eglandular or with sessile glands borne abaxially adjacent to the margin; southeastern Minas Gerais, Rio de Janeiro, and adjacent São Paulo. 76. *S. angustilobum.*
5. Styles all efoliolate; anthers glabrous; lamina with filiform and sometimes also sessile glands borne on the margin.
 6. Laminas abaxially glabrous; Espírito Santo. 8. *S. glabrum.*
 6. Laminas abaxially pubescent or the largest glabrescent (*S. carautae*).
  7. Laminas abaxially pubescent with a mixture of sessile and short-stalked hairs (stalk up to 0.1 mm long), the vesture appressed and sloughed off in patches and then the larger laminas glabrescent; stamens opposing the anterior-lateral sepals with the connective greatly enlarged and the locules reduced; Rio de Janeiro. 21. *S. carautae.*
  7. Laminas abaxially abundantly pubescent with T-shaped hairs (stalk 0.1–0.2 mm long), the vesture spreading and persistent; stamens opposing the anterior-lateral sepals with the connective and locules equally long; Bahia, eastern Minas Gerais. 7. *S. urenifolium.*

4. Larger laminas entire or palmately lobed.
 8. Mature laminas abaxially glabrous to glabrate (commonly with scattered appressed or T-shaped hairs in *S. arenicola*), or sericeous or appressed-tomentulose.
  9. Pedicels glabrous or at most glabrate.
   10. Laminas entire but with the base deeply auriculate and the basal lobes usually overlapping, the margin evenly fringed with filiform glands; umbels solitary, with 3–8 flowers; samara lenticular, the reduced dorsal wing 2–2.5 mm long and encircling the nut, the nut laterally compressed, 3.4–4 mm in diameter; Pernambuco, Bahia, Espírito Santo, Rio de Janeiro, São Paulo, Paraná, Santa Catarina, Rio Grande do Sul, not recorded from Ceará, Rio Grande do Norte, Paraíba, Alagoas, and Sergipe but to be expected there. 55. *S. ciliatum.*
   10. Laminas lobed or entire, the base cordate to auriculate but the basal lobes never overlapping or truncate, hastate, or sagittate; margin of lamina entire or grossly dentate, eglandular or with irregularly spaced filiform glands and sessile glands; umbels in compound inflorescences or sometimes solitary, with 6–15 flowers; samara with the dorsal wing elongate, 3–4.5 cm long, the nut ovoid to spheroid, 6–11 mm in diameter.
    11. Larger laminas palmately to pedately (2–) 5–7 (–9)-lobed, infrequently triangular to elliptical, or rarely suborbicular, 4.2–18.2 cm long, the margin shallowly to grossly dentate, the teeth ending in filiform glands, sometimes also with sessile glands in the sinuses; posterior petal 7–8.5 mm in diameter; mostly along the Río Paraná and Río Uruguay and their tributaries in Paraná, Santa Catarina, and Rio Grande do Sul. 52. *S. jatrophifolium.*
    11. Larger laminas triangular to ovate to elliptical to hastate or sometimes 2–3-lobed, 3.2–10.5 cm long, the margin entire, eglandular or with filiform and/or sessile glands; posterior petal ca. 10 mm in diameter; Ceará, Paraíba, Pernambuco, Bahia, Rio de Janeiro, not recorded from Rio Grande do Norte, Alagoas, Sergipe, and Espírito Santo but to be expected there. 53. *S. auriculatum.*
  9. Pedicels pubescent.
   12. Peduncles rudimentary to 3 mm long (rarely to 4.5 mm long in *S. paralias*), up to 0.2 times as long as the pedicels.
    13. Shrubs; petioles 0.4–1.5 cm long, petiole glands flush with the epidermis, borne at the apex; mature laminas abaxially sparsely to sometimes abundantly sericeous, the margin eglandular; Ceará, Rio Grande do Norte, Paraíba, Pernambuco, Alagoas, Sergipe, Bahia, Espírito Santo, Minas Gerais, Goiás, Rio de Janeiro. 62. *S. paralias.*
    13. Vines; petioles 2.8–7 cm long, petiole glands shallowly cupulate, borne 1–10 mm below the base of the lamina; mature laminas abaxially glabrous but often with T-shaped hairs on the major veins, the margin with irregularly spaced stipitate glands (0.2–0.3 mm in diameter, 0.3–0.6 mm long) and/or filiform glands (up to ca. 2 mm long); São Paulo. 58. *S. bradei.*
   12. Peduncles well developed, (2.5–) 3.5–15 mm long, 0.5–2 times as long as the pedicels.

14. Laminas with the margin grossly dentate, each tooth ending in a filiform gland, the base deeply auriculate; petioles with a pair of shallowly cupulate glands 4–20 mm below the base of the lamina; Rio de Janeiro. 51. *S. vitifolium.*
14. Laminas with the margin entire, eglandular or with sessile or stipitate but never filiform glands, the base attenuate, truncate, or cordate.
    15. Margin of lamina with stipitate glands to 0.7 mm long, borne abaxially adjacent to the margin; anthers pubescent; Pernambuco, Bahia, Espírito Santo, eastern Minas Gerais, Rio de Janeiro, not recorded from Alagoas and Sergipe but to be expected there. 90. *S. salzmannii.*
    15. Margin of lamina eglandular or with sessile glands borne abaxially adjacent to the margin (or in *S. macropodum* sometimes on the margin); anthers glabrous.
        16. Anterior style and posterior styles with only a lateral lip or with tiny folioles less than 1 mm long and wide; anthers of all stamens varying only slightly in size, those of stamens opposing the lateral sepals sometimes with the locules slightly shorter than the connective but not greatly reduced; flowers 15–50 per umbel or pseudoraceme; laminas entire, abaxially abundantly pubescent, the margin eglandular; southern Espírito Santo, southeastern Minas Gerais, Rio de Janeiro, northern São Paulo. 14. *S. lalandianum.*
        16. Anterior and posterior styles with well-developed folioles, 1–3.2 mm long and wide; anthers of stamens opposing the lateral sepals with the connective enlarged and bearing 0–2 reduced locules; flowers 10–25 per umbel or pseudoraceme; laminas entire or lobed, abaxially glabrous to pubescent, the margin eglandular or with sessile glands borne abaxially adjacent to the margin (or sometimes on the margin in *S. macropodum*).
            17. Stems and axes densely (golden-) sericeous, the epidermis hidden by the dense vesture; mature laminas 8.8–27 cm long, 5.8–22.5 cm wide, entire, abaxially densely appressed-tomentulose or sometimes sericeous; nut of samara without air chambers; Bahia. 85. *S. macropodum.*
            17. Stems and axes pubescent when young but soon becoming glabrate to glabrous and the epidermis always visible; mature laminas 5.8–14.5 cm long, 2.7–9.5 cm wide, entire or lobed, abaxially densely sericeous to sparsely so to glabrous (in *S. arenicola* sometimes also with scattered subsessile to T-shaped hairs); nut of samara with air chambers (*S. paraense, S. rotundifolium*) or without them (*S. arenicola*).
                18. Laminas abaxially densely sericeous, the vesture giving the surface a silvery sheen, the margin with glands borne on the margin (or eglandular); limb of all petals with fimbriae up to 1 mm long; Goiás. 71. *S. paraense.*
                18. Laminas abaxially moderately sericeous (never silvery) to sparsely so to glabrous (in *S. arenicola* sometimes also with scattered subsessile to T-shaped hair), the margin with glands borne abaxially adjacent to the margin (or eglandular); limb of petals erose or denticulate/fimbriate, the teeth fimbriae up to 0.3 (–0.5) mm long.
                    19. Limb of anterior-lateral petals ca. 11 mm in diameter, of posterior-lateral petals ca. 8 mm in diameter, of posterior petal ca. 7 mm in diameter, margin of lateral petals erose; anthers of stamens opposing the lateral sepals with the connective enlarged and bearing 2 reduced locules; nut of samara with air chambers surrounding the locule; Bahia, Espírito Santo (?). 84. *S. rotundifolium.*
                    19. Limb of anterior-lateral petals 12–14 mm in diameter, posterior-lateral petals 10–12.5 mm in diameter, of posterior

petal 8–10 mm in diameter, margin of lateral petals denticulate to fimbriate; anthers of stamens opposing the lateral sepals consisting of only a glandular connective without locules; nut of samara without air chambers; coastal Rio de Janeiro, São Paulo, and Paraná. 86. *S. arenicola.*

8. Mature laminas abaxially abundantly pubescent with T-shaped hairs (the stalk 0.1–0.6 mm long), the vesture evenly distributed, spreading (never appressed).
   20. Laminas with sessile to stipitate (nail-like) glands and/or filiform glands borne on the margin; Bahia, Pernambuco, Espírito Santo (?). 61. *S. puberulum.*
   20. Laminas with the margin eglandular or with sessile to stipitate glands borne abaxially adjacent to the margin.
      21. Anterior style with only a narrow lip, up to 0.3 mm wide; anthers of stamens opposing the lateral sepals with the locules equally as long as the connective or slightly shorter, but the connective not greatly enlarged; southern coastal Bahia, southeastern Minas Gerais, and Rio de Janeiro. 15. *S. acuminatum.*
      21. Anterior styles with well-developed folioles; anthers of stamens opposing the lateral sepals with the connective enlarged and bearing 0–2 reduced locules.
         22. Mature laminas adaxially pubescent, velutinous and/or with T-shaped hairs; petioles with a pair of glands borne at the apex or up to 7 mm below the base of the lamina; Rio de Janeiro. 87. *S. gayanum.*
         22. Mature laminas adaxially glabrate to glabrous; petioles with a pair of glands borne at the apex (or sometimes or up to 0.2 mm below the base of the lamina in *S. affine*).
            23. Anthers of stamens opposing the posterior-lateral sepals consisting of only an enlarged glandular connective without locules; anthers pubescent (usually glabrous in *S. tomentosum*).
               24. Margin of petals digitate-fimbriate (fimbriae parallel-sided and obtuse at apex), limb of anterior-lateral petals 8.5–11 mm in diameter; larger laminas narrowly to broadly lanceolate to sometimes ovate, to 12 cm long and 5.5 cm wide; nut of samara 4–7 mm high, on each side with a row of 5–7 lateral winglets; Rio de Janeiro, and adjacent Espírito Santo and Minas Gerais. 77. *S. alternifolium.*
               24. Margin of petals erose to denticulate to fimbriate, the teeth/fimbriae tapered from the base to an acute apex, limb of anterior-lateral petals 11–15 mm in diameter; larger laminas triangular to elliptical to cordate to broadly ovate to orbicular or sometimes 2–5-lobed (rarely lanceolate), to 31 cm long and 19 cm wide; nut of samara 6.5–12 mm high, with 1–3 lateral winglets per side and often also with spurs/crests or only with ridges/tubercles (samara unknown in *S. hatschbachii*).
                  25. Laminas abaxially densely pubescent with golden hairs (trabecula 1.1–2.3 mm long, straight, stalk 0.2–0.4 mm long), the epidermis nearly or completely hidden by the dense vesture; Minas Gerais. 88. *S. hatschbachii.*
                  25. Laminas abaxially densely to sparsely pubescent with white or translucent hairs (trabecula 0.2–1.5 mm long, straight to curled, stalk 0.1–0.2 mm long or to 0.4 mm long in *S. tomentosum*), the epidermis never hidden.
                     26. Laminas with stipitate marginal glands to 0.4 mm long; anthers pubescent; limb of posterior petal without stout gland-tipped fimbriae; trabeculas of abaxial pubescence straight to wavy, adjacent ones touching but not entwined to form a matted vesture, each hair readily discernible; Paraíba, Pernambuco, Alagoas, Sergipe, Bahia, Espírito Santo, Minas Gerais, not recorded from Ceará and Rio Grande do Norte but to be expected there. 89. *S. blanchetii.*
                     26. Laminas with sessile to slightly raised marginal glands, 0.1 (–0.2) mm long; anthers usually glabrous (sometimes with a few hairs); limb of posterior petal at the base often with 1–3

stout gland-tipped fimbriae per side, these 0.3–0.5 mm long, ca. 0.2 mm in diameter; trabeculas of abaxial pubescence straight but more commonly wavy to curled, the adjacent ones overlapping to entwined and commonly forming a matted vesture and each hair not readily discernible; Goiás, Minas Gerais, and adjacent Bahia, Rio de Janeiro, São Paulo, Paraná, Santa Catarina. 79. *S. tomentosum.*

23. Anthers of stamens opposing the posterior-lateral sepals with the connective enlarged and bearing 1–2 reduced locules; unmodified anthers glabrous or sometimes with a few apical hairs (always glabrous in *S. cavernulosum*).

27. Mature laminas abaxially sparsely pubescent, the trabeculas straight or wavy, touching but not greatly overlapping and entwined, the individual hairs readily discernible; nut of samara ca. 20 mm high, 13–15 mm in diameter, the locule surrounded by air chambers; Rio de Janeiro. 82. *S. affine.*

27. Mature laminas abaxially abundantly pubescent, the trabeculas straight to curled, greatly overlapping and entwined and thus the individual hairs not readily discernible (except rarely in *S. tomentosum*); nut of samara 4–12.5 mm high, 5–6.3 mm in diameter, without air chambers (with air chambers in *S. cavernulosum*).

28. Limb of anterior-lateral petals 8–9 mm in diameter, limb of posterior petal 6–7.5 mm in diameter; dorsal wing of samara elongate, 5.5–6.2 cm long, ca. 1.8 cm wide, 3.5–4.7 times as long as wide; Bahia, Espírito Santo, eastern Minas Gerais. 80. *S. saxicola.*

28. Limb of anterior-lateral petals 10.5–15 mm in diameter, limb of posterior petal 8–11 mm in diameter; dorsal wing of samara elongate or triangular to subsquare, 1.6–5.2 cm long, 1.6–2.7 cm wide, 1–2.7 times as long as wide.

29. Umbels borne in dichasia or solitary, these and associated leaves arranged alternately on a primary inflorescence axis; anthers glabrous; nut of samara with an elongate dorsal wing, the locule surrounded by air chambers; eastern Bahia and adjacent Minas Gerais. 83. *S. cavernulosum.*

29. Umbels borne solitary or in dichasia or compound dichasia or thyrses, the units and associated leaves arranged oppositely on a primary inflorescence axis; anthers glabrous or pubescent; nut of samara with an elongate dorsal wing or with a triangular to subsquare dorsal wing encircling the nut, the nut without air chambers.

30. Limb of anterior-lateral petals 10.5–12 mm in diameter, limb of posterior petal without basal stout gland-tipped fimbriae; anthers of stamens opposing the lateral sepals with the connective enlarged and bearing 2 reduced locules; flowers 8–25 per umbel; samara with the dorsal wing triangular to subsquare, encircling the nut, 1.6–2.9 cm high, 1.7–2.7 cm wide; mostly along the Río Paraná and Río Uruguay and their tributaries in São Paulo, Paraná, and Rio Grande do Sul (occasionally adventive in eastern São Paulo). 81. *S. bonariense.*

30. Limb of anterior-lateral petals 11–15 mm in diameter, limb of posterior petal near the base often with 1–3 stout gland-tipped fimbriae per side, these 0.3–0.5 mm long, ca. 0.2 mm in diameter; anthers of stamens opposing the lateral sepals with the connective enlarged and bearing 1 (–2) reduced locules; flowers 10–40 per umbel or pseudoraceme; samara with an elongate dorsal wing 3.6–5.2 cm long, 1.6–2.1 cm wide; Goiás, Minas Gerais, and adjacent Bahia, Rio de Janeiro, São Paulo, Paraná, Santa Catarina,. 79. *S. tomentosum.*

NUMERICAL LIST OF SPECIES

1. S. pseudopuberum
2. S. bogotense
3. S. sarmentosum
4. S. suffruticosum
5. S. stenophyllum
6. S. yungasense
7. S. urenifolium
8. S. glabrum
9. S. peruvianum
10. S. bannisterioides
11. S. harleyi
12. S. cordatum
13. S. romeroi
14. S. lalandianum
15. S. acuminatum
16. S. orientale
17. S. puberum
18. S. maynense
19. S. hypargyreum
20. S. herbaceum
21. S. carautae
22. S. velutinum
23. S. tarapotense
24. S. columbicum
25. S. goudotii
26. S. singulare
27. S. adenophorum
28. S. adenodon
28a. S. adenodon var. adenodon
28b. S. adenodon var. macropterum
29. S. lacunosum
30. S. cuzcanum
31. S. ecuadorense
32. S. nudiflorum
33. S. sinuatum
34. S. convolvulifolium
35. S. macedoanum
36. S. angulosum
37. S. emarginatum
38. S. microphyllum
39. S. sagraeanum
40. S. diversifolium
41. S. floribundum
42. S. florosum
43. S. aberrans
44. S. selerianum
45. S. crenatum
46. S. ellipticum
47. S. eggersii
48. S. venulosum
49. S. echitoides
50. S. tergolanatum
51. S. vitifolium
52. S. jatrophifolium
53. S. auriculatum
54. S. jobertii
55. S. ciliatum
56. S. calcaratum
57. S. laciniatum
58. S. bradei
59. S. boliviense
60. S. coloratum
61. S. puberulum
62. S. paralias
63. S. lindenianum
64. S. retusum
65. S. dichotomum
66. S. panamense
67. S. tonduzii
68. S. finlayanum
69. S. cardiophyllum
70. S. argenteum
71. S. paraense
72. S. palmatum
73. S. strigosum
74. S. stylopogon
75. S. matogrossense
76. S. angustilobum
77. S. alternifolium
78. S. alternans
79. S. tomentosum
80. S. saxicola
81. S. bonariense
82. S. affine
83. S. cavernulosum
84. S. rotundifolium
85. S. macropodum
86. S. arenicola
87. S. gayanum
88. S. hatschbachii
89. S. blanchetii
90. S. salzmannii

INDEX TO NUMBERED COLLECTIONS EXAMINED

The numbers in parentheses refer to the corresponding species in the text and in the Numerical List of Species presented above. Collections of putative hybrids are noted by the symbol "×."

Acevedo-Rodríguez, P. 420 (46), 476 (17), 2854 (41), 2855 (41), 2897 (37), 3309 (33), 3506 (33), 3965 (37), 4042 (37), 4938 (33), 5366 (37), 6291 (40).
Acevedo-Rodríguez, P., & Cedeño, J. A. 7244 (69), 7336 (69), 7466 (78), 7627 (78).
Acevedo-Rodríguez, P., et al. 6794 (46).
Acosta Solís, M. 7571 (2).
Acuña, J. 3780 (40).
Adams, C. D. 11827 (10), 14311 (68).
Agostini, G. 299 (33), 1504 (33), 2711 (33).
Agostini, G., & Fariãs, M. 57 (33).
Aguilar, R. 658 (63).
Aguilar G., J. I. 432 (46).
Ahumada, O., et al. 3852 (81).
Alain, Bro. 3502 (38), 3703 (39), 4453 (39), 9374 (37).
Albert de Escobar, L., & Brand, J. 2067 (48).
Albert de Escobar, L., & Loaiza, C. A. 81-1 (2).
Albert de Escobar, L., et al. 3124 (65), 4263 (2).
Albuquerque, B., et al. 45 (33), 696 (33).
Alexander, E. J. 150 (46).
Allard, H. A. 12185 (37), 13128 (37), 13316 (37), 13684 (37), 14247bis (37), 18093b (37).
Allemão, Fr. R19560 (53), R19561 (53), R19562 (62), R34648 (62), R71974 (62), R72135 (62).
Allemão, Fr., & de Cysneiros 219 (62).
Allen, C. 820 (65).
Allen, P. H. 267 (65), 908 (17), 1777 (46), 2015 (46), 2448 (46), 4514 (46), 4165 (46), 17278 (19).
Almeda, F. 3245 (17), 3383 (24), 5154 (63), 5543 (46), 6585 (46), 7074 (63).
Almeda, F., et al. 3225 (46).
Almeida 3256 (89).
Almeida, E. de F. 206 (79).
Almeida de Jesus, J. 1564 (86), 1667 (86), 1892 (77).
Alston, A. H. G. 6090 (65), 7939 (2).
Amaral, I. L., et al. 95 (33), 961 (71), 1230 (71).
Amorim, A. M. 1278 (15), 1495 (89), 1650 (85).
Amorim, A. M., et al. 367 (55), 372 (55), 378 (85), 872 (85), 988 (89), 1066 (85), 1439 (85), 1533 (62).
Ancuash, E. 1134 (69), 1262 (69×?).
Anderson, W. R. 9208 (7), 11184 (55), 11190 (62), 11197 (86), 11198 (53), 11202 (14), 11610 (14), 11656 (55), 11666 (14), 11711 (14), 11715 (79), 11728 (14), 11729 (14), 11736 (62), 11738 (53), 11756 (62), 12371 (52), 12383 (52), 12389 (52), 13219 (46), 13670 (14), 13677 (14), 13684 (79), 13748 (33).
Anderson, W. R., & Anderson, C. 5511 (63), 5555 (44), 5559 (46), 5564 (46), 5570 (46).
Anderson, W. R., & Lakowski, C. W. 4216 (46), 4221 (46).
Andrade, L. M. 321 (89), 322 (14), 717 (79).
André, E. 237 (65), 358 (2), K903 (2), K906 (2), K973 (65), 2142 (2), 4063 (46).
Angeli 47 (14), 475 (86).
Antonio, T. 1261 (46), 1774 (63), 1921 (46), 2283 (46), 3724 (63), 4359 (46), 4632 (46).
Antonio, T., & Hahn, W. 4427 (17).
Araque Molina, J., & Barkley, F. A. 18S265 (65), 19An038 (2), 19An091.
Araquistain, M., & Castro, D. 1868 (46), 1828 (64), 1909 (64).
Araquistain, M., & Moreno, P. P. 1083 (64), 1505 (64), 1507 (64), 1568 (64), 1630 (64).
Araujo, D. 118 (62), 212 (53), 250 (90), 361 (62), 509 (62), 657 (86), 688 (79), 1115 (82), 1350 (55), 1376 (55), 2290 (79), 3955 (51); 5696 (79), 5831 (82), 6133 (86), 6155 (79), 6552 (86), 7305 (86), 7630 (62), 8295 (53), 8951 (53).
Arbeláez S., G., et al. 1236 (2), 1819 (22).
Arbo, M. M., et al. 2133 (81), 5567 (90), 5626 (62), 6045 (81).
Archer, W. A. 211 (49), 1257 (2), 2463 (33), 2481 (17), 2658 (33), 2662 (33), 2761 (33), 2798 (33), 3385 (2), 8282 (72).
Argeñal, F. 171 (46).
Arias, J. C. 64 (48).
Arias, L. A. 57 (20).
Aristeguieta, L. 1524 (65), 2285 (33), 4774 (10), 4901 (2), 4959 (65).
Aristeguieta, L., & Lizot, J. 7371 (33).
Aristeguieta, L., et al. 7240 (17).
Armond, N. 307 (90), R72128 (79), R72189 (79).
Armond R., 78 (79).
Arnason, T. 17920 (63).
Arnason, T., & Lambert, J. 17548 (46).
Arnoldo, M. 1251 (37), 3433 (40), 3193 (37).
Arrigo, R. 198 (63).
Asplund, E. 5203 (46), 5447 (46), 7608 (2), 9114 (69), 9379 (69×?), 9502 (28), 10233 (18), 12103 (69), 14639 (28a), 16568 (46), 19221 (2).
Atwood, J. T. 3513 (64), 4274 (63).
Aulestia, M. 208 (29).
Austin, D. F. 7257 (33).
Austin, D. F., & Cavalcante, P. 4136 (28a), 4144 (17).
Austin, D. F., et al. 7010 (33), 7016 (28a), 7199 (34), 7328 (10).
Aviles, S. 111 (46), 9609 (46).
Ayala, F. 437 (33), 1851 (42), 2010 (33), 2016 (33), 2295 (69), 2320 (28a), 2357 (33), 2579 (33).
Ayala, F., et al. 2618 (29), 2913 (17), 2938 (28a).
Aymard C., G. 4644 (33).
Aymard C., G., & Cuello, N. 3355 (65), 6577 (33), 7663 (65).
Aymard C., G., & Stergios, B. 925 (28a).

191 (37), 686 (37), 791 (37), 922 (37), 980 (37), 1042 (37), 1192 (37), 2407 (37), 2557 (37), 2629 (37), 3171 (37), 3286 (40), 3398 (37), 3452 (37), 3513 (40).

Bond, F. E., et al. 105 (10).

Bondar, G. 1758 (55).

Boom, B. M. 6991 (37), 7165 (10).

Boom, B., M., & Beardsley, D. 8436 (33).

Boom, B. M., & Devoe, N. 6824 (37).

Boom, B. M., & Grillo, M. 6487 (33).

Boom, B. M., et al. 7867 (2).

Boon, H. 1048 (34), 1104 (34).

Bordenave, B. 100 (33), 195 (34), 281 (34), 242 (33).

Botero, F. FM315 (2).

Botteri, M. 2026 (46).

Bourdeth, J. 57 (46).

Bourgeau, E. 2292 (64).

Bovell, J. R., & Freeman, W. G. 210 (10).

Box, H. E. 612 (37), 684 (40), 791 (37), 857 (40).

Brace, L. J. K. 4872 (39), 5200 (39), 6742 (39), 6971 (39), 7008 (39), 7075 (39), 7124 (39).

Brade, A. C. 218 (53), 230 (62), 7898 (86), 7945 (55), 7987 (58), 9099 (55), 9100 (86), 9101 (79), 9675 (87), 9841 (77), 10041 (53), 10489 (76), 10533 (87), 10638 (14), 10985 (62), 11186 (87), 11359 (53), 11363 (14), 11429 (14), 12018 (62), 12819 (51), 14611 (76), 16094 (86), 11745 (79), 18504 (14), 18700 (77), 20569 (53), R19556 (53), R20868 (77), R78328 (51), RB26238 (53).

Braga, P. I. S. 2438 (51).

Brandbyge, J., & Asanza C., E. 30109 (33), 31600 (18).

Brandbyge, J., et al. 30296 (33).

Brant, A. E., et al. 1727 (65).

Bravo H., H. 11 (64), 18 (63).

Breckon, G., & Deghan, B. 3184 (37).

Bredemeyer, F. 206 (33).

Breedlove, D. 6240 (1), 6565 (46), 9027 (44), 10169 (1), 10395 (1), 12035 (1), 20724 (63), 21417 (1), 23317 (1), 23570 (63), 24135 (44), 24584 (63), 24605 (46), 25190 (46), 26420 (63), 26498 (64), 26648 (64), 27359 (46), 27595 (1), 28620A (46), 28814 (64), 29060 (46), 29285 (1), 30293 (46), 33811 (46), 34225 (46), 34226 (63), 34509 (63), 34869 (46), 34871 (63), 35023 (64), 36643 (46), 37092 (1), 37185 (46), 37480 (46), 39938 (46), 41498 (46), 41958 (64), 42304 (46), 44393 (1), 46460 (44), 46804 (46), 46936 (64), 47263 (63), 47326 (64), 47336 (64), 47959 (46), 48374 (64), 48714 (44), 48850 (1), 49001 (46), 49130 (64), 49288 (1), 49361 (1), 49492 (64), 50177 (44), 50307 (63), 50487 (46), 50725 (46), 51018 (63), 51460 (1), 52070 (1), 52669 (1), 52810 (46), 52968 (1), 53280 (1), 53435 (1), 53814 (46), 54654 (46), 57199 (64), 57642 (64), 57845 (64), 57852 (64), 58471 (64).

Brenes, A. M. 98 (46), 3857 (46), 3892 (46), 4610 (46), 4774 (46), 5245 (46), 5250 (46), 5877 (46), 6547 (46), 6611 (46), 12147 (63), 12281 (46), 12292 (17), 12297 (46), 14337 (46), 15135 (46), 15629 (63), 21466 (45), 22450 (46).

Bresolin, A. 926 (79), 937 (79).

Breteler, F. J. 3109 (2), 3141 (2), 3678 (33), 3757 (33), 4721 (33), 4923 (26), 5104 (33).

Bristan, N. 1154 (17), 1171 (17), 1172 (63), 1988 (39), 6882 (39).

Bristol, M. L. 56 (2), 431 (2), 516 (2).

Brito, H. S. 305 (62).

Brito, H. S., & da Vinha, S. G. 222 (89), 252 (89).

Britton, N. L. 1168 (37), 1897 (40), 1913 (37), 1963 (40), 2249 (40), 2252 (38), 2355 (40), 2549 (37), 4069 (10), 5477 (40), 7145 (37).

Britton, N. L., & Britton, E. G. 2171 (55), 7459 (41), 9005 (10), 9095 (41), 9171 (41), 9459 (37), 9871 (41).

Britton, N. L., & Broadway, W. E. 2624 (68).

Britton, N. L., & Cowell, J. F. 750 (40), 1269 (37), 9897 (40).

Britton, N. L., & Marble, D. W. 433 (37).

Britton, N. L., & Fishlock, W. C. 1045 (37).

Britton, N. L., & Hazen, T. E. 804 (10).

Britton, N. L., & Hollick, A. 1811 (37), 1944 (37), 2159 (37), 2359 (37).

Britton, N. L., & Marble, D. W. 433 (37).

Britton, N. L., & Millspaugh, C. F. 6232 (39).

Britton, N. L., & Shafer, J. A. 378 (37), 515 (37), 871 (37), 1501 (37).

Britton, N. L., & Wheeler, W. M. 44 (37), 288 (41).

Britton, N. L., & Wilson, P. 60 (40), 182 (40), 274 (39), 4568 (40), 5446 (39), 5472 (40), 5525 (39), 5528 (40), 5530 (39), 5622 (40), 5696 (39).

Britton, N. L., et al. 71 (39), 104 (39), 165 (37), 167 (40), 247 (40), 628 (40), 496 (39), 529 (39), 566 (39), 682 (39), 2550 (41), 4663 (40), 4831 (39), 5021 (37), 5550 (39), 5780 (41), 5793 (37), 5880 (40), 5960 (37), 5993 (39), 5995 (39), 6052 (39), 6056 (39), 6230 (39), 6261 (39), 6632 (40), 12947 (38), 13028 (38), 13032 (40).

Broadway, W. E. 113 (68), 280 (10), 481 (72), 596 (10), 602 (55), 627 (68), 630 (28a), 2166 (68), 2172 (10), 2277 (17), 2573 (34), 2623 (55), 4086 (17), 4396 (10), 4474 (10), 5831 (10), 7356 (28a), 7618 (17), 8052 (34), 9091 (17).

Brown, S. 329 (37).

Buchtien, O. 1922 (42).

Bueno, E. A. 156 (45).

Bunbury, C. J. F. 2 (14).

Bunting, G. 4597 (33), 6223 (33), 6475 (33), 6903 (33), 7328 (33), 8009 (46), 8657 (65), 8993 (65), 9147 (65), 9968 (46), 10138 (65), 10610 (65), 10816 (33), 11254 (65), 12883 (65).
Bunting, G., & Licht 500 (46), 837 (63), 1292 (64).
Burandt, C., Jr., & Smith, R. F. V0075 (33).
Burch, D. 389 (46), 1442 (37), 3474 (37), 6666 (37).
Burch, D., et al. 1046 (46).
Burchell, W. J. 1267 (51), 1341 (51), 1505 (51), 1581 (55), 2782 (14), 9532 (17), 9560 (10), 9728 (33), 10025 (72).
Burger, W., & Baker, R. 103 (63), 9877 (63).
Burger, W., & Gentry, J. L., Jr. 8879 (63).
Burger, W., & Liesner, R. 6649 (46).
Burger, W., et al. 10415 (63), 12361 (46).
Burkart, A. 3253 (81), 8906 (81), 20510 (81), 20972 (81).
Burkart, A., & Bacigalupo, N. M. (56).
Burkart, A., & Gamerro, J. A. 21782 (52).
Burkart, A., & Crespo, S. 23009 (81).
Burkart, A., & Troncoso, N. S. 27861 (52).
Burkart, A., et al. 26986 (56), 26987 (81), 26988 (56), 29951 (81), 29954 (81), 30906 (81).
Burke 1035 (63).
Busey, P. 475 (63), 615 (63).
Cabral, E., & Zamudio, C. 510 (52).
Cabrera, A. L. 1718 (81), 1994 (81), 6347 (81).
Cabrera, E. 1367 (46), 1380 (46), 1426 (46), 1967 (64), 2072 (46), 2130 (46), 2200 (46), 2268 (63), 2440 (63), 3044 (1), 8204 (63), 4538 (46), 11193 (63), 11265 (63), 11347 (63), 15030a (63), 15049 (63).
Cabrera R., I. 3473 (2), 4410 (2), 4454 (2).
Calderón, S. 234 (46), 759 (46).
Callejas, R. 2140 (2), 2151 (5).
Callejas, R., & Balslev, H. 1041 (2).
Callejas, R., & Fonnegra G., R. 7317 (48).
Callejas, R., and Marulanda, O. 5892 (78).
Callejas, R., et al. 4465 (20), 4690 (20), 4742 (49), 5019 (10), 5027 (17), 5608 (25), 5615 (49), 5745 (17), 8510 (2), 8930 (49), 9311 (24).
Calzada, J. I. 309 (63), 699 (46), 753 (46), 950 (46), 1458 (46), 1689 (46), 2340 (63), 2754 (63), 4414 (63), 4746 (46), 6424 (63).
Camacho Durán, R. C-12 (2), 105 (2).
Camp, W. H. E853 (69), E3337 (2), E3620 (46).
Campbell et al. P20920 (33).
Campêlo, C. R., et al. 1569 (89).
Campos, J. 295 (18).
Campos Novaes, J. SP1975 (14).
Campos Porto, J. 22 (79).
Campos Porto, P. 961 (62), 1418 (62), 1421 (62).
Cano, A. 1547 (3).
Cantú R., J. 193 (63).
Capell, P. RB85161 (77).
Capus, F. 179 (33).
Carauta, J. P. P. 177 (53), 573 (14), 1876 (14), 2025 (82), 2372 (77), 2401 (14), 2435 (82), 3289 (87), 3531 (87), 4309 (21), 4647 (79), 4780 (77), 5815 (79).
Carballo, G. 403 (63).
Cárdenas L., D. 1347 (65).
Cárdenas L., D., et al. 2603 (24), 2915 (20), 2939 (20).
Cárdenas de Guevara, L., et al. 2565 (33).
Cardona, F. 589 (10).
Cardozo, A., et al. 65 (33), 91 (65).
Carleton, M. A. 635 (17).
Carlson, M. C. 63 (46).
Carnevali, R. 3899 (81).
Casari, M. B., et al. 107 (53), 337 (87), 376 (51), 448 (44).
Casas, J. F., & Molero, J. 5785 (81).
Castañeda, R. R. 2717 (46).
Castellanos, A. 23041 (62), 23589 (14), 24838 (52), 24979 (80).
Castillo C., G. 3880 (63).
Cavalcante, P. 801 (34), 1401 (33), 1600 (33), 1651 (72), 1770 (72), 1839 (10), 1923 (10), 1924 (10), 2260 (28a), 3284 (42).
Cavalcante, P., & Silva, M. 2780 (33).
Cerón M., C. E. 1688 (18), 2038 (69), 2723 (33).
Cerón M., C. E., & Cerón, M. 3081 (18), 3110 (28).
Cerón M., C. E., & Iguago, C. 5511 (18), 5592 (69), 5665 (2).
Cerón M., C. E., & Neill, D. A. 2367 (69×?).
Cerón M., C. E., & Ocampo, L. 11865 (3).
Cerón M., C. E., et al. 2879 (33).
Cerón, V. H. FM316 (2).
Chacón G., I. A. 607 (63), 1292 (17).
Chagas, J. SP74964 (33), SP74975 (33), SP79177 (33).
Chambers, K. L. 2508 (40), 2509 (40), 2659 (40).
Chanek, M. 164 (63), 186 (63).
Chargas, F., & Silva 1171 (56).
Charpin, A., & Jacquemoud, F. AC13277 (2).
Chase, M. W. 90163 (68).
Chavarría, U. 128 (17), 338 (46).
Chavelas, P., et al. ES-372 (46), 2800 (63).
Chaves, D. 323 (64).
Chávez, R. 839 (18).
Chickering, A. M. 64 (63).
Chindoy B., P. J. 104 (2).
Choussy, F. 3 (46).
Cid F., C. A. 9169 (33).
Cid, C. A., & Lima, J. 3434 (33), 3894 (33).
Cid F., C. A., & Ramos, J. 1043 (33).
Cid F., C. A., et al. 592 (33), 945 (33), 1793 (33), 4648 (74), 5218 (33).
Clare, T. 1453 (55).
Clark & Cave M82-011 (64).
Clarke, O. F, 56 (63), 341 (1).

Claussen, P. 1 (14), 15 (79), 51 (79), 103A (79), 105A (14), 192 (55), 541 (14), 2062 (76).
Clemente, Bro. 2507 (40), 2584 (40), 5177 (39), 6047 (39).
Clewell, A., & Cruz, G. 4076 (64).
Clewell, A., & Tyson, E. L. 3267 (17).
Coêlho, L. S. SP74951 (33), SP79193 (33), SP79235 (33).
Coêlho, L. S., & Mello, F. SP74952 (33).
Coêlho, L. S., et al. 262 (17).
Coêlho de Moraes, J. 736 (53), 1990 (53).
Coello, F. 171 (33).
Cogollo, A. 932 (65), 710 (20).
Coker, W. C. 223 (39), 502 (39).
Collela, M., & Molina, G. 1433 (33).
Collela, M., et al. 1286 (37).
Combs, R. 14 (39, 40), 678 (40), 756 (40).
Contreras, E. 82 (46), 742 (46), 1881 (46), 2636 (64), 3214 (46), 3373 (63), 3564 (46), 3584 (63), 6244 (63), 6569 (64), 7371 (46), 7759 (64), 7824 (64), 9138 (64), 9155 (17), 9561 (46), 9595 (46), 10129 (17), 10192 (63), 10825 (64), 11144 (64).
Conzatti, C. 1680 (44), 2422 (46), 3437 (64), 3514 (46), 3789 (46).
Cook, M. T. 37 (39).
Cook, O. F., & Collins, G. N. 515 (37), 702 (37).
Cook, O. F., & Gilbert, G. B. 939 (30).
Cook, O. F., & Griggs, R. F. 775 (46).
Cook, O. F., & Martin, R. D. 28 (10).
Coradin, L., & Cordeiro, R. 943 (33), 1004 (33).
Cordeiro, J. 38 (79).
Cordeiro, M. R., et al. 1363 (33).
Córdoba, W. A., & García, F. 379 (17).
Core, E. L. 1117 (2), 1301 (2).
Cornejo, F., & Rubia, A. 1607 (18).
Correa, M. D., & Dressler, R. L. 386 (46).
Correll, D. S. 43461 (39), 44863 (39), 48173 (39), 48540 (37).
Correll, D. S., & Godfrey, R. K. 41259 (39).
Correll, D. S., & Hill, S. R. 45231 (39).
Correll, D. S., & Proctor, G. R. 47893 (39).
Correll, D. S., & Sauleda, R. 20438 (39).
Costa, E. 228 (53).
Coulon 294 (17).
Courbon, A. 102 (81), 491 (81).
Cowan, C. 2024 (63), 2056 (63), 2492 (63), 2679 (63), 3189 (63), 3442 (64).
Cowan, R. S. 1648 (40), 38056 (34), 38206 (34), 38473A (34), 38473B (33), 38566 (33), 38929 (34).
Cowan, R. S., & Maguire, B. 38086 (33), 38112 (33).
Cowell, J. F. 70 (63), 380 (63), 577 (41), 752 (37).
Crawford, J. 476 (63), 633 (37).
Cremers, G. 81 (33), 5119 (34), 5991 (33), 7190 (33), 7348 (34), 7531 (34), 7672 (34), 7806 (72), 7812 (10), 7979 (34), 8095 (34), 8369 (10), 9439 (34), 9522 (10).
Cremers, G., & Hoff, M. 11246 (33), 11345 (34).
Cremers, G., et al. 12441 (33), 12496 (34).
Cristóbal, C. L., et al. 1527 (81), 2072 (81).
Croat, T. B. 27 (37), 4707 (46), 4850 (46), 4946 (63), 5107 (63), 6090 (46), 6238 (46), 6357 (46), 6438 (46), 7021 (63), 7040 (19), 7226 (19), 7233 (46), 7385 (63), 7607 (28a), 7901 (46), 9325 (19), 9861 (10), 10073 (10), 10146 (63), 11169 (46), 11303 (17), 12033 (46), 12780 (63), 12843 (63), 13097 (46), 13476 (17), 15530 (17), 17999 (70), 19640 (70), 20868 (33), 21989 (46), 21910 (46), 22501 (46), 22983 (63), 23446 (64), 33077 (46), 33836 (46), 34435 (63), 35010 (46), 39734 (64), 40296 (64), 40534 (46), 42568 (63), 51688 (60), 58052 (70), 60848 (37), 60930 (41), 63553 (46), 64144 (64), 64655 (63), 64766 (63), 70653 (2), 72860 (2).
Crosby, G. T. 42 (10).
Cruz, M. E. 181 (46).
Cuadros V., H. 1086 (65), 2407 (46), 3276 (65), 3287 (65), 3767 (65).
Cuatrecasas, J. 3235 (2), 8061 (2), 8283 (24), 9131 (4), 9629 (2), 10803 (17), 11102 (28a), 11437 (2), 11531 (2), 13875 (2), 15352 (47, 48), 15940 (46), 17663 (46), 17791 (48), 18362 (2), 18429 (2), 18671 (48), 18695 (2), 19514 (2), 19559 (48), 21744 (2), 22032 (20), 22489 (2), 22602 (48), 22702 (48).
Cuatrecasas, J., & García Barriga, H. 1874 (2).
Cuatrecasas, J., & Rodríguez, L. 27967 (49).
Cuatrecasas, J., & Soderstrom, T. 27115 (33).
Cuatrecasas, J., &Willard, L. 26251 (2).
Cucalón S., H., & Estévez G., O. 3 (48).
Cuello, N., & Cuello, E. 59 (33).
Cuello, N., & Fernández, Y. 508 (33).
Curran, H. M. 13 (86), 53 (65), 151 (90), M-619 (33), 777 (37).
Curran, H. M., & Haman, M. 1279 (10), 1289 (10), 1342 (10), 1358 (10).
Curtiss, A. H. 213 (39).
Cuasto, L. 969 (52).
Czerwenka, J., & Paláez F., S. 406 (37).
da Costa, R. 149 (33).
Daly, D. C., et al. 319 (71), 648 (69), 736 (33), 1435 (33), 4051 (33), 4442 (33).
Daniel, Bro. 117 (2), 947 (49), 960 (2), 1534 (2), 2710 (2).
Daniel, T. F. 1286 (46), 1290 (63), 3748 (46), 5489 (63), 5501 (66), 5814 (63).
Daniel, T. F., & Bartholomew, B. 4995 (64).
da Fonseca, W. N. 40 (62), 300 (62).
D'Arcy, W. 793 (37), 1975 (37), 4000B (46), 4863 (37), 9376 (46), 9593 (46), 9733 (46), 10054

8951 (33), 9709 (10), 11146 (33), MG4754 (10), MG4780 (17).
Dugand G., A. 405 (65), 429 (65), 650 (65), 931 (65), 2958 (24), 5172 (65), 4717 (46), 6747 (65).
Dugand G., A., & García Barriga, H. 2306 (65), 2460 (65).
Dugand G., A. & Jaramillo, R. 2993 (48), 2793 (65), 2819 (65), 3015 (2), 3025 (2), 3064 (2), 3835 (2), 3840 (2), 4065 (65), 4493 (48).
Duke, J. A. 340 (17), 4240 (46), 4361 (46), 4883 (17), 4910 (46), 5712 (46), 5863 (63), 5976 (46), 6021 (46), 7446 (37), 8474 (17), 8551 (46), 8609 (63), 8935 (46), 8939 (46), 9294 (65), 9328 (19), 9648 (46), 10083 (63), 10151 (63), 10199 (63), 10341 (66), 10365 (66), 10382 (66), 10560 (65), 10636 (46), 10662 (46), 11577 (46), 11668 (46), 11937 (63), 12311 (46), 13710 (46), 14611 (17), 15497 (63).
Dunlap, V. C. 248 (63), 524 (63).
Dunn 21982 (64).
Duque Jaramillo, J. M. 1620 (2), 2325 (29), 3747 (2), 3802 (2), 4044A (20), 4106-B (49), 4153 (2), 4464 (2).
Durán, R., & Olmsted, I. 894 (46).
Dusén, P. 3 (53), 4 (55), 240 (14), 782a (79), 4365 (79), 4416 (55), 6631 (79), 8858 (55), 11517 (79), 13694 (79), 13699 (55), 13744 (55), 17212 (79).
Duss, A. 166 (17), 172 (40), 249 (40), 437 (40), 438 (40), 439 (37), 1414 (10), 1472 (17), 1473 (34), 2413 (40), 2414 (17), 2893 (10), 4407 (17).
Dwyer, J. D. 1602A (46), 2235 (46), 2362 (46), 2438 (46), 2827 (46), 4002 (66), 6858 (46), 6862 (17), 7076 (46), 7223 (63), 7313 (19), 8433 (46), 9962 (63), 10945 (63), 12492 (46), 12655 (63), 13710 (46).
Dwyer, J. D., & Coomes, R. 12985 (17).
Dwyer, J. D., & Liesner, R. 12259 (63).
Dwyer, J. D., & Nee, M. 11993 (63).
Dwyer, J. D., et al. 491 (63), 575 (63).
Dyer, F. J. A87 (63), A106 (17).
Earle, F. S. 80 (40), 618 (39).
Ebinger, J. E. 85 (17), 242 (63), 434 (46).
Eggers, H. F. A. 110 (37), 125 (37), 157 (37), 215 (37), 258 (37), 352 (37), 390 (37), 407 (10), 646 (41), 651 (17), 689 (10), 813 (17), 1530 (36), 2672 (10), 3259 (41), 5726 (28a), 7617 (39), 14335 (47), 14893 (46).
Egler, W. A. 1367 (33).
Egler, W. A., & Irwin, H. S. 46066 (34).
Egler, W. A., & Murça Pires, J. 4771 (34).
Eiten, G., & Clayton, W. D. 6113 (55), 6233 (55).
Eiten, G., & Eiten, L. T. 2542 (86), 2800 (86), 4102 (62), 6202 (86), 7858 (55).
Ekman, E. L. 115 (39), 286 (39), 849 (40), 1084 (39), 1110 (39), 1283 (40), H1468 (37), 1527 (52), 1528 (81), 1529 (81), H2668 (10), 2819 (39), 2863 (40), 4009 (39), 4106 (10), 4107 (39), H5124 (37), H5252 (37), 6453 (39), 6466 (39), 8451 (40), H8487 (37), 8607 (38), 8648 (40), 8670 (57), H8820 (57), 8984 (39), H11148 (17), H11532 (17), H11880 (36), H12508 (17), 12625 (38), 13385 (40), H14357 (10), H14796 (17), H15123 (37), 15126 (40), 17212 (39).
Elias, Bro. 503 (65), 719 (65), 1289 (65).
Elias, J. 223 (33).
Elias, T. S. 1661 (46).
Ellenberg, H. 2588 (73).
Elliot, K. 94 (46).
Elliott, G. F. S. 4098 (10).
Elliott, W. R. 41 (37).
Elmore, F. H. F19 (63), 42 (46).
Emmerich, M. 856 (28a), 717 (53).
Emygdio, L. 420 (87), 1714 (62).
Encarnación, F. 986 (28a), 1295 (28a), 26052 (69).
Endres, A. R. 6 (46), 88 (46).
Enrich, K. 18 (52).
Enríquez, O. G. 518 (63).
Erlanson, C. O. 36 (46), 520 (46).
Ernst, W. R. 1299 (17), 1358 (17), 1413 (40), 1711 (17).
Ervendberg, C. F. L. 323 (64).
Espina, J., & Arias, L. A. 1182 (20).
Espina, J., et al. 2455 (46), 2703 (17), 2710 (46), 3676 (46).
Estrada, U., et al. 89 (46).
Euponino, A. 99 (89), 557 (62).
Euponino, A., & da Vinha, S. G. 526 (89).
Evans, R., & Lewis, G. 1883 (34).
Everaarts, A. P. 519 (33).
Ewernberg 323 (81).
Ewan, J. A. 15930 (49).
Eyerdam, W. J. 63 (57), 219 (57), 517 (36).
Eyerdam, W. J., et al. 23365 (81).
Fagerlind, F., & Wibom, G. 439 (46), 2701 (46).
Fairchild, D. 3815a (37), 3850 (37), 3850a (40).
Falcão, J. I. A., et al. 776 (89), 1145 (89).
Farias, G. L. 158 (89), 418 (89).
Faris, J. A. 40 (37).
Fassett, N. C. 25827 (2).
Feddema, C. 1894 (17), 2032 (55), 2033 (10).
Fendler, A. 47 (63), 48 (46), 49 (10), 250 (10), 251 (68).
Fernandes, A. EAC20118 (10).
Fernandes, A., & Matos EAC4066 (71).
Fernández, A. 1072 (33), 2515 (33), 2667 (33), 2950 (33), 3748 (68), 3846 (68), 3849 (68).
Fernández, A., & Jaramillo 5400 (65), 7820 (65), 7826 (65).
Fernández, A., et al. 72 (52), 7913 (17).

Fernández C., J., & Molero, J. FC5724 (52).
Fernández P., A. 172 (2), 241 (46), 361 (46).
Ferrari, G. 745 (33), 1901 (17).
Ferrari, G., & Trujillo, B. 1595 (65).
Ferreyra H., R. 1130 (69), 1184 (33), 2196 (69), 4464 (69), 4489 (69), 4632 (42), 4767 (42), 4897 (33), 7932 (69), 11200 (2), 18273 (69), 18949 (33).
Ferrucci, S., et al. 365 (52).
Feuillet, C. 79 (33), 222 (33), 225 (34), 541 (72), 749 (34), 898 (10), 1578 (10).
Fevereiro, P. C. 282 (89).
Fiebrig, K. 536 (52), 1327 (56), 1327a (56), 5459 (81), 5862 (81), 6120 (52).
Figueroa P., A. 896 (2).
Finlay 1 (68), 109 (10), 195 (55).
Fischer, W. H. 305 (82).
Fisher, G. L. 3784 (64).
Fisher, M. J. 25 (41).
Fisher-Meerow, L. L. 768 (36), 5746 (37).
Fishlock, W. C. 68 (37), 196 (37), 319 (41).
Flaster, B. 68 (53), 1146 (87).
Fleury, M. 334 (33), 661 (33).
Flora Falcón 343 (33).
Florschütz, J., & Florschütz, P. A. 897 (17), 1654 (34), 1727 (34).
Florschütz, J., & Maas, P. J. M. 2502 (33).
Focke, H. C. 715 (17), 554 (17), 683 (33).
Foldats, E. 147A (33).
Folsom, J. P. 3730 (63), 4926 (46).
Folsom, J. P., & Collins 6544 (46)
Folsom, J. P., et al. 7005 (46).
Fonnegra, R., et al. 2303 (24), 2385 (46).
Forero, E. 1035 (17).
Forero, E., & Gentry, A. 796 (46).
Forero, E., & Jaramillo M., R. 2000 (55).
Forero, E., & Wrigley 7056 (42), 7084 (33).
Forero, E., et al. 981 (17), 4415 (46), 6401 (42), 9236 (17), 9727 (2).
Forest Dept. Br. Guiana 2971 (17), 4935 (17).
Fosberg, F. R. 19230 (50), 28875 (33), 55341 (37), 57572 (37), 59047 (37).
Foster, R. B. 710 (63), 762 (63), 1033 (17), 1320 (46), 1645 (46), 2648 (69), 3041 (69), 3094 (69), 3347 (69), 6443 (69), 6506 (69), 9704 (69), 12012 (18), 12576 (18), 13335 (69).
Foster, R. B., & Fitzpatrick, J. 5164 (18), 5255 (18).
Foster, R. B., & Terborgh, J. 6169 (18).
Foster, R. B., et al. 7668 (2), 7781 (2).
Fox, H. S. 16 (52), 267 (81), 295 (55).
Frame, D. 189 (33).
Franco, P. 3005 (46).
Freire, C. V. 359 (53).
Freire, C. V., & Brade, C. R26443 (53).
Freire F., A., et al. 1022 (3).
Friedrichsthal 865 (46), 1638 (46).
Fróes, L. 1779 (10), 20018 (89), 20021 (85), 29378 (69), 30185 (33).
Fróes, L., & Black, G. A. 27456 (17), 27632 (10).
Fromm, E., et al. 1318 (79), 1320 (79).
Fryxell, P. A., & Anderson, W. R. 3485 (64), 3522 (64).
Fryxell, P. A., & Lott, E. 3249 (46), 3264 (46).
Fryxell, P. A., et al. 4400 (65).
Fuchs, H. P., et al. 21712 (46).
Fuertes, J., et al. 405 (17).
Fuertes, M. 17 (36), 21 (37), 313 (10).
Galeotti, H. 29 (39), 4341 (46), 4344 (64), 4352 (39).
Gandoger 28 (33).
Garber, A. P. 36 (41), 125 (37).
García, R., et al. 1922 (37).
García, R., & Alba, N 32 (37).
García Barriga, H. 4402 (2), 7657 (65), 7703 (24), 7747 (2), 11550 (2), 11561 (2), 12608 (2).
García Barriga, H., & Jaramillo, R. 21181 (2).
García Barriga, H., & Lozano C., G. 18248 (33).
García Barriga, H., et al. 12973 (2), 12977 (2).
García Zorrón, N. 1656 (81).
Gardner, G. 1257 (62), 1427 (62), 1487 (62), 4474 (7).
Garnier, A. 76 (64), 103 (33), 942 (64), 1289 (64), 3006 (64).
Garwood, N. C. 987 (17).
Garwood, N. C., et al. 158 (19).
Gasche, J., & Desplats, J. 59 (33).
Gaudichaud, C. 258 (55), 553 (14), 581 (79), 962 (62), 1185 (52).
Gaumer, G. F. 408 (63), 23264 (63), 23477 (63), 23538 (63), 23900 (63), 24075 (63), 24118 (63), 24418 (63).
Geay, E. 187770 (33).
Gehriger, W. 389 (2), 432 (2).
Gehrt, A. SP5485 (79), SP8237 (79), SP12036 (79).
Gentle, P. 276 (46), 1425 (63), 1462 (63), 2224 (63), 2253 (46), 2481 (63), 2490 (46), 2490A (64), 2867 (63), 3567 (17), 4018 (55), 4665 (46), 4816 (46), 5451 (64), 5525 (64), 6123 (46), 6486 (46), 7312 (63), 7492 (46), 7776 (17), 8019 (55), 8385 (64), 8733 (64), 8975 (64), 9053 (63).
Gentry, A. 1950 (63), 1968 (63), 2415 (17), 3391 (46), 5151 (63), 5585 (63), 6065 (17), 6408 (17), 6433 (46), 7522 (64), 7676 (63), 7927 (63), 9165 (33), 9451 (10), 10028 (46), 10685 (33), 11142 (33), 12473 (33), 13166 (33), 13478 (46), 15353 (20), 15377 (20), 15507 (33), 17209 (46), 19654 (69), 19977 (20), 20010 (20), 20293 (2), 21261 (28a), 22784 (9), 23623 (69), 25532 (33), 28132 (28a), 28374 (37), 28800 (2), 29481 (69), 29929 (69), 29968 (33), 30734 (46), 30844 (2), 34796A (24), 37633 (69), 37688 (23), 37915

(33), 38624 (46), 40101 (43), 40716 (46), 40879 (2), 41312 (17), 41413 (70), 42673 (33), 43015 (42), 47202 (33), 47713 (2), 49365 (86), 49414 (62), 50146 (62), 50498 (37), 50726 (37), 54823 (32), 56181 (28a), 58206 (46), 59522 (49), 59536 (2), 60518 (2), 63198 (72), 69969 (2), 72667 (47), 77559 (73).
Gentry, H. S. 12285 (64).
Gerbin, W. C134 (39).
Gereau, R. E., et al. 1920 (44), 2227 (64).
Giacometto, J. 1062 (65).
Gibert, E. J. 121 (81), 2608 (10).
Gillespie 2991 (33).
Gillett, J. M., & Dickerson, V. 16521 (33).
Gillis,W. T. 5825 (37), 6116 (39), 6240 (39), 6262 (39), 9098 (46), 10296 (64), 12015 (39), 12519 (39), 13069 (39).
Gillis, W. T., & Plowman, T. C. 10178 (46).
Gilly, C. L., & Hernández X., E. 84 (63), 210 (63), 433 (64).
Ginés, Bro. 69 (26), 496R (65), 1925 (65), 2096 (65), 4519 (10), 4908 (28a), 5096 (10).
Ginzbarg, S. 776 (90).
Ginzberger, A. 587 (56), 589 (33), 1600 (55).
Giraldo C., D. A. 745 (24).
Glassman, S. F. 1596 (46), 1699 (46).
Glaziou, A. 227 (82), B594 (33), 715 (14), 717 (55), 1061 (79), 1146 (51), 1146a (52), 2115 (79), 2141 (76), 4935 (76), 4936 (82), 5755 (62), 5767 (82), 8463 (76), 8580 (15), 9674 (28a), 10360 (76), 10361 (79), 10362 (79), 10372 (51), 10373 (53), 12481 (79), 12482 (7), 13604 (7, 86).
Gleason, H. A. 66 (33), 437 (33), 488 (33), 809 (33).
Gleason, H. A., & Cook, M. T. T-8 (17).
Glocker 187 (10).
Goeldi, A. 3969 (69), 7764 (33).
Goés, L. E. RB73296 (51).
Goés, G. C. 1121 (77).
Goés, G. C., & Constantino, D. 203 (87), 614 (76), 1059 (77).
Goldman, E. A. 577 (63), 749 (44).
Goll, G. P. 380 (41).
Gómez A., P. 509 (28).
Gómez P., L. D. 14927 (63), 20486 (63), 22769 (63), 24127 (17).
Gómez P., L. D., et al. 20311 (46).
Gómez Pompa, A. 4882 (46).
Gómez Pompa, A., & Nevling, L. I., Jr. 1541 (63), 5203 (63).
González L., L. A., & Pérez J., L. A. 4261 (63).
González Q. 642 (64), 1854 (64).
González R., G. FM313 (2).
Gooding, E. G. B. 335 (55).
Goodwin, G. G., & Goodwin, B. W. 21 (37).
Gordon, B. L. 84C (17).
Gottsberger, G., & Döring, J. G14-26888 (73).
Gottsberger, I., & Gottsberger, G. 27-2273 (62).
Graham, S. A. 232 (46), 918 (40), 919 (39).
Grández, C., & Chiquispama, A. 916 (28).
Grant, J. R. 91-01551 (46).
Grant, M. L. 9576 (3), 9638 (3).
Granville, J. J. de 60 (72), 95 (34), 1324 (33), 2465 (33), 4834 (34), 5036 (34), 5119 (34), 5139 (34), 5193 (34), 7010 (33), 8211 (17), 8353 (10), 10235 (34), 11945 (34).
Granville, J. J. de, & Poncy, O. 11705 (10).
Grayum, M. 2799 (63), 2995 (63), 8682 (17), 8702 (63), 9516 (33).
Grayum, M., & Sleeper, P. 5891 (46).
Greenman, J. M., & Greenman, M. T. 5088 (46), 5196 (66), 5683 (64).
Grenand 518 (17), 1867 (72), 1906 (33), 2136 (33).
Grijalva, A., & Vanegas, O. 3416 (64).
Grijalva, A., et al. 286 (18).
Guánchez, F. 656 (33).
Gudiño, E., et al. 1827 (33), 2113 (28a).
Guedes, M. 1819 (10), 1914 (10).
Guerra, F. RB48267 (87).
Gues, E. P. 45 (89).
Guillemin, J. B. A. 51 (79), 267 (55), 650 (14), 661 (79), 755 (79).
Guimarães, E. F. 128 (86).
Gutiérrez V., G. 35647 (24), 35657 (24).
Gutiérrrez V., G., & Schultes, R. E. 505 (33), 778 (33), 819 (33).
Haber, W. A. 3565 (46), 3722 (46), 5655 (63), 5987 (46), 6925 (63), 8354 (63).
Haenke, T. 2255 (46).
Hage, J. L. 12 (55), 154 (89), 2160 (83).
Hage, J. L., & dos Santos, E. B. 1416 (83).
Hahn, L. 1132 (17).
Hahn, W. 189 (65), 3575 (72), 3581 (72), 3826 (10), 4814 (10), 5118 (33), 5622 (33).
Halle, F. 4040 (33).
Hamilton, C., et al. 873 (46), 981A (46).
Hamilton, S. H. 188 (39), 189 (39), 196 (40), 219 (39).
Hammel, B. 1348 (65), 4388 (46), 4574 (46), 4583 (17), 8398 (63), 8461 (63), 9017 (63), 9325 (17), 10016 (63), 10661 (63), 11640 (63), 12283 (63), 17952 (17), 18261 (46), 18822 (67).
Hammel, B., & D'Arcy, W. 4965 (17), 5029 (19), 5030 (46).
Hammel, B., & Trainer, J. 12784 (63), 12899 (63), 13221 (63).
Hammel, B., et al. 6883 (46).
Hans, D. 50 (14).
Harley, R. M. 150117 (89), 15057 (62), 16098 (62), 16411 (62), 16479 (53), 17124 (62), 17244 (89), 17469 (55), 17486 (85), 17573 (85),

(37), 13411 (36), 13581 (37), 14045 (37), 14099 (10), 15597 (37), 15693 (37).
Leoni, L. S. 998 (62), 1072 (14), 1827 (14), 2257 (79).
Leprieur, F. R. 304 (33).
Lescure 2165 (28b), 3225 (17).
Levy, P. 188 (64), 1320 (64).
Lewis, W. H. 1685 (46), 12691 (78).
Lewis, W. H., & Vásquez, R. 4044 (28a).
Lewis, W. H., et al. 86 (46), 198 (19), 200 (17), 238 (46), 389 (46), 1621 (46), 1685 (46), 1817 (63), 5202 (63), 5299 (66), 5463 (63), 10083 (33), 10386 (28a), 10651 (33).
Lewton, F. L. 252 (46), 329 (63), 430 (55).
Liebmann, F. 5399 (46), 5400 (46).
Liesner, R. 156 (63), 2214 (46), 2910 (17), 3042 (46), 4341 (46), 4372 (46), 15966 (33), 24471 (33).
Liesner, R., & Dwyer, J. D. 1574 (64), 1631 (64).
Liesner, R., & González, A. 9267 (33).
Liesner, R., et al. 15031 (63).
Lima, J. C. A. 37 (89), 47 (53), 616 (33).
Lima, J. C. A., et al. 128 (42).
Lima, F. MG10772 (10), MG10776 (10).
Lindeman, J. C. 4455 (34), 5001 (34), 5440 (34).
Lindeman, J. C., & Haas, H. 1980 (79), 2360 (79), 3217 (79).
Lindeman, J. C., et al. 47 (33), 278 (33).
Linden, J. J. 907 (46), 1724 (39).
Linder 89 (33).
Lino, A. M. 68 (89).
Liogier, A. H. 10322 (41), 10753 (41), 10785 (37), 10804 (37), 10905 (37), 11138 (37), 11787 (37), 11863 (37), 12442 (37), 13606 (37), 14473 (10), 14820 (37), 14992 (17), 15044 (37), 15169 (37), 15180 (37), 15683 (37), 15944 (37), 16215 (37), 16222 (37), 16568 (37), 16630 (37), 17597 (37), 18094 (37), 18246 (37), 20878 (37), 21308 (37), 27428 (37), 31178 (37), 35045 (37).
Liogier, A. H., & Liogier, P. 19308 (37), 20034 (36), 22015 (36), 23035 (36), 23326 (37), 25595 (36), 26625 (36), 26695 (36), 26900 (37), 28999 (10).
Liogier, A. H., & Martorell, L. F. 35173 (41).
Liogier, A. H., et al. 28354 (41), 28408 (41), 29367 (37), 29562 (41), 30089 (37), 30076 (17), 31327 (10), 31567 (37), 32547 (41), 32870 (37), 33469 (37), 33781 (41), 34457 (17), 34931 (37).
Lisbôa, A. 47 (33).
Lisbôa, P. 8 (87), 115 (33), 739 (33).
Lisbôa, P., et al. 1203 (72), 1296 (62), 1503 (72).
Llatas Q., S. 770 (46), 1418 (3).
Lloyd, F. E. 831 (40).
Lobato, L. C. B., & Oliveira, J. 108 (10).
Lobato, L. C. B., et al. 230 (10).
Lobo, M. G. A., et al. 166 (33).
Löfgren, A. 238 (81), 277 (62), 632 (62), 641 (14), 785 (14), 1077 (53), 3116 (86), 4160 (86).
Löfgren, A., & Edwall 1663 (55).
López, C. 51 (2).
López Ch., L. 582 (64).
López Figueiras, M. 268 (39), 8001 (2).
López M., A. 8757 (46), 8806 (2).
López M., A., & Sagástegui A., A. 7878 (46).
López Palacios, S. 9 (2), 395 (33).
López Palacios, S., & Bautista B., J. A. 3456 (2).
Lorence, D. H. 3071 (46).
Lorentz, P. G. 107 (52), 159 (81), 266 (81), 577 (81), 578 (52).
Lot, A. 1643 (63).
Lourteig, A. 1800 (34).
Lowe, J. 4003 (33), 4284 (33), 4301 (33).
Lowrie, S. R. 223 (69), 646 (69).
Lowrie, S. R., et al. 292 (73), 384 (42).
Lugo S., H. 37 (2), 168 (69×?), 1327 (2), 1807 (2), 1981 (69×?), 2001 (69×?), 2027 (69×?), 2273 (69×?), 2578 (33), 2917 (33), 3093 (28), 3228 (33), 3366 (28), 3718 (33), 3899 (69), 4184 (33), 4238 (33), 4942 (33), 5032 (28), 5047 (28), 5108 (28), 5446 (33), 5519 (33), 5876 (33), 5916 (2), 6069 (2), 6129 (28).
Luna, A. 935 (39).
Lund, P. W. 313 (55), 508 (62), 561 (82), 569 (82).
Lundell, C. L. 975 (46), 3128 (46), 4086 (63), 4087 (10), 4089 (10), 15523 (46), 15601 (46), 15992 (46), 17185 (46), 17253 (46), 17771 (63).
Lundell, C. L., & Contreras, 20819 (63).
Lundell, C. L., & Lundell 7668 (46).
Lurvey, E. 473 (52), 657 (52).
Luteyn, J. L. 11530 (41), 11550 (37).
Luteyn, J. L., et al. 5081 (2).
Lutz, A. 657 (55), 1093 (53), 1148 (55).
Lutz, B. 463 (79), 1537 (55), 1652 (55), 1721 (51), 1722 (55), R147049 (76), R15909 (82).
Lützelburg, P. 12511 (77), 25866 (62), 25881 (62), 25965 (62), 25972 (62), 26158 (62).
Maas, P. J. M., & Maas, H. 362 (33).
Maas, P. J. M., & Westra, L. Y. T. 2213 (33), 3537 (33).
Maas, P. J. M., et al. 247 (33), 5543 (34), 5574 (17), 7203 (33), 12815 (33).
MacBride, F. J. 2719 (63).
MacBryde, B. 695 (2), 950 (2), 1465 (33).
MacBryde, B., & Herrera-MacBryde, O. 685 (46), 689 (46).
MacDougal, J. M. 634GR (46).
MacDougal, J. M., & Rodán, F. J. 3635 (2).
MacDougal, J. M., & Velásquez S., M. P. 4113 (2).
MacDougall, T. H323 (46), H326 (63).
Macêdo, A. 5057 (35), 5486 (35).
Machado, O. RB62061 (62), RB76247 (62).

Smith, A. 2466 (46).
Smith, A. C. 2821 (33), 3166 (33), 10440 (40), 10494 (40), 10579 (41).
Smith, C. 4844 (86).
Smith, C. L. 1107 (63).
Smith, D. N. 250 (63), 267 (63), 3794 (18), 3885 (18), 4002 (70).
Smith, G. L. 10010 (37), 10453 (37).
Smith, G. W. 109 (28a).
Smith, H. H. 341 (46), 1525 (65), 2070 (2).
Smith, H. H., & Smith, G. W. 220 (17), 418 (32, 66), 1261 (17).
Smith, L. B. 3291 (40), 2000 (86), 2033 (55).
Smith, L. B., & Brade, A. C. 2335 (51).
Smith, L. B., & Hodgdon, A. R. 3171 (40).
Smith, L. B., & McWilliam, E. L. 15354 (86).
Smith, L. B., & Reitz, R. 6081 (55).
Smith, L. C. 531 (44).
Smith, R. F. V1270 (65), V3134 (2).
Snedaker, S. C. C-19 (46).
Sobel, G. L., & Strudwick, J. 2108 (33), 2274 (33).
Sobel, G. L., et al. 4739 (10).
Sodiro, L. 202 (46).
Soejarto, D. D. 270 (2).
Soeprata 5H (17), 8G (17), 20 (17).
Solomon, J. C. 6099 (42), 6256 (73), 7682 (33).
Soria, N. 2969 (52), 3392 (52).
Soria, N., & Basualdo, I. 2084 (52).
Soto, C. 388 (17).
Soukup, J. 2211 (18).
Soule, J. A. 2958 (64).
Sousa, M. 129 (63), 597 (64), 767 (64), 1320 (64), 1521 (64), 2070 (46), 2291 (63), 2393 (46).
Sousa, A. B. RB208623 (62).
Spada, J. 69 (62), 142 (89).
Sparre, B. 13922 (2), 18227 (20), 19633 (46), 19752 (46).
Spellman et al. 151 (63).
Sperling, C. R. 5994 (33), 6057 (33), 6322 (71).
Splitgerber, F. L. 1706 (17).
Sprague, T. A. 320 (2).
Spruce, R. 202 (10), 243 (10), 767 (33), 1001 (33), 1644 (69), 1880 (33), 3277 (33).
Stafford, D. 9 (30).
Stahl, A. 1068 (10).
Standley, P. C. 2303 (46), 7929 (63), 8757 (64), 11437 (64), 11730 (46), 15211 (64), 16139 (46), 17535 (64), 17563 (64), 19468 (46), 19643 (64), 19847 (46), 20347 (64), 20832 (46), 20961 (46), 21272 (64), 21465 (46), 21973 (46), 22158 (46), 23013 (46), 24055 (63), 24059 (17), 24140 (63), 24300 (63), 24828 (63), 24987 (63), 25132 (10), 25614 (66), 26328 (46), 26934 (46), 27075 (66), 28355 (46), 28531 (19), 28635 (63), 29019 (19), 29599 (46), 30355 (63), 30987 (63), 30991 (63), 31218 (46), 31568 (63), 31923 (66), 36051 (46), 40926 (63), 53309 (63), 53697 (63), 54029 (63), 54986 (63), 55042 (63), 55453 (63), 56510 (46), 56653 (63), 58157 (46), 62209 (46), 70619 (63), 70828 (46), 72187 (10), 72347 (46), 74973 (46), 75218 (46), 78475 (64), 78536 (46), 78593 (64), 78707 (64), 78839 (64), 78842 (46), 79473 (46), 79488 (64), 79594 (64), 81345 (12), 81349 (12), 88189 (64), 88789 (64).
Standley, P. C., & Chacón, P. 5307 (63), 5619 (63), 5877 (46), 6067 (46), 6068 (63), 6095 (46), 6421 (46), 6906 (64), 7567 (63).
Standley, P. C., & Padilla V., E. 1924 (64), 2737 (64), 3207 (46).
Standley, P. C., & Valerio, J. 48355 (63).
Stannard, B. L., & Arrais, M. G. M. 835 (33).
Starry, D. E. 237 (46).
Stehlé, H. 121 (37), 194 (40), 273 (40), 410 (17), 448 (40), 451 (10), 594 (40), 595 (40), 632 (40), 912 (37), 985 (17), 1577 (40), 1689 (40), 1691 (40), 2000 (40), 2868 (40), 2946 (55), 2979 (55), 6302 (37), 6775 (37).
Stehlé, H., & Stehlé 4477 (10), 4506 (17), 4897 (17), 5958 (17), 7098 (17), 7253 (40).
Stehlé, H., et al. 5515 (10).
Stein, B. A. 1378 (19).
Stein, B. A., & McDade, L. 3326 (24).
Stein, B. A., et al. 2563 (33).
Steinbach, J. 7931 (60).
Steiner, K. E. 83 (63), 129 (17), 151 (17), 173 (63), 217 (17), 218 (17), 219 (31), 275 (69), 277 (42), 280 (42).
Stergios, B. 3220 (33); 11159 (33); 11814 (33); 11864 (33).
Stergios, B., & Aymard, G. 8833 (65), 8942 (33).
Stergios, B., & Delgado, L. 13493 (33); 13580 (33).
Stergios, B., et al. 3563 (33), 3615 (33), 3681 (33), 5322 (33), 5926 (33), 6224 (33), 8902 (33).
Stern, W. L., & Wasshausen, D. 2422 (17), 2539 (40), 114920 (10).
Stern, W. L., et al. 886 (46), 1740 (46), 1741 (46).
Stevens, W. D. 1395 (64), 3642 (46), 4188 (46), 5842 (64), 6198 (64), 7256 (64), 7729a (46), 11983 (64), 12282 (64), 12445 (64), 12532 (64), 12705 (64), 12970 (64), 15505 (64), 16709 (64), 17465 (46), 17672 (46), 17741 (55), 19188 (46), 19272 (64), 19912 (63), 20026 (46), 20824 (10), 20931 (64), 21357 (64), 21427 (64), 21475 (46), 22858 (64), 24601 (63).
Stevens, W. D., et al. 25345 (46).
Stevenson, J. A. 279 (41), 1712 (37).
Steward, W. et al. 9 (33).
Steyermark, J. A. 29379 (46), 30312 (64), 30363 (46), 38072 (64), 38399 (46), 38458 (63), 38685 (63), 38884 (64), 38925 (63), 39213 (64), 44074 (64), 44396 (64), 45796 (63),

45778 (46), 45848 (46), 46294 (63), 46340 (64), 51840 (1), 52520 (2), 54021 (46), 54395 (3), 55960 (2), 61019 (33), 61506 (33), 87299 (33), 87417 (17), 87495 (17), 87775 (17), 87783 (28a), 87917 (33), 91317 (10), 95736 (33), 97226 (2), 101761 (65), 109847 (33), 110996 (65), 114285 (10), 114297 (10), 114536 (10), 114709 (10), 114833 (33), 116443 (33), 118529 (2), 119249 (65), 120332 (65), 125252 (2), 126178 (33), 126838 (2).
Stier, F. 40 (46), 133 (17).
Stoddart, D. R. 3013 (40), 3020 (40).
Stoffers, A. L. 2276 (37), 2383 (37), 2405 (37), 2569 (37), 2673 (37), 2701 (37), 2712 (37).
Stoffers, A. L., et al. 35 (33), 207 (33).
Stone, D. E. 3173 (46).
Stork, H. E. 3199 (46).
Stork, H. E., & Horton, O. B. 10209 (46).
Stork, H. E., et al. 8742 (46).
Stoutamire, W. 2046 (19).
Strudwick, J. J., & Sobel, G. L. 3353 (17), 3474 (33).
Sturrock 270 (10).
Sucre, D. 1079 (55), 1799 (55), 1924 (53), 1926 (53), 2227 (15), 4059 (87), 5165 (76), 6167 (82), 6982 (86), 7935 (53), 8386 (62), 8879 (77), 9167 (87).
Sucre, D., et al. 6433 (53).
Sullivan, G. A. 122 (10), 260 (46).
Svensson, B., & Bremer, K. 709 (39).
Sytsma, K. 1023 (46).
Sytsma, K., & D'Arcy, W. G. 3469 (65).
Tamayo, F. 340 (2), 1523 (33), 1821 (2).
Tapajoz SP12034 (79).
Tate, G. H. H. 15 (33).
Tate, R. 282 (64), 413 (46).
Tavares, S. 481 (62), 846 (62).
Taylor, C. 707 (46), 1385 (63).
Taylor, D. 18 (46).
Taylor, N. 14 (39), 148 (40), 377 (37).
Téllez, O. 824 (64), 1656 (46), 1678A (63), 1695 (46), 1893 (63), 2013 (63), 4208 (63), 4950 (17), 21880 (63).
Tenorio L., P. 5186 (46), 5688 (63).
ter Steege, J. 249 (33), 330 (33).
Terbogh, J. 155 (37).
Terry, M. E., & Terry, R. A. 1376 (65).
Tessmann, G. 3156 (18), 3412 (69), 3966 (69), 4298 (28a), 4661 (28a), 4859 (18), 4916 (28a), 5003 (28a), 5518 (17).
Teunissen, P., & Teunissen-Werkhoven, M. 12617 (34).
Teixeira, L. O. A. et al. 826 (33), 980 (33), 1340 (33).
Thiebaut, C. 272 (40).
Thieme, C. 5168 (46), 5164 (63).
Thieret, J. W. 889 (39).
Thomas, W. W. 3362 (33).
Thomas, W. W., et al. 4949 (33), 5034 (33), 8997 (90), 9424 (90), 9442 (89), 9738 (90), 9847 (90), 9854 (89), 9898 (89), 9960 (90), 9979 (62), 10004 (89), 10113 (89), 10193 (85).
Thompson, J. B. 1040 (37).
Thompson, S. A. 7286 (36).
Thorne, R. F., & Lathrop, E. 40247 (1), 41490 (1).
Tipaz, G., & Aulestia, C. 1679 (2).
Tipaz, G., et al. 453 (33).
Todzia, C. 2230 (33).
Tomás, Bro. 2139 (2).
Ton, A. S. 2643 (1), 2743 (1), 2867 (1), 3469 (64), 4262 (64), 5169 (64), 5868 (63).
Tonduz, A. 6985 (17), 8943 (46), 9365 (63), 9942 (17), 11454 (63), 11455 (63), 12803 (63), 13479 (67), 13480 (46), 13810 (46), 13824 (67), 14008 (67), 17862 (46).
Toriola-Marbot, D., & Hoff, M. 50 (34), 344 (10), 354 (10).
Toro, R. A. 317 (48), 354 (48), 747 (2), 809 (25), 847 (2).
Torrend, C. 294 (62).
Torres C., R. 4418 (46), 4517 (63), 6413 (46).
Torres M., J. 206 (33), 279 (28a).
Torres, J. H., et al. 1364 (2).
Townsend, C. H. T. A84 (3), A87 (46).
Tredwell, S. 751 (41).
Tresling, J. 35 (34), 230 (17), 364 (34).
Tressens, S. G., et al. 1497 (52), 2479 (52), 3915 (52).
Triana, J. J. 3411 (78).
Trinta, Z. A. 471 (79), 502 (14), 738 (79), 745 (14), 882 (86), 917 (87).
Troncoso, N. S., et al. 1341 (52).
Trujillo, B. 1489 (10), 1890 (65), 4127 (33), 4577 (10), 9230 (68), 13079 (10), 14137 (10).
Trujillo, B., & Ferrari, G. 14049 (65).
Tsugaru, S., & Sano, Y. B-511 (17), B-689 (33).
Tsugaru, S., et al. B-1365 (89), B-1367 (62).
Tuberquia, D., et al. 54 (2).
Tún Ortíz, R. 373 (46), 1467 (63), 2137 (63), 2315 (64).
Tunqui, S. 454 (18), 491 (18), 708 (69), 717 (69), 889 (18).
Tutin, T. G. 1310 (30), 1328 (30).
Tyson, E. L. 1642 (46), 1734 (63), 2228 (46), 6878 (46).
Tyson, E. L., & Blum, K. E. 2614 (66), 2616 (46), 3697 (63).
Tyson, E. L., & Lazor, R. L. 6164 (19).
Tyson, E. L., et al. 2939 (19), 4507 (17), 4589 (17).
Ule, E. 30x (79), 776 (51), 3996 (14), 3998 (55), 4681 (79), 4707 (62), 5539 (69), 6438 (69), 7253 (53), 7948 (33), 8185 (33), 9484 (69), 9485 (73).
Underwood, L. M. 6a (37).

Underwood, L. M., & Earle, F. S. 98 (39), 127 (39), 200 (39), 212 (10), 1351 (39), 1353 (39), 1359 (39), 1379 (10).
Underwood, L. M., & Griggs, R. F. 344 (41), 437 (37), 458 (41), 582 (37), 708 (37).
Urgent, D. 87 (63).
Uribe U., L. 1148 (24), 1181 (49), 1377 (2), 1419 (2), 1434 (2), 1439 (19), 1440 (49), 1574 (2), 1632 (24), 1686 (24), 1856 (2), 2169 (2), 2428 (2), 3185 (10), 3419 (2), 3798 (2), 4033 (2), 4034 (24), 4309 (65), 4845 (2), 5050 (2), 5621 (2), 5666 (2), 5885 (20), 6017 (2), 6241 (2).
Usteri, P. A. SP12030 (14).
Utley, J., & Utley, K. 1102 (63), 3114 (46).
Valerio, M. 437 (63), 956 (46).
Valeur, E. J. 66 (36), 416 (37), 961 (37).
Valverde, F. M. 880 (47), 1857 (32).
Valverde, L. 205 (2).
van der Hammen 1284 (33).
van der Werff, H. 3610 (2).
van der Werff, H., & Ortíz, R. 5897 (2).
van der Werff, H., & Ortega, F. 6112 (2).
van Doesburg, P. H., Jr. 70 (33).
van Hermann 189a (39), 834 (39), 2628 (39).
Vanni et al. 1944 (52), 2873 (81), 3264 (81).
van Severen 29 (63), 39 (63).
Vara, A., & Arias, L. M. 320 (63).
Vareschi, V. 8691 (33).
Vargas C., C. 814 (30), 14670 (73), 16941 (18), 18668 (69), 22750 (30).
Vargas, E. 93 (46).
Vargas, E., & Seidel, R. 2036 (60).
Vasques, A., et al. 27 (42).
Vásquez, R. 3847 (69).
Vásquez, R., & Jaramillo, N. 894 (28a), 8624 (29), 9319 (33), 11736 (28a), 12072 (69).
Vauthier 6 (55).
Vázquez, M. 390 (63), 433 (63).
Vázquez, R., et al. 366 (42), 6970 (33).
Vázquez Y., C. 665 (46).
Velasco, L. V. 8850 (46).
Vélez, I. 3815 (39).
Vélez, P. 2818 (32).
Velloso, H. P. 41 (55), 717 (83), 719 (85), 732 (85), 823 (83), 826 (89).
Ventura A., F. 935 (46), 1296 (64), 2624 (46), 3541 (46), 4892 (46), 8079 (46), 9081 (46), 11015 (64), 11045 (46), 11275 (64), 13897 (46), 14106 (64), 19341 (46), 19628 (64), 19661 (64), 20047 (63), 20174 (46), 20404 (10), 20801 (63), 20822 (46), 21325 (46).
Ventura V., E. 1223 (64).
Venturi, S. 6 (81).
Vera Santos, J. 2796 (64), 2853 (46),
Verboom, W. V. 5154 (63).
Versteg, G. M. 47 (34), 861 (34).
Vervoorst, S. 1613 (52).
Vianna, E. RB150317 (87).
Vianna, M. C. 49 (79), 109 (14), 115 (79), 125 (77), 131 (62), 633 (79), 1633 (87).
Vidal, J. R147046 (79), R147065 (87), R34645 (55), R71955 (53), R72076 (53).
Vidal, J., & Valle, M. H. I60 (55).
Viégas, A. P., 6815 (81).
Viégas, A. P., & Costa A. S. 2410 (79).
Vieira, M. R72086 (14).
Villacorta, R., & Lemus, P. 254 (46).
Vincent, M. A. 6222 (55).
Vivaldi, J. L. 356 (10).
Vogel, S. 55 (62), 196 (17).
Vogelsang, E. 10 (33).
Vogl, C. 181 (33).
von Sneidern, K. 428 (2), 660 (2), 1362 (33), 2362 (2), 4852 (2).
von Türckheim, H. 2521 (36), 2587 (37), 2673 (37), 3033 (36), 3186 (36), 3246 (37), 3501 (37), 3626 (37), 3734 (10), 3798 (1), 7830 (64), 7831 (64), 7832 (64), 8385 (1), II1356 (55), II2359 (63).
von Wedel, H. 82 (46), 716 (17), 787 (17), 836 (63), 847 (17), 891 (46), 995 (17), 995A (46), 1176 (46), 1249 (17), 1300 (46), 1525 (63), 1618 (46), 1649 (63), 1686 (63), 1794 (46), 2044 (63), 2654 (46), 2662 (17), 2754 (63), 2761 (17), 2828 (46), 2944 (46).
Votava, F., & Liogier, A. H. 143 (37).
Wachenheim, G. 86 (33), 87 (33), 242 (34), 341 (33).
Wagner, R. J. 136 (37), 359 (17), 1468 (37), 1712 (41).
Waide, R. E. 62312 (20).
Walker, M. M. 92-038 (40).
Wall, E. 25 (37).
Wallnöfer, B. 11-31588 (70).
Warming, E. 593 (37), 926 (37).
Watson, S. 35 (64).
Wawra, H. 387 (55).
Wawra, H., & Maly 373 (85), 444 (55), 570 (90), 871 (64).
Weberbauer, A. 1963 (2), 3591 (42), 4207 (2), 4989 (30), 6216 (9).
Webster, G. L. 13175 (40), 20975 (63), 22401 (46), 25035 (89), 25064 (15), 30202 (2), 31114 (2).
Webster, G. L., & Lynch, 17700 (64).
Webster, G. L., & Miller 9785 (10).
Webster, G. L., & Wilson 5229 (10).
Webster, G. L., et al. 8229 (37), 12715 (17).
Weddell, H. A. 22 (14), 198 (14), 456 (87), 511 (62), 540 (51), 586 (14), 597 (82).
Weir, J. 7 (87).

INDEX TO SCIENTIFIC NAMES

Accepted names are in roman type; the main entry for each is in **boldface**. Synonyms are in *italics*.